A PHOTOGRAPHIC ATLAS OF ACUPUNCTURE POINTS

Antoine **Bereder**, M. D.

The Photographic **Atlas** of Acupuncture

FINDHORN PRESS

Published by Findhorn Press, Scotland, in 2011 under licence from
Éditions Quintessence, Rue de la Bastidonne, 13678 Aubagne Cedex, France
www.editions-quintessence.com

ISBN 978-1-84409-538-4

A CIP record for this title is available from the British Library.

Front cover and title page calligraphy: Jung-Shih Chou
Proofread by Jason Bussell, MSOM, L.Ac

Printed and bound in China

1 2 3 4 5 6 7 8 9 17 16 15 14 13 12 11

Published by
Findhorn Press
117-121 High Street,
Forres IV36 1AB,
Scotland, UK
Tel +44 (0)1309 690582
Fax +44 (0)131 777 2711
email: info@findhornpress.com
www.findhornpress.com

In memory of Alain Carpentier.

I would have liked to do this with you, but you left before…

To Raquel, who didn't think twice before offering her body to Science!

To my Wu Shen, Flori, Raquel, Mavi, Fini, ... not to forget Jorge.

To Maria Teresa, thank you for this beautiful face...

To all those who helped me make this project come true.

Foreword

The aim of this atlas of acupunctural anatomy is to break away from the traditional acupuncture drawings which are far from real life and from the problems encountered when localizing the acupuncture points.

Instead of following the traditional order of the channels, I have prefered to take another course and study the human body, beginning with the three Yin, then the three Yang, upper channels (meridians), and likewise for the lower channels (meridians).

The first part of the book will give a general view of the channels, with a description of the tendino-muscular channel, the Main channel with its internal branches and the Luo vessels and, finally, the distinct channel and the points of the Main channel.

This first part also comprises the description of the Marvellous vessels.

The second part studies the different anatomical parts of the body, indicating the points of the Main channels as well as the other Extra channel points, such as the curious points, the new points, and also the points pertaining to the system of Master Tong.

The last part is a numerical and alphabetical index of all the points, also giving the numbers of the pages where these points can be found.

This book doesn't give the indications of the points, nor the way to use them, this being outside the scope of an anatomical atlas.

I hope that this book will help you to know the acupuncture points better, and that using it will be enjoyable as well as useful.

Introduction

The human body can be divided into proportional units, called Cun or Distances, which permit to localise the points with a certain precision.

In my opinion, it is better to use this system rather than the one based on the breadth of fingers, much less precise.

As a matter of fact, a tall person can have long thin fingers, as a consequence, the distance will be underestimated. Conversely, a small man with very thick fingers will overestimate it.

We will therefore use the traditional measures given by *Ling Shu*, chapter IV, paragraph 14, of which this is a reminder.

For the upper limb:

- from the end of the anterior axillary crease to the cubital crease: 9 Cun,
- from the cubital crease to the anterior crease of the wrist: 12 Cun.

For the lower limb:

- from the great trochanter to the inferior border of the patella: 19 Cun,
- from the inferior border of the patella to the extremity of the external malleolus: 16 Cun.

For the trunk and the abdomen:

- from the sternoxyphoidian joint to the umbilicus: 8 Cun,
- from the umbilicus to the superior border of the pubis: 5 Cun.

For the back:

- between the internal borders of the scapullas: 6 Cun.

For the cranium:

- between the eye brows and the frontal hairline: 3 Cun,
- from the frontal hairline to the posterior hairline: 12 Cun.

For the anterior part of the body we will use particular lines:

- the anterior midline, vertical axis of the body which it divides into two equal parts,
- the mammary line, from the scapula to the pubis, by way of the nipple,
- the intermammary line, between the anterior midline and the mammary line,
- the paraxillary line, between the mammary line and the axillary line,
- the axillary line that goes down vertically on the lateral side of the chest, from the centre of the axillary fossa to the apex of the iliac spine.

For the posterior part of the body:

- the posterior midline that passes by the tip of the spinous processes,
- the line that passes between the internal border of the scapula, 3 Cun from the midline,
- the line that passes between the midline and the one passing by the internal border of the scapula (1.5 Cun).

Chapter 1

A general view of the channels (meridians)

The 3 upper Yin

- Shu Tai Yin (Lung)
- Shu Jue Yin (Master of the Heart or Pericardium)
- Shu Shao Yin (Heart)

Shu Tai Yin system (Lung)

- Tendino-muscular channel
- Main channel
- Internal branches
- Luo vessels
- Distinct channel
- Points of the Main channel

Tendino-muscular of the Shu Tai Yin (Lung)

***Ling Shu,* chapter IV, paragraph 13**

The tendino-muscular of the Lung starts
at the upper part of the thumb.

It goes up along that finger and spreads behind
the thenar eminence.

It passes by the external side of the Deku
(foramen of the pulse).

It goes up on the forearm and spreads
at the midpoint of the ulnar crease.

It keeps ascending on the internal side of the
humerus, penetrates under the axillary fold,
and goes out by the supra-clavicular fossa.

It goes back in front of the shoulder joint
and goes up to get together again
in the supra-clavicular fossa.

It goes up and spreads inside the chest.

It unites with the cardia.

It finally goes down to the area of the 11th
and 12th ribs (floating ribs).

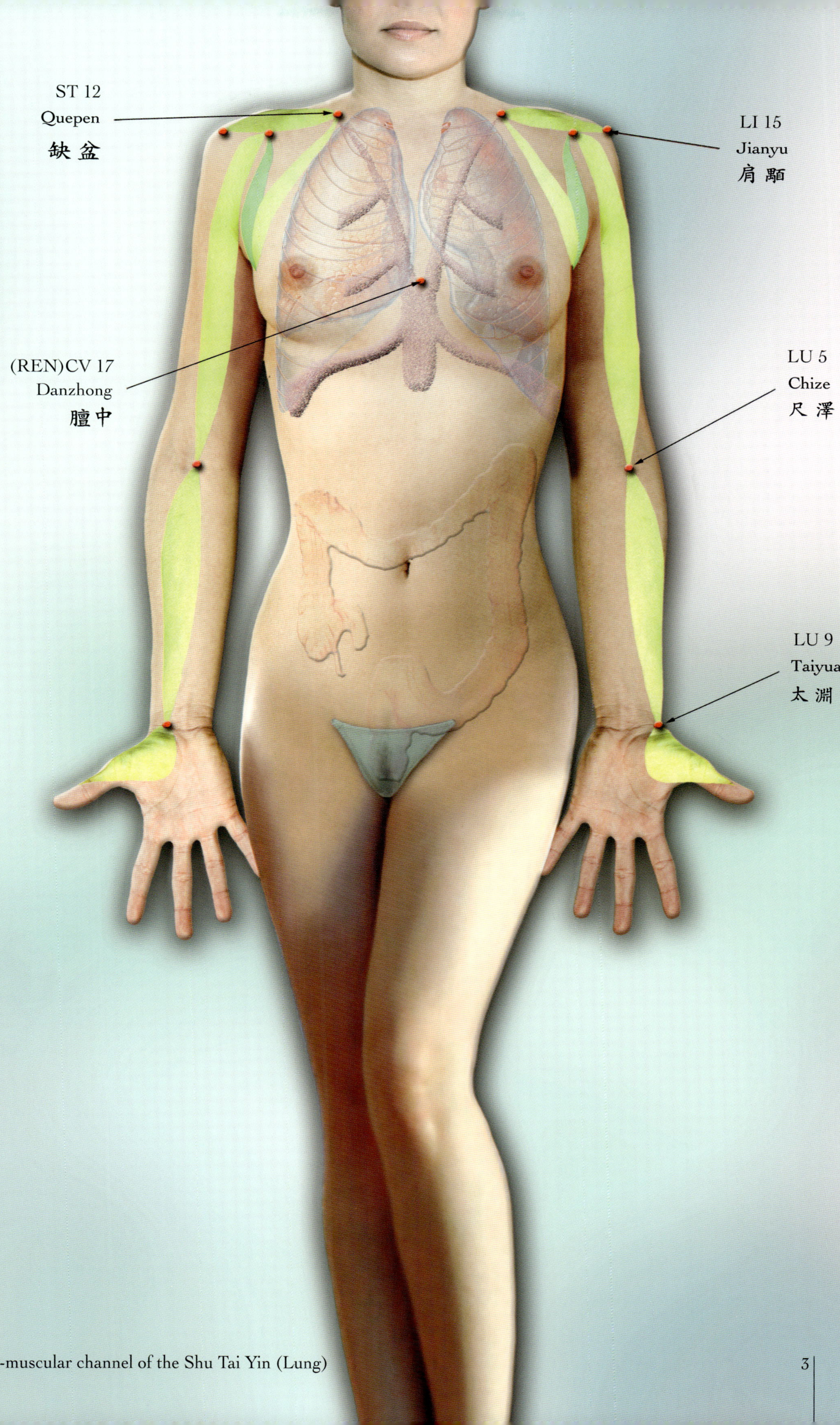

Tendino-muscular channel of the Shu Tai Yin (Lung)

Main channel of the Shu Tai Yin (Lung)

Ling Shu, chapter III, paragraph 10

The Lung vessel, Shu Tai Yin (strong Yin) of the hand, starts in Zhongjiao (Middle Burner) and going down, joins the Large Intestine.

It passes by the foramina of the Stomach and ascends in the diaphragm that belongs to the Lungs.

From the 'duct of the Lungs', it reaches the armpit on a transversal course.

It goes down on the internal side of the arm.

It passes in front of the Shu Shao Yin vessel, called 'Heart.'

It goes down by the middle of the elbow.

It follows the internal border of the forearm and the lower tip of the 'high bone' (process).

It penetrates inside the Cun Kou (foramen or entrance of the pulse).

It goes up again on the thenar eminence and goes out by the tip of the thumb.

The branch, behind the wrist, goes out by the tip of the internal border of the forefinger.

Commentaries

The superficial course of the Lung channel starts in the chest at point LU 1 Zhongfu, in the 1st intercostal space on the paraxillary line.

Then, it goes up vertically to the inferior border of the scapula where LU 2 Yunmen is located.

It goes down again by the antero – external side of the arm (delto-pectoral groove) to reach the cubital crease on the radial side of the tendon of the biceps where LU 5 Chize is.

From LU 5, it goes down following a line that joins LU 5 to the flex crease of the wrist on the side of the radial styloid.

Getting closer to the crease of the wrist, it deviates, following an exterior course, on the external side of the styloid process where LU 7 Lieque is.

LU 9 Taiyuan is to be found in the flex crease of the wrist.

From that point, it moves to the external border of the thenar eminence passing by LU 10 Yugi.

It then goes to its final point, LU 11 Shaoshang, situated on the radial border of the thumb, at 0.1 Cun from the external ungueal angle.

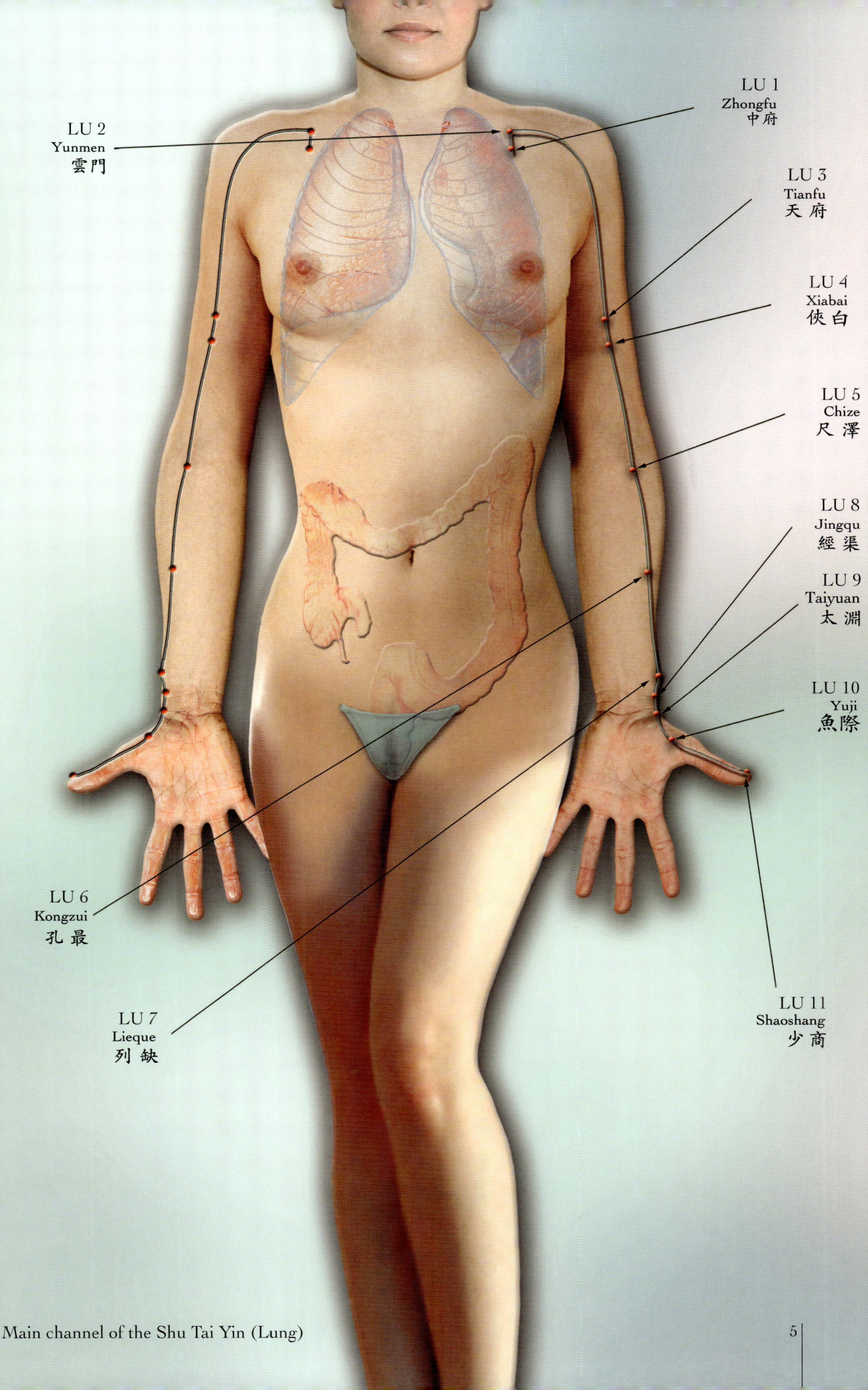

Main channel of the Shu Tai Yin (Lung)

Internal branches of the Shu Tai Yin (Lung)

Ling Shu, chapter III, paragraph 10

The 'vessel of the Lungs', Shu Tai Yin (strong Yin) of the hand starts in the Zhongjiao (Middle Burner) and going down, joins the Large Intestine.

It passes by the foramina of the stomach and goes up in the diaphragm that pertains to the Lungs.

From the 'duct of the Lungs', it transversally joins the armpit.

Commentaries

The Lung channel originates in the middle burner or Zhongjiao.

It runs downward to unite with the Large Intestine (viscus).

It goes up again and goes to the Stomach, passing by its upper foramen or cardia.

It then goes through the diaphragm to reach the Lung, its corresponding organ.

It goes up again and passes by the trachea up to the larynx.

From the larynx, it deviates its course and goes transversally to its first point, LU 1 Zhongfu, where its superficial course starts.

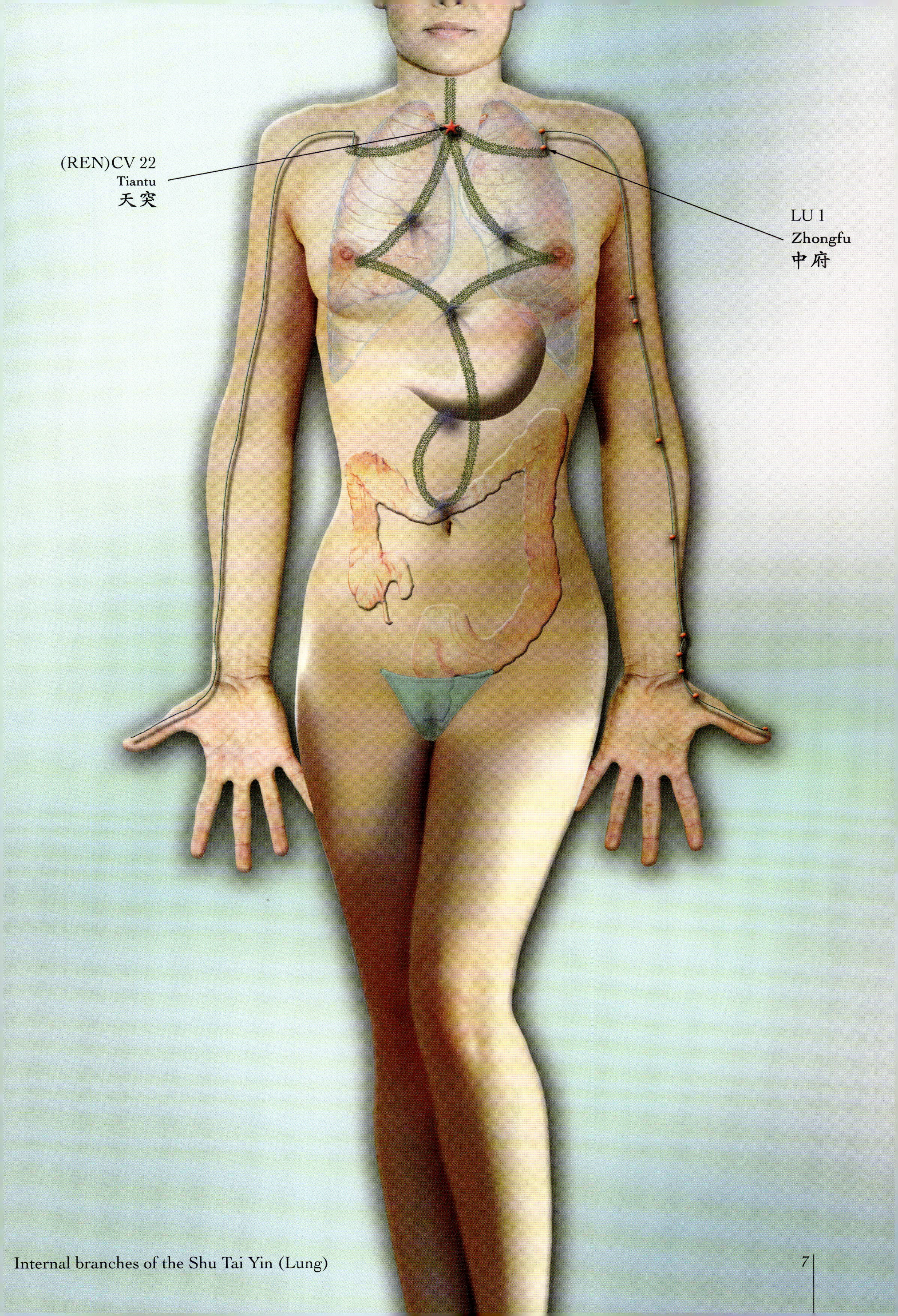

Internal branches of the Shu Tai Yin (Lung)

Luo vessels of the Shu Tai Yin (Lung)

***Ling Shu*, chapter III, paragraph 10**

The Luo of the Shu Tai Yin of the hand (Lung) is called Lieque.

It goes out above the wrist at point LU 7 Lieque.

It unites with the Shu Tai Yin channel of the hand.

It enters directly the palm of the hand (thenar eminence) at point LU 10 Yuji.

This Luo joins the Yang Ming of the hand (Large Intestine).

Commentaries

The Luo system is constituted by two branches going out of LU 7 Lieque.

- A longitudinal Luo vessel.
- A transversal Luo vessel.

The longitudinal Luo vessel goes out of LU 7 Lieque and reaches the thenar eminence where it spreads, then keeps going, following the radial border of the forefinger and finally ends its course at point LI 1 Shangyang, 2 mm behind the ungueal angle of the forefinger, on the side of the thumb.

The transversal Luo vessel also goes out of LU 7 Lieque and goes to point LI 4 Hegu, situated between the 1st and the 2nd metacarpal bone, at the midpoint of its radial border.

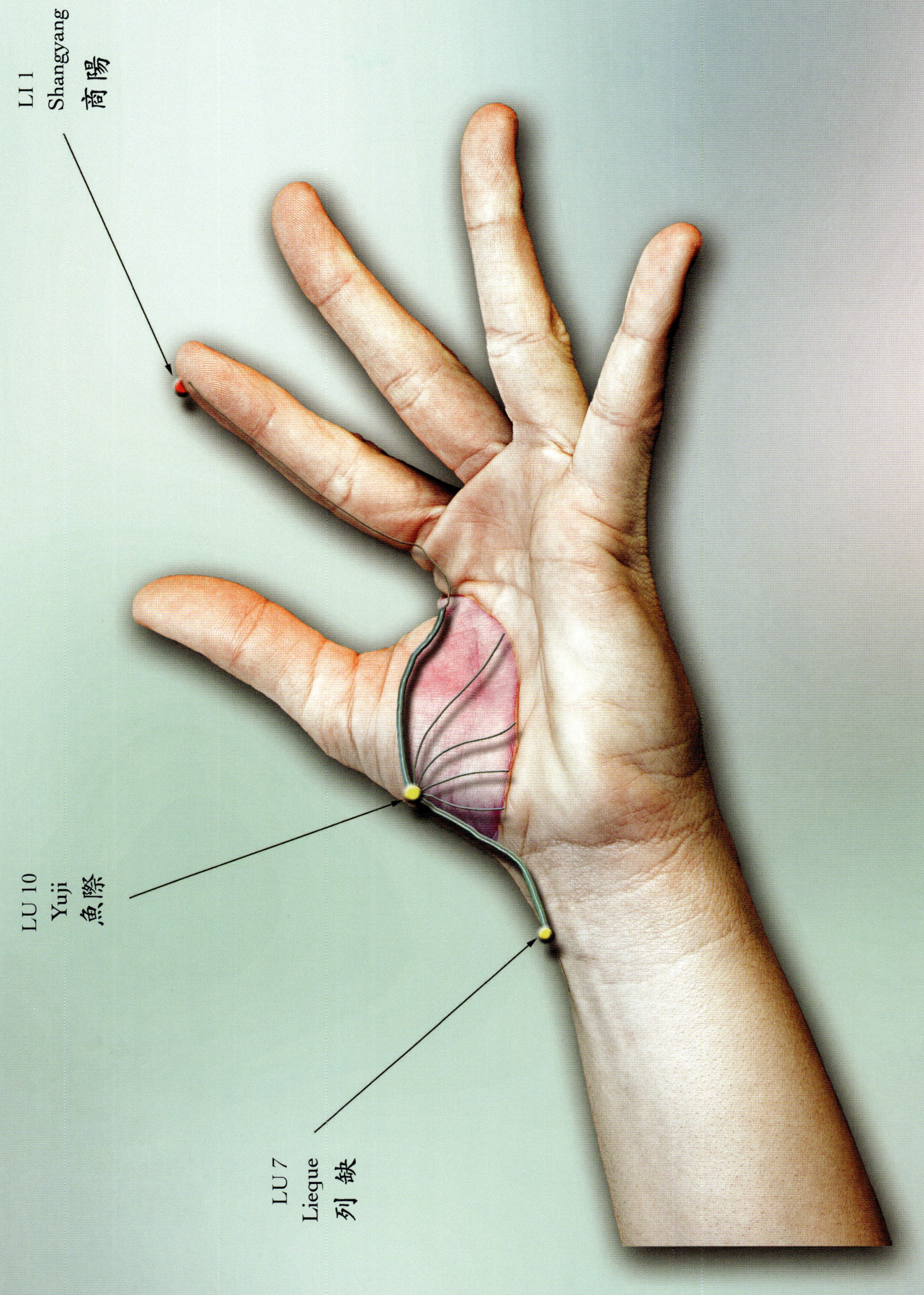

Luo vessels of the Shu Tai Yin (Lung)

Distinct channel of the Shu Tai Yin (Lung)

Ling Shu, chapter III, paragraph 11

The distinct channel of the Lung channel penetrates into Yuanye (deep in the axillary fold) in front of the channel of Shu Shao Yin (Heart).

It reaches the Lungs.

It spreads in the Large Intestine.

Goes up to exit by the supra-clavicular fossa.

Goes up along the throat and unites with Yang ming.

Commentaries

The distinct channel of the Lung leaves the Main channel on the level of the armpit.

It goes to point GB 22 Yuanye located at the crossing between the 4th intercostal space and the middle axillary line.

It goes into the chest and spreads in the Lungs.

It passes by the Large Intestine.

It goes up in the supra-clavicular fossa on the level of point ST 12 Quepen, then goes up along the throat to unite with Yang Ming of the hand (Large Intestine).

N.B.

For some authors, the meeting with the Large Intestine occurs on the level of point ST 12 Quepen and forms the 6th union.

For others, the distinct channel goes to point LI 18 Futu where it unites with the Large Intestine to form the 6th union.

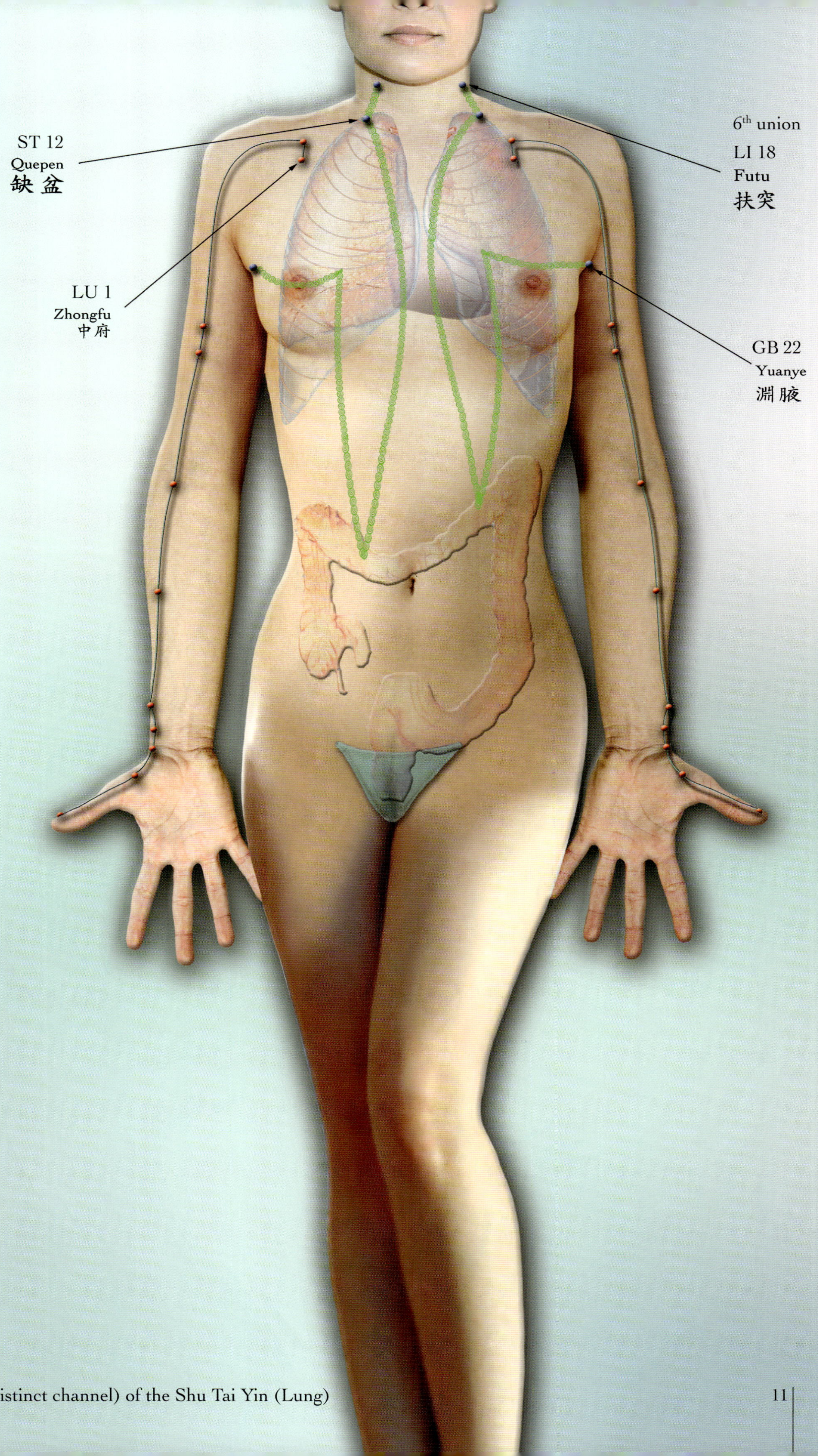

Jing Bie (distinct channel) of the Shu Tai Yin (Lung)

Points of the Main channel Shu Tai Yin (Lung)

LU 1 Zhongfu
In the superior lateral part of the anterior thoracic wall, 1 Cun below LU 2 Yunmen, on the level of the 1st intercostal space, 6 Cun lateral to the anterior midline.

LU 2 Yunmen
In the superior lateral part of the anterior thoracic wall, superior to the coracoid process of the scapula, in the depression of the infraclavicular fossa, 6 Cun lateral to the anterior midline.

LU 3 Tianfu
On the medial side of the upper arm and on the radial border of the biceps muscle of the arm, 3 Cun below the anterior end of the axillary fold.

LU 4 Xiabai
On the medial side of the upper arm and on the radial border of the biceps muscle of the arm, 4 Cun below the anterior end of the axillary fold, or 5 Cun above the cubital crease.

LU 5 Chize
In the cubital crease, in the depression of the radial side of the tendon of the biceps muscle of the arm.

LU 6 Kongzui
On the radial side of the palmar surface of the forearm, and on the line connecting LU 5 Chize and LU 9 Taiyuan, 7 Cun above the cubital crease.

LU 7 Lieque
On the radial side of the forearm, proximal to the styloid process of the radius, 1.5 Cun above the crease of the wrist, between the brachioradial muscle and the tendon of the long abductor muscle of the thumb.

LU 8 Jingqu
On the radial side of the palmar surface of the forearm, 1 Cun above the crease of the wrist, in the depression between the styloid process of the radius and radial artery.

LU 9 Taiyuan
On the radial end of the crease of the wrist, where the pulsation of the radial artery is palpable. (Do NOT needle directly into the artery!)

LU 10 Yuji
On the palm, in the middle of the first metacarpal, at the junction between white and red flesh.

LU 11 Shaoshang
Radial side of thumb, 2 mm behind the corner of the nail.

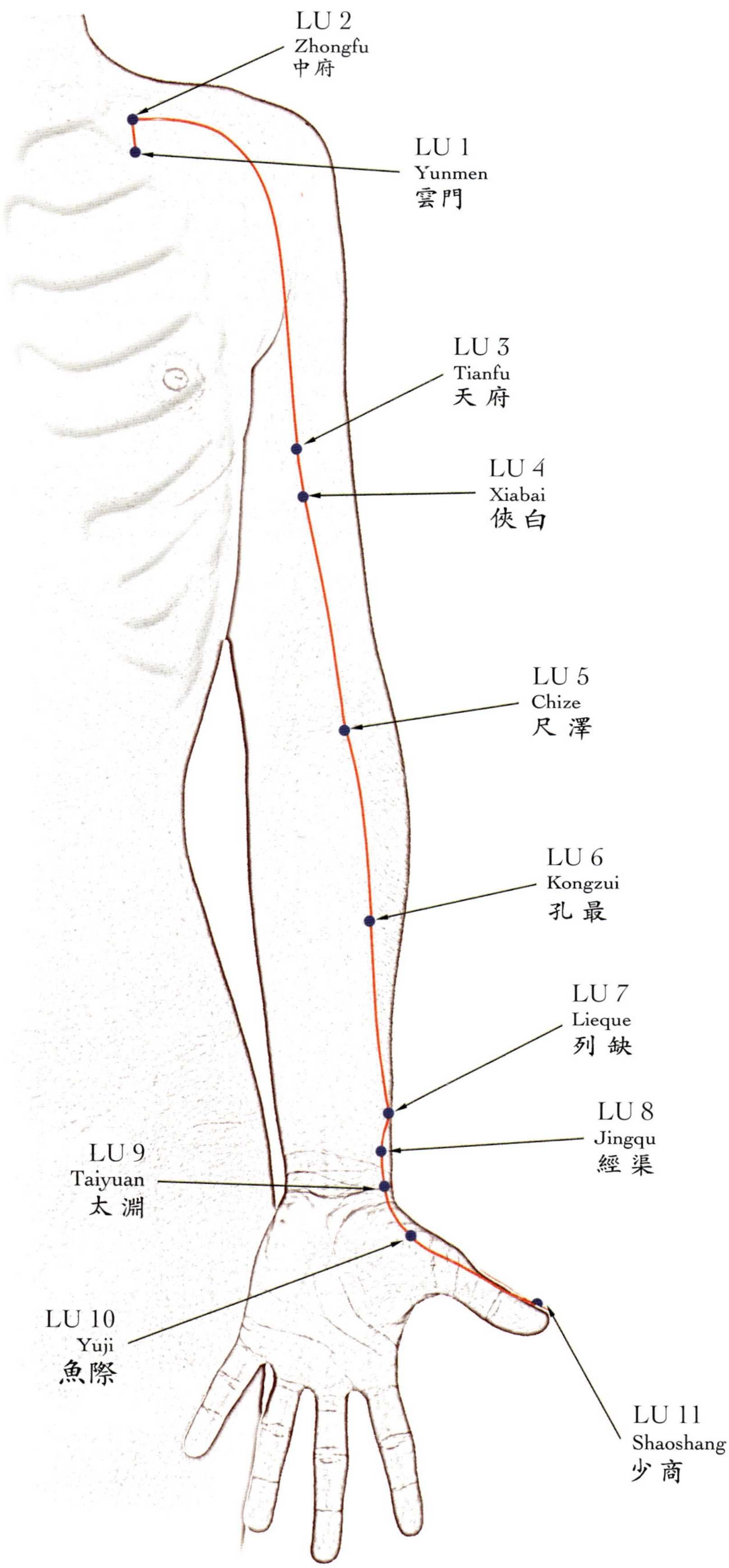

Summary of the Shu Tai Yin channel (Lung)

Shu Jue Yin system (Master of the Heart or Pericardium)

- Tendino-muscular channel
- Main channel
- Internal branches
- Luo vessels
- Distinct channel
- Points of the Main channel

Tendino-muscular of the Shu Jue Yin (Master of the Heart or Pericardium)

Ling Shu, chapter IV, paragraph 13

The tendino-muscular of the Master of the Heart or Pericardium starts at the middle finger and follows the course of the tendino-muscular of the Lung.

It joins the internal border of the elbow.

It goes up by the internal border of the arm.

It gets together under the armpit.

It spreads and goes down, passing in front and behind the ribs.

The branch penetrating under the armpit scatters in the middle of the chest and joins the cardia.

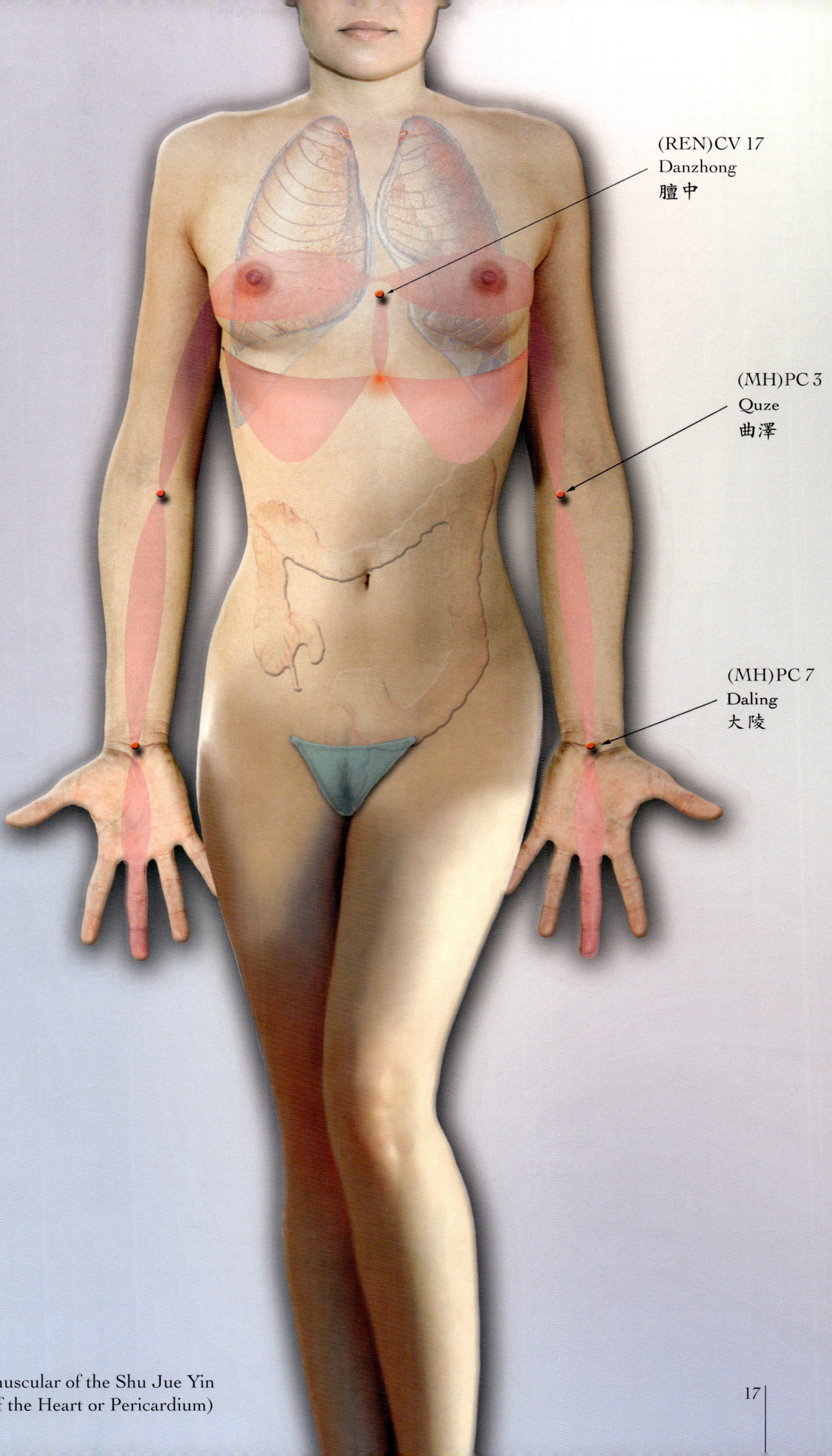

Tendino-muscular of the Shu Jue Yin
(Master of the Heart or Pericardium)

Main channel of the Shu Jue Yin (Master of the Heart or Pericardium)

***Ling Shu*, chapter III, paragraph 10**

The vessel of the 'Master of the Heart', Shu Jue Yin of the hand, or 'Pericardium', starts at the midpoint of the chest.

It goes out and belongs to the Pericardium.

It goes down to the diaphragm, joins and passes by the Triple Heater or Triple Burner.

A branch goes along the chest and exits by the ribs.
It goes down 3 Cun below the armpit.

It goes up again under the armpit and goes along the internal side of the arm.

It moves between Shu Tai Yin (channel of the Lungs) and Shao Yin (channel of the Heart).

It penetrates at the midpoint of the elbow, goes down on the forearm and circulates between the 2 tendons.

It enters at the midpoint of the palm, goes along the middle finger and goes out at its tip.

Another branch goes out of the middle of the palm, goes along the ring finger on the side of the little finger and goes out at the tip.

Commentaries

The superficial branch of the Master of the Heart or Pericardium channel starts from the chest 3 Cun under the axillary crease, beside and outside the nipple, at point (MH)PC 1 Tianchi, situated in the 4th intercostal space.

It goes up to the armpit.

Then, it goes down to the anterio-internal part of the biceps, where it crosses point (MH)PC 2 Tianquan, between the Lung channel outside, and the Heart channel, inside.

It passes by the cubital crease on the internal side of the tendon of the biceps, where point (MH)PC 3 Quze is.

It goes down again on the forearm, between the tendons of the long palmar and the radial flexor of the carpus.

It reaches the midpoint of the flex crease of the wrist where point (MH)PC 7 Daling is situated.

Passing by point (MH)PC 8 Laogong, it gets inside the palm of the hand in the 3rd intermetacarpian space.

It follows the anterior part of the middle finger.

It reaches the tip of this finger at point (MH)PC 9 Zhongchong.

A branch goes out of (MH)PC 8 Laogong and runs to the internal tip of the ring finger to join the channel Shu Shao Yang (Triple Heater or Triple Burner).

N.B.

(MH)MC 9 Zhongchong, three different locations:

- In the middle of the base of the nail of the middle finger.
- Approximately 1 fen behind and outside the outer corner of the middle finger nail.
- At the tip of the middle finger, currently the most common location.

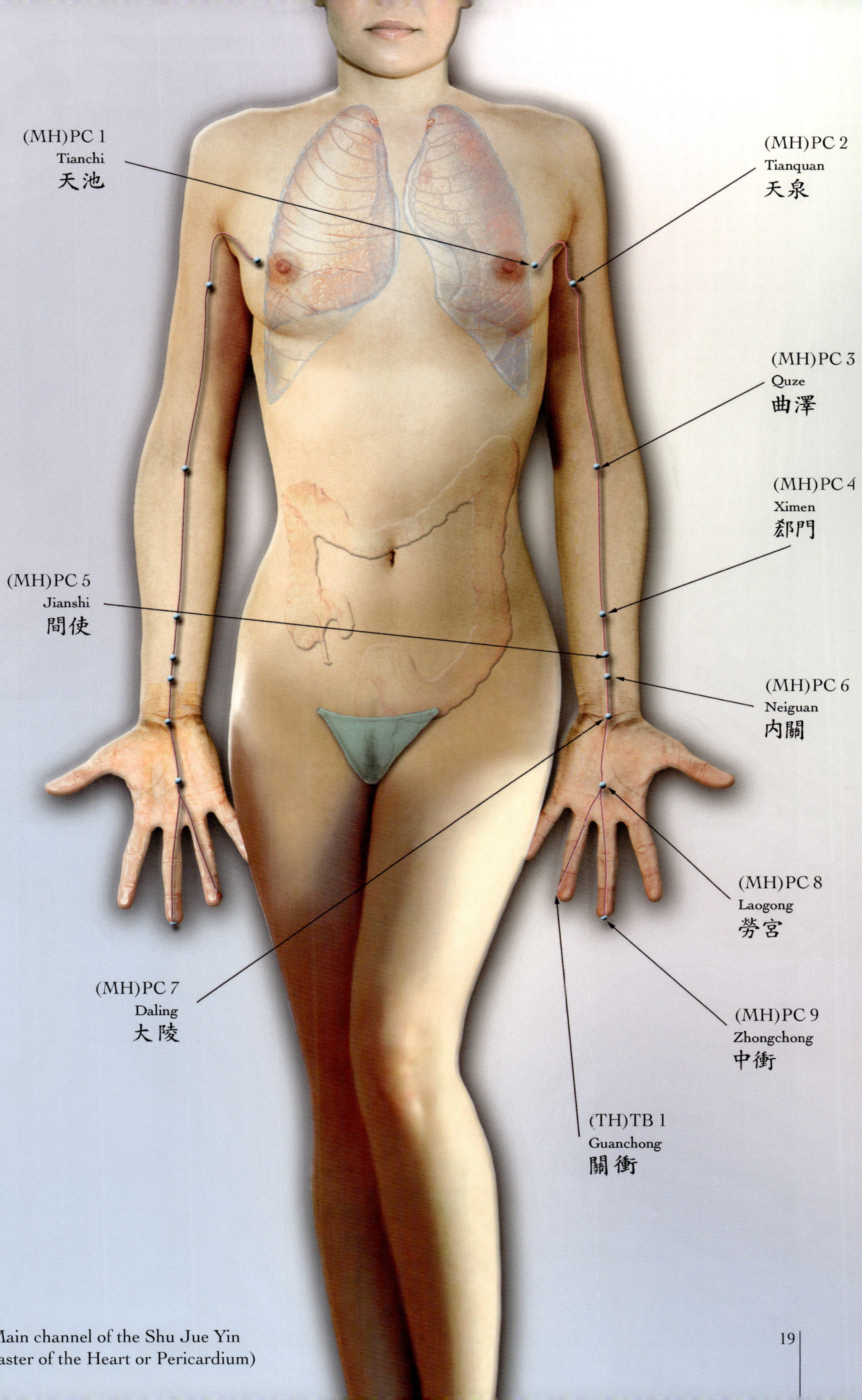

Main channel of the Shu Jue Yin
(Master of the Heart or Pericardium)

Internal branches of the Shu Jue Yin (Master of the Heart or Pericardium)

Commentaries

The vessel of the 'Master of the Heart', Shu Jue Yin of the hand, or 'Pericardium' starts at the midpoint of the chest.

It goes out and belongs to the Pericardium.

It goes down on the diaphragm, joins and passes by the Triple Heater or Triple Burner.

A branch runs along the chest and goes out by the ribs. It goes down 3 Cun below the axilla.

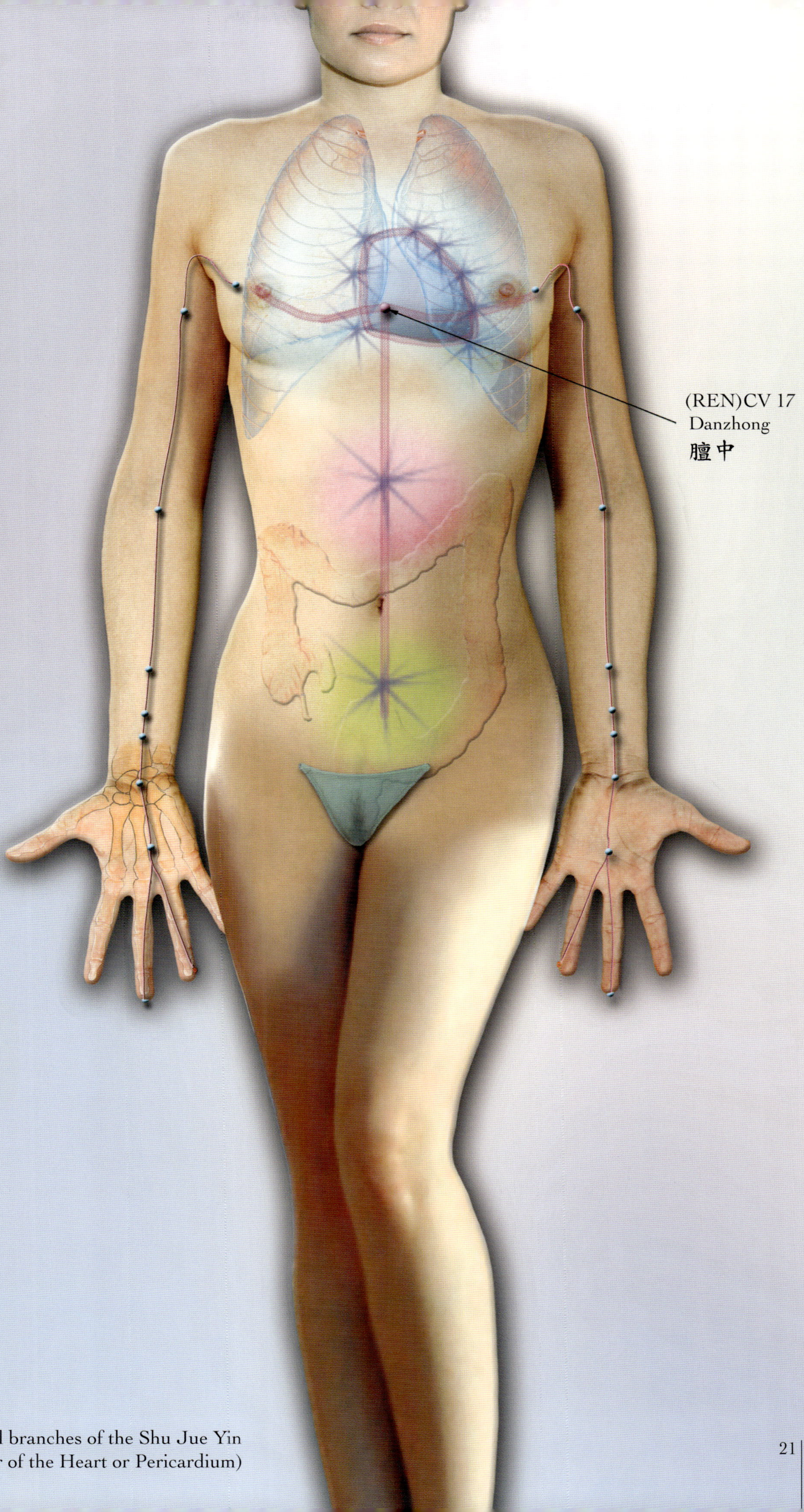

Internal branches of the Shu Jue Yin
(Master of the Heart or Pericardium)

Luo vessels of the Shu Jue Yin (Master of the Heart or Pericardium)

Ling shu, chapter III, paragraph 10

The Luo vessel of the Master of the Heart or Pericardium, (MH)PC 6 is called Neiguan.

It starts 2 Cun above the wrist and goes out between the tendons.

It goes up along the Master of the Heart or Pericardium channel and passes by the envelope of the Heart.

Commentaries

The Luo vessels of the Master of the Heart or Pericardium start at point (MH)PC 6 Neiguan, situated 2 Cun above the anterior flex crease of the wrist, and divides into 2 branches:

- The longitudinal Luo vessel that follows the course of the Master of the Heart or Pericardium channel. It then runs to the chest where point (REN)CV 17 Danzhong is, then passes by the envelope of the Heart and enters the Heart.
- The transversal Luo vessel unites with Shu Shao Yang channel at point (TH)TB 4 Yangchi, located at the midpoint of the dorsal crease of the wrist.

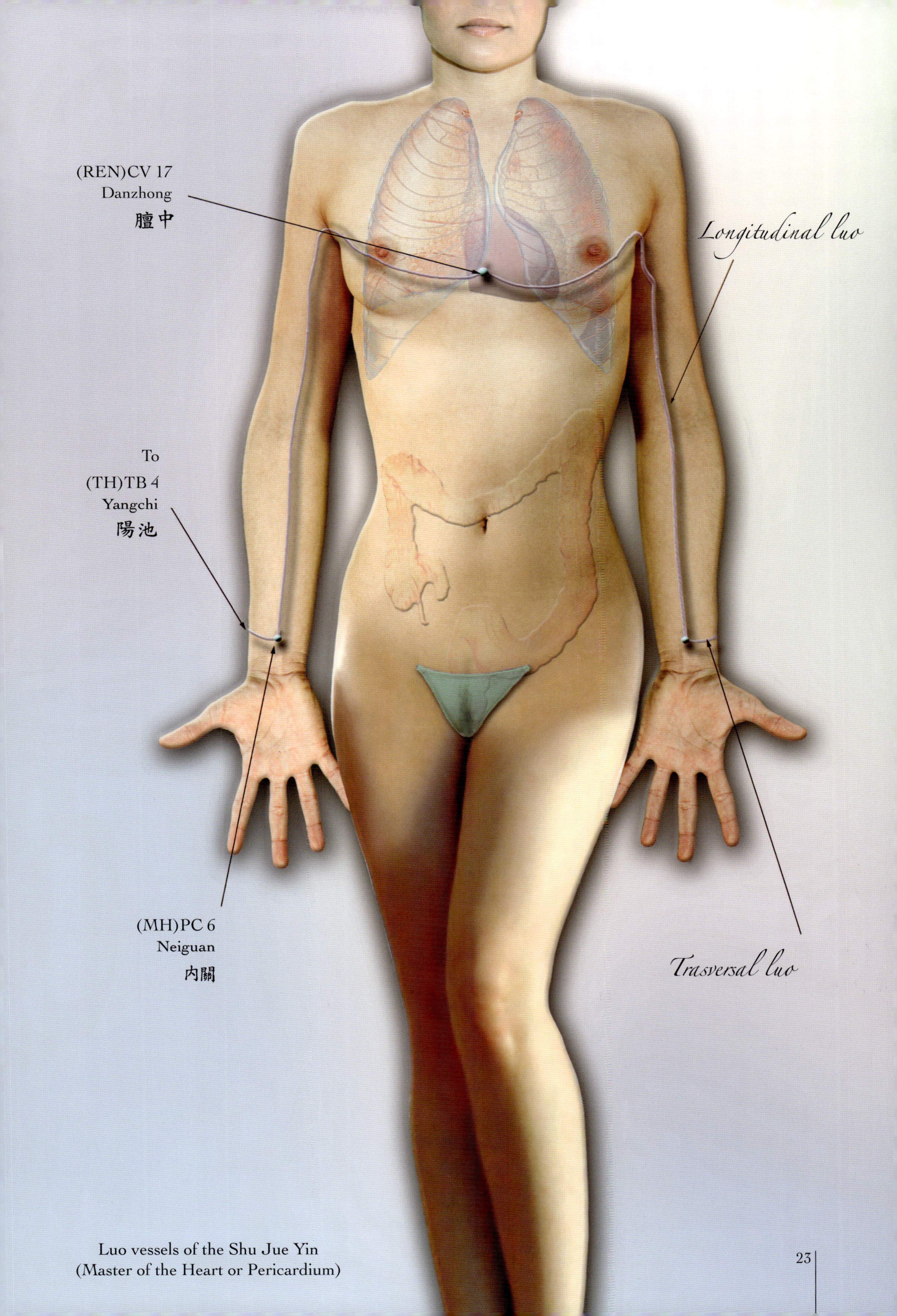

Luo vessels of the Shu Jue Yin
(Master of the Heart or Pericardium)

Distinct channel of the Shu Jue Yin (Master of the Heart or Pericardium)

***Ling Shu,* chapter III, paragraph 11**

The distinct channel goes out of GB 22 Yuanye 3 Cun below the armpit and enters at the middle of the chest.

It unites with the Triple Heater or Triple Burner, flows up along the throat and passes behind the ear.

It joins Shao Yang below the mastoid.

Commentaries

The distinct channel of the Master of the Heart or Pericardium leaves the Main channel in the area of (MH)PC 1 Tianchi.

It joins GB 22 Yuanye, on the middle axillary line, 3 Cun below the armpit in the 4th intercostal space.

It penetrates into the chest and into the Heart out of which 2 branches go out:

- The first one joins the three Heaters.
- The second one goes up to the throat passing by (REN)CV 23 Lianquan, passes by the lateral side of the neck and joins the Main channel and the distinct channel of Shu Shao Yang (TH)TB below the mastoid, where point (TH)TB 16 Tianyu is located, to form the fifth union.

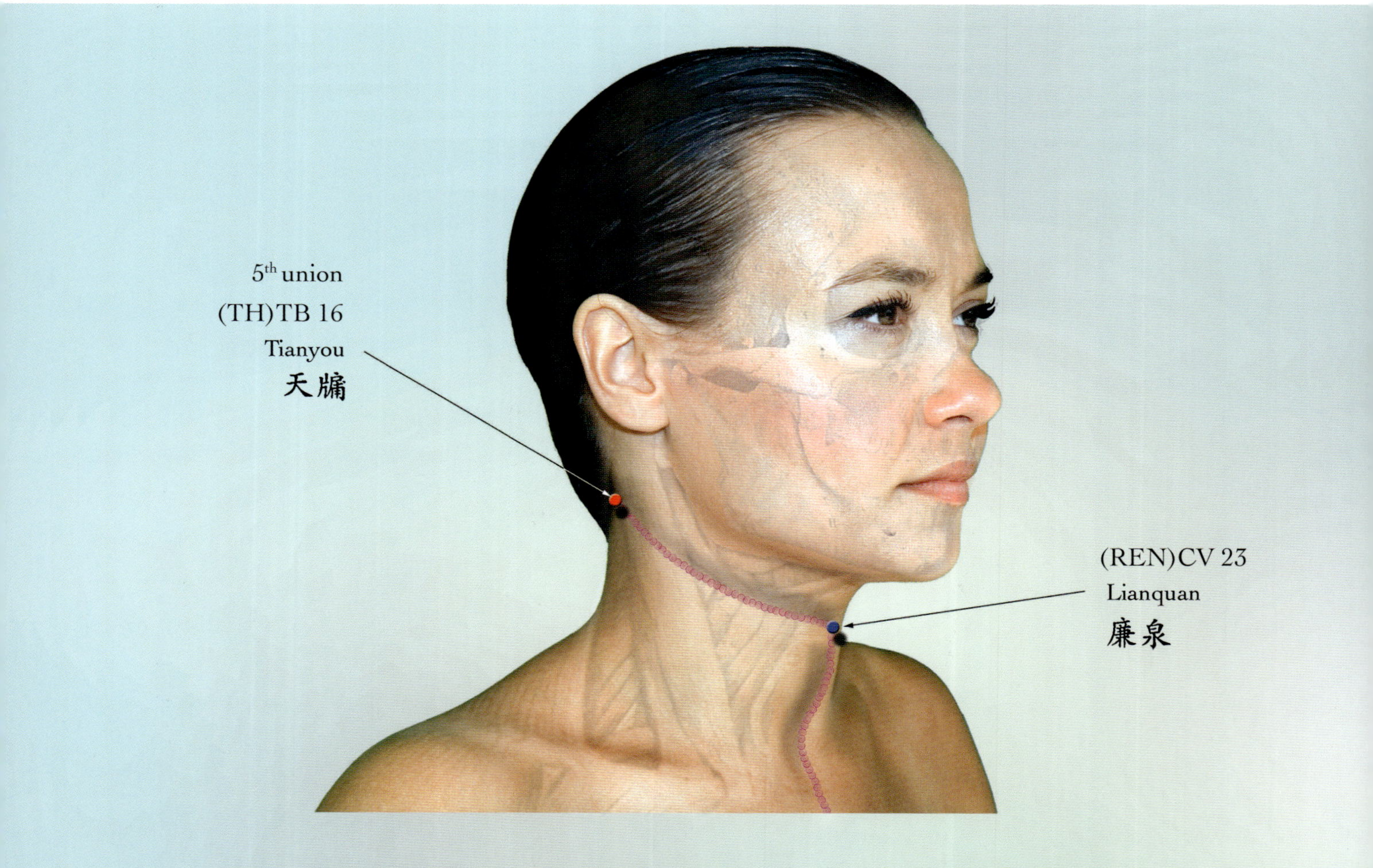

Cephalic route of the Shu Jue Yin meridian (Master of the Heart or Pericardium)

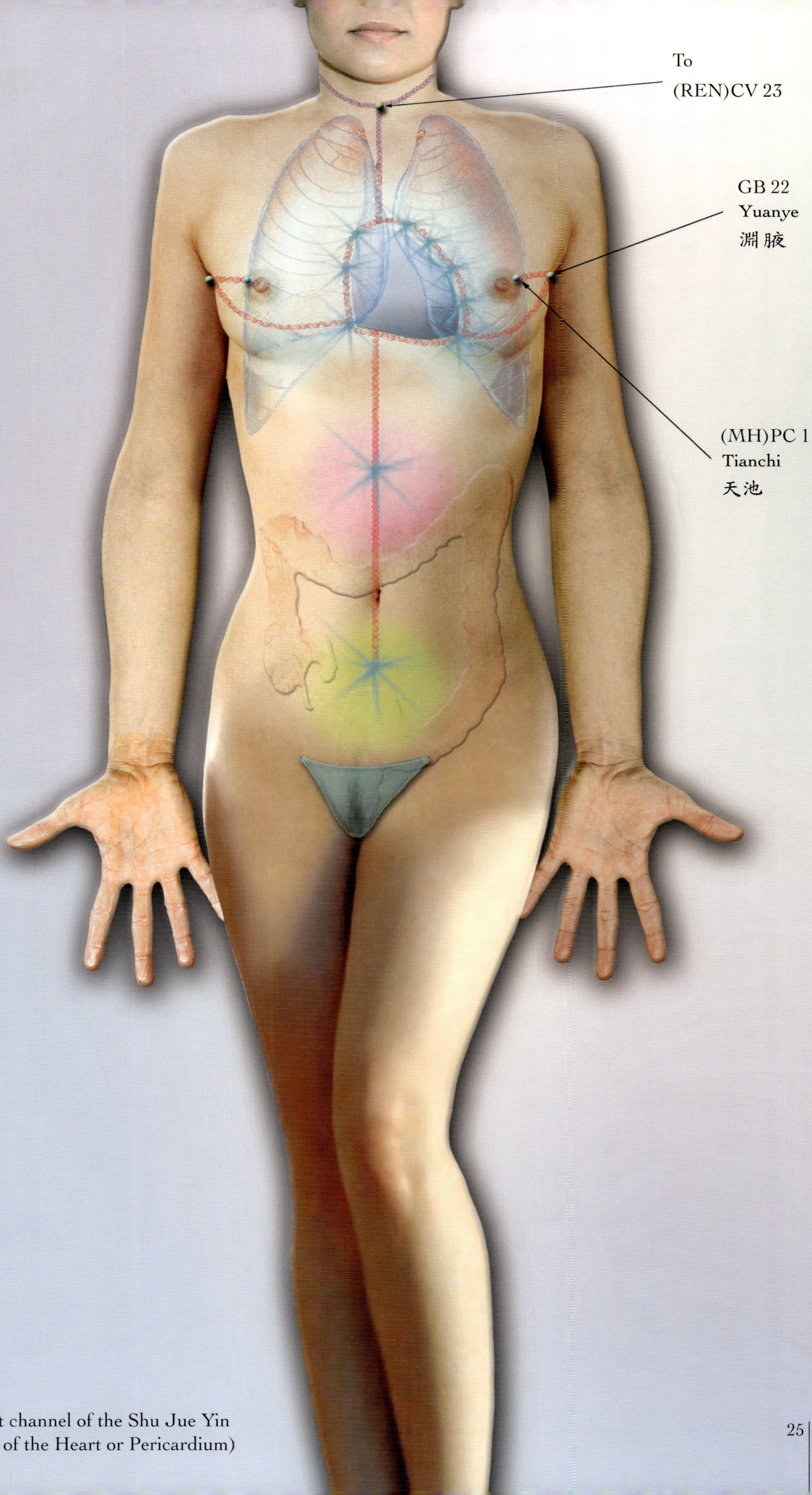

Distinct channel of the Shu Jue Yin
(Master of the Heart or Pericardium)

Points of the Main channel Shu Jue Yin (Master of the Heart or Pericardium)

(MH)PC 1 Tianchi
On the chest, in the 4th intercostal space, 1 Cun lateral to the nipple and 5 Cun lateral to the anterior midline.

(MH)PC 2 Tianquan
On the medial side of the arm, 2 Cun below the anterior end of the axillary fold, between the long and short heads of the biceps muscle of the arm.

(MH)PC 3 Quze
At the midpoint of the cubital crease, on the ulnar side of the tendon of the biceps muscle of the arm

(MH)PC 4 Ximen
On the palmar side of the forearm and on the line connecting (MH)PC 3 Quze and (MH)PC 7 Daling, 5 Cun above the crease of the wrist.

(MH)PC 5 Jianshi
On the palmar side of the forearm and on the line connecting (MH)PC 3 Quze and (MH)PC 7 Daling, 3 Cun above the crease of the wrist, between the tendons of the long palmar muscle and radial flexor muscle of the wrist.

(MH)PC 6 Neiguan
On the palmar side of the forearm and on the line connecting (MH)PC 3 Quze and (MH)PC 7 Daling, 2 Cun above the crease of the wrist, between the tendons of the long palmar muscle and radial flexor muscle of the wrist.

(MH)PC 7 Daling
At the mid point of the crease of the wrist, between the tendons of the long palmar muscle and radial flexor muscle of the wrist.

(MH)PC 8 Laogong
At the centre of the palm, between the 2nd and 3rd metacarpal bones, but close to the latter, and in the part touching the tip of the middle finger when a fist is made.

MC 9 Zhongchong
Three different locations:

- In the middle of the base of the nail of the middle finger.
- About 1 fen behind and outside the outer corner of the middle finger nail.
- At the tip of the middle finger, currently the most common location.

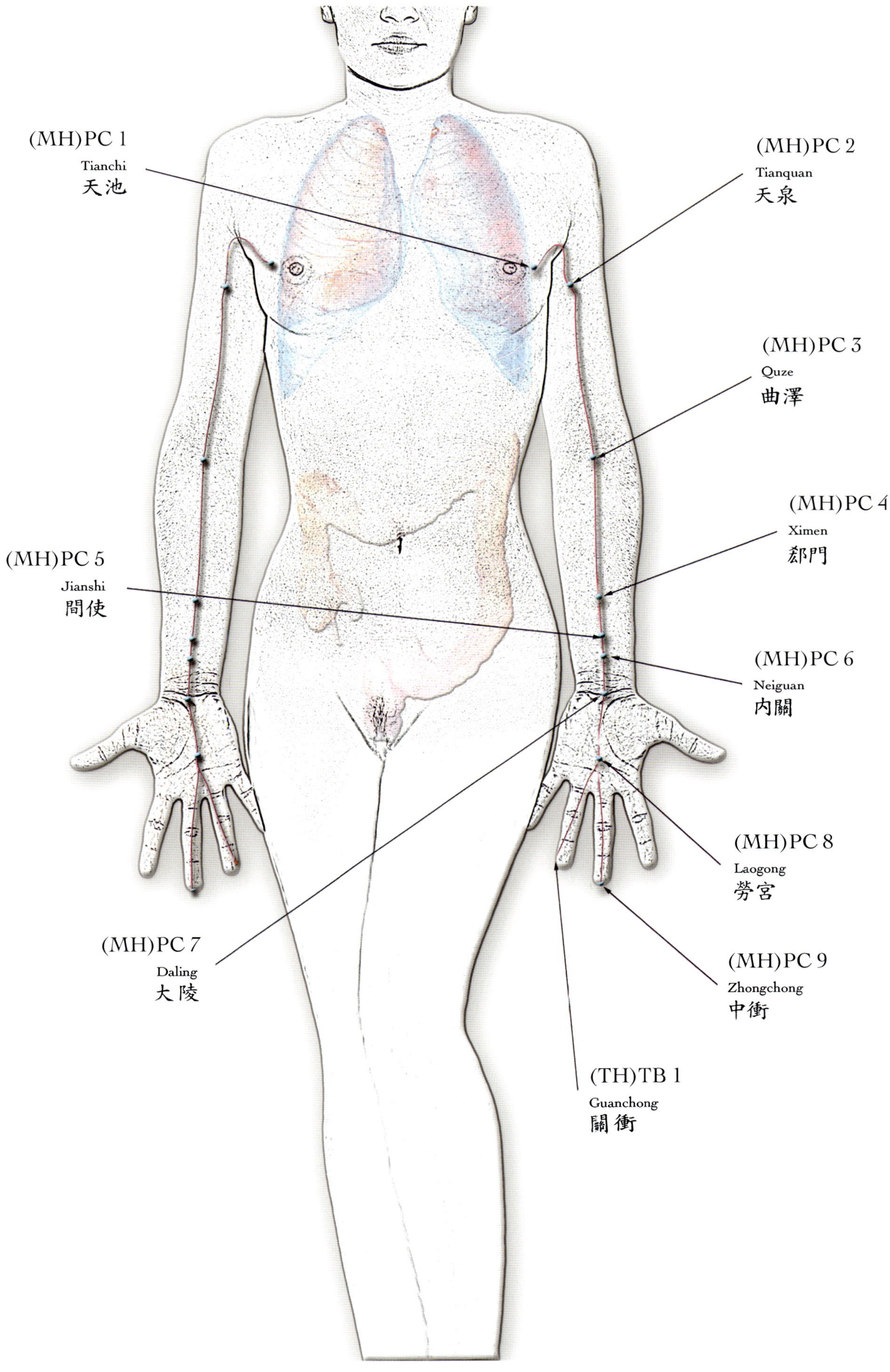

Points of the Main channel Shu Jue Yin (Master of the Heart or Pericardium)

Shu Shao Yin system (Heart)

- Tendino-muscular channel
- Main channel
- Internal branches
- Luo vessels
- Distinct channel
- Points of the Main channel

Tendino-muscular of the Shu Shao Yin (Heart)

***Ling Shu*, chapter IV, paragraph 13**

The tendino-muscular of the Heart originates on the internal border of the little finger of the hand.

It unites with the radial styloid of the cubitus.

It goes up and joins the internal border of the elbow.

Still going up, it penetrates the axillary fossa.

Crosses the Lung channel, enters the breast and gathers at midchest.

It passes by the cardia and going down, unites with the umbilicus.

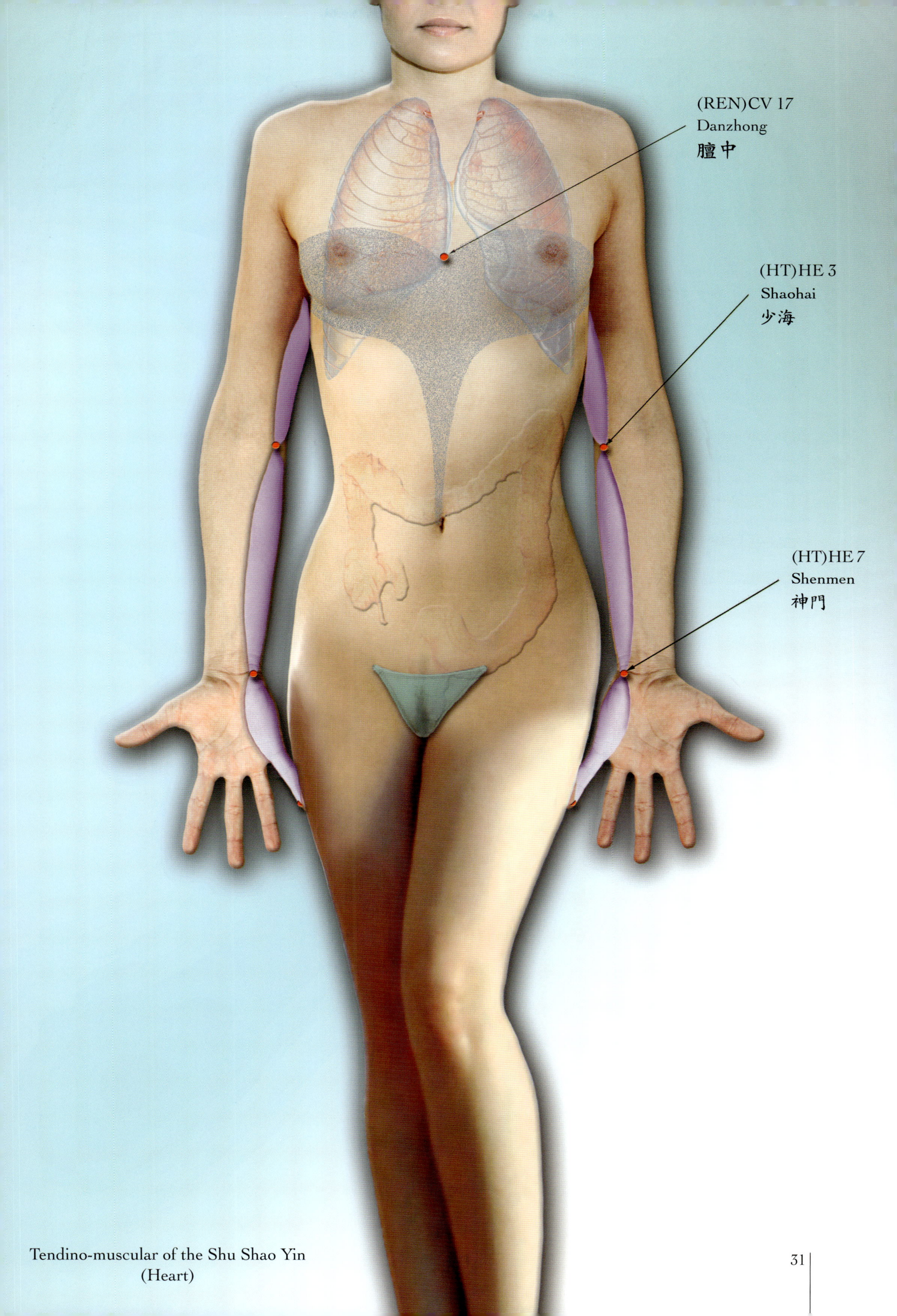

Tendino-muscular of the Shu Shao Yin
(Heart)

Main channel of the Shu Shao Yin (Heart)

Ling Shu, chapter III, paragraph 10

The vessel of the Heart, Shu Shao Yin of the hand, originates in the middle of the Heart and when it goes out, belongs to the 'connections of the Heart'.

It runs down to the diaphragm and joins the Small Intestine.

A branch leaves the 'connections of the Heart', ascends close to the throat and joins the 'eye connections'.
The direct branch also starts from the 'connections of the Heart', goes up by way of the Lungs and goes down under the armpits.

It runs down alongside the internal aspect of the arm.

It passes behind the Shu Tai Yin or 'Pericardium'.

It goes down on the internal side of the elbow, follows the internal side of the forearm and goes down to the extremity of the cubital process.

It enters the palm through the internal border.

It runs along the internal side of the little finger and goes out at the tip.

Commentaries

The Main channel of Shu Shao Yin (Heart) starts under the armpit at point (HT)HE 1 Jiquan.

It runs down on the internal side of the arm to the internal border of the elbow where (HT)HE 3 Shaohai is to be found.

Of the three upper Yin channels, this is the one with the innermost course.

It then passes by the cubital border at the anterio-internal part of the forearm, and reaches the cubital styloid.

It passes by the inferior border of the pisiform, where (HT)HE 7 Shenmen is.

It passes on the palm of the hand, at point (HT)HE 8, on the distal palmar flex crease, between the fourth and the fifth metacarpal bone.

It terminates at point (HT)HE 9 Shaoshong, on the radial side of the little finger, 0.1 Cun posterior to the ungueal angle

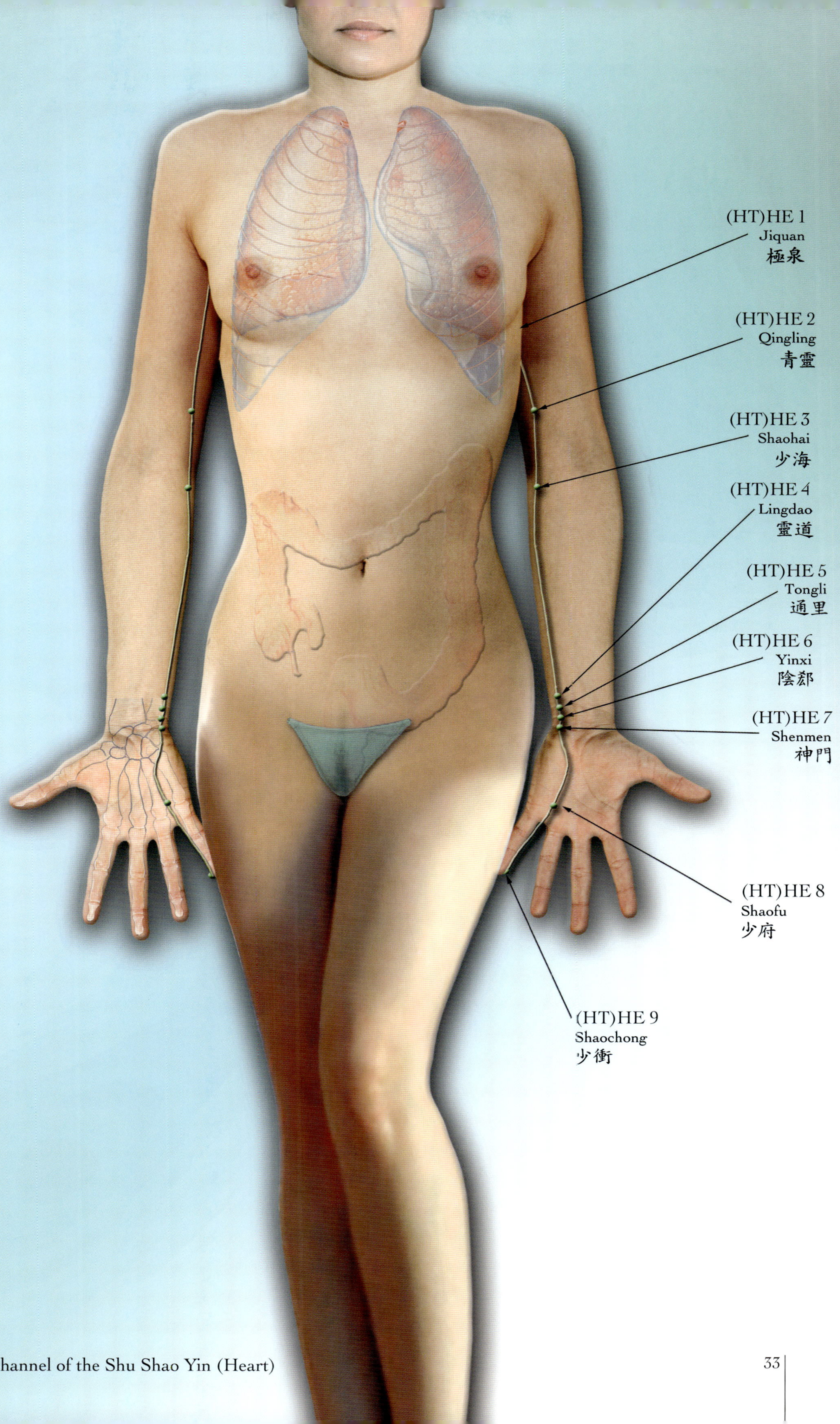

Main channel of the Shu Shao Yin (Heart)

Internal branches of the Shu Shao Yin (Heart)

***Ling Shu*, chapter III, paragraph 10**

The vessel of the Heart, Shu Shao Yin of the hand, originates in the middle of the Heart and going out, belongs to the 'connections of the Heart'.

It runs down on the diaphragm and joins the Small Intestine.

A branch starts from the 'connections of the Heart', goes up close to the throat and joins the 'eye connections'.

The straight branch also starts from the 'connections of the Heart', goes up linking the Lungs and goes down under the armpits.

Commentaries

There are three internal branches that originate, all three of them, inside the Heart.

- the first one goes through the diaphragm and joins the Small Intestine,
- the second one passes by the throat, the eyes and ends in the brain,
- the third one runs through the Lung and exits at point (HT)HE 1 Jiquan, where the Main channel starts.

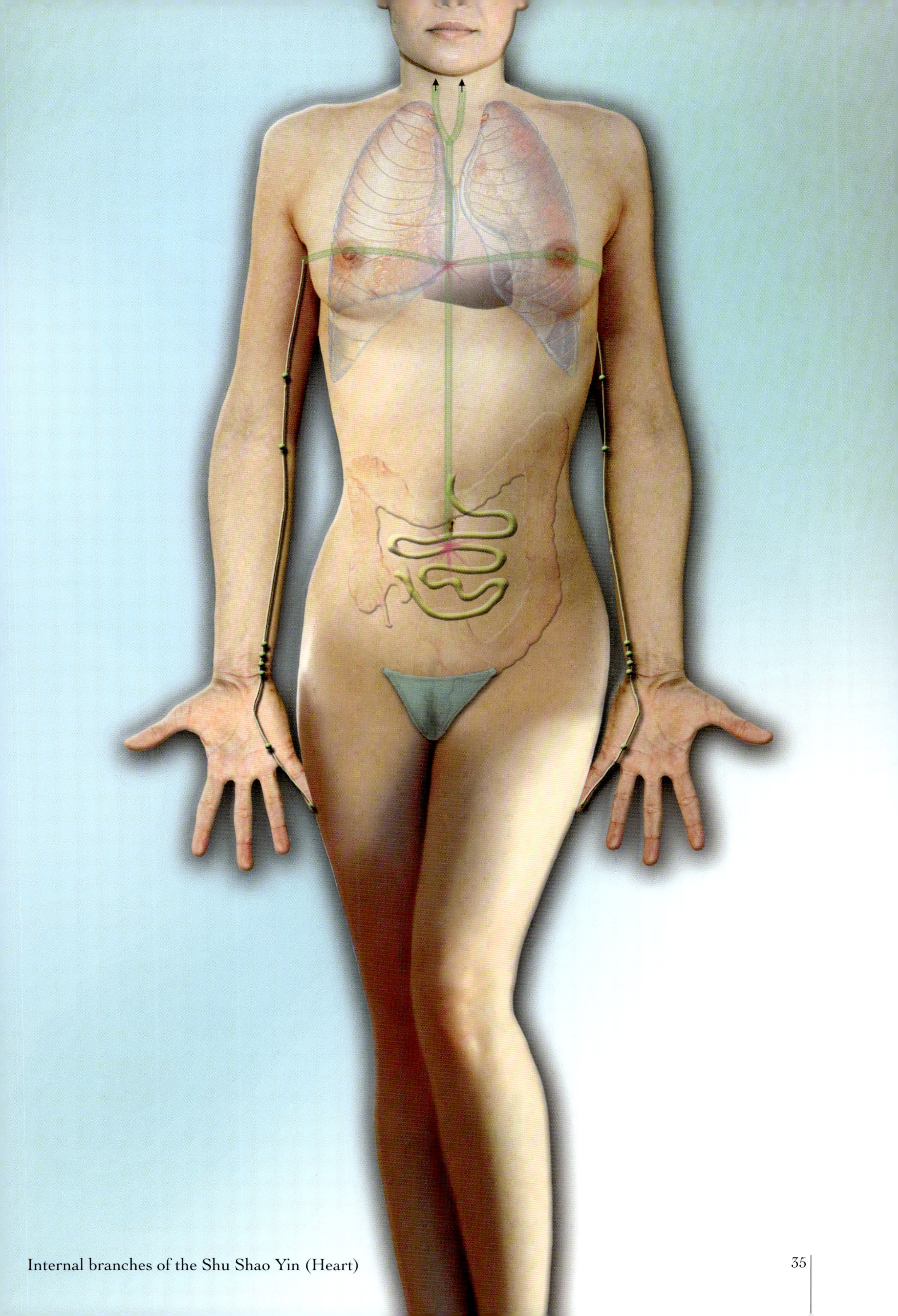

Internal branches of the Shu Shao Yin (Heart)

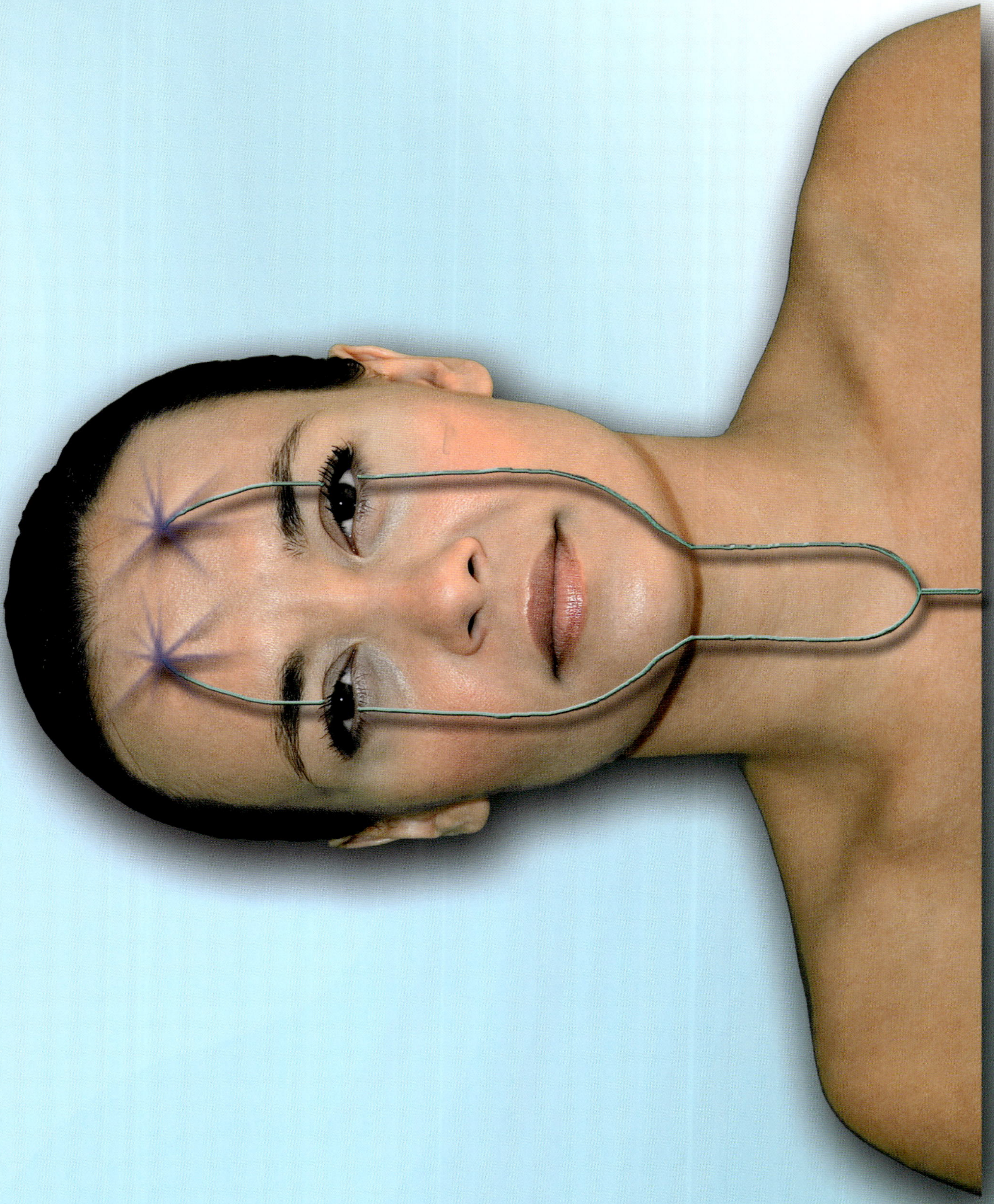

Internal branches of the Shu Shao Yin, cephalic course

Internal branches of the Shu Shao Yin, cephalic course

Luo vessels of the Shu Shao Yin (Heart)

Ling Shu, chapter III, paragraph 10

The Luo vessel of Shu Shao Yin of the hand is called Tongli and corresponds to point (HT)HE 5.

It starts 1.5 Cun above the wrist.

It goes out and travels all along Shu Shao Yin then enters the Heart.

It joins ' the root of the tongue' and the 'eye connections'.

Commentaries

The Luo vessel of the Heart goes out at point (HT)HE 5 Tongli, vertically to the styloid cubital, 1 Cun above the flex crease of the wrist.

The longitudinal Luo follows the course of the Main channel of the Heart.

It then passes by the axillary fossa and the Heart.

From there, it goes up and passes by the eye, then enters the brain.

The transversal Luo goes out from the same point, (HT)HE 5 Tongli, joins with Shu Tai Yang (Small Intestine) at point SI 4 Wangu, on the cubital side of the hand, in the depression between the base of the 5th metacarpal bone and the unciform bone.

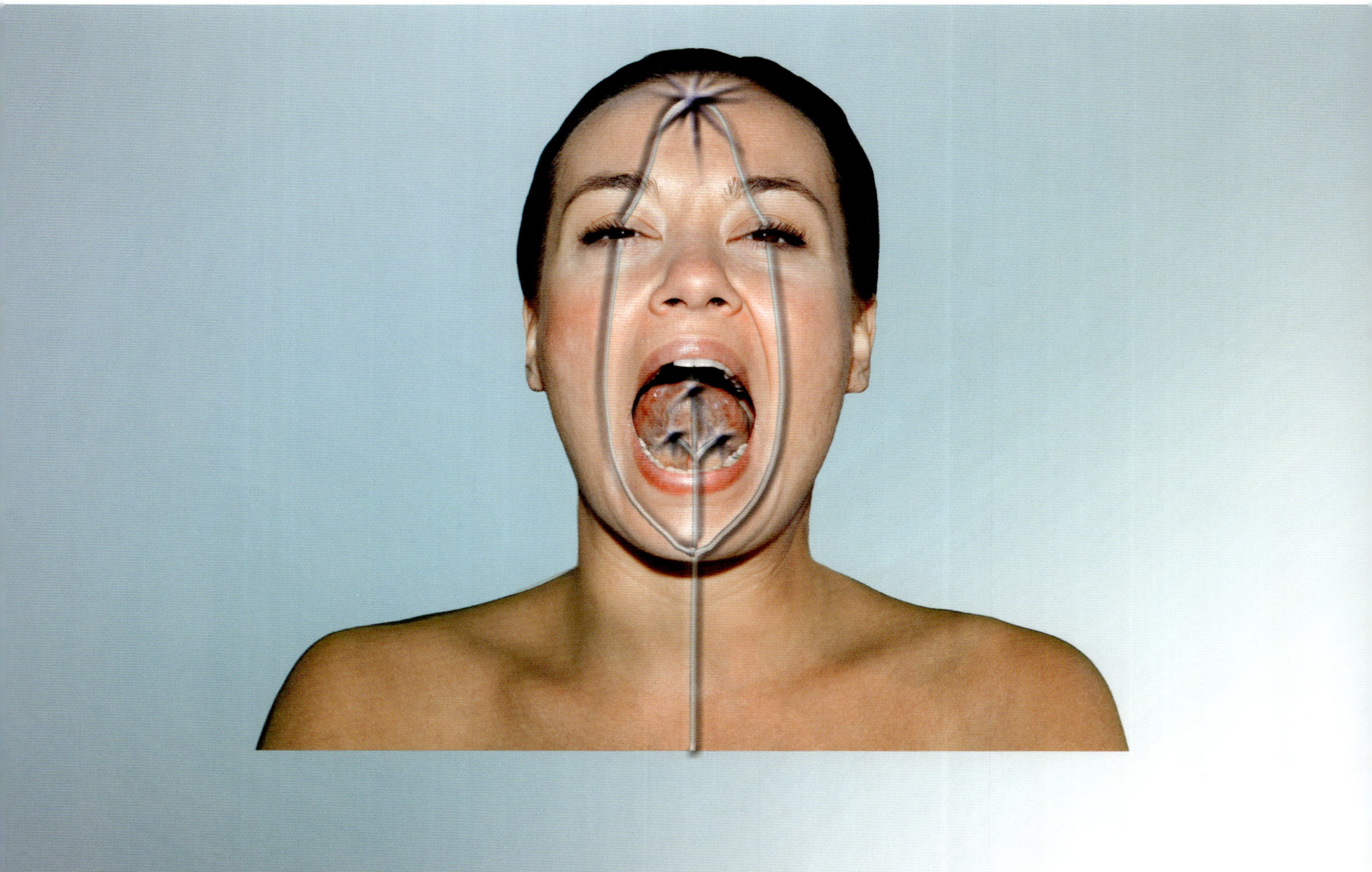

Luo vessels of the Shu Shao Yin, cephalic course

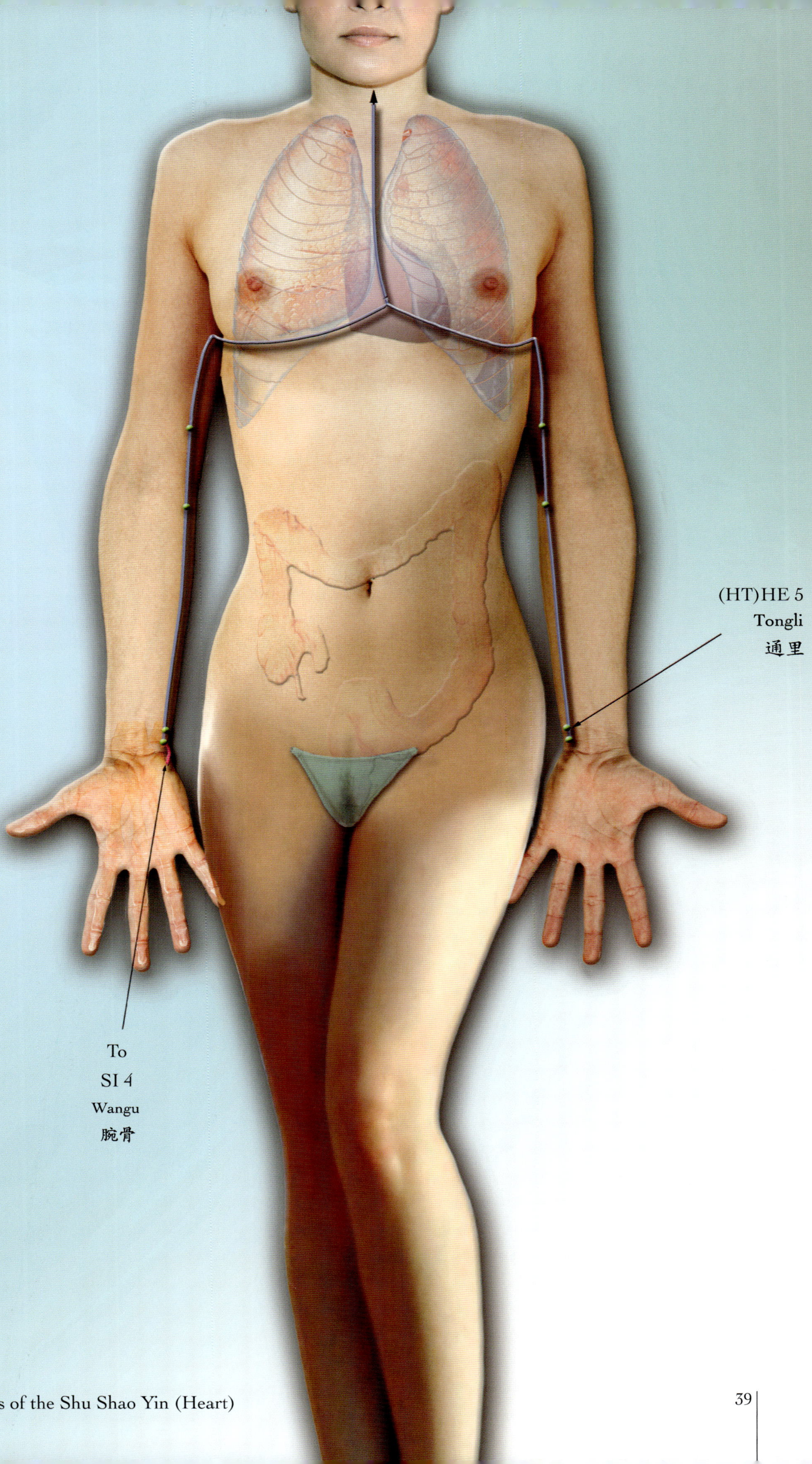

Luo vessels of the Shu Shao Yin (Heart)

Distinct channel of the Shu Shao Yin (Heart)

Ling Shu, chapter III, paragraph 11

The distinct channel of the Main channel Shu Shao Yin of the hand (Heart) enters the GB 22 Yuanye (deep in the armpit) between two tendons.

It belongs to the Heart.

It runs up by the throat, exits by the face and joins the inner canthus.

Commentaries

The distinct channel of Shu Shao Yin of the hand gets out of the axillary area at point (HT)HE 1 Jiquan.

It reaches GB 22 Yuanye at the 4th intercostal space and enters the chest.

It reaches the Heart and joins the distinct channel of Shu Tai Yang (Small Intestine) to form the 4th union.

It goes up and passes by the throat.

It reaches the surface of the face on the level of the cheek and ends its course in the internal ungueal angle of the eye.

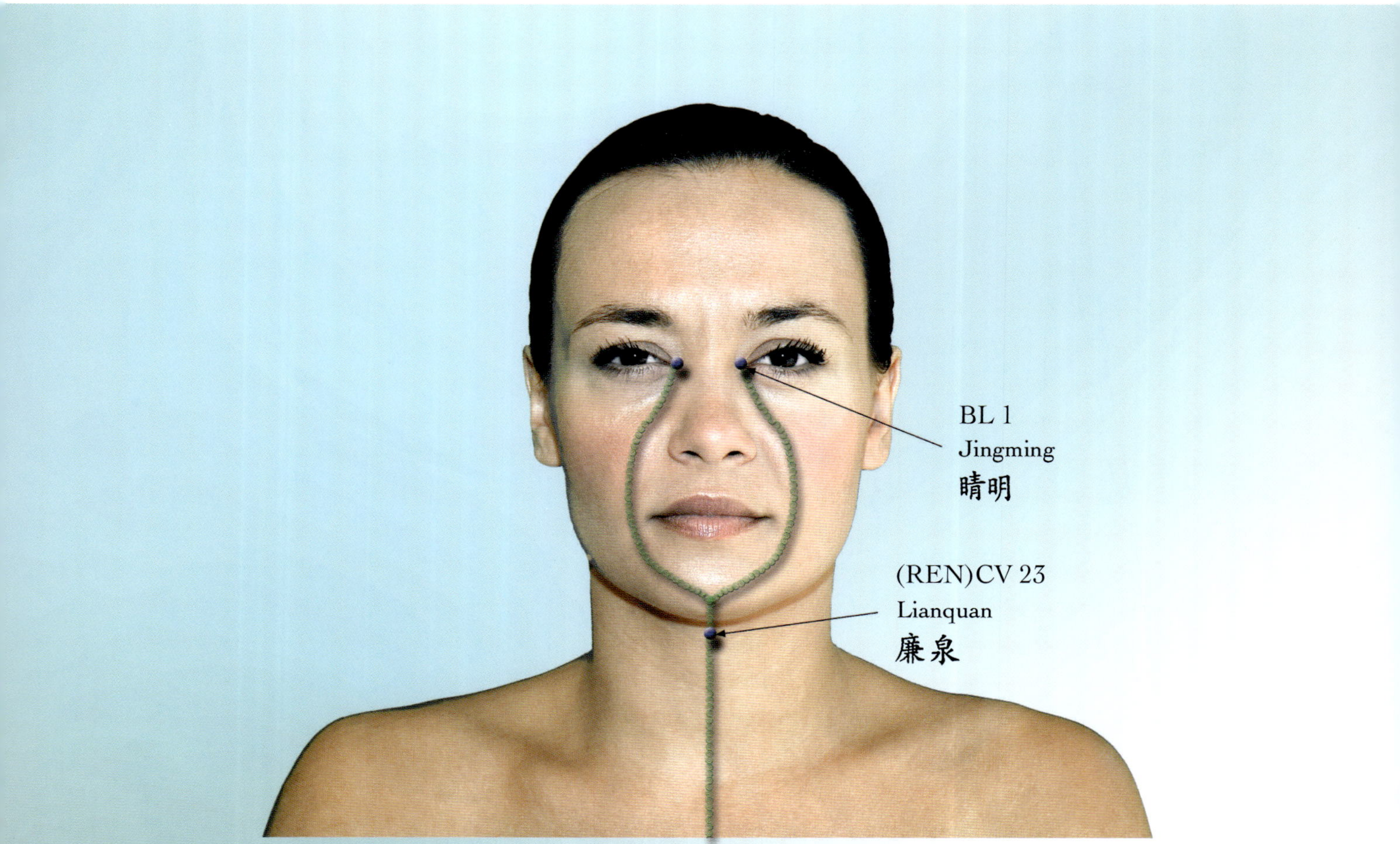

Distinct channel of the Shu Shao Yin, cephalic course

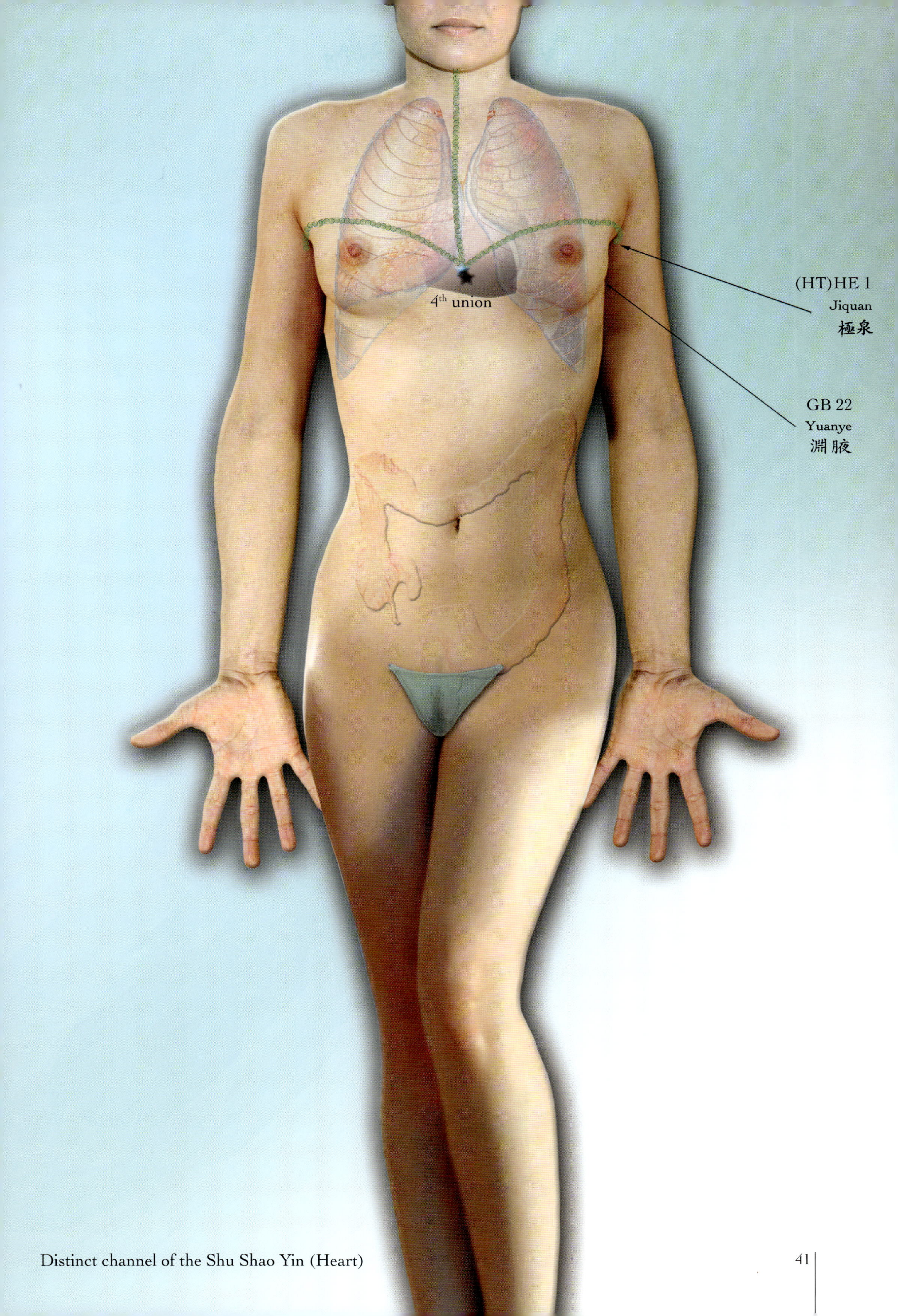

Distinct channel of the Shu Shao Yin (Heart)

Points of the Main channel Shu Shao Yin (Heart)

(HT)HE 1 Jiquan
At the apex of the axillary fossa, where the pulsation of the axillary artery is palpable. (Do NOT needle directly into the artery!)

(HT)HE 2 Qingling
On the medial side of the arm and on the line connecting (HT)HE 1 Jiquan and (HT)HE 3 Shaohai, 3 Cun above the cubital crease, in the groove medial to the biceps muscle of the arm.

(HT)HE 3 Shaohai
With the elbow flexed, at the midpoint of the line connecting the medial end of the cubital crease and the medial epicondyle of the humerus.

(HT)HE 4 Lingdao
On the palmar side of the forearm and on the radial side of the tendon of the ulnar flexor muscle of the wrist, 1.5 Cun proximal to the crease of the wrist.

(HT)HE 5 Tongli
On the palmar side of the forearm and on the radial side of the tendon of the ulnar flexor muscle of the wrist, 1 Cun proximal to the crease of the wrist.

(HT)HE 6 Yinxi
On the palmar side of the forearm and on the radial side of the tendon of the ulnar flexor muscle of the wrist, 0.5 Cun proximal to the crease of the wrist.

(HT)HE 7 Shenmen
On the wrist, at the ulnar end of the crease of the wrist, in the depression of the radial side of the tendon of the ulnar flexor muscle of the wrist.

(HT)HE 8 Shaofu
In the palm, between the 4th and 5th metacarpal bones, at the part of the palm touching the tip of the little finger when a fist is made.

(HT)HE 9 Shaochong
On the radial side of the distal segment of the little finger, 0.1 Cun from the corner of the nail.

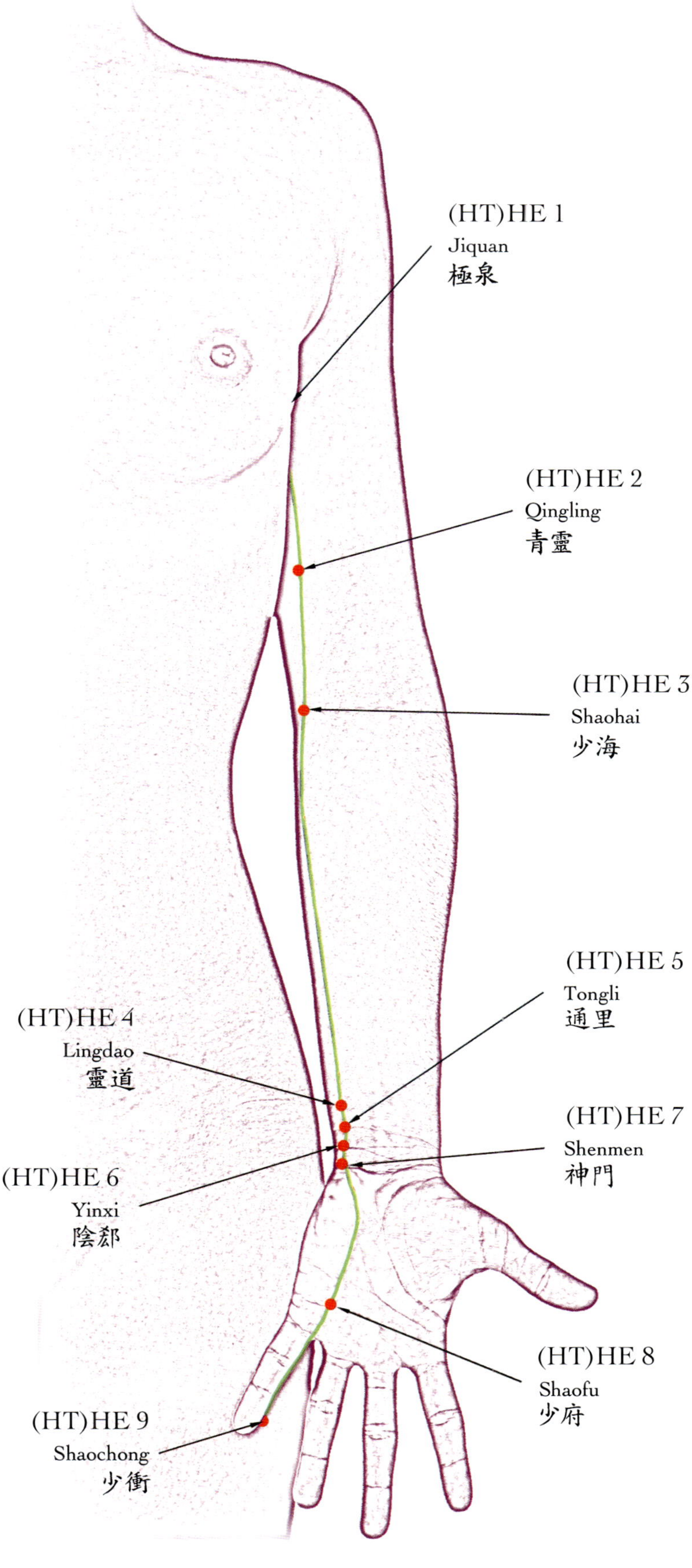

Points of the Main channel Shu Shao Yin (Heart)

The 3 upper Yang

- Shu Yang Ming (Large Intestine)
- Shu Tai Yang (Small Intestine)
- Shu Shao Yang
  (Triple Heater or Triple Burner)

Shu Yang Ming system (Large Intestine)

- Tendino-muscular channel
- Main channel
- Internal branches
- Luo vessels
- Distinct channel
- Points of the Main channel

Tendino-muscular of the Shu Yang Ming (Large Intestine)

***Ling Shu*, chapter IV, paragraph 13**

The tendino-muscular of the Large Intestine starts at the tip of the forefinger.

It unites with the wrist, goes up along the forearm and unites,while going up, with the external side of the elbow.

It goes up on the arm and unites with the acromio-clavicular joint.

A branch goes round the scapula and unites with the backbone.

The vertical branch, going out of point LI 15 Jianju, runs up along the neck.

A branch passes by the cheek bone and unites with the zygomatic.

The tendino-muscular goes up and goes out in front of the tendino-muscular of the Triple Heater or Triple Burner.

It goes up to the left angle of the forehead, passes by the head and goes down to the right mandible.

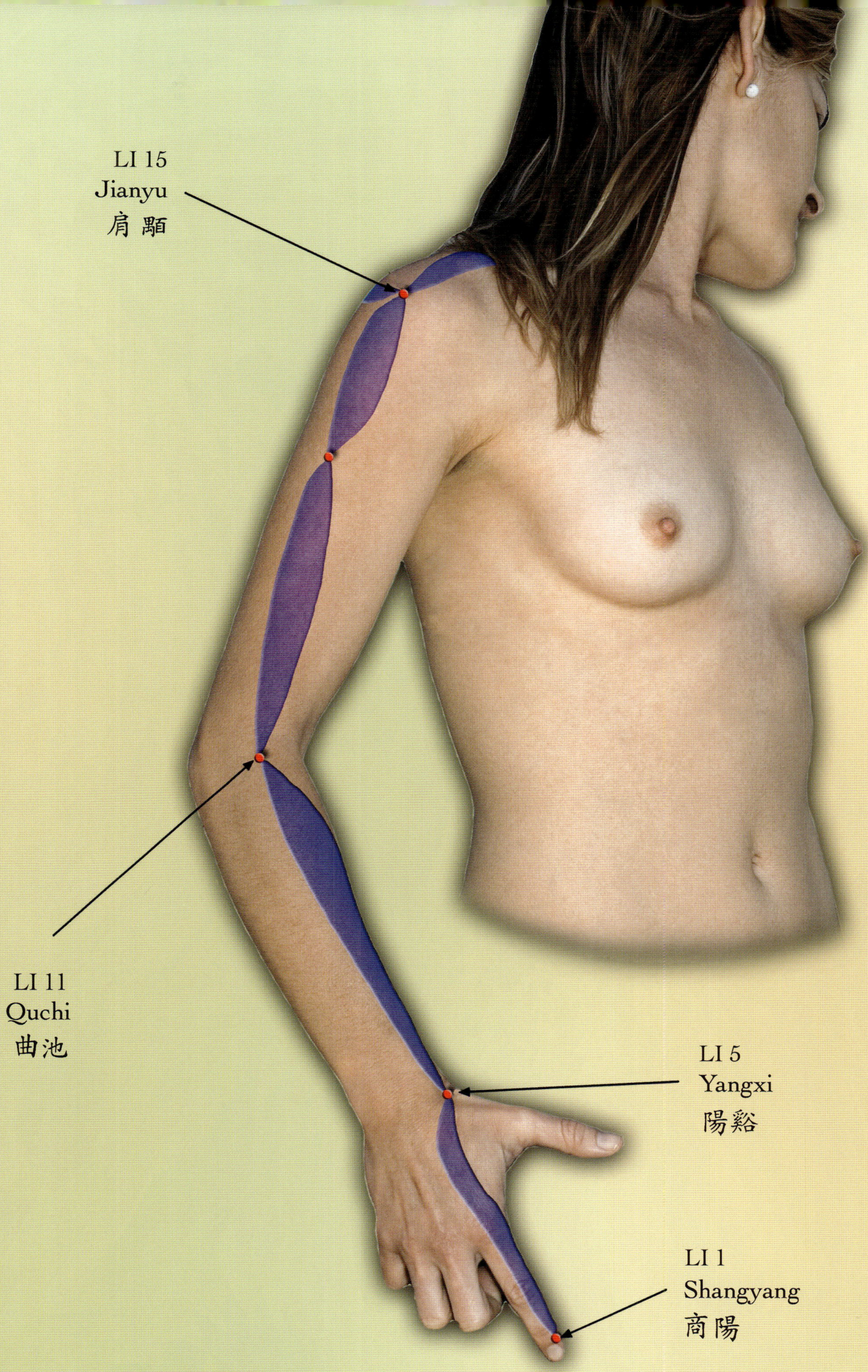
LI 15
Jianyu
肩髃
LI 11
Quchi
曲池
LI 5
Yangxi
陽谿
LI 1
Shangyang
商陽

Tendino-muscular of the Shu Yang Ming (Large Intestine), cephalic course

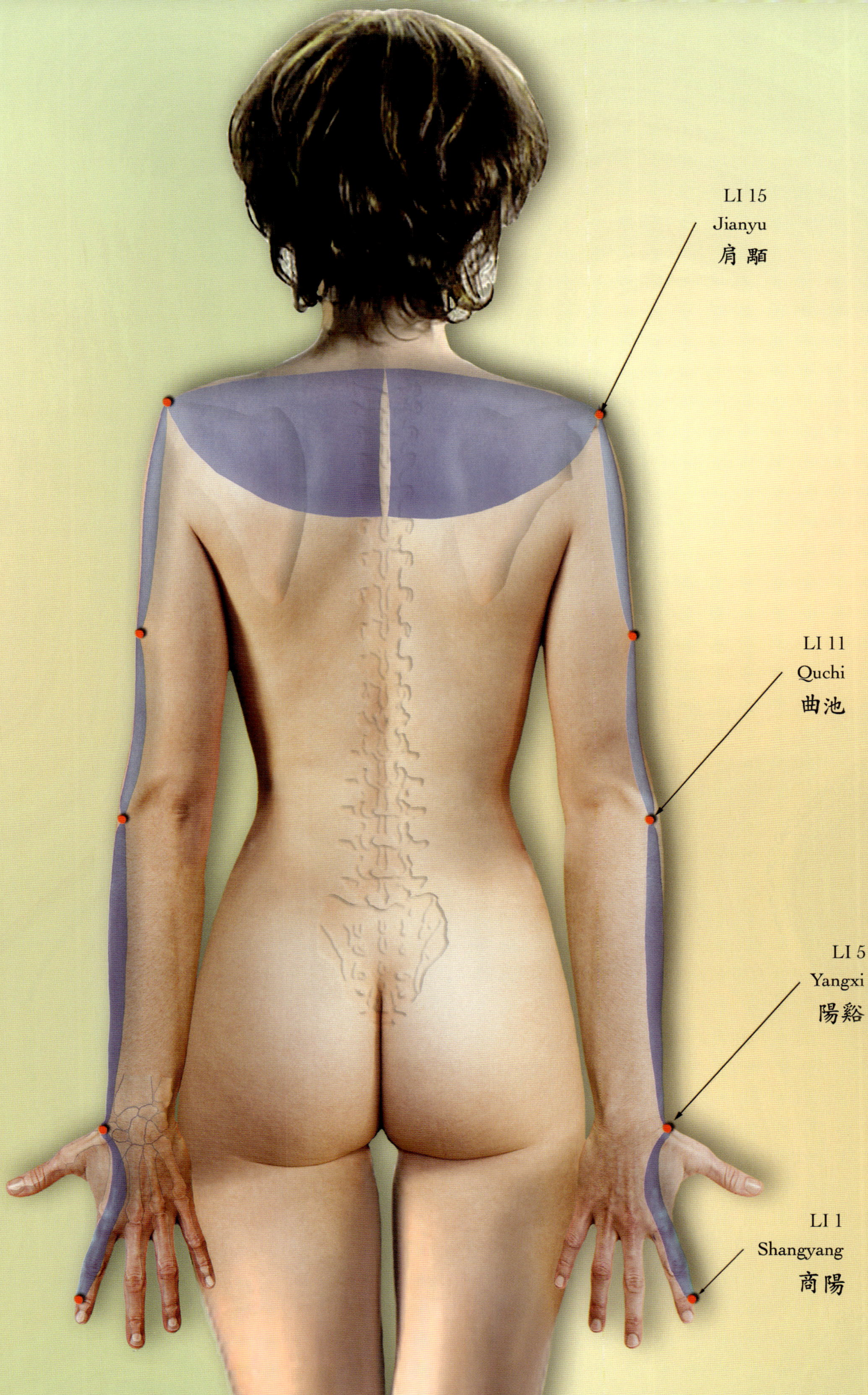

Tendino-muscular of the Shu Yang Ming (Large Intestine), posterior view

MAIN CHANNEL OF THE SHU YANG MING (LARGE INTESTINE)

***Ling Shu*, chapter III, paragraph 10**

The course of the Large Intestine, Shu Yang Ming of the hand, starts at the tip of the forefinger.

It goes along the upper part of the finger (the external side of the forefinger).

It goes out by the Hegu in the interval of the 2 bones (in the angle formed by the 1st and 2nd metacarpal bones).

It goes up and gets inside at the midpoint of the 2 tendons (the long and the short extensor of the thumb).

It goes along the upper part of the forearm (the external side of the forearm).

It goes inside on the external side of the elbow.

It goes up on the antero-external border of the arm. It goes up on the shoulder and goes out by the anterior border of the acromion.

It goes up and goes out by the meeting-point of the backbone (below the process of the 7th cervical).

It goes down again in the supraclavicular fossa, joins the Lungs, goes down on the diaphragm and belongs to the Large Intestine.

The branch starting from the supra-clavicular fossa runs up to the neck, passes by the cheek and penetrates the lower gum.
It goes out encircling the mouth (running round the upper lip) and crossing the Shuigou (Middle of the man, in the philtrum).

The left-hand branch moves to the right, the right-hand branch moves to the left, surrounding the 2 nostrils.

Commentaries

The channel of the Large Intestine starts at the external ungueal angle of the forefinger, at point LI 1 Shangyang.

It goes up, passing by the radial border of the forefinger and between the 1st and 2nd metacarpal bones at point LI 4 Hegu.
It enters the depression between the tendons of the long extensor and of the short palmar.

It goes up by the anterior border of the external border of the forearm, passing by the external part of the elbow where LI 11 Quchi, and after, follows the anterior border of the external side of the arm to reach the shoulder at point LI 15 Jianyu.

From there, it follows the anterior border of the acromion, passes behind in the back at point (DU)GV 14 Dazhui, under the 7th cervical, and runs again to the anterior part of the chest in the supra-clavicular fossa where point ST 12 Quepen is.

From ST 12 Quepen, a branch communicates with the Lung, follows a descending course to go through the diaphragm and ends in the Large Intestine.

From the same point, a branch goes up along the neck to the cheek and penetrates the gum of the lower teeth.

It keeps going, surrounding the upper lip, then crosses its symetrical channel at point (DU)GV 26 Shuigou.

It goes up in direction of the opposite side of the nostril at point LI 20 Yingxiang to connect with the channel of the Stomach, Yang Ming of the foot.

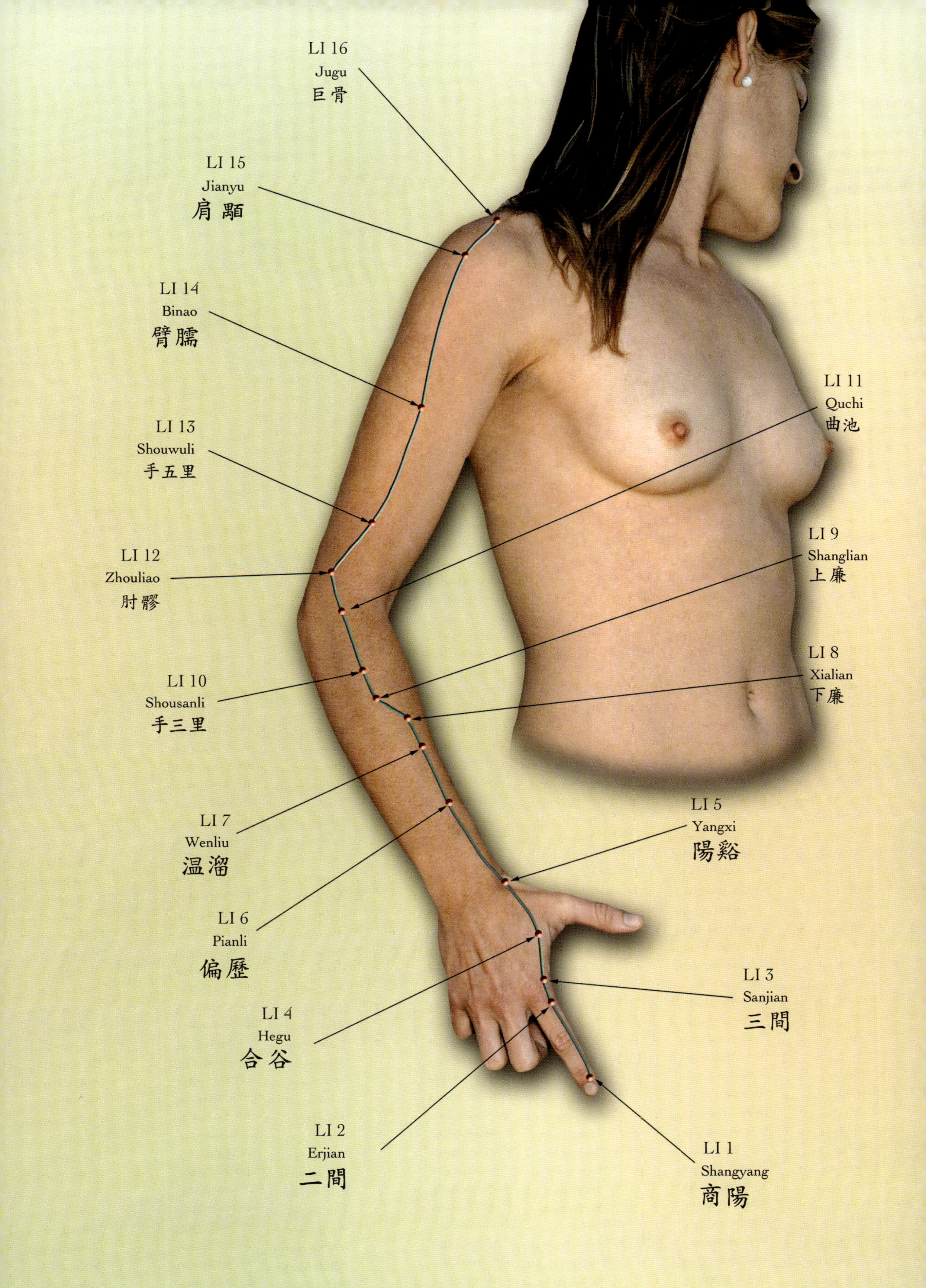

Main channel of the Shu Yang Ming (Large Intestine)

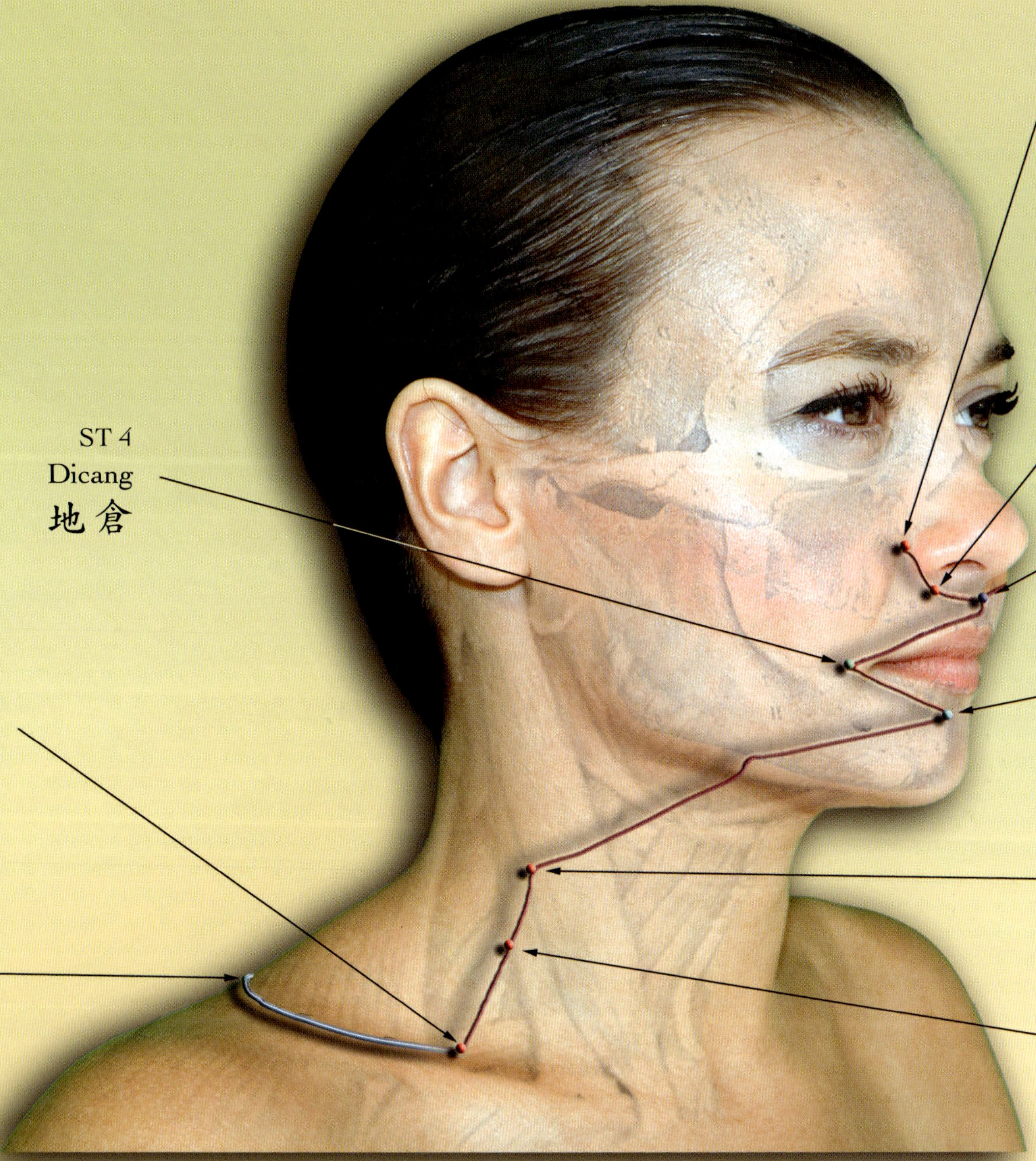

Main channel of the Shu Yang Ming (Large Intestine), cephalic course

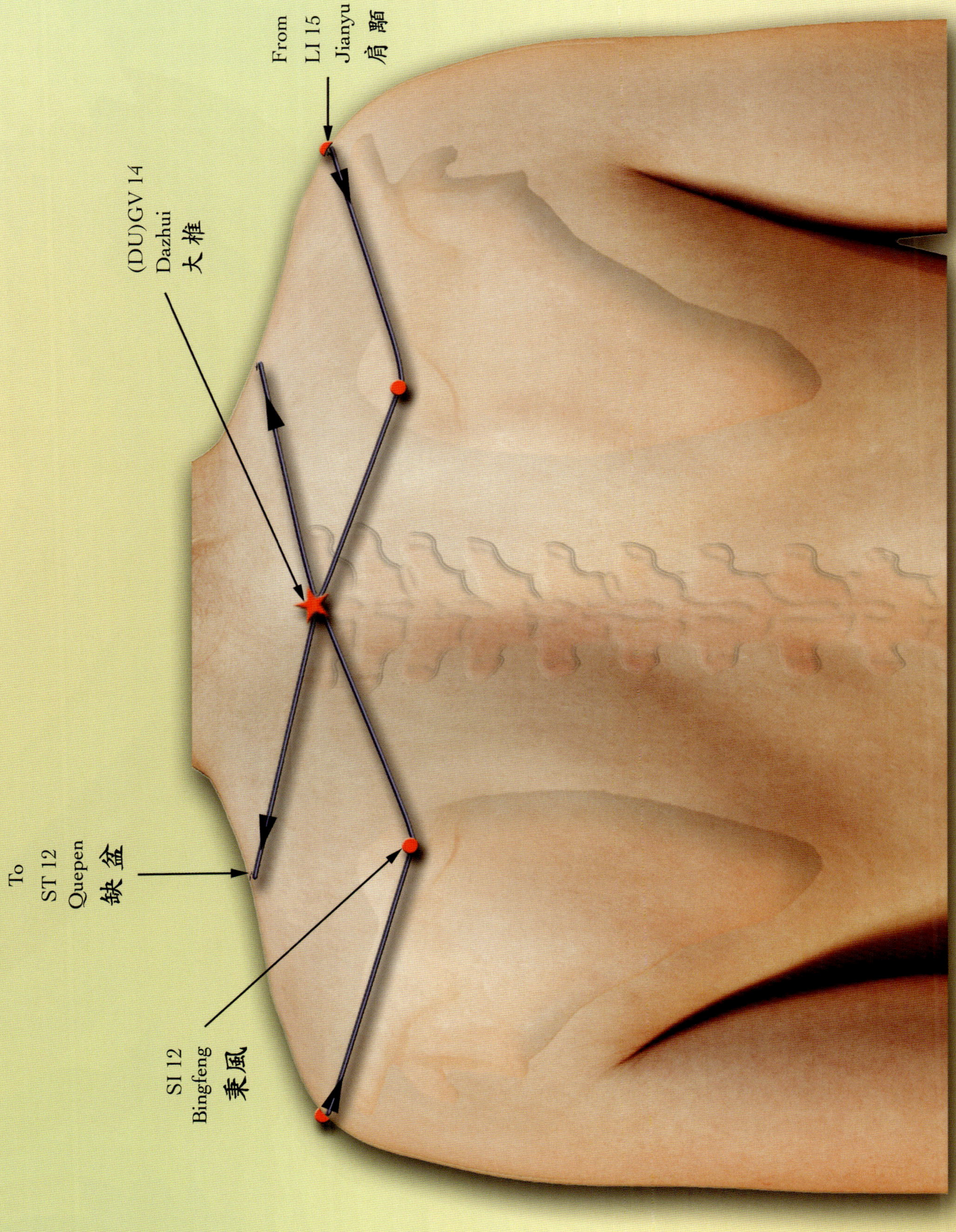

Main channel of the Shu Yang Ming (Large Intestine), dorsal route

Internal branches of the Shu Yang Ming (Large Intestine)

Ling Shu, chapter III, paragraph 10

The internal branch starts from ST 12 Quepen and communicates with the Lung, follows an descending course to go through the diaphragm and ends in the Large Intestine.

Commentaries

The internal branch of the Shu Yang Ming channel starts in the supra-clavicular fossa at point ST 12 Quepen.

From that point, it penetrates the chest to unite with the lungs.

It goes down and goes through the diaphragm.

It ramifies with the Large Intestine on the level of point ST 25 Tianshu.

It continues its descending course to relate with point ST 37 Shangjuxu, point He of specific action of the Large Intestine, located 3 Cun below ST 36 Zusanli.

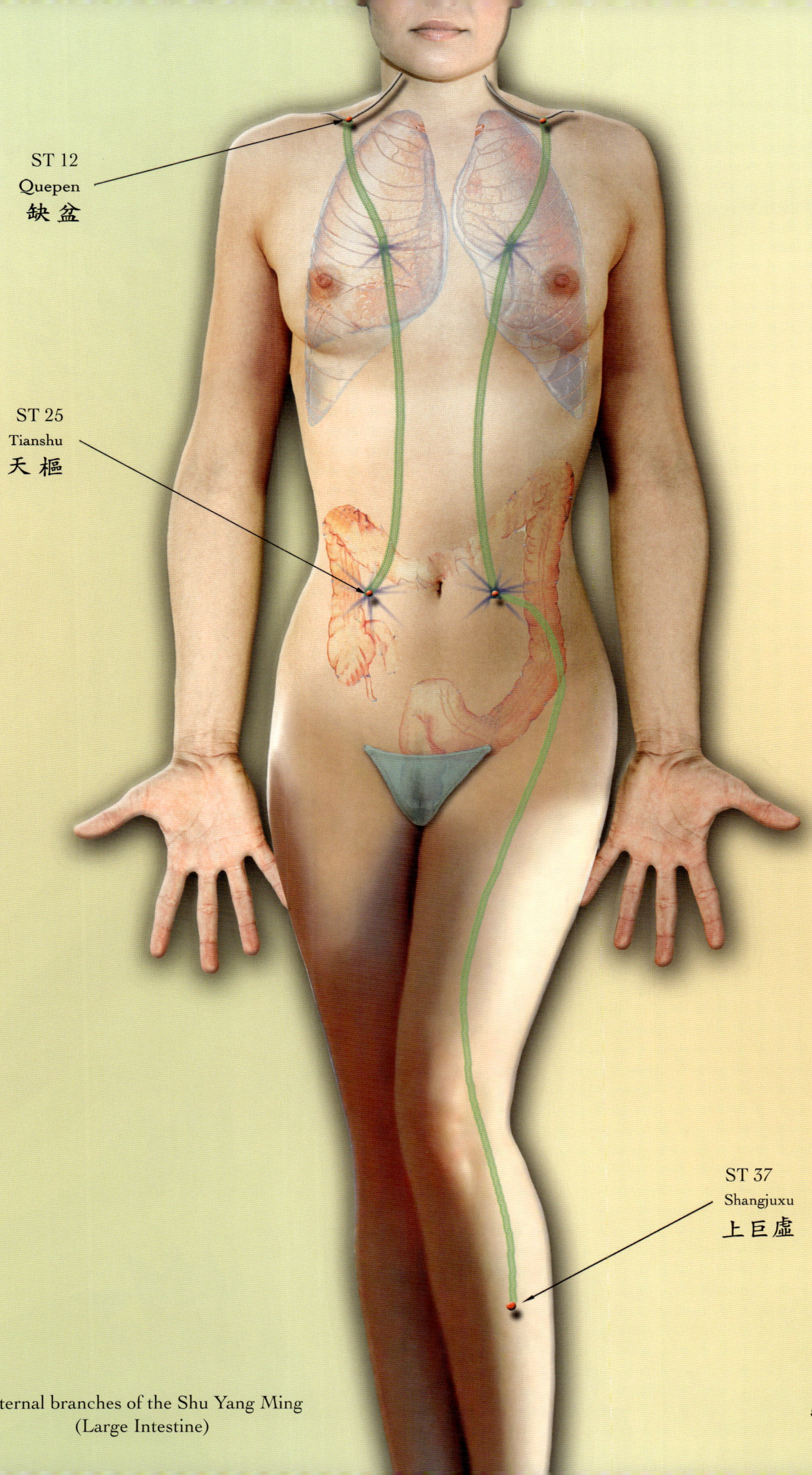

Internal branches of the Shu Yang Ming
(Large Intestine)

Luo vessels of the Shu Yang Ming (Large Intestine)

Ling Shu, chapter III, paragraph 10

The Luo vessel of the Shu Yang Ming of the hand is called Pianli.

It starts 3 Cun above the crease of the wrist and penetrates the Shu Tai Yin.

A branch runs up passing by the forearm, above the acromio-clavicular joint and goes to the inferior maxillary angle.
It unites with the teeth laterally.

A branch penetrates into the ear and joins the 'ancestral vessel'.

Commentaries

- The Luo vessels go out of point LI 6 Pianli 3 Cun above the crease of the wrist.

- The longitudinal vessel follows the course of the Large Intestine channel to reach point LI 15 Jianyu.
  Then it passes by the supra-clavicular fossa to reach point ST 12 Quepen.
  From there, a branch enters the chest to join the Lung.
  From the same point, another branch goes up along the neck to the inferior maxilla then unites with the root of the teeth and penetrates into the ear to join the Main channel.

- The transversal Luo joins the Shu Tai Yin channel of the Lung at point LU 9 Taiyuan.

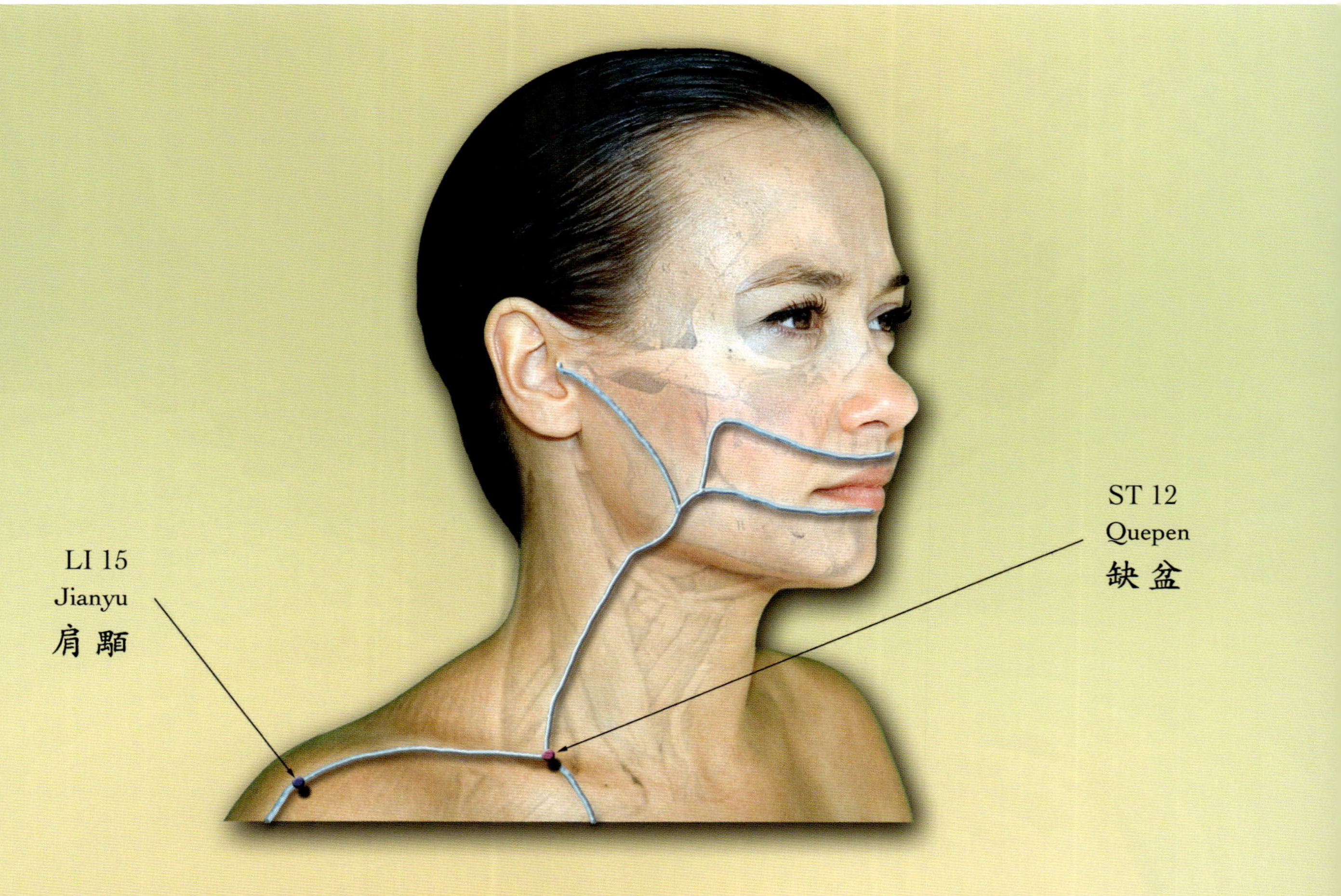

Luo vessels of the Shu Yang Ming (Large Intestine), cephalic course

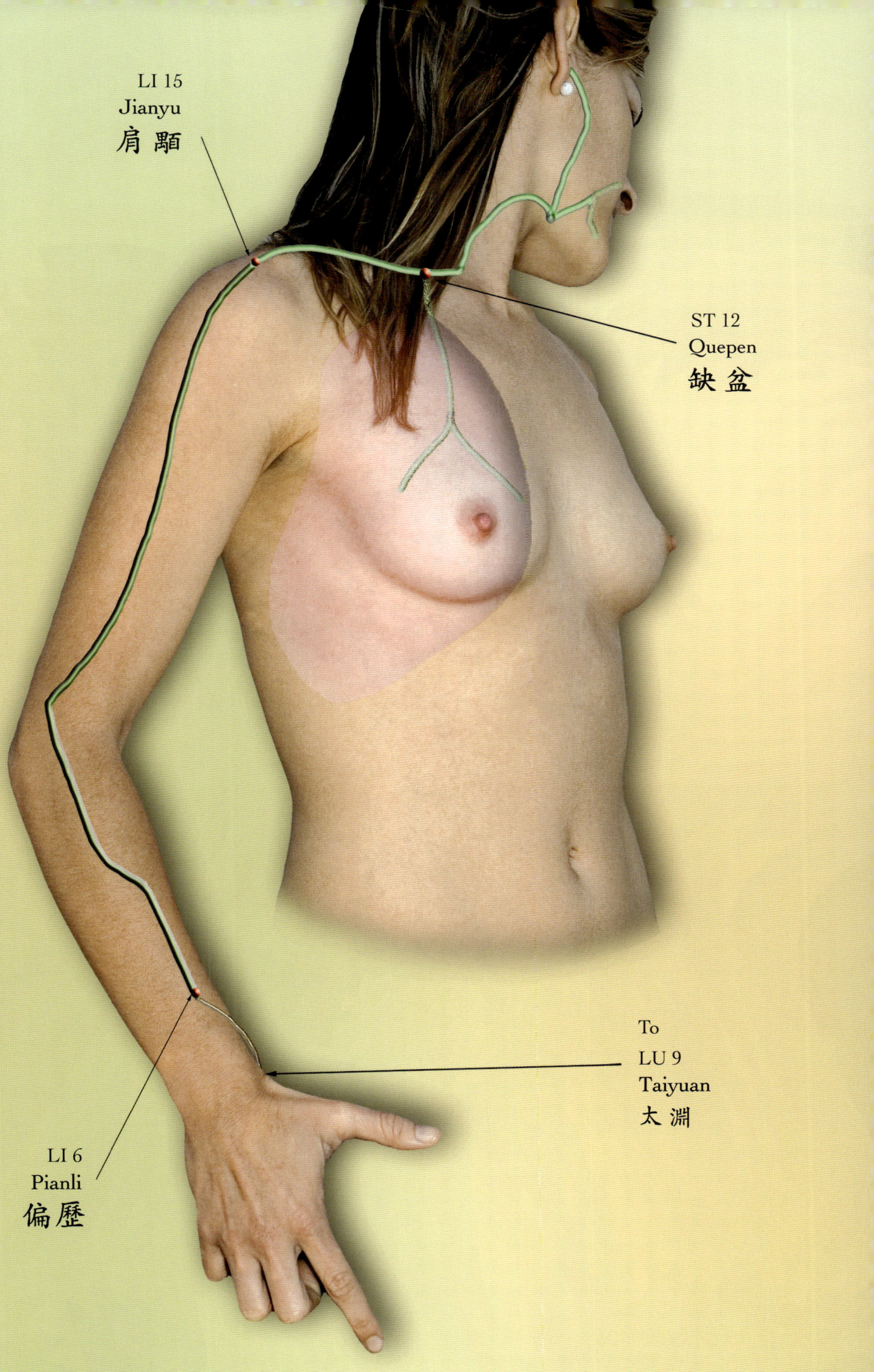

Luo vessels of the Shu Yang Ming (Large Intestine)

Distinct channel of the Shu Yang Ming (Large Intestine)

Ling Shu, chapter III, paragraph 11

The distinct channel of the Shu Yang Ming of the hand goes out of the arm, passes by the nipple and exits on the level of the acromio-clavicular joint.

It enters the Zhugu (bone of the column) that is to say the spinous process of the 7th cervical.

Commentaries

The distinct channel goes out of the Main channel by the hand, the elbow or the shoulder, according to the authors.

It follows the Main channel to the shoulder, at point LI 15 Jianyu.

From that point starts a branch that goes to the chest and unites with the breast.

Another branch goes back to the middle of the supra-clavicular fossa, goes down into the Large Intestine, passes by the Lung, goes out again at point ST 12 Quepen, runs along the neck to join the Main channel on the level of point LI 18 Futu to form the sixth union.

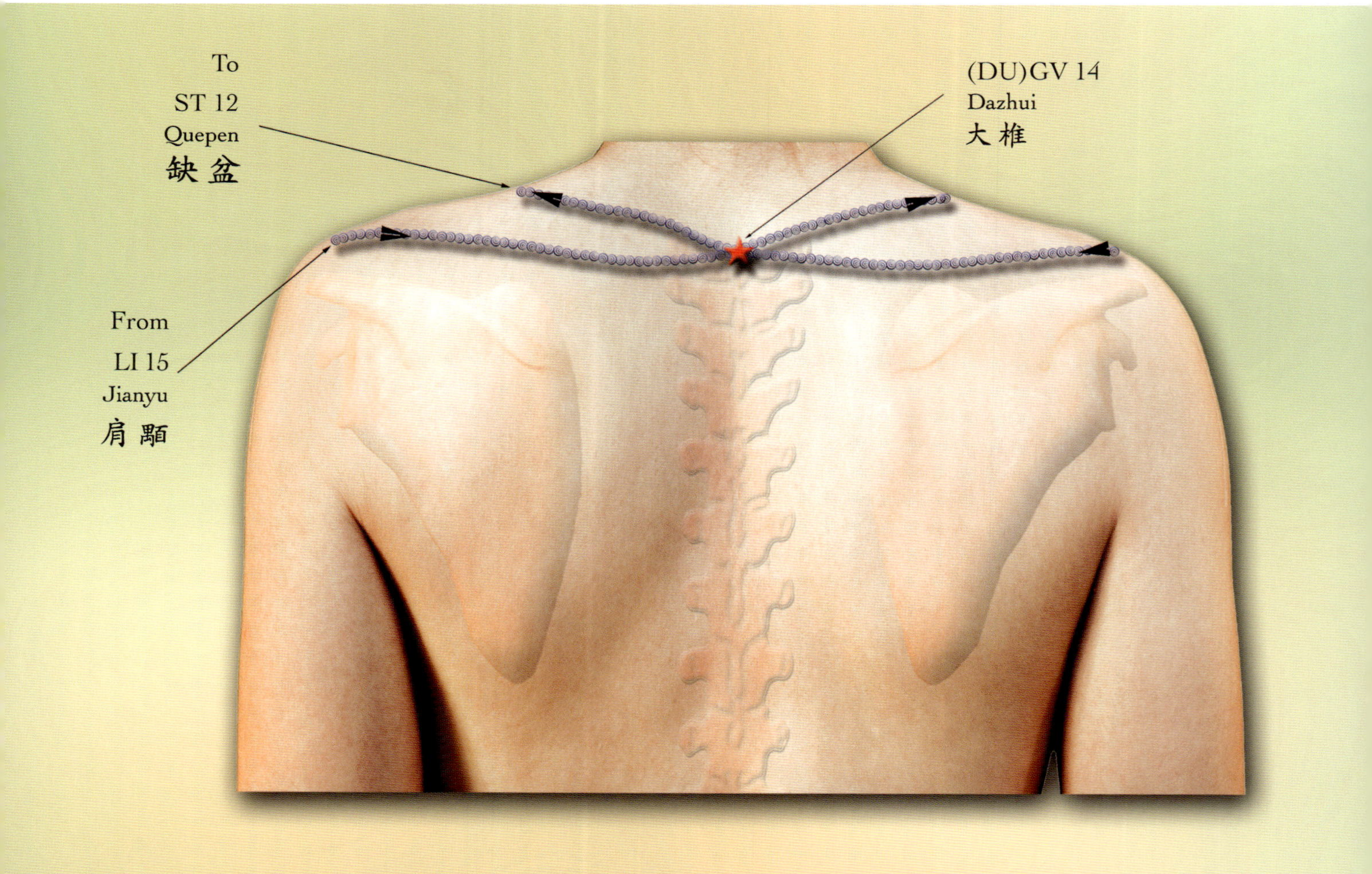

Distinct channel of the Shu Yang Ming (Large Intestine), posterior course

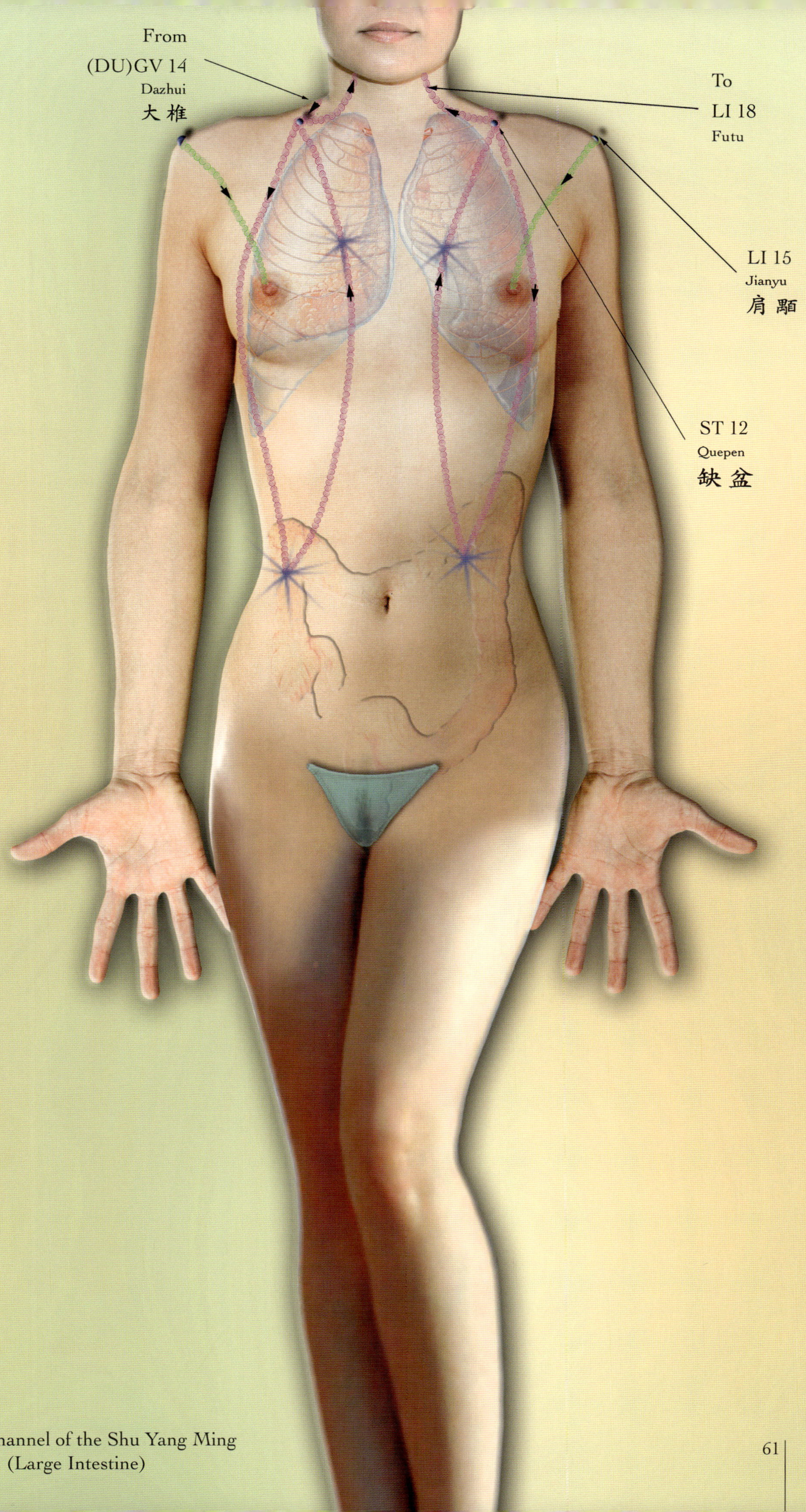

Distinct channel of the Shu Yang Ming
(Large Intestine)

Points of the Main channel of the Shu Yang Ming (Large Intestine)

LI 1 Shangyang
On the radial side of the distal segment of the indes-finger, 0.1 Cun from the corner of the nail.

LI 2 Erjian
In the depression of the radial side, distal to the 2nd metacarpophalangeal joint when a loose fist is made.

LI 3 Sanjian
In the depression of the radial side, proximal to the 2nd metacarpophalangeal joint when a loose fist is made.

LI 4 Hegu
On the back of the hand between the first and the second metacarpal bone, in the middle of the radial edge of the second metacarpal.

LI 5 Yangxi
At the radial end of the crease of the wrist, in the depression between the tendons of the short extensor and long extensor muscles of the thumb when the thumb is tilted upwards.

LI 6 Pianli
With the elbow slightly flexed, on the radial side of the dorsal surface of the forearm and on the line connecting LI 5 Yangxi and LI 11 Quchi, 3 Cun above the crease of the wrist.

LI 7 Wenliu
With the elbow flexed, on the radial side of the dorsal surface of theforearm and on the line connecting LI 5 Yangxi and LI 11 Quchi, 5 Cun above the crease of the wrist.

LI 8 Xialian
On the radial side of the dorsal surface of the forearm and on the line connecting LI 5 Yangxi and LI 11 Quchi, 4 Cun below the cubital crease.

LI 9 Shanglian
On the radial side of the dorsal surface of the forearm and on the line connecting LI 5 Yangxi and LI 11 Quchi, 3 Cun below the cubital crease.

LI 10 Shousanli
On the radial side of the dorsal surface of the forearm and on the line connecting LI 5 Yangxi and LI 11 Quchi, 2 Cun below the cubital crease.

LI 11 Quchi
With the elbow flexed, at the lateral end of the cubital crease, at the midpoint of the line connecting LU 5 Chize and the external humeral epicondyle.

LI 12 Zhouliao
With the elbow flexed, on the lateral side of the upper arm, 1 Cun above LI 11 Quchi, on the border of the humerus.

LI 13 Shouwouli
On the lateral side of the upper arm and on the line connecting LI 11 Quchi and LI 15 Jianyu, 3 Cun above LI 11 Quchi.

LI 14 Binao
On the external side of the arm at the lower end of the deltoid and a line connecting LI 11 Quchi and LI 15 Jianyu, 7 Cun above LI 11 Quchi.

LI 15 Jianyu
On the shoulder, in the depression located in front of the lower edge of the acromion in the upper deltoid when the arm is abducted or raised in front at the shoulder.

LI 16 Jugu
On the shoulder, in the depression between the acromial extremity of the clavicle and scapular spine.

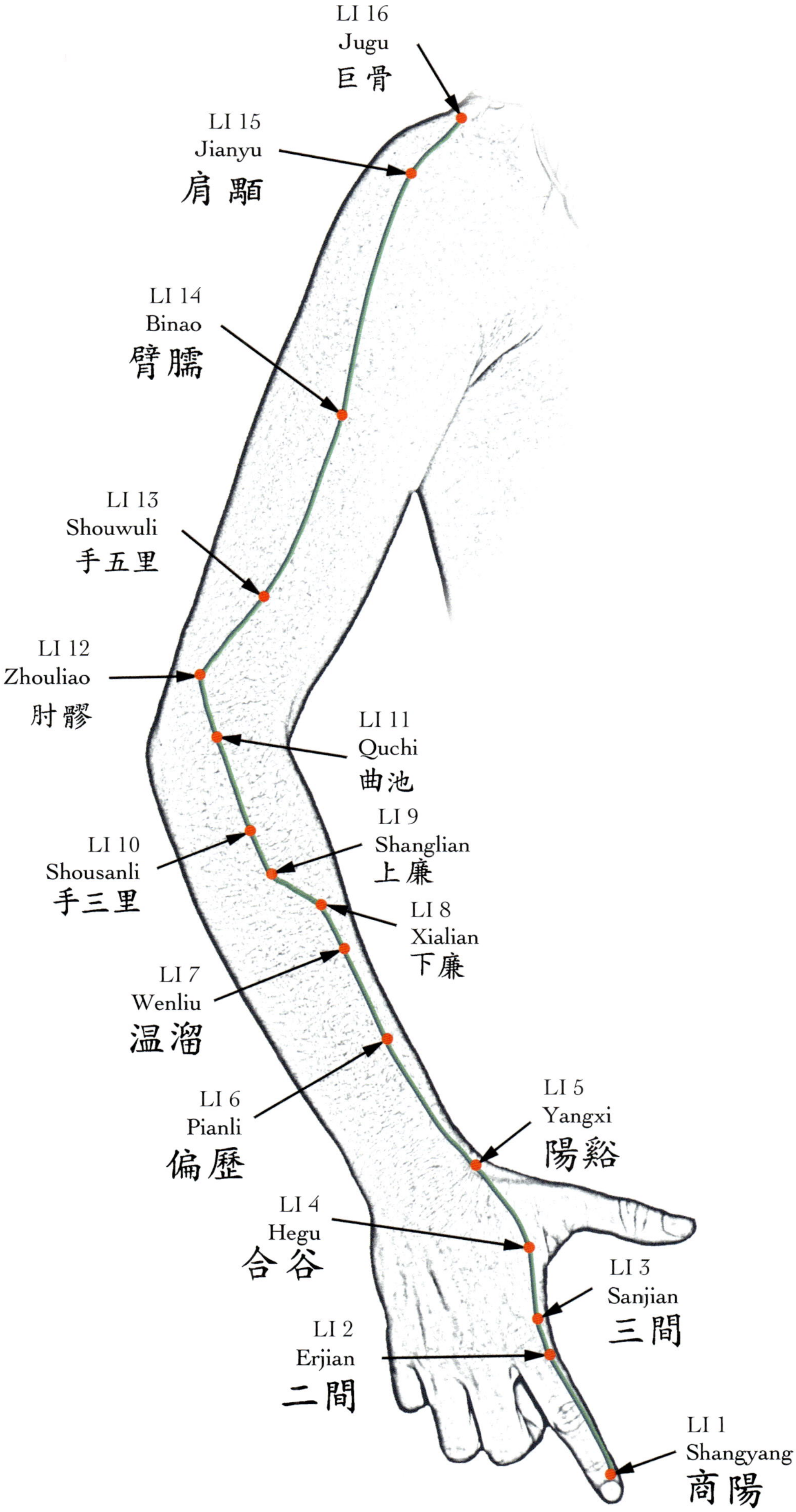

Points of the Main channel of the Shu Yang Ming (Large Intestine)

LI 17 Tianding
On the lateral side of the neck, at the posterior border of the sterno-cleido-mastoid muscle beside the laryngeal protuberance, at the midpoint of the line connecting LI 18 Futu and ST 12 Quepen.

LI 18 Futu
On the lateral side of the neck, beside the laryngeal protuberance, between the anterior and posterior borders of the sternocleidomastoid muscle.

LI 19 Zouliao
On the upper lip, directly below the lateral border of the nostril, on the level of (DU)GV 26 Shuigou (Renzhong).

LI 20 Yingxiang
In the nasolabial groove, beside the midpoint of the lateral border of the nasal ala.

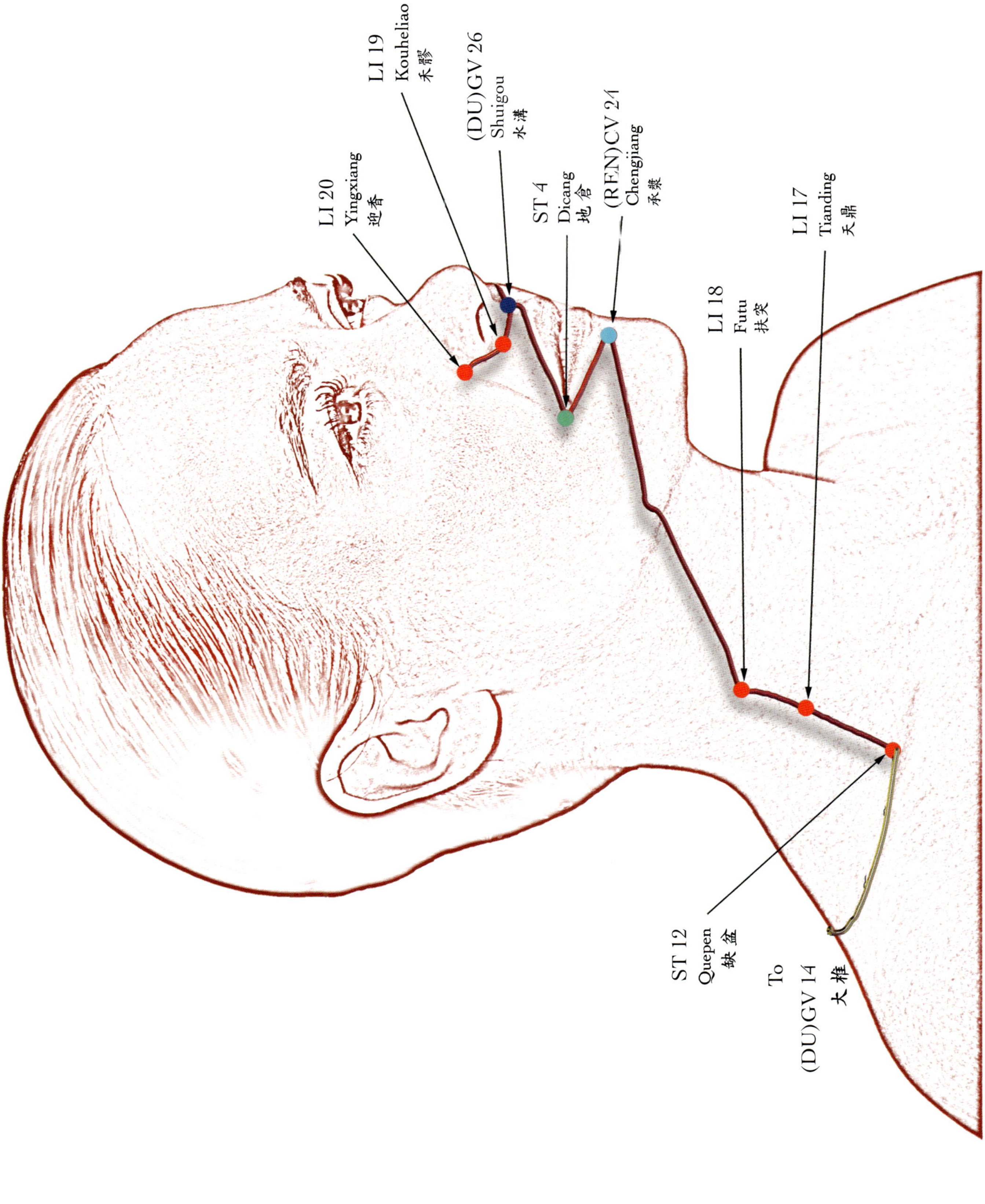

Points of the Main channel of the Shu Yang Ming (Large Intestine)

Shu Tai Yang system (Small Intestine)

- Tendino-muscular channel
- Main channel
- Internal branches
- Luo vessels
- Distinct channel
- Points of the Main channel

Tendino-muscular of the Shu Tai Yang (Small Intestine)

***Ling Shu*, chapter IV, paragraph 13**

The tendino-muscular of the Small Intestine starts above the little finger of the hand.

It unites with the wrist and goes up along the internal side of the forearm.

It passes behind the olecranon.

It penetrates under the armpit where it spreads.

This branch continues, backward, to the posterior border of the armpit and goes round the shoulder-blade.

It passes along the neck and goes out in front of the tendino-muscular of the Bladder.

It unites behind the ear at point GB 12 Wangu.

Another branch enters at the midpoint of the ear.

The vertical branch goes out above the ear and unites with the part of the maxilla which is situated under the ear.

It goes up and arrives at the outer canthus.

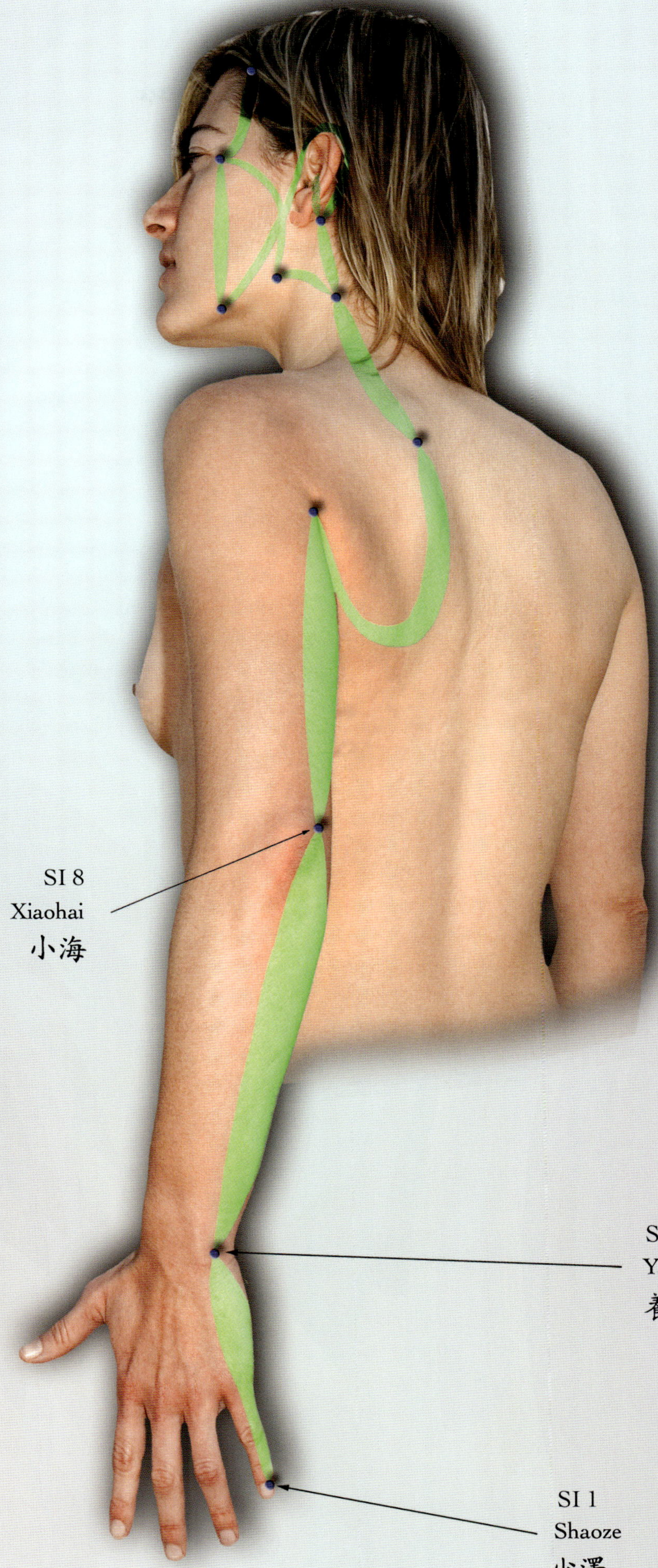

Tendino-muscular of the Shu Tai Yang (Small Intestine)

Tendino-muscular of the Shu Tai Yang (Small Intestine)

Commentaries

On that level, the tendino-muscular divides into 2 branches.

– The first one passes under the armpit.

– The second one passes by the posterior part of the shoulder blade, goes up by the lateral part of the neck in front of the tendino muscular of the Bladder and unites with the tendino muscular of Shu Shao Yang (Triple Heater or Triple Burner).

From that point start 2 branches, one anterior, the other posterior.

– The anterior branch passes by the angle of the lower maxilla, follows the vertical branch of the maxilla, goes up in front of the ear to the zygomatic arch, and reaches the edge of the orbital cavity.

– The posterior branch unites with the mastoid process and sends out a branch penetrating the ear. It goes up to the posterior part of the ear, goes round it, runs down to join the lower maxilla, goes up to reach the external border of the orbital cavity and unites with it.

It finally gets to the frontal area and ends at point GB 13 Benshen where it unites with the other Yang tendino-musculars of the upper limb.

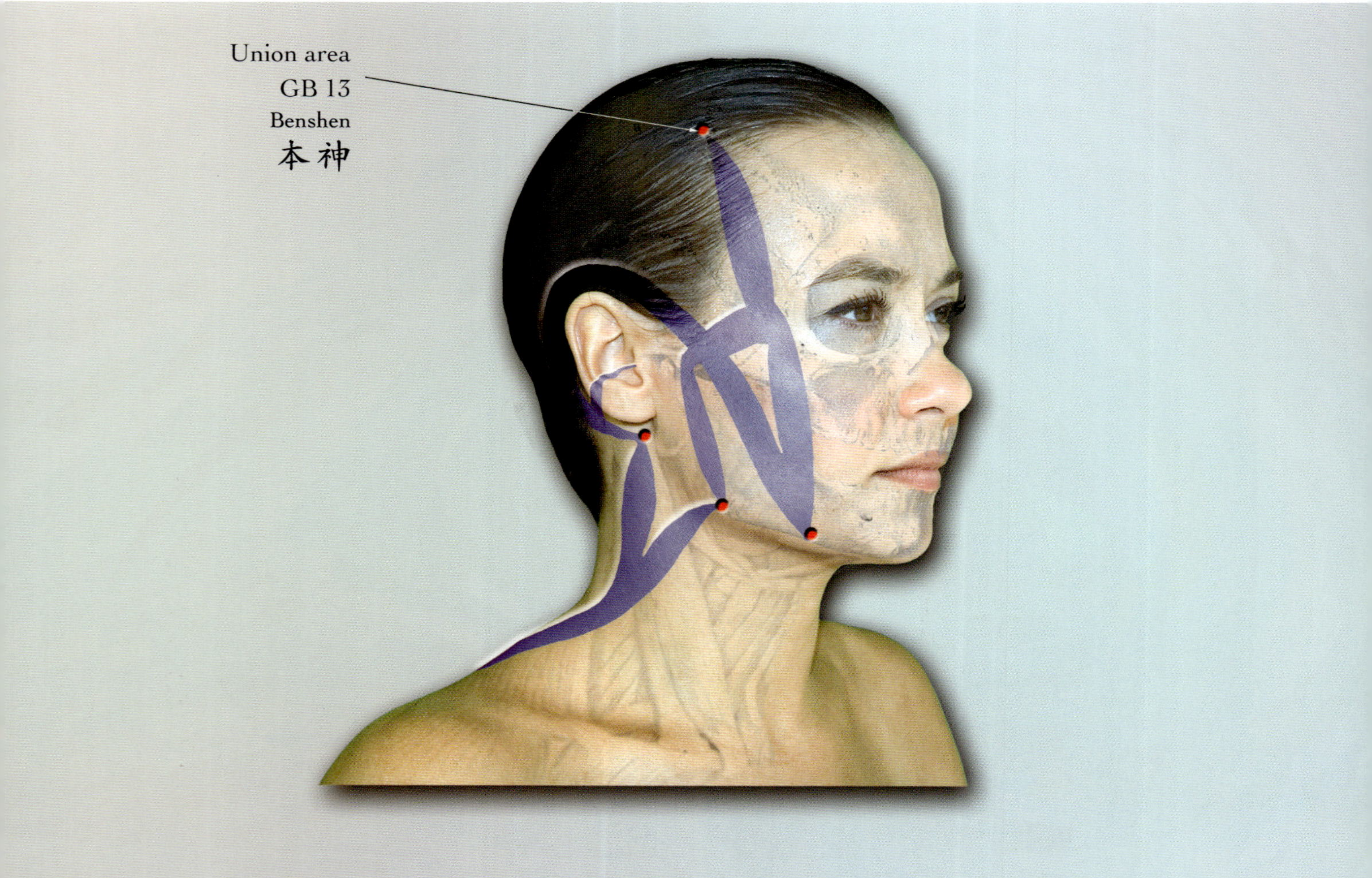

Tendino-muscular of the Shu Tai Yang (Small Intestine), cephalic course

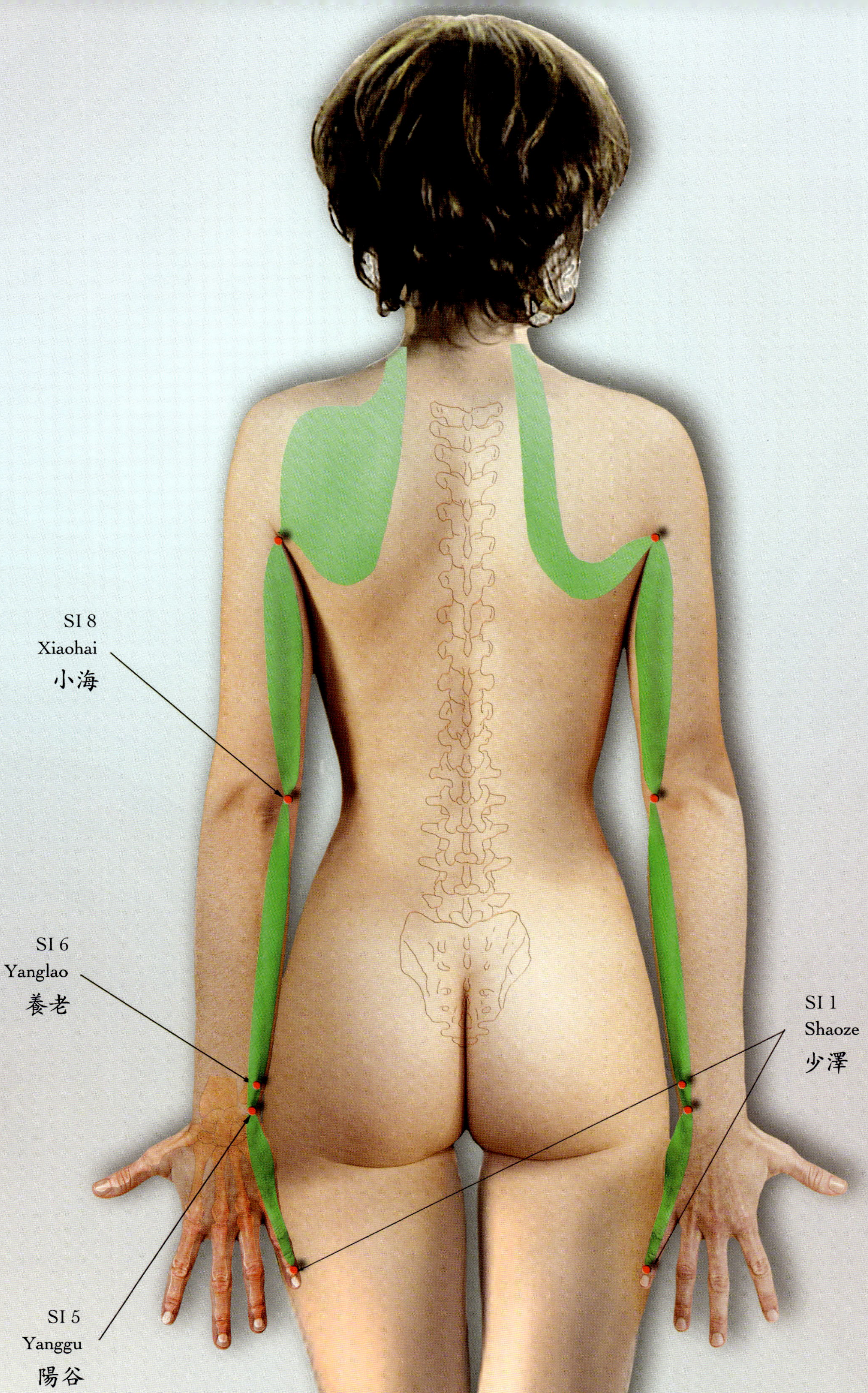

Tendino-muscular of the Shu Tai Yang (Small Intestine), posterior view

Main channel of the Shu Tai Yang (Small Intestine)

***Ling Shu,* chapter III, paragraph 10**

The vessel of the Small Intestine, Shu Tai Yang of the hand, starts at the tip of the little finger, follows the external side of the hand (we are dealing with the internal border of the hand) and goes up on the wrist.

It exits by the process and goes up directly on the forearm by the inferior extremity of the cubitus.

It goes out by the posterio-internal side of the elbow between the tendons.

It goes up on the posterio-internal part of the arm and goes out by the joint of the shoulder, enters the supra-clavicular fossa and joins the Heart.

It goes along the oesophagus and goes down on the diaphragm, reaches the Stomach and belongs to the Small Intestine.

The branch that starts from the supra-clavicular fossa goes along the neck, goes up on the malar and reaches the outer canthus.

It goes back 'to the midpoint of the ear' (in reality, to the centre of the tragus).

A separate branch goes up from the lower maxilla under the inferior orbital edge, passes by the nose and reaches the inner canthus.

An oblique branch joins the malar area.

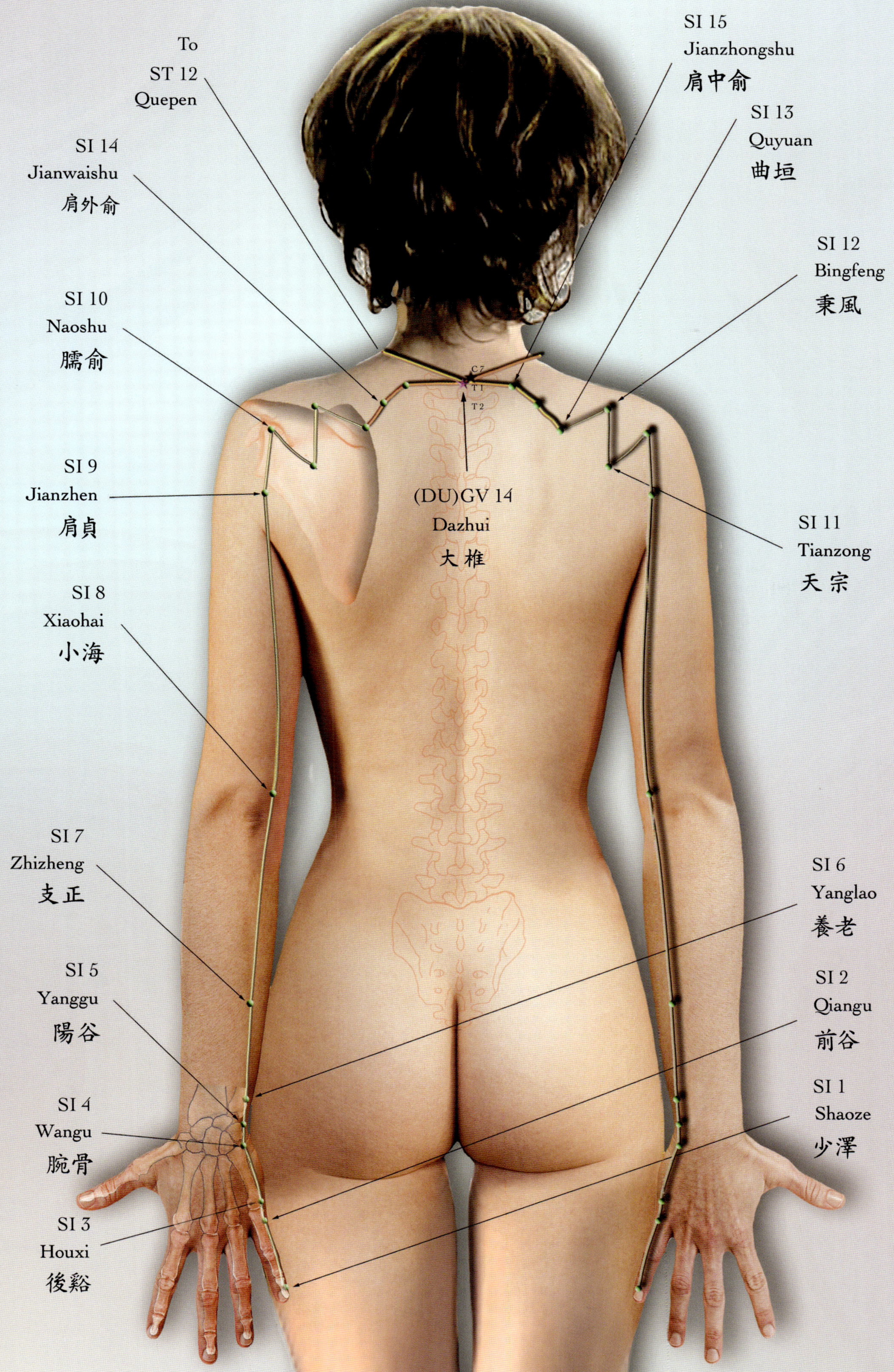

Main channel of the Shu Tai Yang (Small Intestine)

From the shoulder, the branch goes to the shoulder-blade and joins with point (DU)GV 14 Dazhui, at the tip of the spinous process of (HT)HE 7 Shenmen, passing, or not, according to the authors, by points BL 41 Fufen and BL 11 Dazhu.

It then goes back to the supra-clavicular fossa at point ST 12 Quepen.

From that point, the channel goes up by the external side of the neck, passes by point SI 17 Tianrong in front of the anterior border of the sterno-cleido-mastoïd.

It reaches the cheek at point SI 18 Quanliao, on the lower edge of the zygomatic arch.

From point SI 18 Quanliao, a branch joins point BL 1 Jingming, above the inner canthus while another branch joins the external edge of the orbital cavity and crosses point GB 1 Tongziliao.

This branch flows forward from the ear and crosses point (TH)TB 22 Erheliao.

It enters the ear at point SI 19 Tinggong where the channel ends.

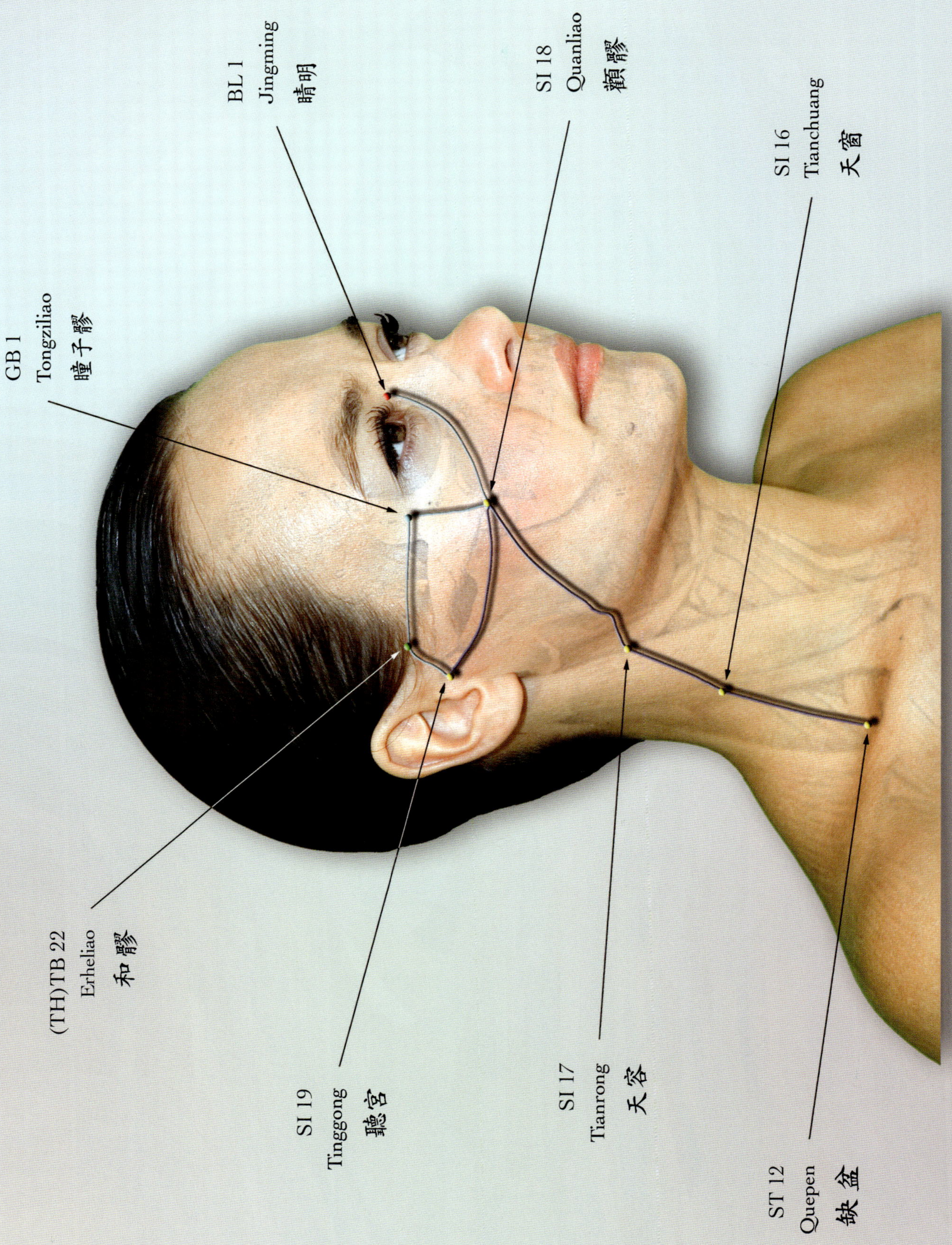

Main channel of the Shu Tai Yang (Small Intestine), cephalic course

Internal branches of the Shu Tai Yang (Small Intestine)

***Ling Shu*, chapter III, paragraph 10**

The internal branch goes up again
on the posterio-internal part of the arm
and goes out by the joint of the shoulder,
enters the supra-clavicular fossa
and joins the Heart.

It goes along the oesophagus and descends
on the diaphragm, reaches the Stomach
and belongs to the Small Intestine.

Commentaries

The internal branch starts at the supra-clavicular fossa, at point ST 12 Quepen.

It penetrates into the chest to unite with the Heart.

It goes down by the oesophagus, goes through the diaphragm and unites with the Stomach.

On that level, it crosses point (REN)CV 12 Zhongwan and point (REN)CV 13 Shangwan deep inside the body, then keeps descending to finally end in the Small Intestine.

A branch connects with point ST 39 Xiajuxu, 6 Cun below ST 36 Zusanli.

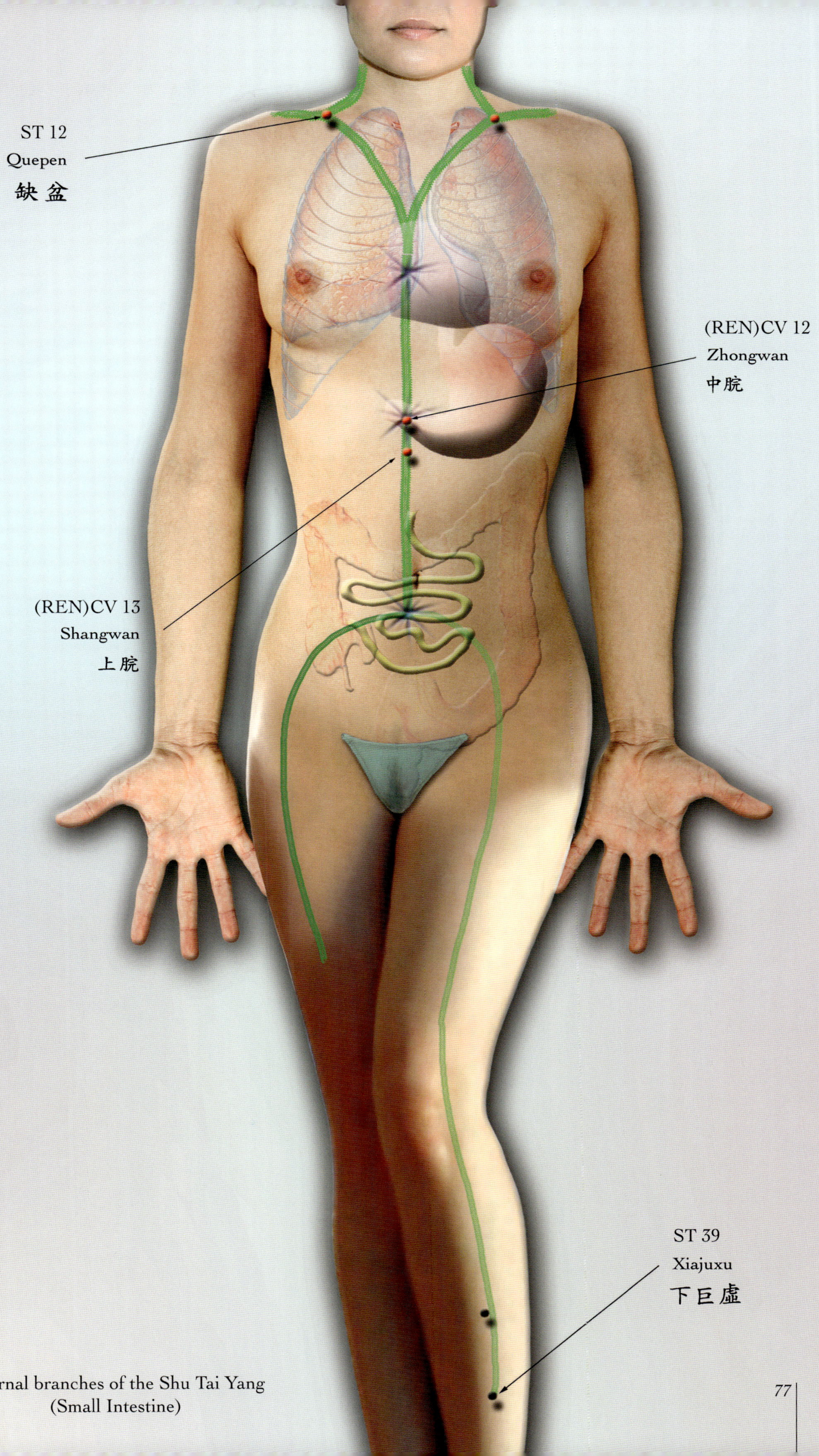

Internal branches of the Shu Tai Yang
(Small Intestine)

Luo vessels of the Shu Tai Yang (Small Intestine)

***Ling Shu,* chapter III, paragraph 10**

The Luo vessel of the Shu Tai Yang of the hand is called Zhizheng and goes out 5 Cun above the inside of the wrist.

It unites with Shu Shao Yin (Heart) channel.

A distinct branch goes up from the elbow and joins the acromio-clavicular joint at point LI 15 Jianyu.

Commentaries

It goes out from point SI 7 Zhizheng, 5 Cun above the crease of the wrist, outside the posterior border of the cubitus, and divides into two branches:

- a longitudinal branch that reaches point LI 15 Jianyu, then passes by point ST 12 Quepen and penetrates into the chest to finally end its course in the Heart,
- a transversal branch that joins Shu Shao Yin (Heart) channel at point (HT)HE 7 Shenmen.

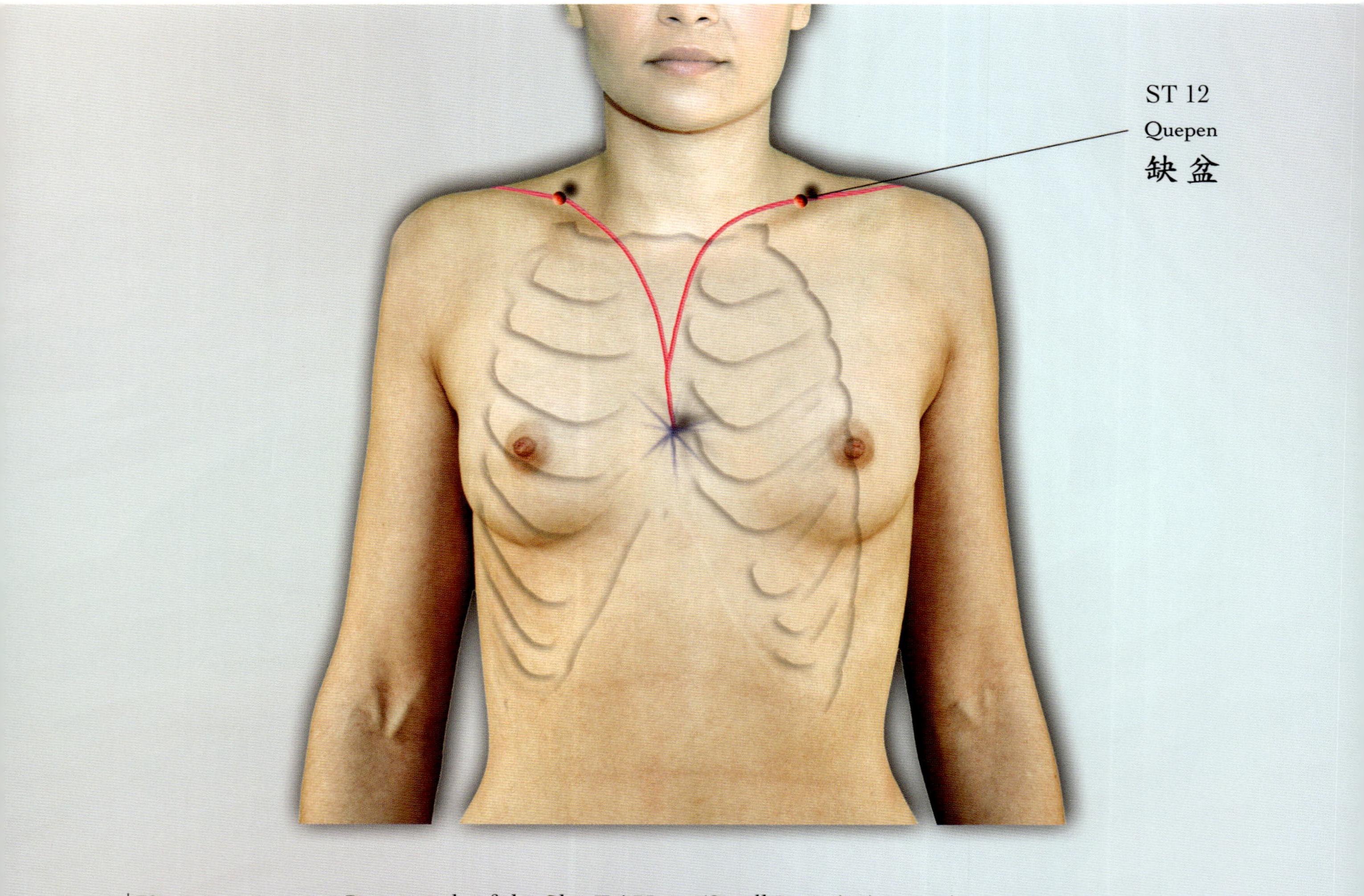

Luo vessels of the Shu Tai Yang (Small Intestine), anterior course

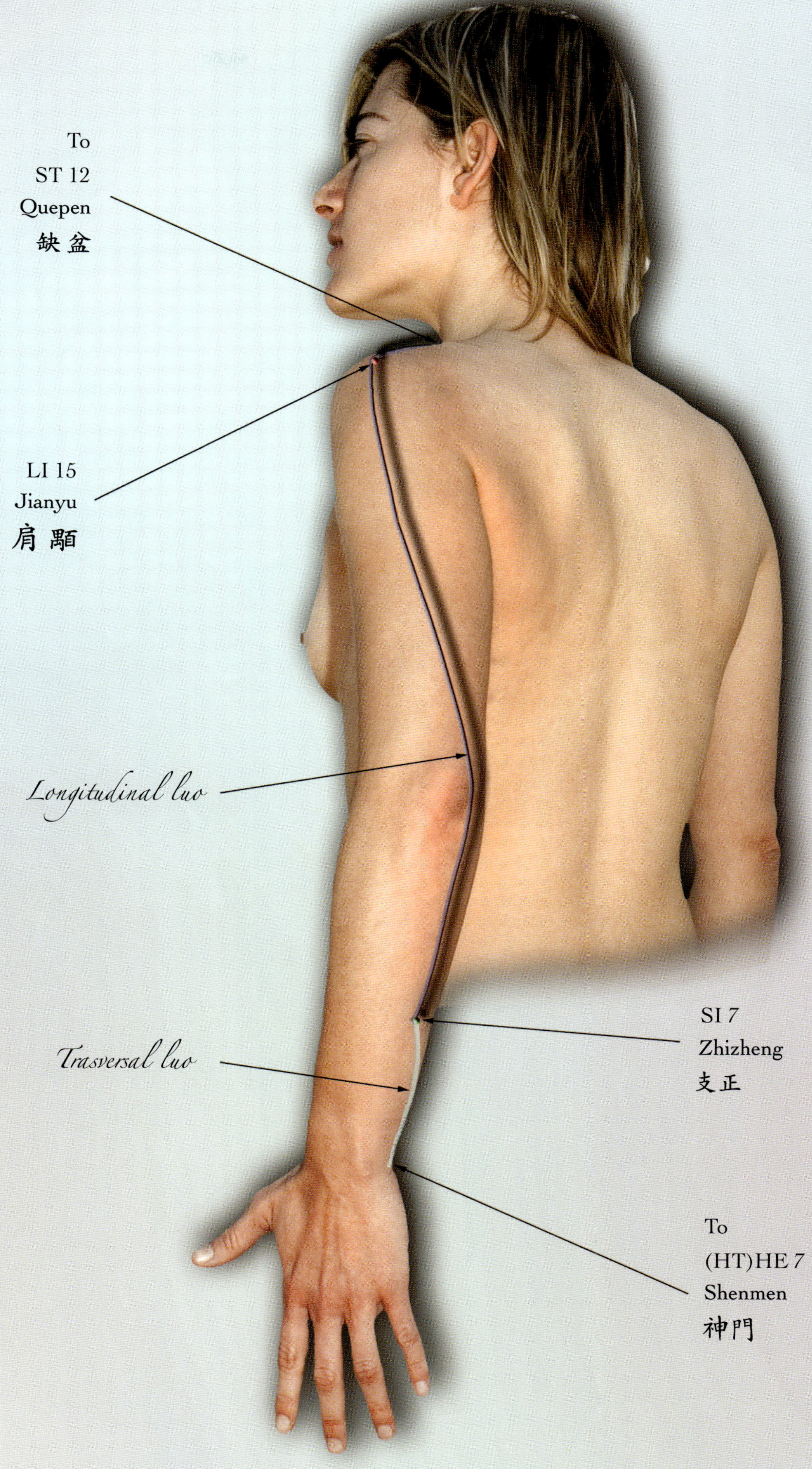

Luo vessels of the Shu Tai Yang (Small Intestine)

Distinct channel of the Shu Tai Yang (Small Intestine)

***Ling Shu*, chapter III, paragraph 11**

The distinct channel of the Main channel Shu Tai Yang of the hand (Small Intestine) symbolizes the Earth.

It goes out of the shoulder joint, enters the armpit and goes to the Heart.

It joins with the distinct channel of the Shu Shao Yin (Heart) to form the 4th union.

It goes down through the diaphragm, then keeps going down to join the Small Intestine.

Commentaries

It goes out of its Main channel on the level of point SI 12 Bingfeng, at the midpoint of the supra-spinal fossa.

It goes to the axillary hollow at point GB 22 Yuanye, penetrates into the chest and ramifies in the Heart.

It joins the distinct channel of Shu Shao Yin (Heart) to form the 4th union.

It goes down through the diaphragm, then keeps going down to unite with the Small Intestine.

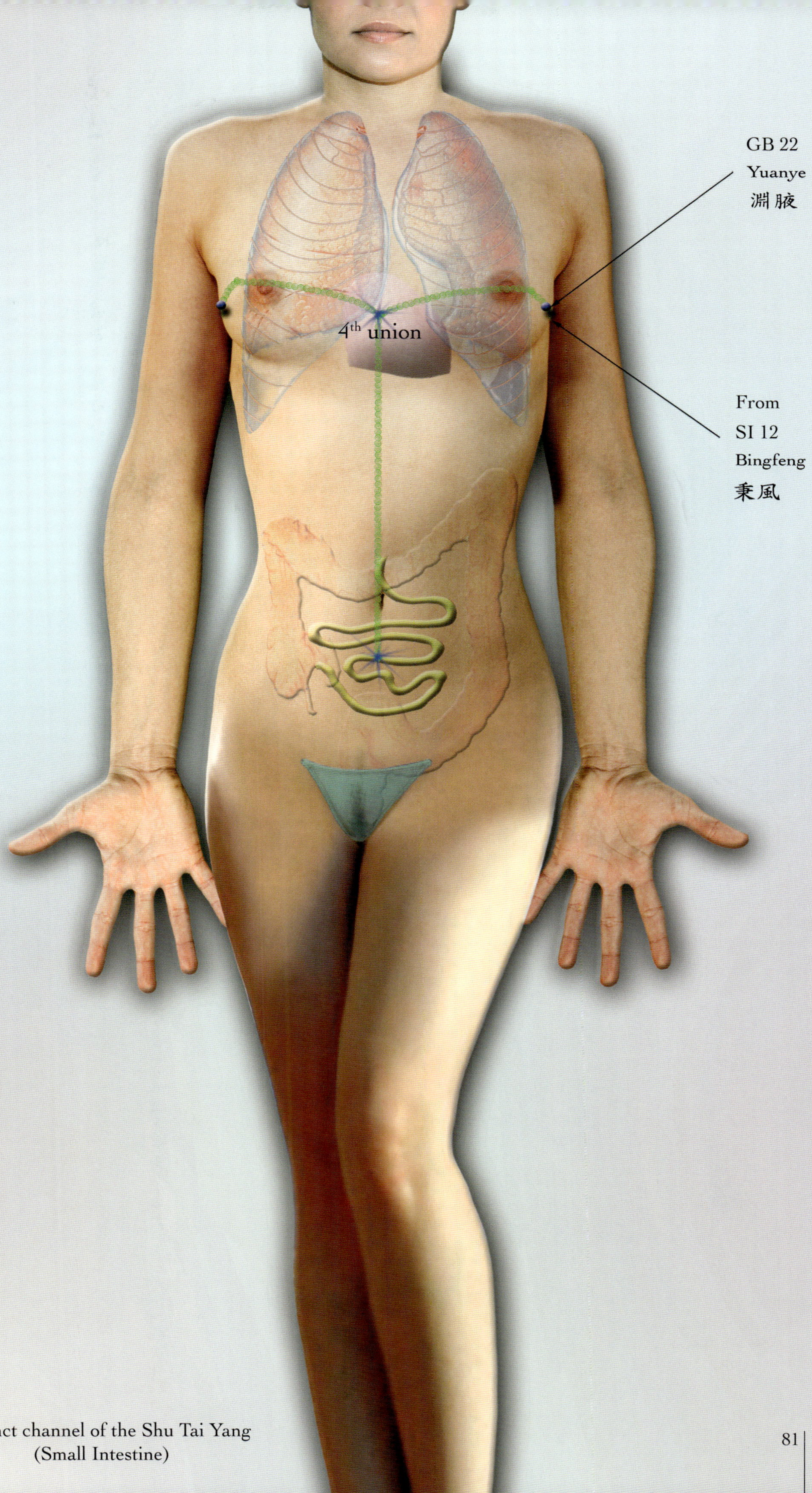

Distinct channel of the Shu Tai Yang
(Small Intestine)

Points of the Main channel Shu Tai Yang (Small Intestine)

SI 1 Shaoze
On the ulnar side of the distal segment of the little finger, 0.1 Cun from the corner of the nail.

SI 2 Qiangu
At the 'junction of the red and white skin', along the ulnar border of the hand, at the ulnar end of the crease of the 5th metacarpo-phalangeal joint when a loose fist is made.

SI 3 Houxi
At the 'junction of the red and white skin' along the ulnar border of the hand, at the ulnar end of the distal palmar crease, proximal to the 5th metacarpo-phalangeal joint when a hollow fist is made.

SI 4 Wangu
On the ulnar border of the hand, in the depression between the proximal end of the 5th metacarpal bone and hamate bone, and at the 'junction of the red and white skin'.

SI 5 Yanggu
On the ulnar border of the wrist, in the depression between the styloid process of the ulna and triangular bone.

SI 6 Yanglao
On the ulnar side of the posterior surface of the forearm, in the depression proximal to and on the radial side of the head of the ulna.

SI 7 Zhizheng
On the ulnar side of the posterior surface of the forearm and on the line connecting SI 5 Yanggu and SI 8 Xiaohai, 5 Cun proximal to the dorsal crease of the wrist.

SI 8 Xiaohai
On the medial side of the elbow, in the depression between the olecranon of the ulna and the medial epicondyle of the humerus.

SI 9 Jianzhen
Posterior and inferior to the shoulder joint, 1 Cun above the posterior end of the axillary fold with the arm abducted.

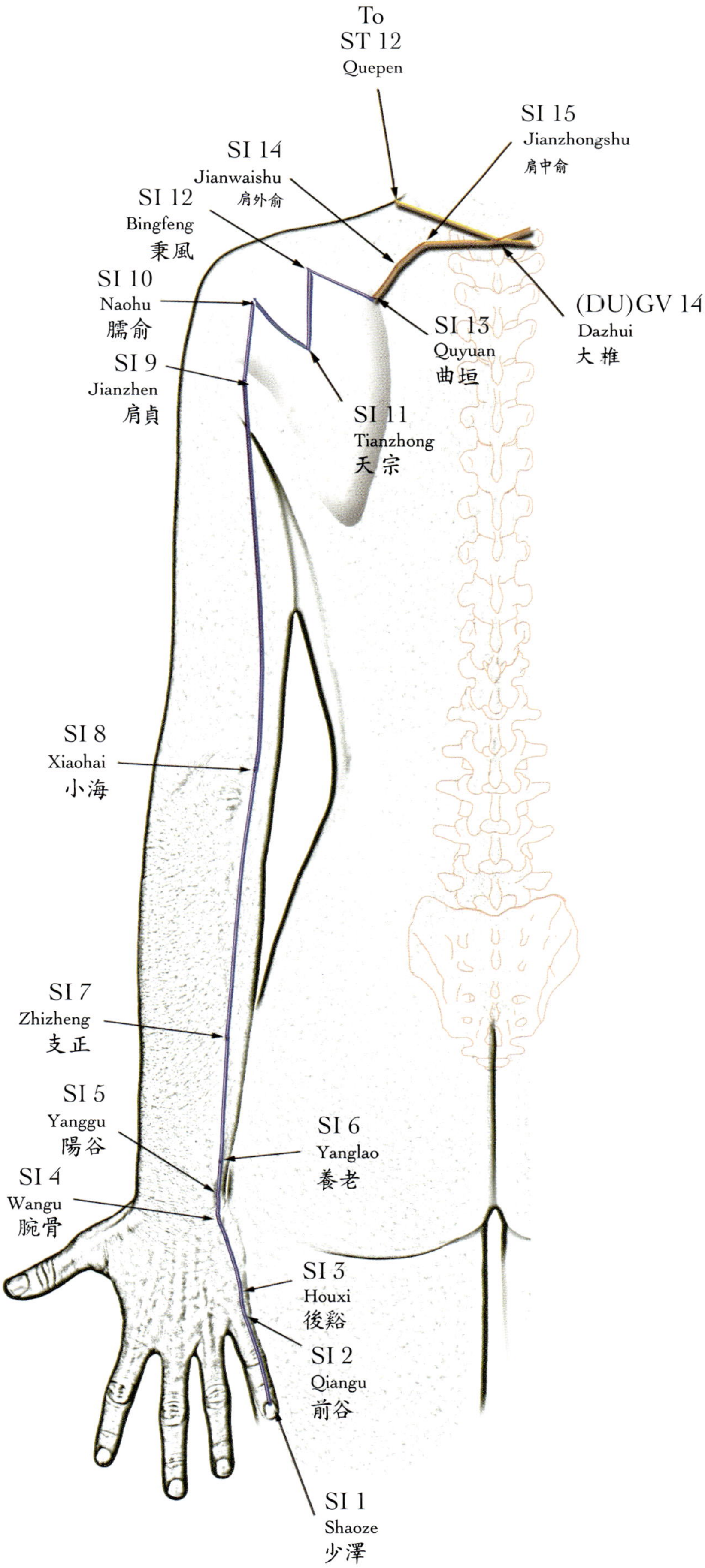

Points of the Main channel Shu Tai Yang (Small Intestine)

SI 10 Naoshu
On the shoulder, above the posterior end of the axillary fold, in the depression below the lower border of the scapular spine.

SI 11 Tianzong
On the scapula, in the depression of the centre of the sub-scapular fossa, and on the level of the 4th thoracic vertebra.

SI 12 Bingfeng
On the scapula, at the centre of the suprascapular fossa, directly above SI 11 Tianzong, in the depression found when the arm is raised.

SI 13 Quyuan
On the scapula, at the medial end of the suprascapular fossa, at the midpoint of the line connecting SI 10 Naoshu and the spinous process of the 2nd thoracic vertebra.

SI 14 Jianwaishu
On the back, below the spinous process of the 1st thoracic vertebra, 3 Cun lateral to the posterior midline.

SI 15 Jianzhongshu
On the back, below the spinous process of the 7th cervical vertebra, 2 Cun lateral to the posterior midline.

SI 16 Tianchuang
On the lateral side of the neck, posterior to the sterno-cleido-mastoid muscle and LI 18 Futu, on the level of the laryngeal protuberance.

SI 17 Tianrong
On the lateral side of the neck, posterior to the mandibular angle, in the depression of the anterior border of the sterno-cleido-mastoid muscle.

SI 18 Quanliao
On the face, directly below the outer canthus, in the depression below the zygomatic bone.

SI 19 Tinggong
On the face, anterior to the tragus and posterior to the mandibular condyloid process, in the depression found when the mouth is open.

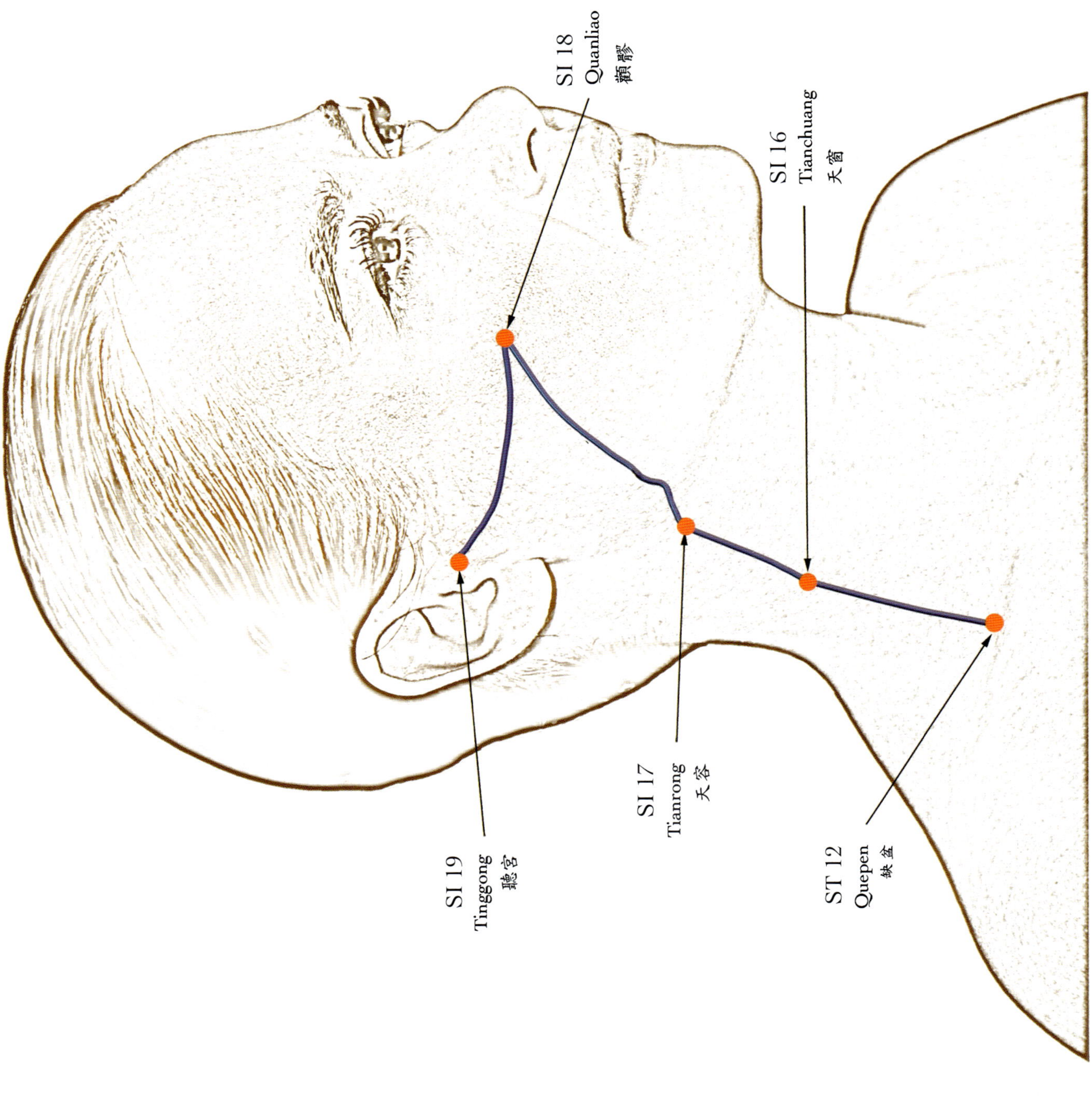
SI 18
Quanliao
顴髎
SI 16
Tianchuang
天窗
SI 17
Tianrong
天容
SI 19
Tinggong
聽宫
ST 12
Quepen
缺盆

Shu Shao Yang system (Triple Heater or Triple Burner)

- Tendino-muscular channel
- Main channel
- Internal branches
- Luo vessels
- Distinct channel
- Points of the Main channel

Tendino-muscular of the Shu Shao Yang (Triple Heater or Triple Burner)

Ling Shu, Chapter IV, paragraph 13

The tendino-muscular of the Triple Heater or Triple Burner starts at the tip of the ring-finger.

It joins the wrist and follows the midline of the forearm, then ramifies in the elbow.

It goes up along the external border of the forearm, follows its course passing by the shoulder, runs on the neck and joins the tendino-muscular of the Small Intestine.

A branch emerges from the angle of the mandible, goes inside and joins the root of the tongue.

Another branch goes up to point ST 6 Jiache, passes in front of the ear to reach the outer canthus.

It goes up again on the mandible to end at the angle of it.

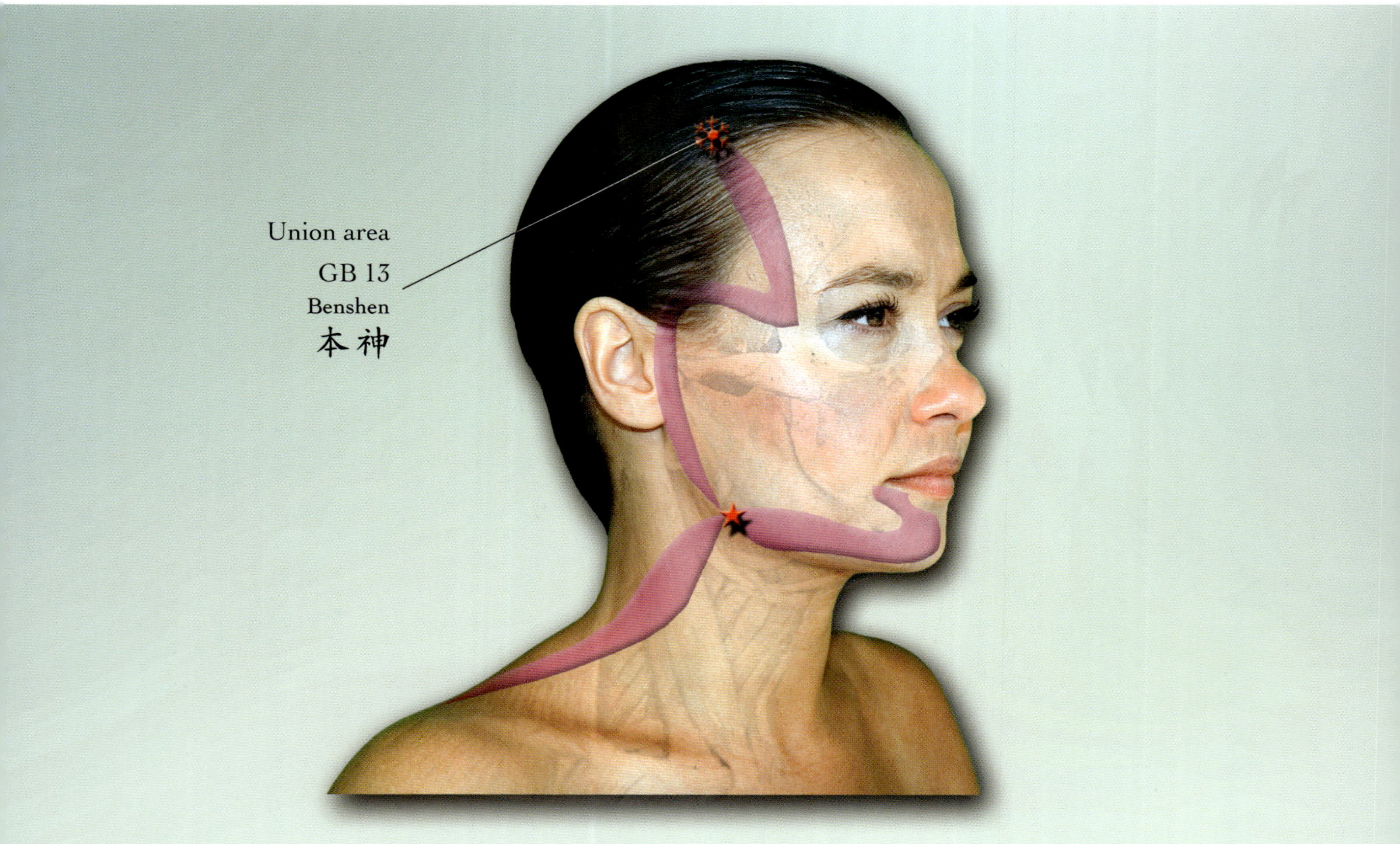

 Tendino-muscular of the Shu Shao Yang (Triple Heater or Triple Burner), cephalic course

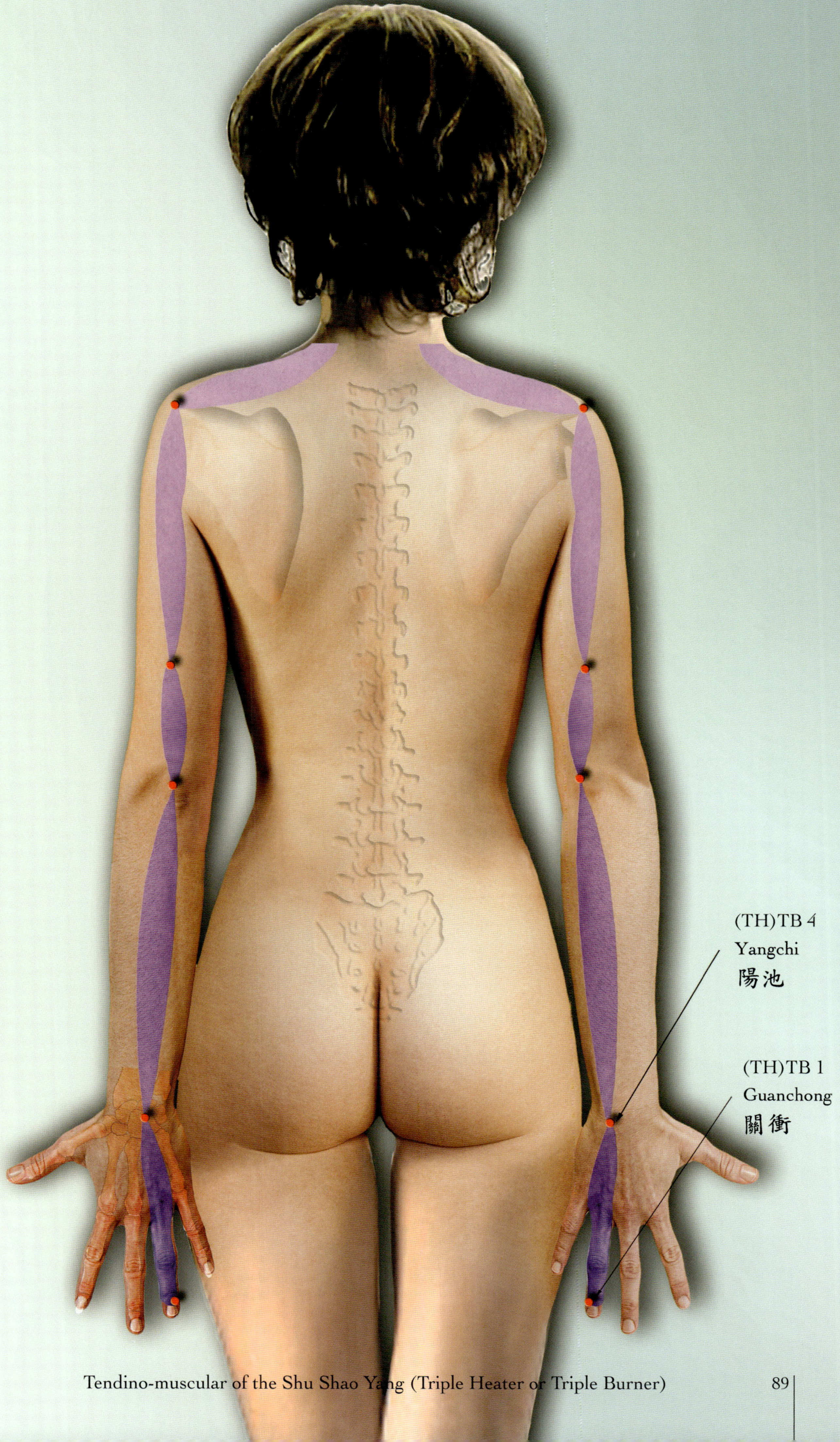

Tendino-muscular of the Shu Shao Yang (Triple Heater or Triple Burner)

Main channel of the Shu Shao Yang (Triple Heater or Triple Burner)

Ling Shu, chapter III, paragraph 10

The vessel of the Triple Heater or Triple Burner, Shu Shao Yang of the hand, starts at the tip of the ring finger, on the side of the little finger.

It goes up and emerges between the two fingers.

It runs along the posterior side of the hand and goes over the dorsal side of the wrist.

It emerges on the external side of the forearm between the bones (between the radius and the ulna).

It goes up following the elbow, goes along the external side of the arm and goes up to the shoulder.

It emerges crossing the Shao Yang of the foot from behind and penetrates in the supra-clavicular fossa.

It exits from point Danzhong, at the midpoint of the chest, and ramifies joining the Pericardium.

It descends on the diaphragm and belongs to the Triple Heater or Triple Burner while running along it.

The branch leaving Danzhong ascends and goes out by the supra-clavicular fossa.

It rises by the neck and joins the posterior ear (the posterior part of the auricle).

It goes up directly on the ear and emerges by the apex (apex of the auricle).

It goes down, running round the cheek so as to reach the lower part of the orbital cavity.

The branch that starts from the posterior part of the ear penetrates at the midpoint of the ear (centre of the tragus) and passing before Kezhuren (GB 3 Shangguan), it runs in front of the auricle.

It crosses the branch of the area of the cheek and reaches the canthus.

Commentaries

The channel reaches the elbow at point (TH) TB 10 Tianjing; it runs on the posterio internal part of the arm until it reaches the shoulder at point (TH)TB 14 Jianliao.

It passes under the posterior border of the acromion, then by the trapezius to join point SI 12 Bingfeng and (DU)GV 14 Dazhui.

It goes back to the midpoint of the supra clavicular fossa, at point ST 12 Quepen, crossing the channel of the Gall Bladder at point GB 21 Jianjing.

It then goes up on the lateral part of the neck, passing behind the nape and the ear.

It reaches the temple and joins points GB 6 Xuanli and GB 4 Hanyan.

It goes down to the lower maxilla, goes up on the lower part of the orbital cavity to join the Gall Bladder channel.
It crosses point SI 18 Quanliao.

From the posterior part of the ear, a branch penetrates there. It goes out in front of the tragus where it crosses point SI 19 Tinggong and GB 3 Shangguan or Kezhuren.

It reaches point GB 1 Tongziliao, at the outer canthus.

It ends at the extremity of the eyebrow, at point (TH)TB 23 Sizhukong.

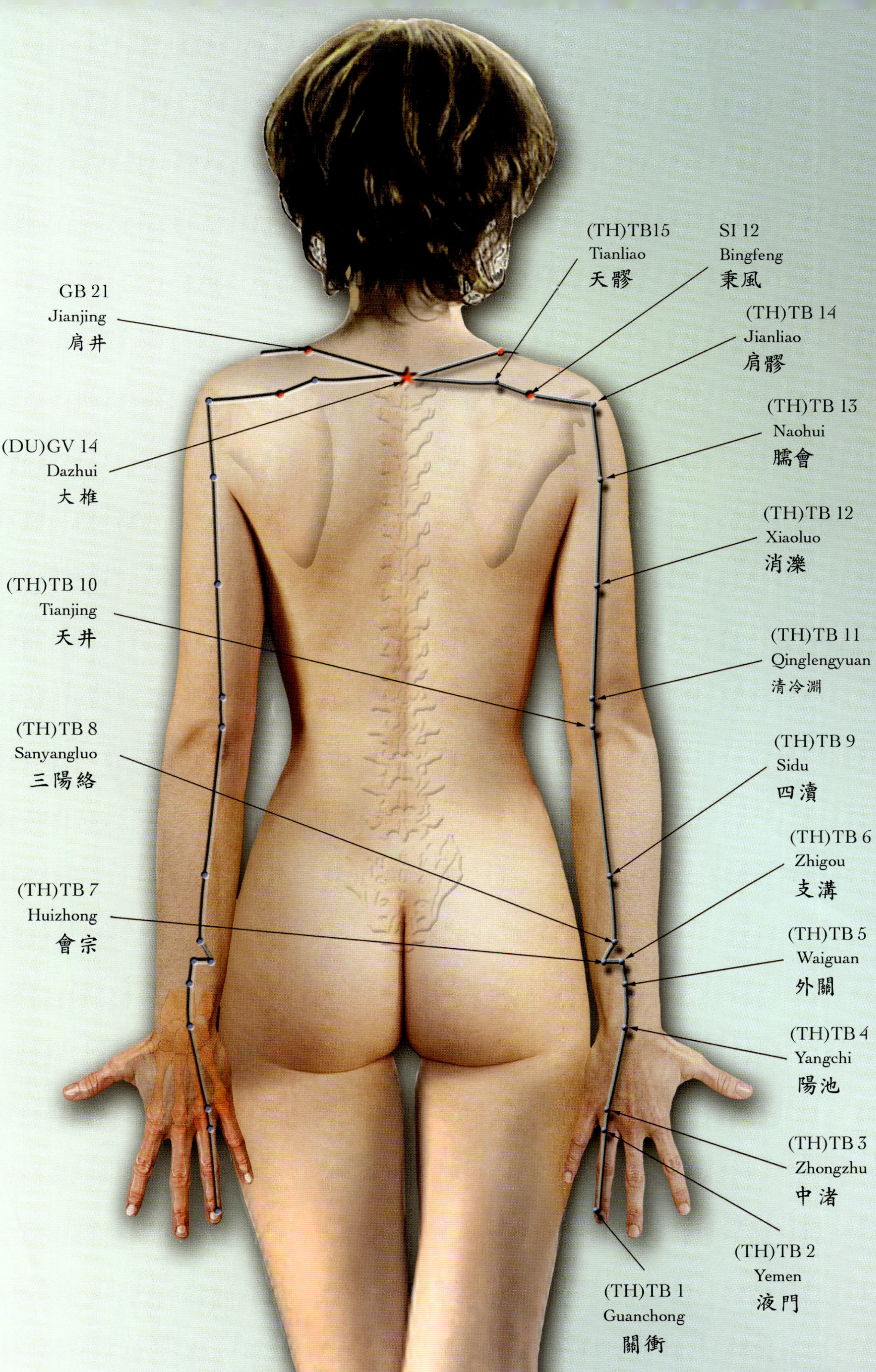

Main channel of the Shu Shao Yang (Triple Heater or Triple Burner)

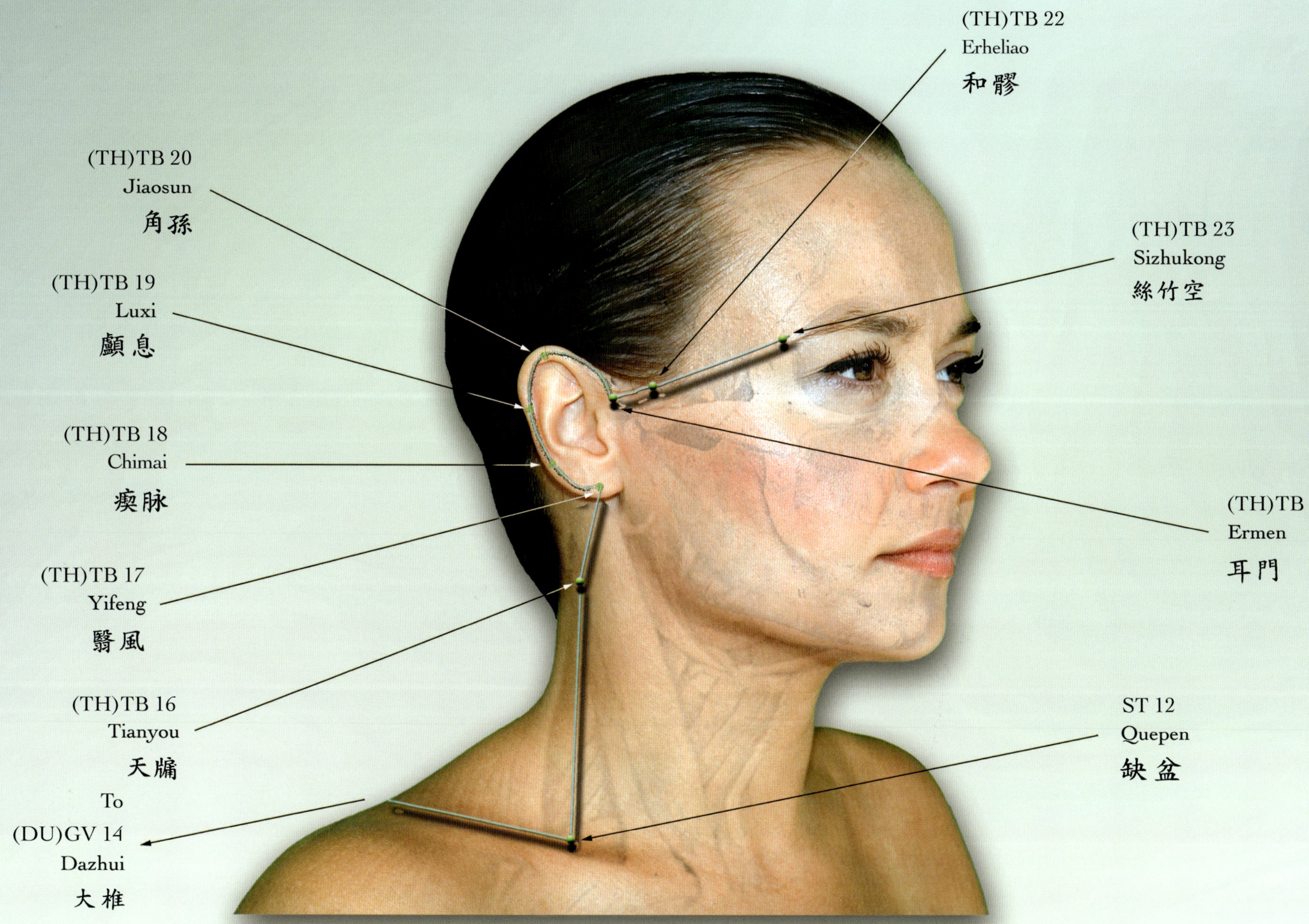

Main channel of the Shu Shao Yang (Triple Heater or Triple Burner), cephalic course n° 1

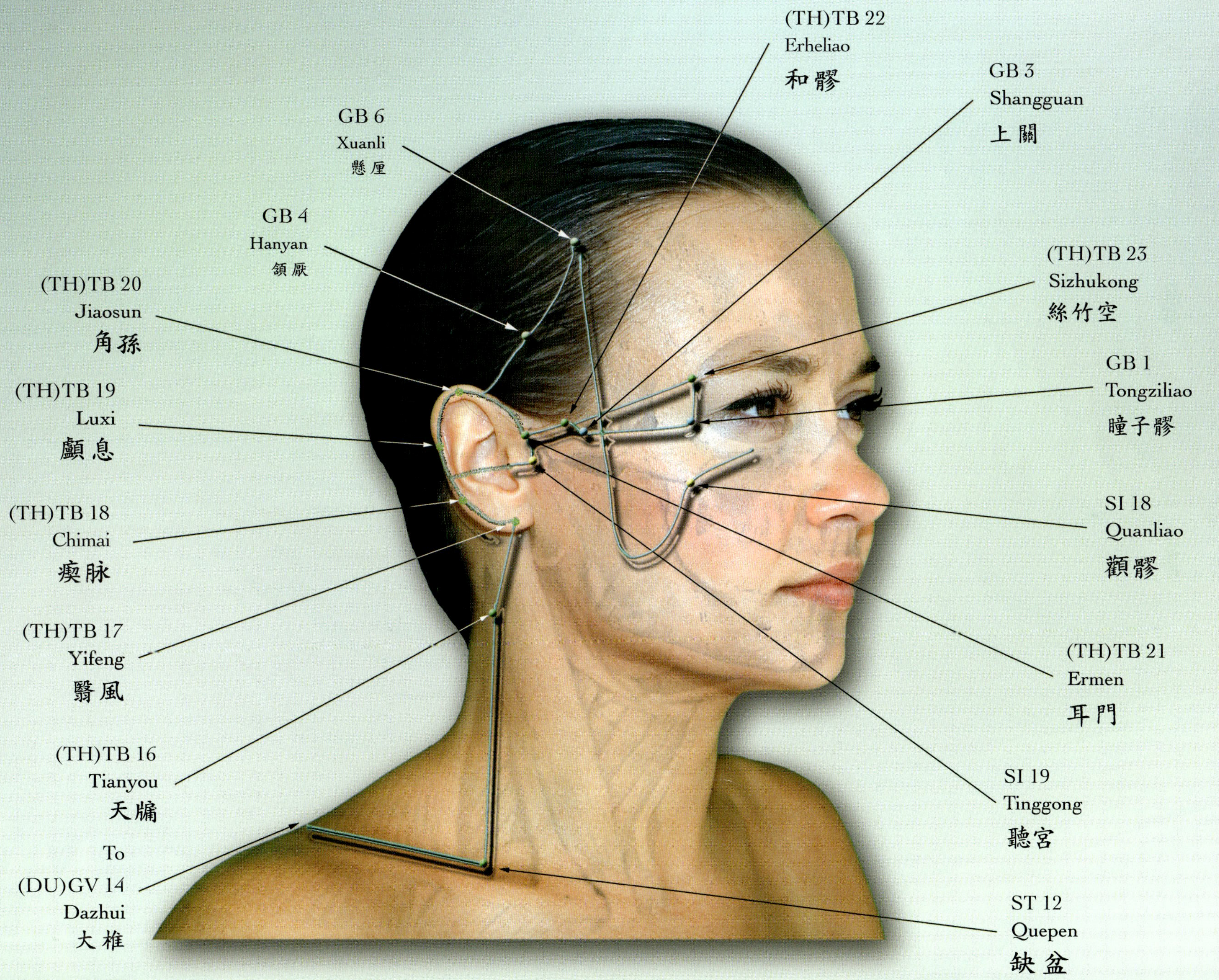

Main channel of the Shu Shao Yang (Triple Heater or Triple Burner), cephalic course n°2

Internal branches of the Shu Shao Yang (Triple Heater or Triple Burner)

***Ling Shu*, chapter III, paragraph 10**

(...)It emerges from point Danzhong at the midpoint of the chest and spreads joining the Pericardium.

It descends on the diaphragm and belongs to the Triple Heater or Triple Burner while running along it.

The branch that leaves Danzhong goes up and out by the supra-clavicular fossa.

Commentaries

From point ST 12 Quepen, the channel enters the chest to join the Heart Protector (the Pericardium) and the channel of the Master of the Heart at point (REN)CV 17 Danzhong.

It then goes down to cross the diaphragm and join the Three Heaters.

From the lower burner, a branch descends to join point BL 39 Weiyang at the external extremity of the popliteal crease.

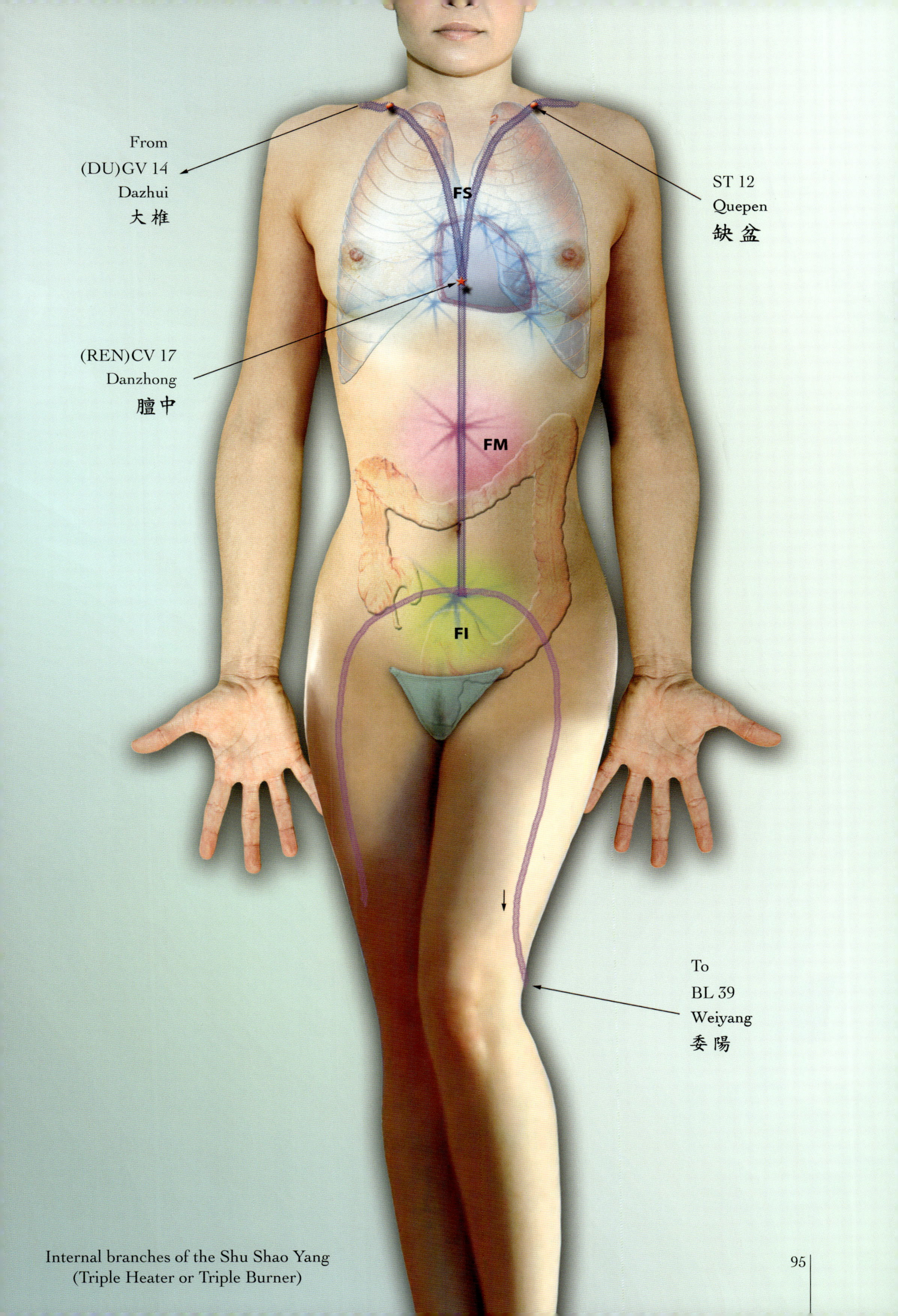

Internal branches of the Shu Shao Yang
(Triple Heater or Triple Burner)

Luo vessels of the Shu Shao Yang (Triple Heater or Triple Burner)

***Ling Shu*, chapter III, paragraph 10**

The Luo vessel of the Shao Yang of the hand (Triple Heater or Triple Burner) is called Waiguan.

It starts 2 Cun above the wrist.

It goes externally round the forearm and unites at the midpoint of the chest.

It joins the Master of the Heart or Pericardium.

Commentaries

It goes out from point (TH)TB 5 Waiguan situated 2 Cun above the posterior crease of the wrist and divides into 2 branches:

- The longitudinal Luo follows the posterior part of the forearm and of the arm.
  It passes on the posterior part of the shoulder, goes back to point ST 12 Quepen, and enters the chest to join the channel of the Master of the Heart or Pericardium, and then spreads in the Three Heaters.

- The transversal Luo joins the Main channel of the Master of the Heart or Pericardium at point (MH)PC 7 Daling located in a depression, at the midpoint of the transversal crease of the wrist, between the tendons of the big and small palmar.

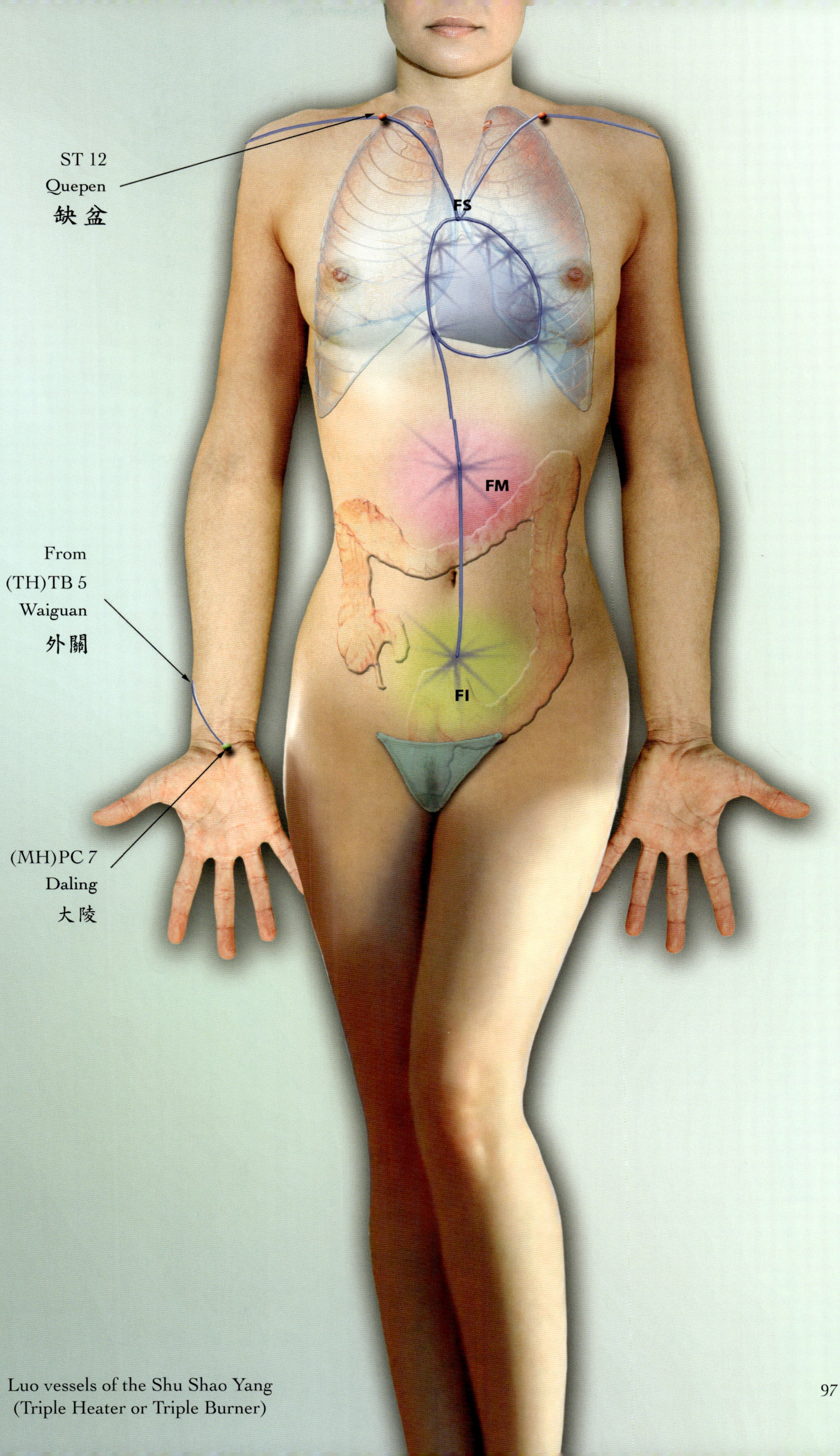

Luo vessels of the Shu Shao Yang
(Triple Heater or Triple Burner)

Distinct channel of the Shu Shao Yang (Triple Heater or Triple Burner)

***Ling Shu*, chapter III, paragraph 11**

The distinct channel of the Main channel Shu Shao Yang of the hand (Triple Heater or Triple Burner) symbolises the sky.

It emerges from the vertex, enters the supra clavicular fossa, descends in the Triple Heater or Triple Burner and disperses at the midpoint of the chest.

Commentaries

The distinct channel starts at point (DU)GV 20 Baihui. A branch unites that point with point (TH)TB 20 Jiaosun.

It runs down behind the ear, then passses by point (TH)TB 16 Tianyu where it joins its Main channel and the distinct channel of the Master of the Heart or Pericardium, to form the 5th union.

It descends to point ST 12 Quepen, penetrates in the chest, joins the 'Pericardium' and disperses in the Three Heaters.

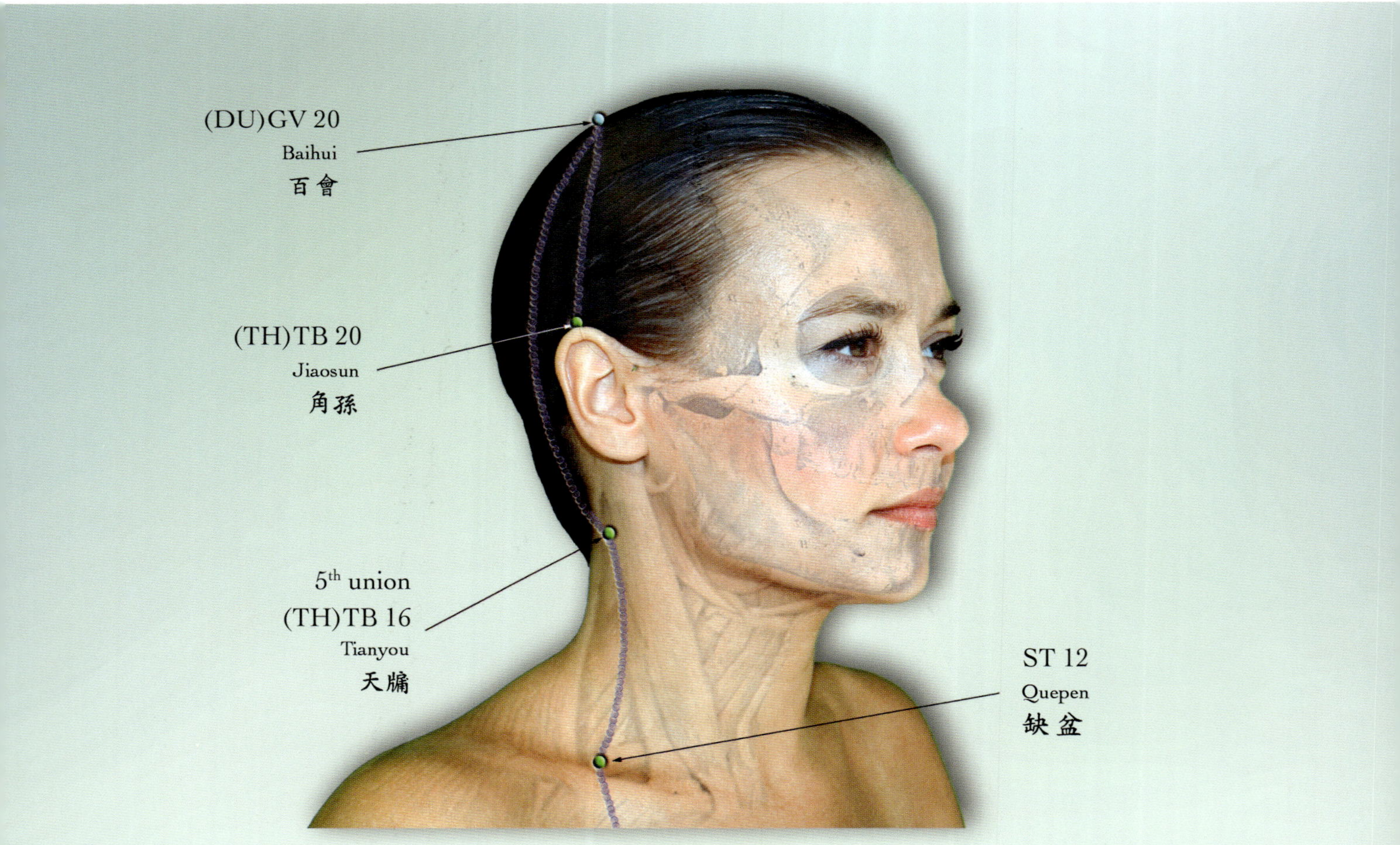

 Distinct channel of the Shu Shao Yang (Triple Heater or Triple Burner), cephalic course

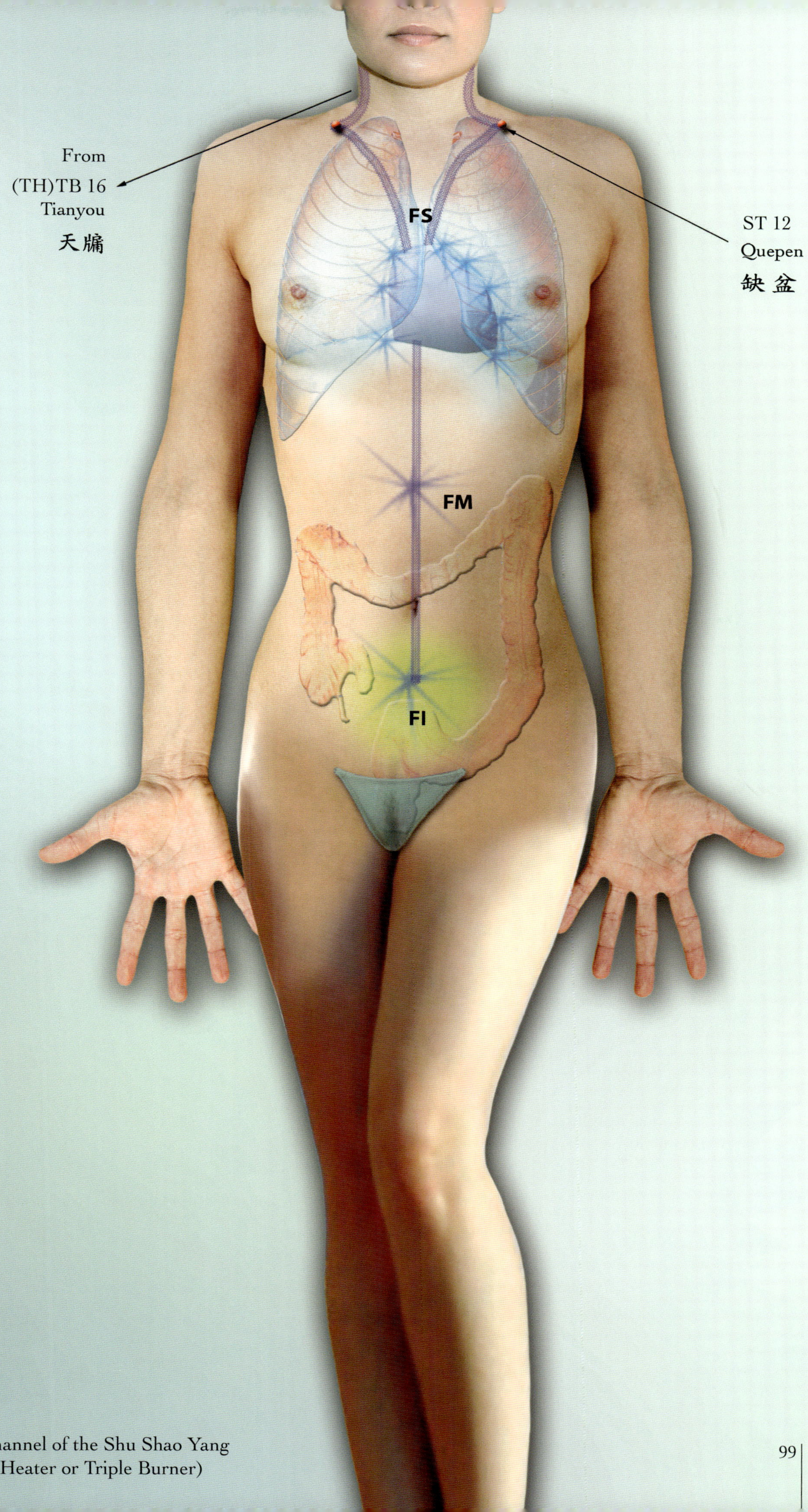

Distinct channel of the Shu Shao Yang
(Triple Heater or Triple Burner)

Points of the Main channel Shu Shao Yang (Triple Heater or Triple Burner)

(TH)TB 1 Guangchong
On the ulnar side of the distal segment of the 4th finger, 0.1 Cun from the corner of the nail.

(TH)TB 2 Yemen
On the dorsum of the hand, between the 4th and 5th fingers, at the 'junction of the red and white skin', proximal to the margin of the web.

(TH)TB 3 Zhongzhu
On the dorsum of the hand, proximal to the 4th metacarpo-phalangeal joint, in the depression between the 4th and 5th metacarpal bones.

(TH)TB 4 Yangchi
At the midpoint of the dorsal crease of the wrist, in the depression on the ulnar side of the tendon of the extensor muscle of the fingers.

(TH)TB 5 Waiguan
On the dorsal side of the forearm and on the line connecting (TH)TB 4 Yangchi and the tip of the olecranon, 2 Cun proximal to the dorsal crease of the wrist, between the radius and ulna.

(TH)TB 6 Zhigou
On the dorsal side of the forearm and on the line connecting (TH)TB 4 Yangchi and the tip of the olecranon, 3 Cun proximal to the dorsal crease of the wrist, between the radius and ulna.

(TH)TB 7 Huizong
On the dorsal side of the forearm, 3 Cun proximal to the dorsal crease of the wrist, on the ulnar side of (TH)TB 6 Zhigou and on the radial border of the ulna.

(TH)TB 8 Sanyangluo
On the dorsal side of the forearm, 4 Cun proximal to the dorsal crease of the wrist, between the radius and ulna.

(TH)TB 9 Sidu
On the dorsal side of the forearm and on the line connecting (TH)TB 4 Yangchi and the tip of the olecranon, 5 Cun distal to the tip of the olecranon, between the radius and the ulna.

(TH)TB 10 Tianjing
On the lateral side of the upper arm, in the depression 1 Cun proximal to the tip of the olecranon when the elbow is flexed.

(TH)TB 11 Qinglengyuan
With the elbow flexed, on the lateral side of the upper arm, 2 Cun above the tip of the olecranon and 1 Cun above (TH)TB 10 Tianjing.

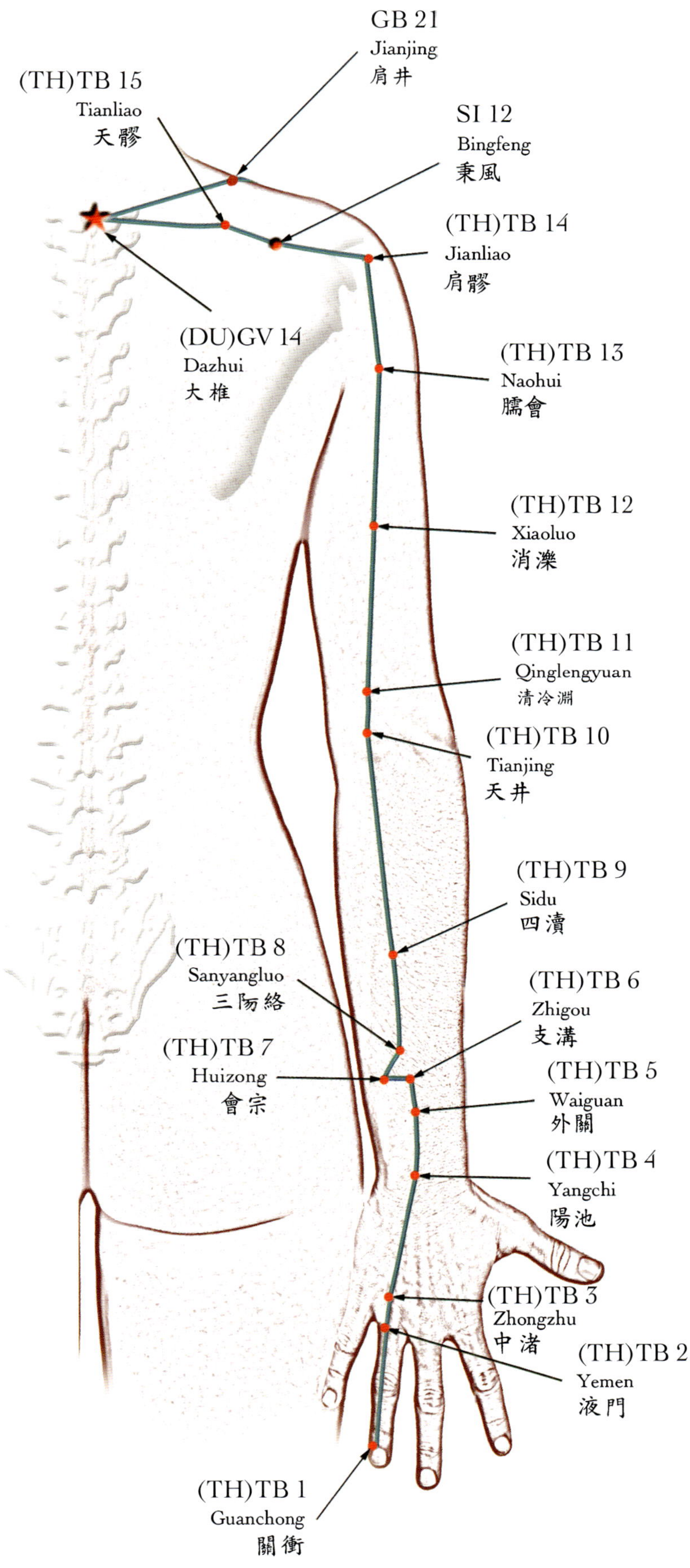

Points of the Main channel Shu Shao Yang (Triple Heater or Triple Burner)

(TH)TB 12 Xiaoluo
On the lateral side of the upper arm, at the midpoint of the line connecting (TH)TB 11 Qinglengyuan et (TH)TB 13 Naohui.

(TH)TB 13 Naohui
On the lateral side of the upper arm and on the line connecting the tip of the olecranon and (TH)TB 14 Jianliao, 3 Cun below (TH)TB 14 Jianliao, and on the posterio-inferior border of the deltoid muscle.

(TH)TB 14 Jianliao
On the shoulder, posterior to LI 15 Jianyu, in the depression inferior and posterior to the acromion when the arm is abducted.

(TH)TB 15 Tianliao
On the scapula, between GB 21 Jianjing and SI 13 Quyuan, at the superior angle of the scapula.

(TH)TB 16 Tianyou
On the lateral side of the neck, directly below the posterior border of the mastoid process, on the level of the mandibular angle, and on the posterior border of the sterno-cleido-mastoid muscle.

(TH)TB 17 Yifeng
Posterior to the ear lobe, in the depression between the mastoid process and mandibular angle.

(TH)TB 18 Chimai
On the head, at the centre of the mastoid process, and at the junction of the middle third and lower third of the line connecting (TH)TB 20 Jiaosun and (TH)TB 17 Yifeng along the curve of the ear helix.

(TH)TB 19 Luxi
On the head, at the junction of the upper third and middle third of the line connecting (TH)TB 20 Jiaosun and (TH)TB 17 Yifeng along the curve of the ear apex.

(TH)TB 20 Jiaosun
On the head, on the scalp, the external ear being bent forward, directly above the top of the fold thus formed.

(TH)TB 21 Ermen
On the face, in front of the notch above the tragus, in the depression located on the posterior condyle of the mandible when the mouth is open.

(TH)TB 22 Erheliao
On the lateral side of the head, on the posterior margin of the temples, anterior to the anterior border of the root of the ear auricle and posterior to the superficial temporal artery.

(TH)TB 23 Sizhukong
On the face, in the depression of the lateral end of the eyebrow.

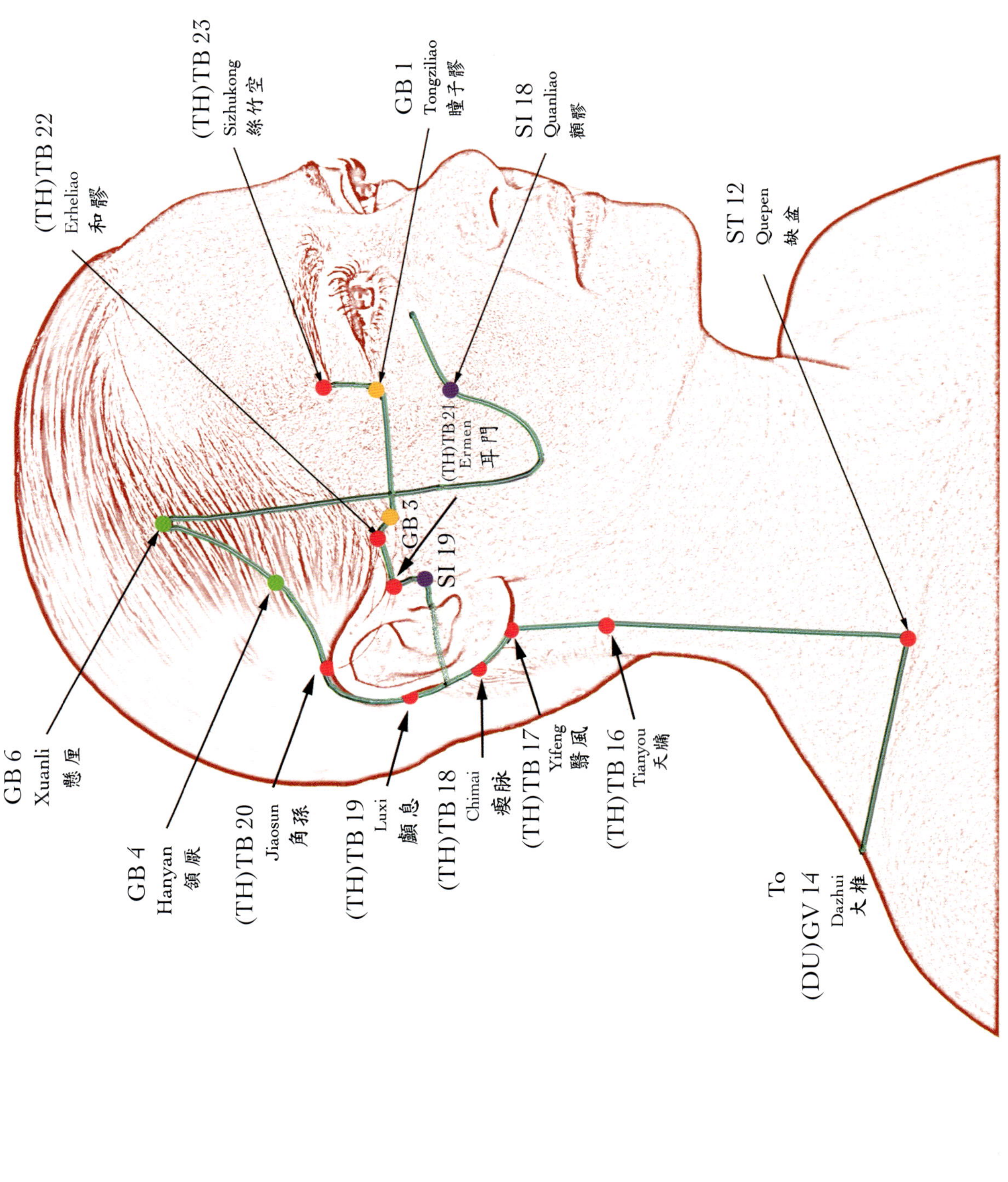

Points of the Main channel Shu Shao Yang (Triple Heater or Triple Burner)

The 3 lower Yin

- Zu Tai Yin (Spleen)
- Zu Jue Yin (Liver)
- Zu Shao Yin (Kidney)

ZU TAI YIN SYSTEM (SPLEEN)

- Tendino-muscular channel
- Main channel
- Internal branches
- Luo vessels
- Distinct channel
- Points of the channel

Tendino-muscular of the Zu Tai Yin (Spleen)

***Ling Shu,* chapter IV, paragraph 13**

The tendino-muscular of the Spleen starts on the internal side of the foot at the tip of the big toe.

It goes up and spreads in the internal malleolus.

It takes a vertical course to the knee and the internal femoral condyle.

It follows the internal side of the thigh, uniting with the hip.

It joins the genitalia and goes up to the abdomen, uniting with the umbilicus.

It travels along the inside of the abdomen and spreads in the ribs.

It spreads at the midpoint of the chest.

Its internal branch unites with the anterior part of the backbone.

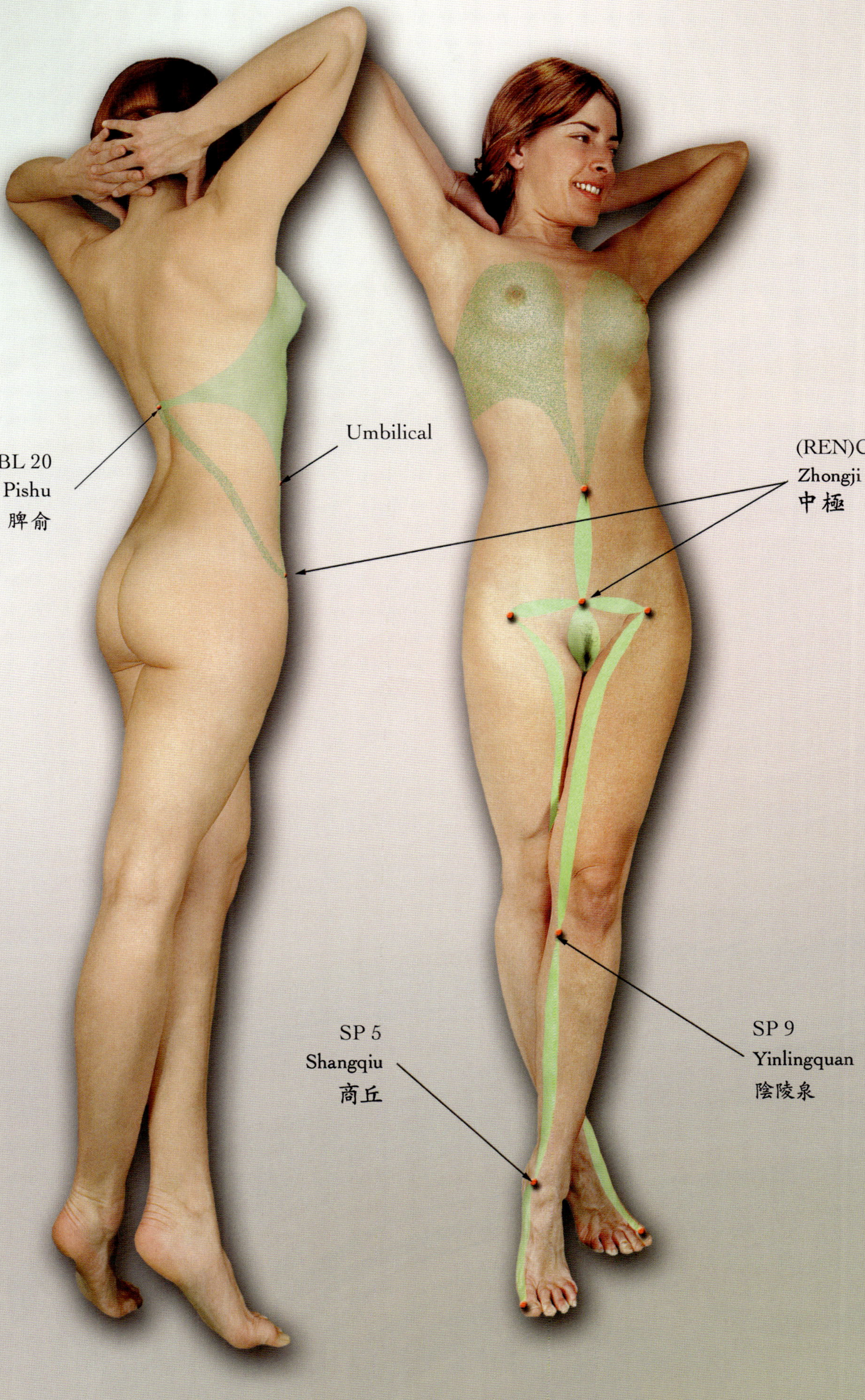

Tendino-muscular of the Zu Tai Yin (Spleen)

Main channel of the Zu Tai Yin (Spleen)

***Ling Shu*, chapter III, paragraph 10**

The vessel of the Spleen, Zu Tai Yin of the foot, starts at the tip of the big toe and goes along this toe, following the edge of the white skin (internal border or Yin face).

It passes behind the 1st metatarsal bone.

It goes up in front of the internal malleolus and by the internal side of the leg.

It follows the posterior side of the tibia, crosses the Jue Yin in front.

It goes up before the internal side of the knee and the internal side of the thigh.

It penetrates in the abdomen, belongs to the Spleen and joins the Stomach.

It goes up by the diaphragm, encircles the throat, joins the 'root of the tongue' and spreads under the tongue.

The branch that leaves the Stomach stands apart by going up in the diaphragm and joining the middle of the Heart.

Commentaries

Going out of its 1st point SP 1 Yinbai, the channel passes by the internal border of the big toe and of the 1st metatarsal bone.

It passes by the depression that is in front of and below the internal malleolus at point SP 5 Shangqiu.

It follows the internal border of the tibia by crossing the channels of the Zu Jue Yin (Liver) and of the Zu Shao Yin (Kidney) at point SP 6 Sanyinjiao, meeting point of the three lower Yin.

It passes by the internal side of the knee by crossing point SP 9 Yinlingquan, and runs along the anterior and internal side of the thigh to reach the inguinal crease.

From there, it goes to point SP 13 Fushe.

It goes back to the midline and passes by points (REN)CV 3 Zhongji and (REN)CV 4 Guanyuan.

It flows to the mammary line to join point SP 14 Fujie, and another point of the Spleen, SP 15 Daheng, on the level of the umbilicus.

From SP 15 Daheng, a branch goes to point (REN)CV 10 Xiawan, penetrates deep inside to join the Spleen and the Stomach.

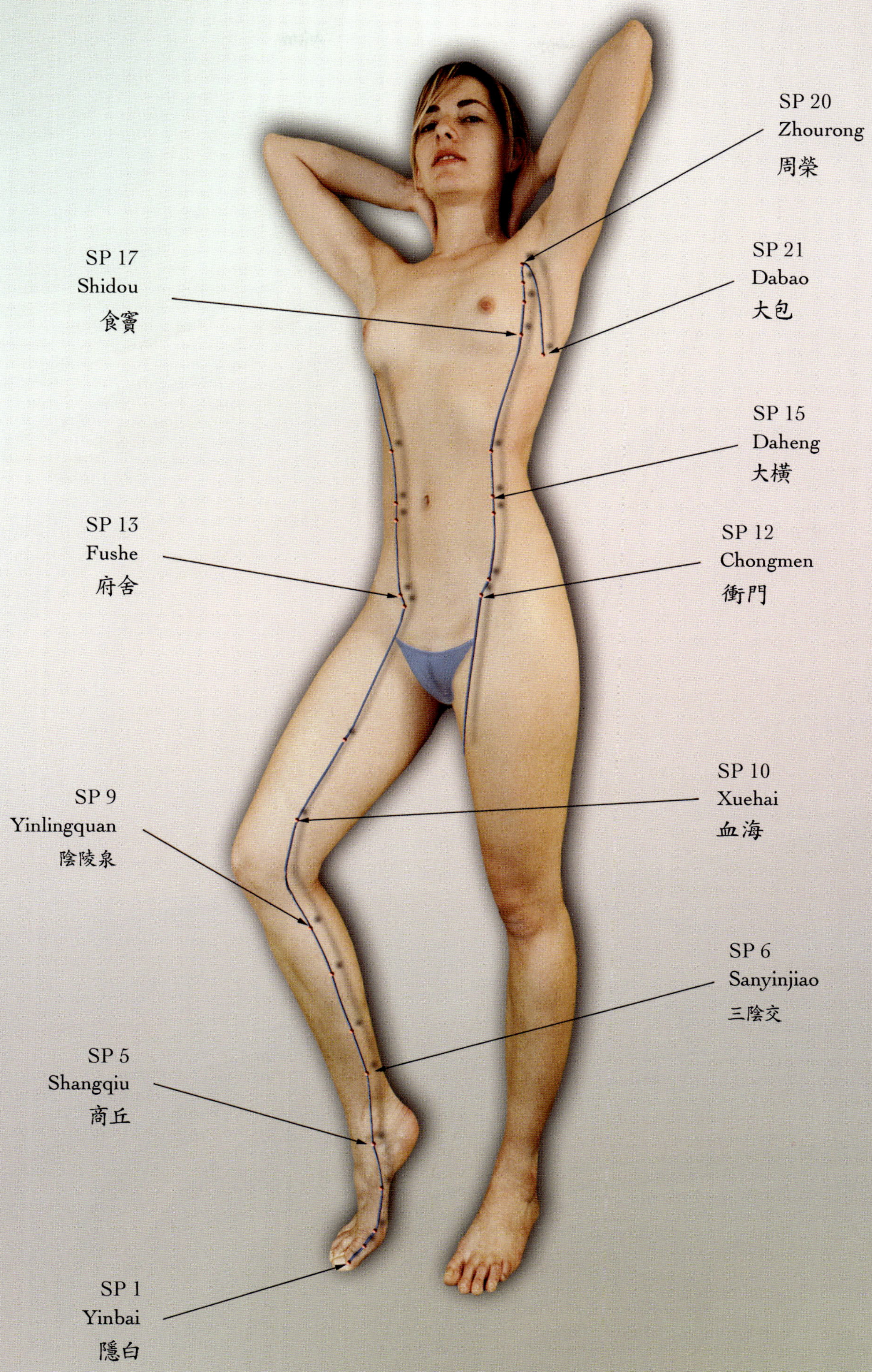

Main channel of the Zu Tai Yin (Spleen), course n° 1

It goes up by going through the diaphragm, passes along the oesophagus and reaches the Heart to unite with its own channel.

From point (REN)CV 10 Xiawan, the channel follows its course, passing by point SP 16 Fuhai, 3 Cun above SP 15, at the lower border of the 9th rib, horizontally to (REN)CV 11 Jianli.

It keeps going up to join point GB 24 Riyue, below the nipple, in the 7th intercostal space, and with (LR)LV 14 Qimen, at the inferior border of the 6th rib, in the 6th intercostal space on the mammary line.

It goes up to reach the level of the 2nd intercostal space, where point SP 20 Zhourong is.

From there, it sends out a branch to point LU 1 Zhongfu, from which another branch goes out and runs to the root of the tongue, then it spreads at its lower part.

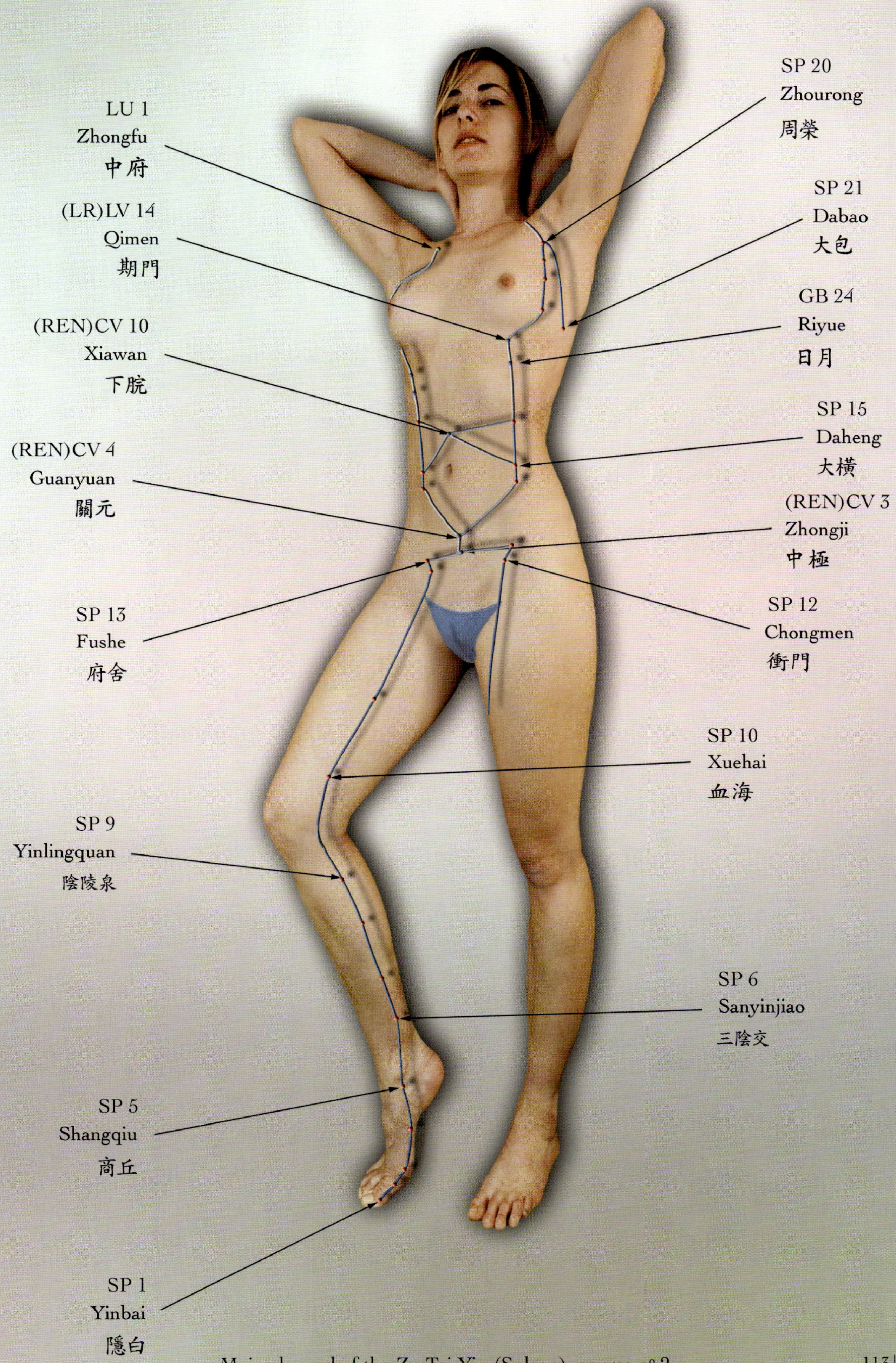

Main channel of the Zu Tai Yin (Spleen), course nº 2

Internal branches of the Zu Tai Yin channel (Spleen)

***Ling Shu,* chapter III, paragraph 10**

From SP 15 Daheng, a branch runs to point (REN)CV 10 Xiawan, enters deep inside to join the Spleen and the Stomach.

It goes up, crossing the diaphragm, passes along the oesophagus and reaches the Heart to unite with its own channel.

From point SP 20 Zhourong, a branch goes to point LU 1 Zhongfu, from which another branch exits to go to the root of the tongue, to then disperse at its lower part.

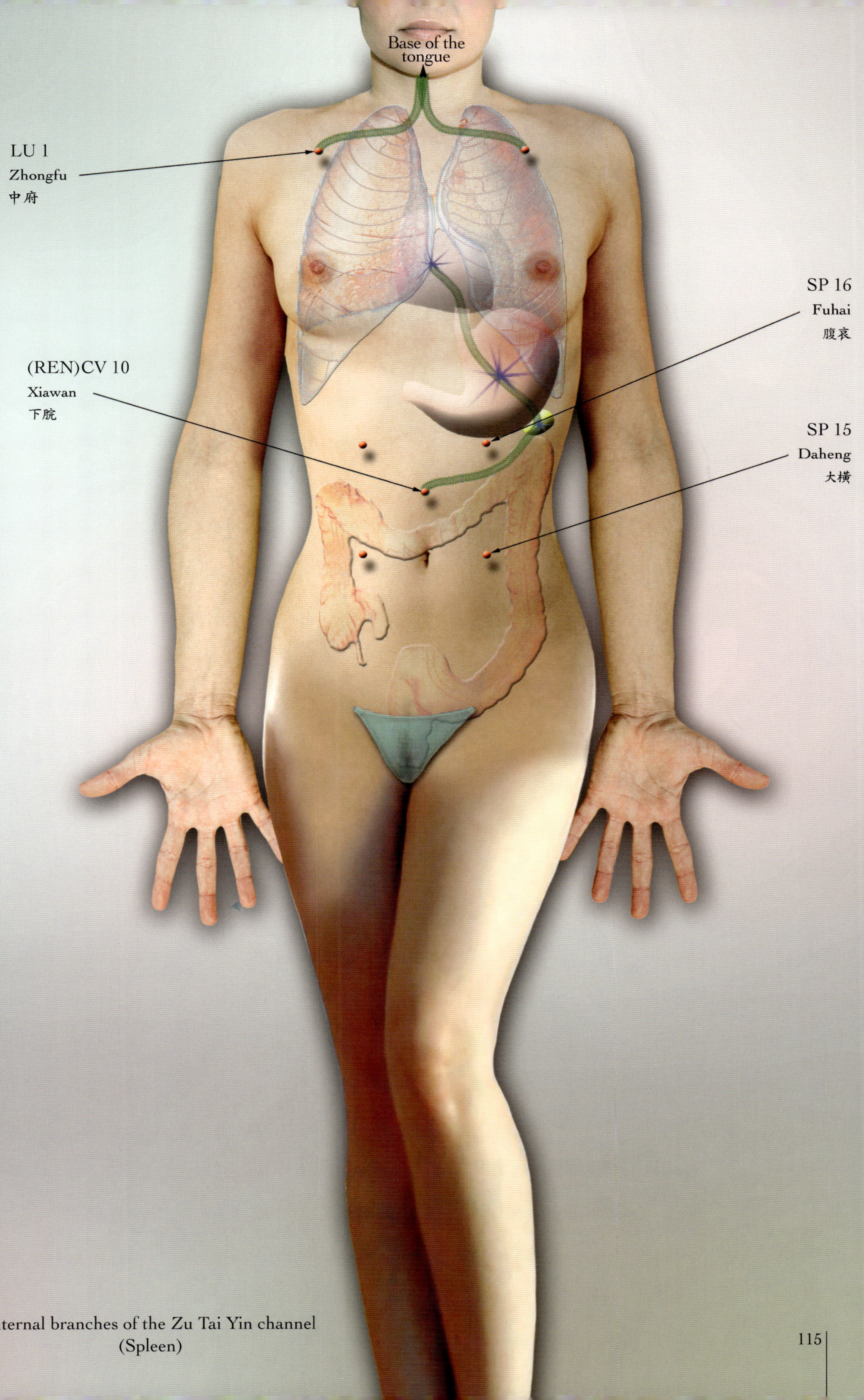

Internal branches of the Zu Tai Yin channel (Spleen)

Luo vessels of the Zu Tai Yin (Spleen)

***Ling Shu*, chapter III, paragraph 10**

The distinct branch of the Zu Tai Yin of the foot (channel of the Spleen) is called Gong Sun (grand-father and grand-son).

It starts 1 Cun behind the original joint (behind the head of the first metatarsal bone).

This distinct vessel runs to Zu Yang Ming (channel of the Stomach).

A distinct branch joins the Intestines and the Stomach.

Commentaries

It goes out of point SP 4 Gongsun and divides into two branches.

- The longitudinal branch goes up to the abdomen where it enters and ramifies with the Large Intestine and the Stomach.
- The transversal branch joins the channel of the Stomach at point ST 42 Chongyang.

Great Luo of the Spleen

It goes out from point SP 21 Dabao and unites in itself the fourteen Luo vessels that envelop the body like a net.

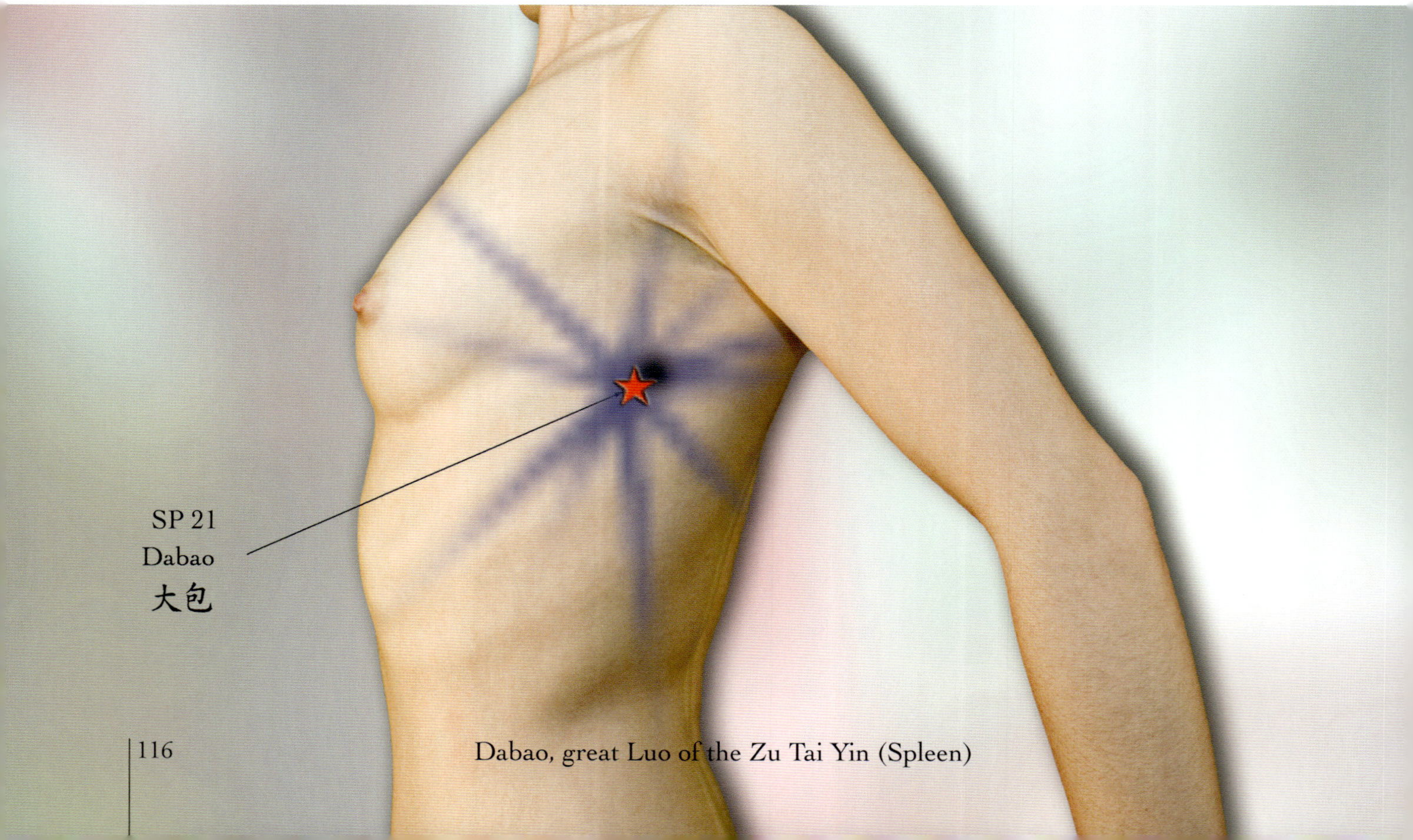

Dabao, great Luo of the Zu Tai Yin (Spleen)

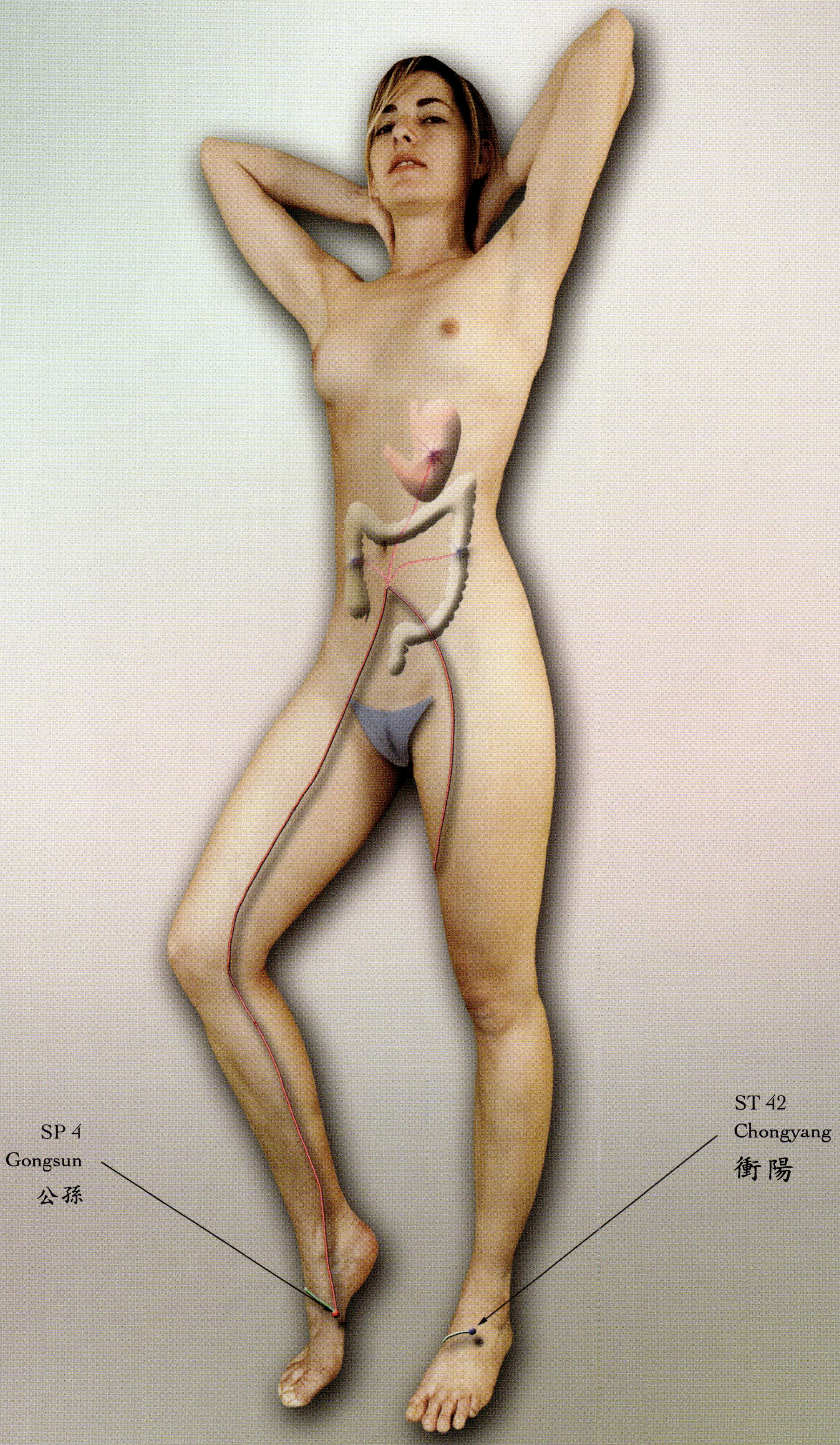

Luo vessels of the Zu Tai Yin (Spleen)

Distinct channel of the Zu Tai Yin (Spleen)

Ling Shu, chapter III, paragraph 11

The distinct channel of the Main channel of the Zu Tai Yin of the foot goes up along the anterio-superior part of the thigh.

It joins Zu Yang Ming passing by the throat and the midpart of the tongue.

Commentaries

The distinct channel of the Zu Tai Yin of the foot goes out of the Main channel at the internal part of the thigh.

It goes to point ST 30 Qichong, where it penetrates.

It goes to the Spleen and spreads in the Stomach.

It crosses the Heart, goes up along the oesophagus and passes by point ST 12 Quepen, at the midpoint of the supra-clavicular fossa.

It goes up to the throat and joins the distinct channel and the Main channel of Zu Yang Ming (Stomach) at point ST 9 Renying to form the third union.

It penetrates deep into the tongue where it ends.

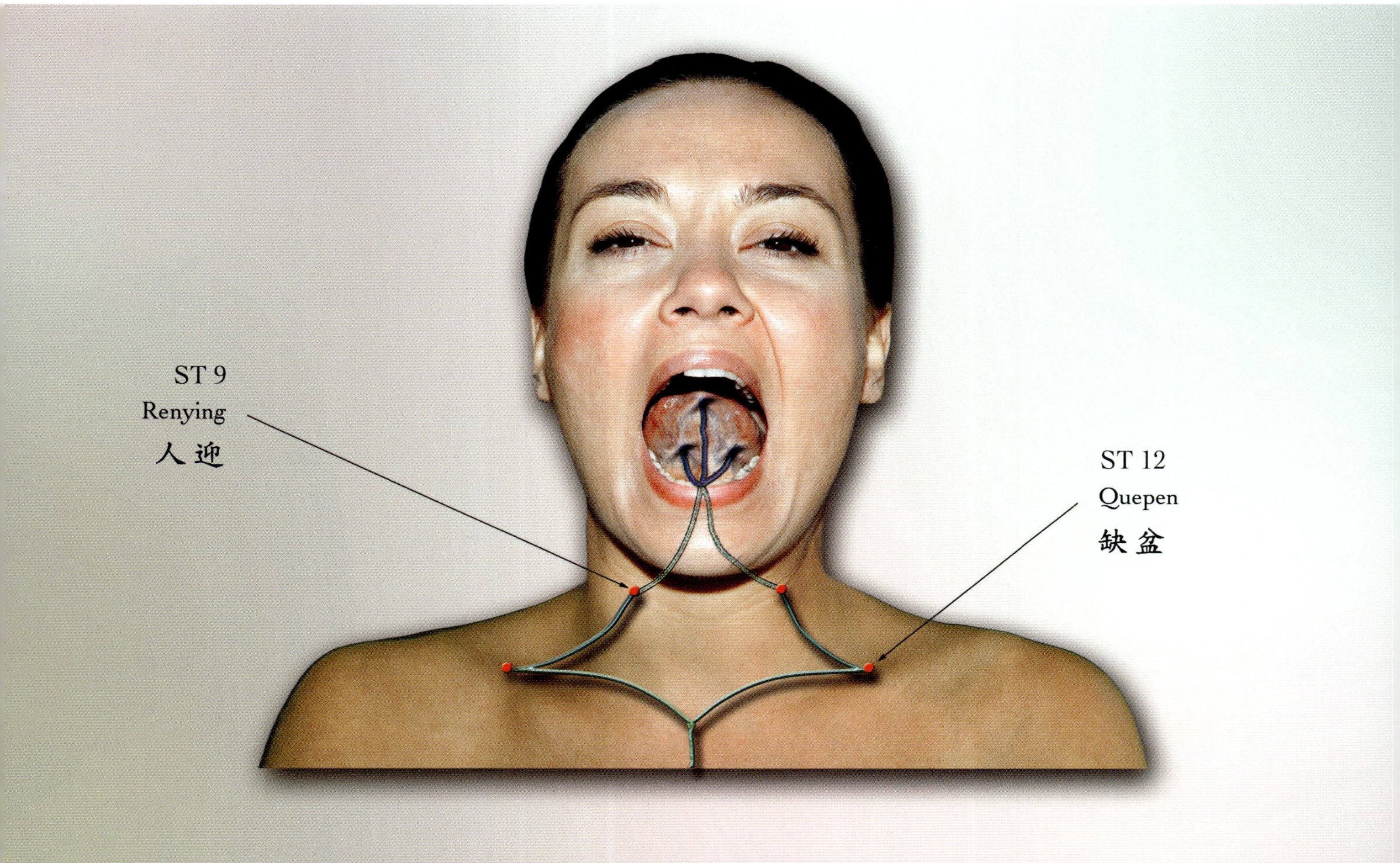

Distinct channel of the Zu Tai Yin (Spleen), cephalic course

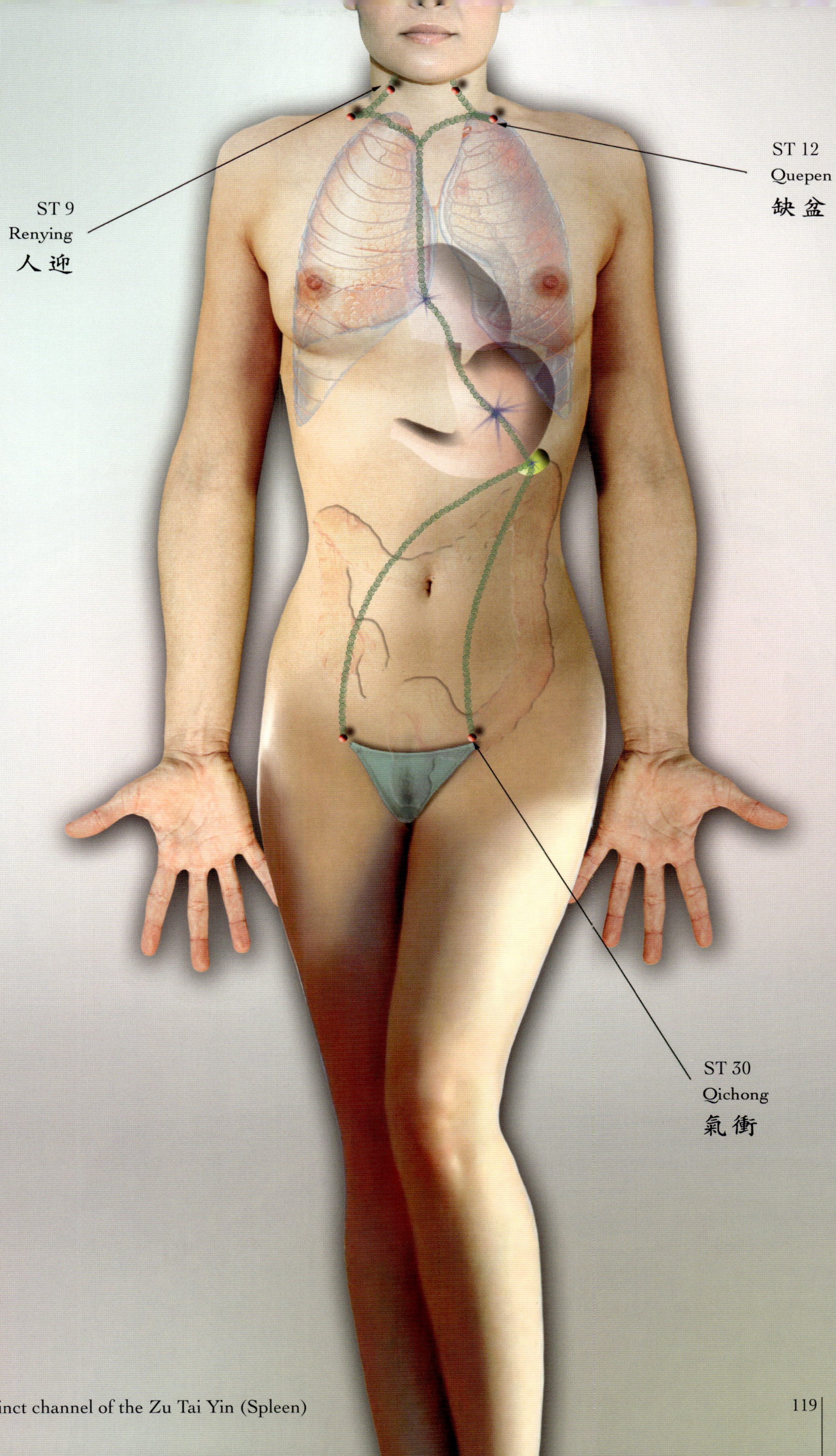
ST 12
Quepen
缺盆
ST 9
Renying
人迎
ST 30
Qichong
氣衝

Points of the Main channel Zu Tai Yin (Spleen)

SP 1 Yinbai
On the medial side of the distal segment of the big toe, 0.1 Cun from the corner of the toenail.

SP 2 Dadu
On the medial border of the foot, in the depression of the 'junction of the red and white skin', anterior and inferior to the 1st metatarso-phalangeal joint.

SP 3 Taibai
On the medial border of the foot, in the depression of the 'junction of the red and white skin', posterior and inferior to the 1st metatarso-phalangeal joint.

SP 4 Gongsun
On the medial border of the foot, anterior and inferior to the proximal end of the 1st metatarsal bone.

SP 5 Shangqiu
In the depression anterior and inferior to the medial malleolus, at the midpoint of the line connecting the tuberosity of the navicular bone and the tip of the medial malleolus.

SP 6 Sanyinjiao
On the medial side of the leg, 3 Cun above the tip of the medial malleolus, posterior to the medial border of the tibia.

SP 7 Lougu
On the medial side of the leg and on the line connecting the tip of the medial malleolus and SP 9 Yinlingquan, 6 Cun from the tip of the medial malleolus, posterior to the medial border of the tibia.

SP 8 Diji
On the medial side of the leg and on the line connecting the tip of the medial malleolus and Yinlingquan, 3 Cun below SP 9 Yinlingquan.

SP 9 Yinlingquan
On the medial side of the leg, in the depression posterior and inferior to the medial condyle of the tibia.

SP 10 Xuehai
With the knee flexed, on the medial side of the thigh, 2 Cun above the superior medial corner of the patella, on the prominence of the medial head of the quadriceps muscle of the thigh.

SP 11 Jimen
On the medial side of the thigh and on the line connecting SP 10 Xuehai and SP 12 Chongmen, 6 Cun above SP 10 Xuehai.

SP 12 Chongmen
At the lateral end of the inguinal groove, 3.5 Cun lateral to the midpoint of the upper border of the symphysis pubis, lateral to the pulsating external iliac artery.

SP 13 Fushe
On the lower abdomen, 4 Cun below the centre of the umbilicus, 0.7 Cun above SP 12 Chongmen, and 4 Cun lateral to the anterior midline.

SP 14 Fujie
On the lower abdomen, 1.3 Cun below SP 15 Daheng, and 4 Cun lateral to the anterior midline.

SP 15 Daheng
On the middle abdomen, 4 Cun lateral to the centre of the umbilicus.

SP 16 Fu'ai
On the upper abdomen, 3 Cun above the centre of the umbilicus, and 4 Cun lateral to the anterior midline.

SP 17 Shidu
On the lateral side of the chest and in the 5th intercostal space, 6 Cun lateral to the anterior midline.

SP 18 Tianxi
On the lateral side of the chest and in the 4th intercostal space, 6 Cun lateral to the anterior midline.

SP 19 Xiongxiang
On the lateral side of the chest and in the 3rd intercostal space, 6 Cun lateral to the anterior midline.

SP 20 Zhourong
On the lateral side of the chest and in the 2nd intercostal space, 6 Cun lateral to the anterior midline.

SP 21 Dabao
On the lateral side of the chest and on the middle axillary line, in the 6th intercostal space.

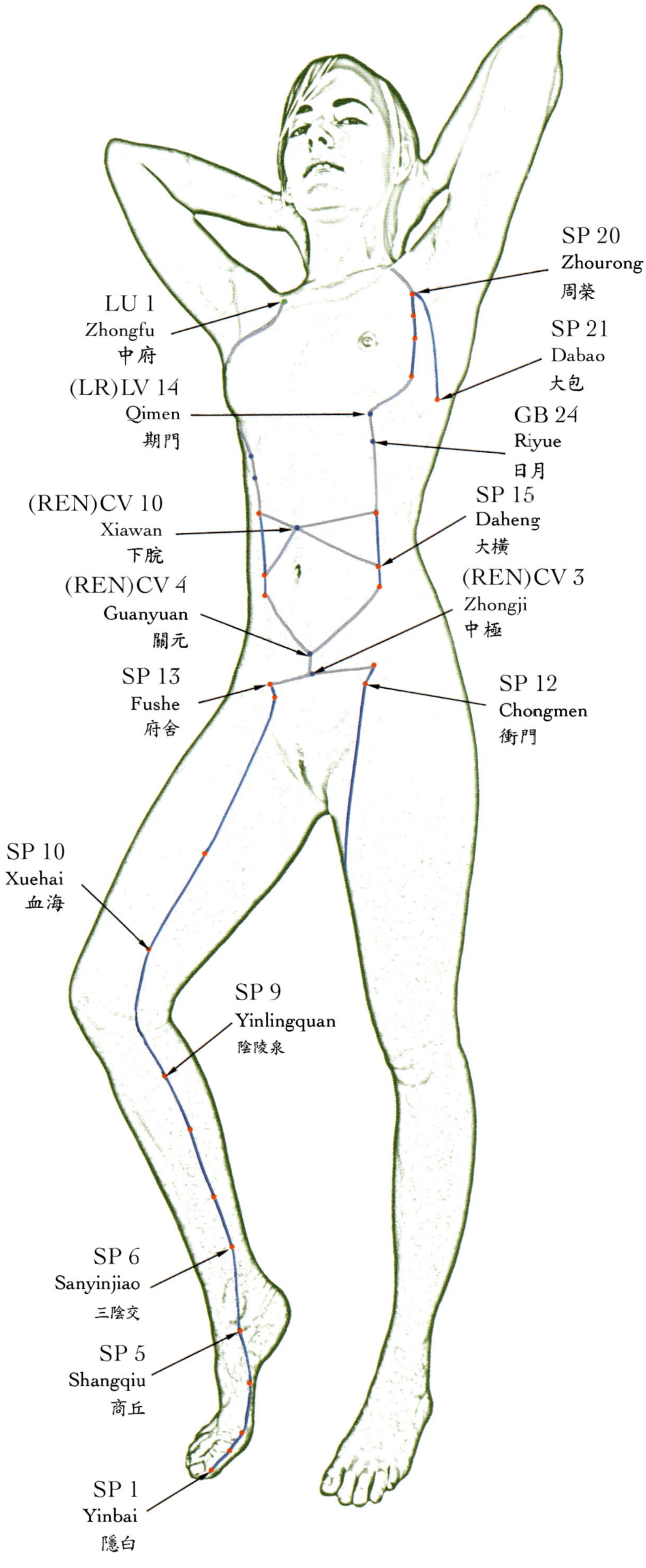

Points of the Main channel Zu Tai Yin (Spleen)

Zu Jue Yin system (Liver)

- Tendino-muscular channel
- Main channel
- Internal branches
- Luo vessels
- Distinct channel
- Points of the Main channel

Tendino-muscular of the Zu Jue Yin (Liver)

***Ling Shu*, chapter IV, paragraph 13**

The tendino-muscular of the liver starts above the big toe.

It goes up and diverges in front of the internal malleolus.

It goes up along the tibia, and keeps going up passing under the internal femoral condyle.

It goes up by the internal side of the thigh.

It passes by the genitalia and unites with the other tendino-muscular channels.

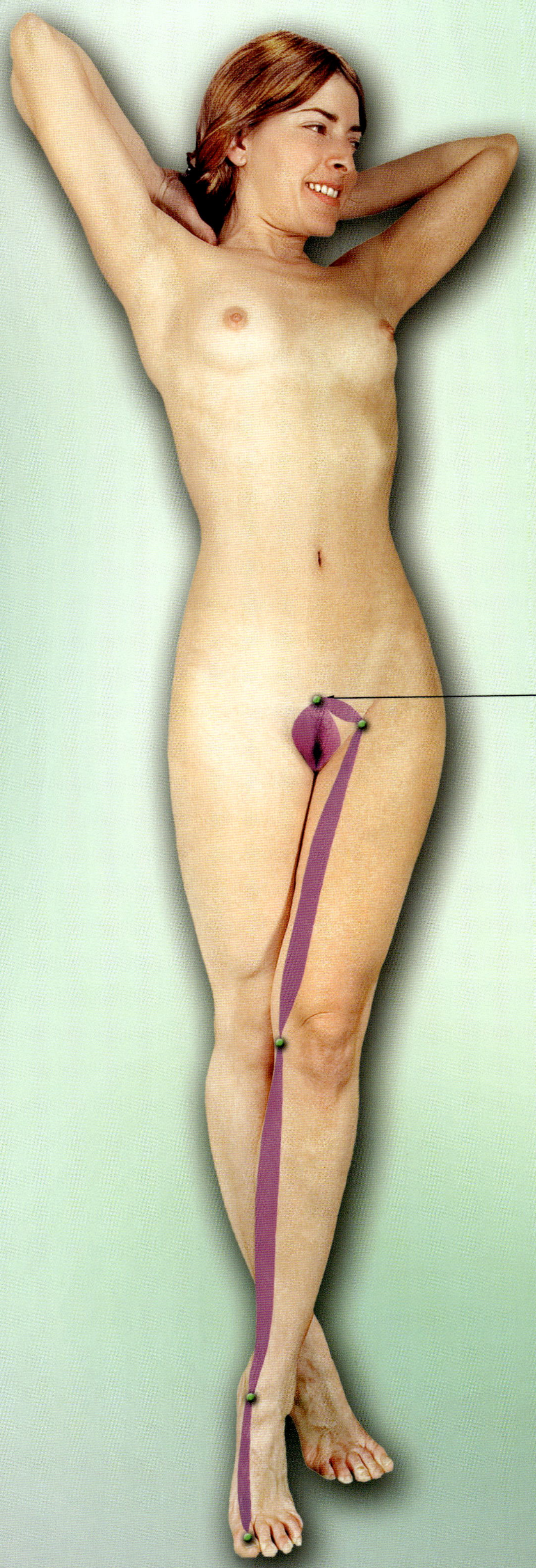
Union area
(REN)CV 3
Zhongji
中極

Main channel of the Zu Jue Yin (Liver)

Ling Shu, chapter III, paragraph 10

The vessel of the Liver, Zu Jue Yin of the foot, starts at the hairline of the big toe (in front of the metatarso-phalangeal joint).

It goes over the dorsal side of the foot by the upper region and gets to 1 Cun of the internal malleolus.

It goes up to 8 Cun above the malleolus.
It crosses the back of the Zu Tai Yin (channel of the Spleen) and runs up to the internal border of the popliteal crease.

It goes along the internal part of the thigh and penetrates into the pubic hairs, going round the genitalia.

It reaches the lower part of the abdomen, runs close to the Stomach, belongs to the Liver and joins the Gall Bladder.

It goes up following the diaphragm and spreads in the costal area.

It goes along the posterior part of the trachea.
It gets inside, going up on the upper palate.

It links the 'eye connections', runs up and goes out by the forehead.

It joins the Du Mai (Governor vessel) at the top of the head (at point Baihui [100 links]).

The branch leaving the 'eye connections' goes down on the cheek and encircles the lips.
The branch leaving the Liver follows a different course: it gets inside the diaphragm and goes up, spreading in the Lungs.

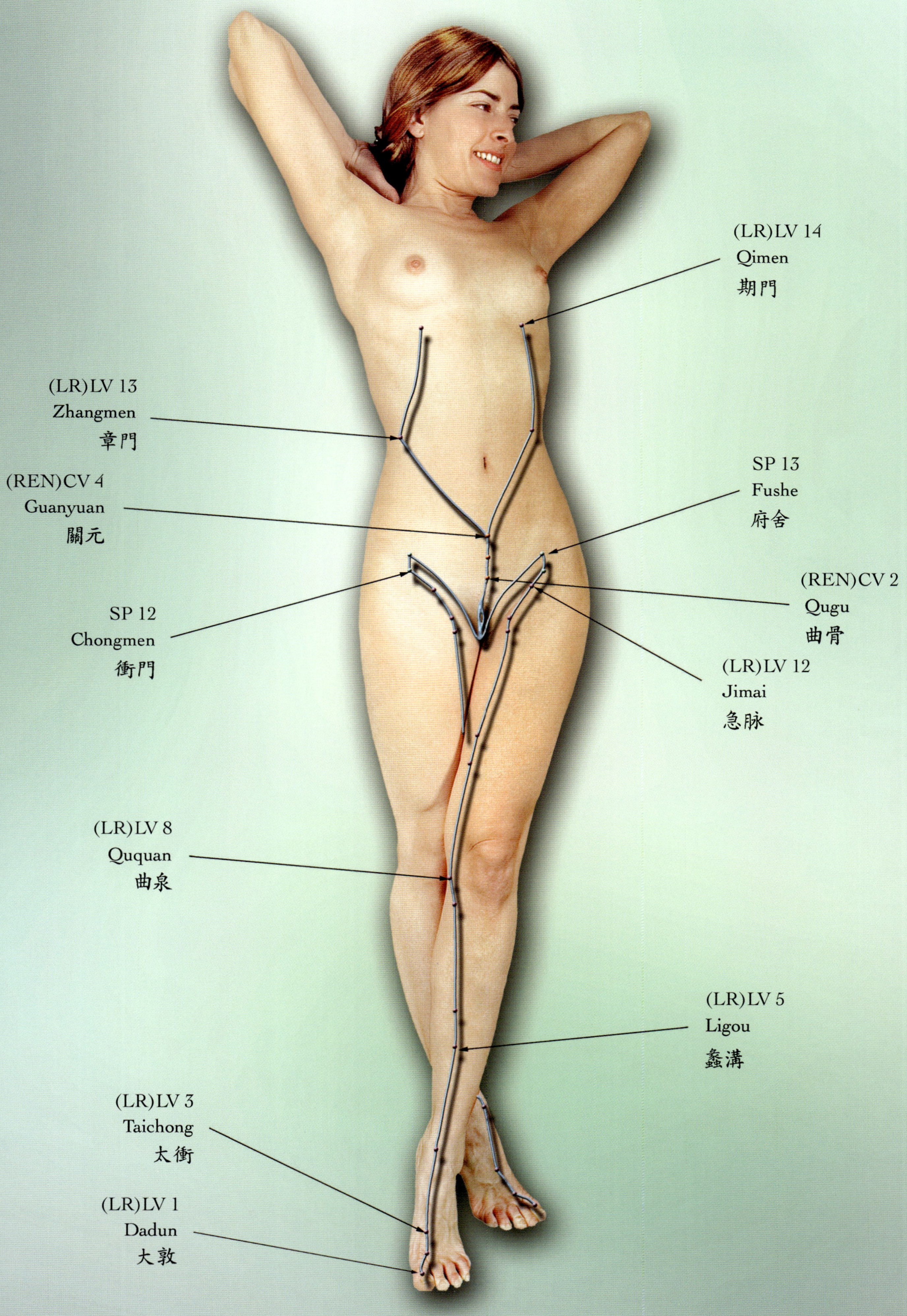

Main channel of the Zu Jue Yin (Liver), course n° 1

Going out of its first point (LR)LV 1 Dadun, the Main channel of the Zu Jue Yin follows the dorsal side of the foot, passes in front of point (LR)LV 4 Zhongfeng and goes up by the posterio-internal border of the tibia to reach point SP 6 Sanyinjiao, where the three lower Yin get together.

It follows the internal border of the tibia to 8 Cun above the internal malleolus and goes inside.

It crosses the channel of the Zu Tai Yin and passes behind it.

It gets to the internal extremity of the flex crease of the knee where point (LR)LV 8 Ququan is located.

It passes by the internal part of the thigh, penetrates the pubis and connects with points SP 12 Chongmen and SP 13 Fushe.

It goes toward the genitalia, runs round them, goes up again, connects on the midline with points (REN)CV 2 Qugu, (REN)CV 3 Zhongji and (REN)CV 4 Guanyuan.

It keeps its ascending course and gets to the tip of the 11th rib where (LR)LV 13 Zhangmen is and finally reaches the 6th intercostal space at point (LR)LV 14 Qimen.

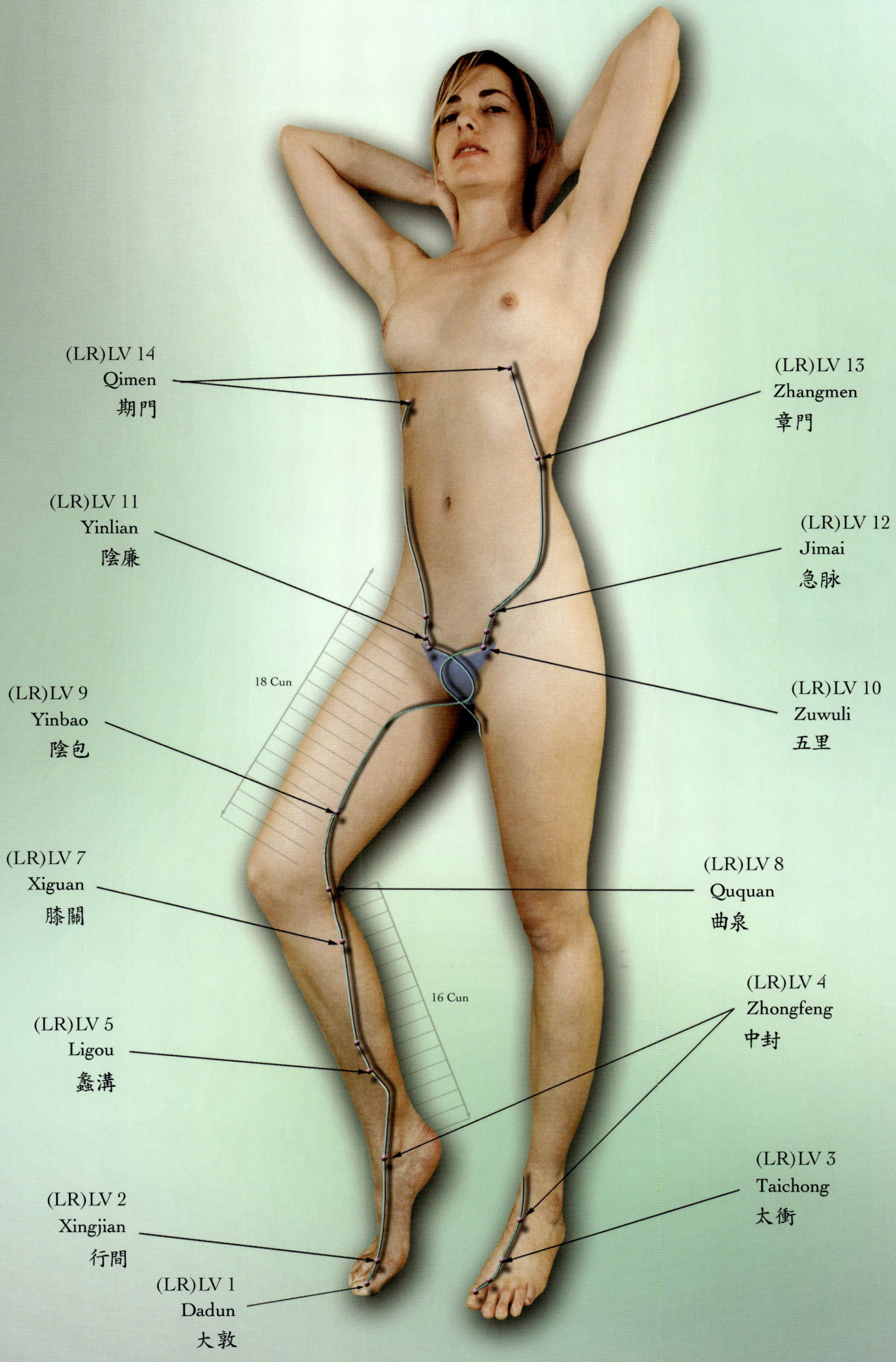

Main channel of the Zu Jue Yin (Liver), course n° 2

Internal branches of the Zu Jue Yin channel (Liver)

Commentaries

From (LR)LV 13 Zhangmen, the channel enters the Stomach, the Liver and the Gall Bladder.

Then, it goes across the diaphragm and ramifies in the ribs and in the Lung, where it unites with the Main channel of Shu Tai Yin (Lung).

Leaving (LR)LV 14 Qimen, a branch penetrates the Liver, and another passes behind the trachea, goes to the pharynx, the maxillary sinus, the eyeball and the brain.

From the maxillary sinus fork out two other branches:

- The first goes up on the forehead and reaches point (DU)GV 20 Baihui.
- The second goes down along the cheeks and goes round the lips on the inside.

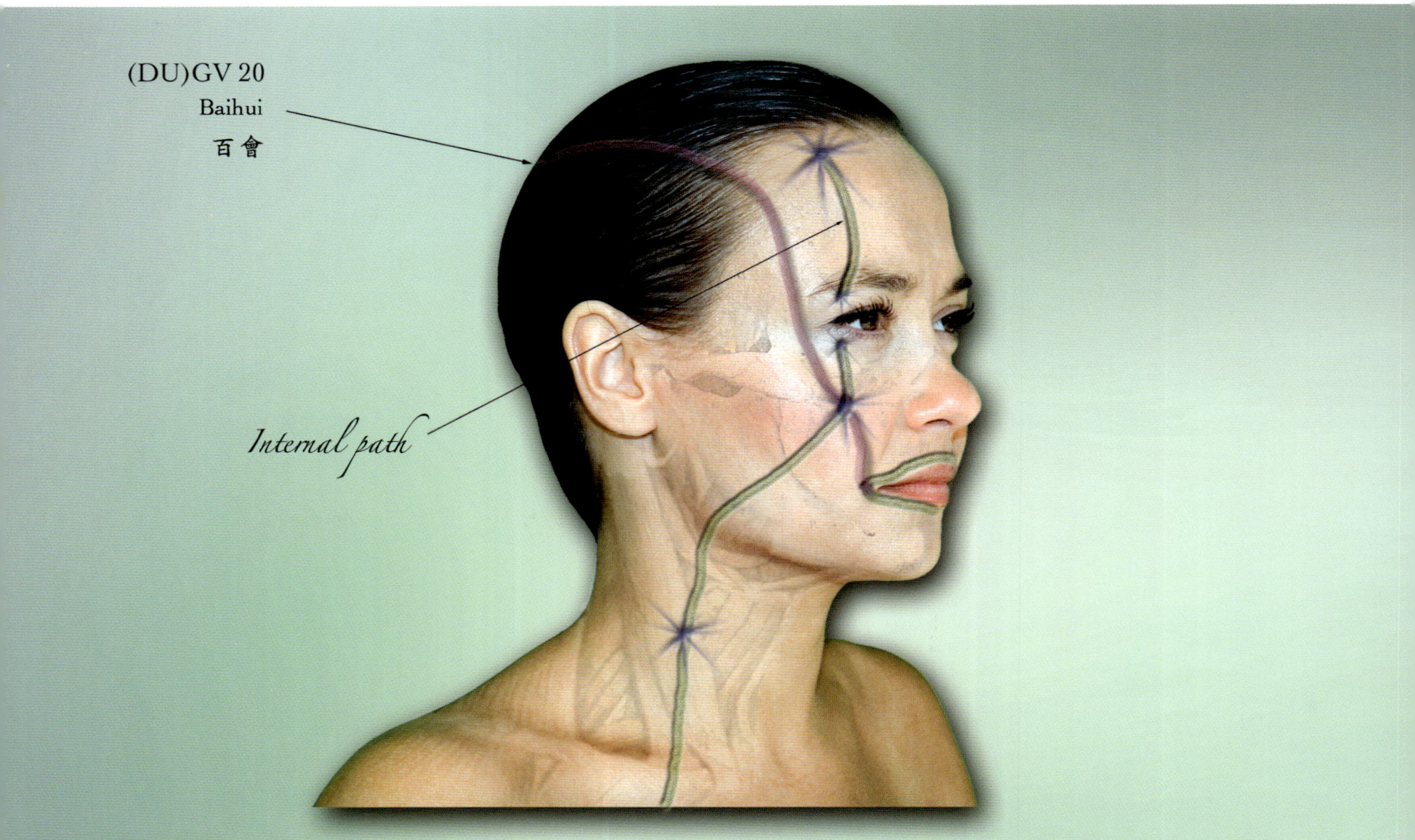

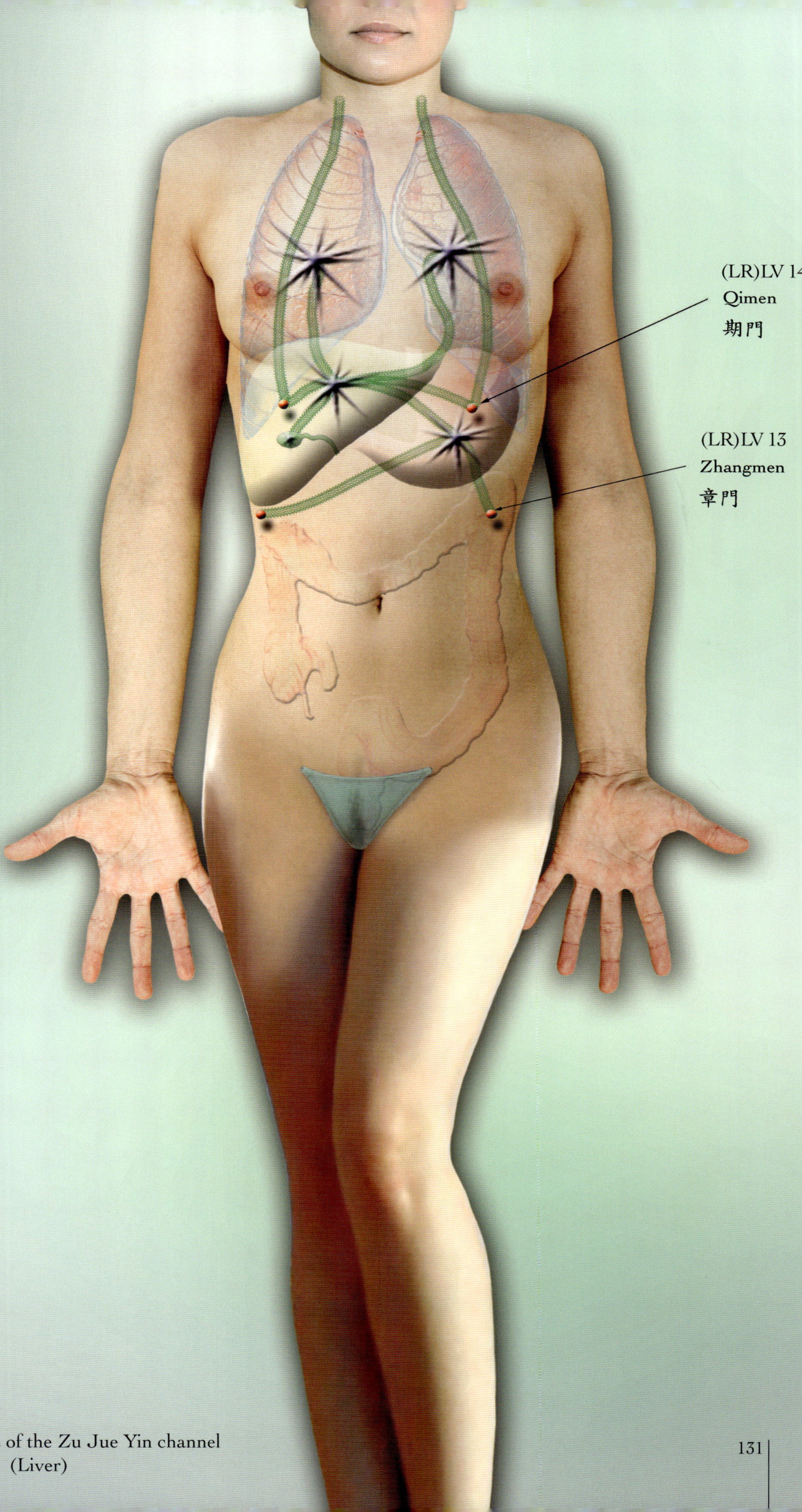

Internal branches of the Zu Jue Yin channel
(Liver)

Luo vessel of the Zu Jue Yin (Liver)

Ling Shu, chapter III, paragraph 10

The distinct branching of the Zu Jue Yin of the foot (channel of the Liver) is called Ligou (woodworm in the groove).

It starts 5 Cun above the internal malleolus.

This distinct vessel runs to the Zu Shao Yang (channel of the Gall Bladder).

A distinct branch passes by the tibia (the leg), goes up to the testes and joins the penis.

Commentaries

The Luo vessel leaves point (LR)LV 5 Ligou 5 Cun above the internal malleolus and divides into two branches.

- The longitudinal branch passes along the internal part of the calf and reaches the external genitalia where it ramifies.
- The transversal branch unites with the channel of the Gall Bladder at point GB 40 Qiuxu in the interspace between the calcaneum and the cuboid.

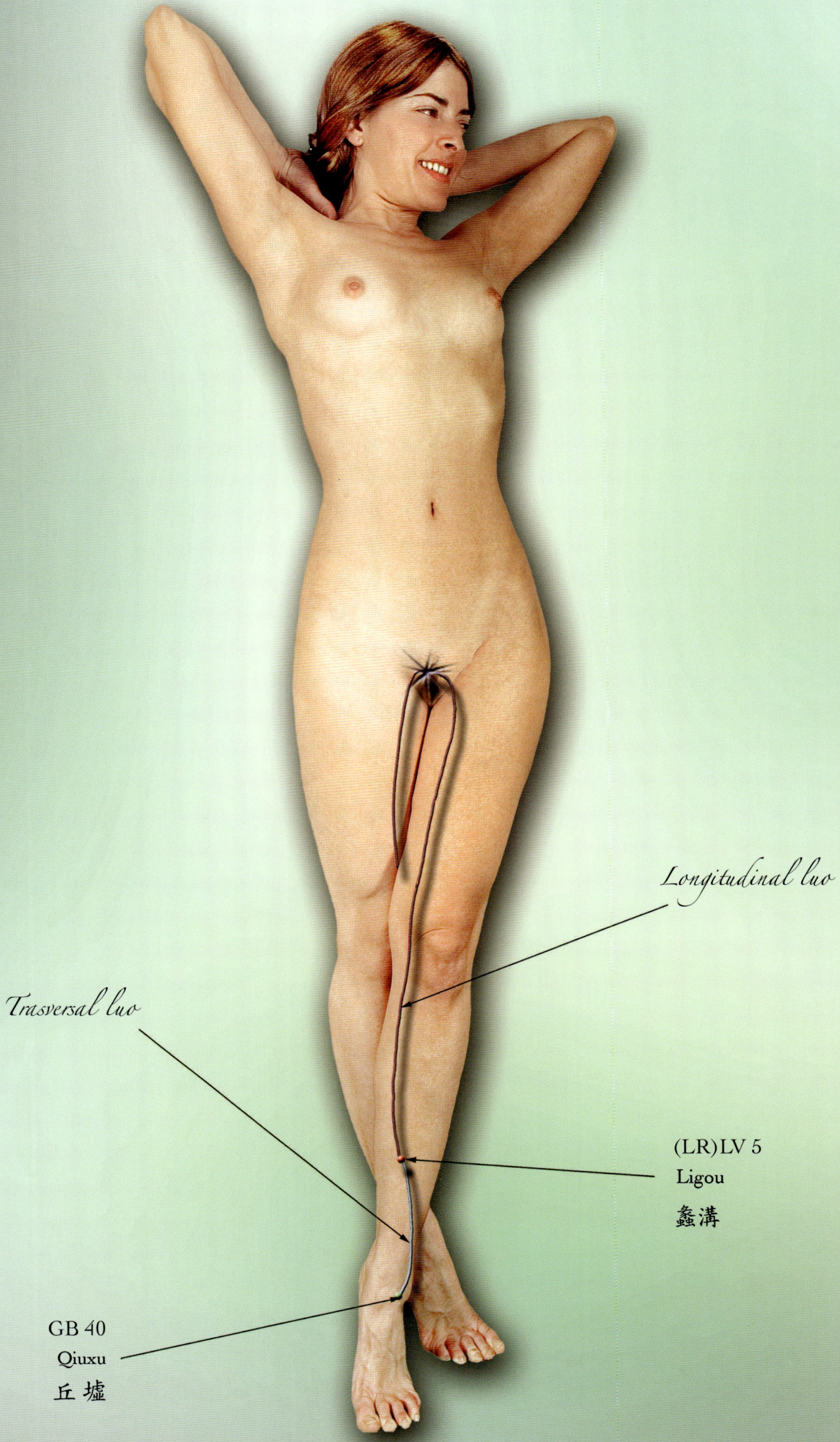

Luo vessel of the Zu Jue Yin (Liver)

Distinct channel of the Zu Jue Yin (Liver)

Ling Shu, chapter III, paragraph 10

The distinct channel of the Zu Jue Yin of the foot (Liver) goes out on the dorsal side of the foot.

It goes up to the pubic area, joins the channel of the Shao Yang (Gall Bladder) and circulates with it to form the second union.

Commentaries

From the dorsal side of the foot, the distinct channel goes up to the pubis and unites with the genitalia.

It joins the channel of the Gall Bladder in the region of (REN)CV 2 Qugu.

It runs up to the abdomen to reach point (LR) LV 13 Zhangmen, penetrates deep inside and unites with the Liver, with the Gall Bladder and goes through the Heart.

It ascends to the throat passing on each side of the trachea and spreads on the face.

It joins the distinct channel and to the Main channel of the Gall Bladder to form the second union in the vicinity of point GB 1 Tongziliao.

It finally joins the 'eye connections' and penetrates into the brain.

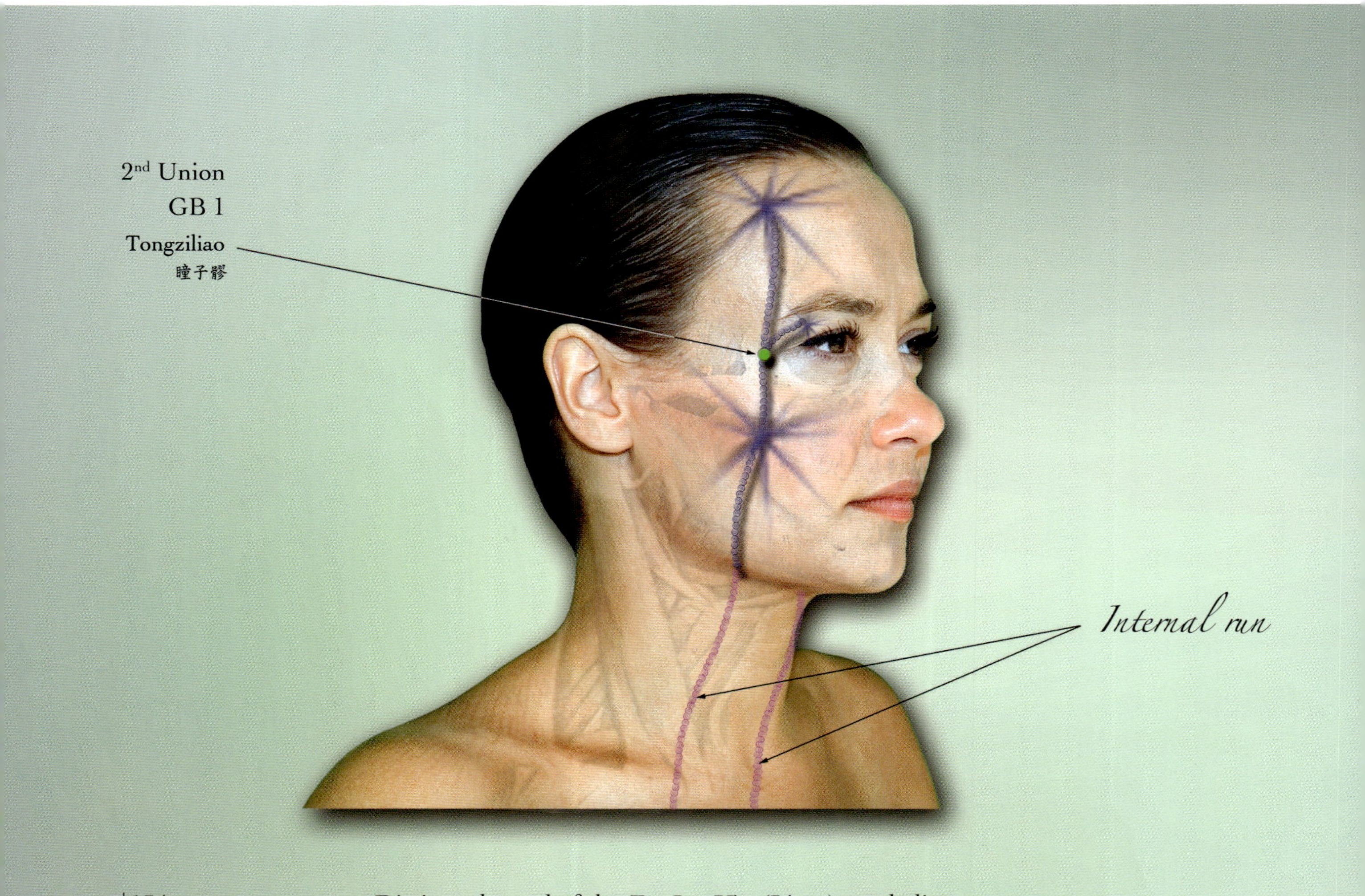

Distinct channel of the Zu Jue Yin (Liver), cephalic course

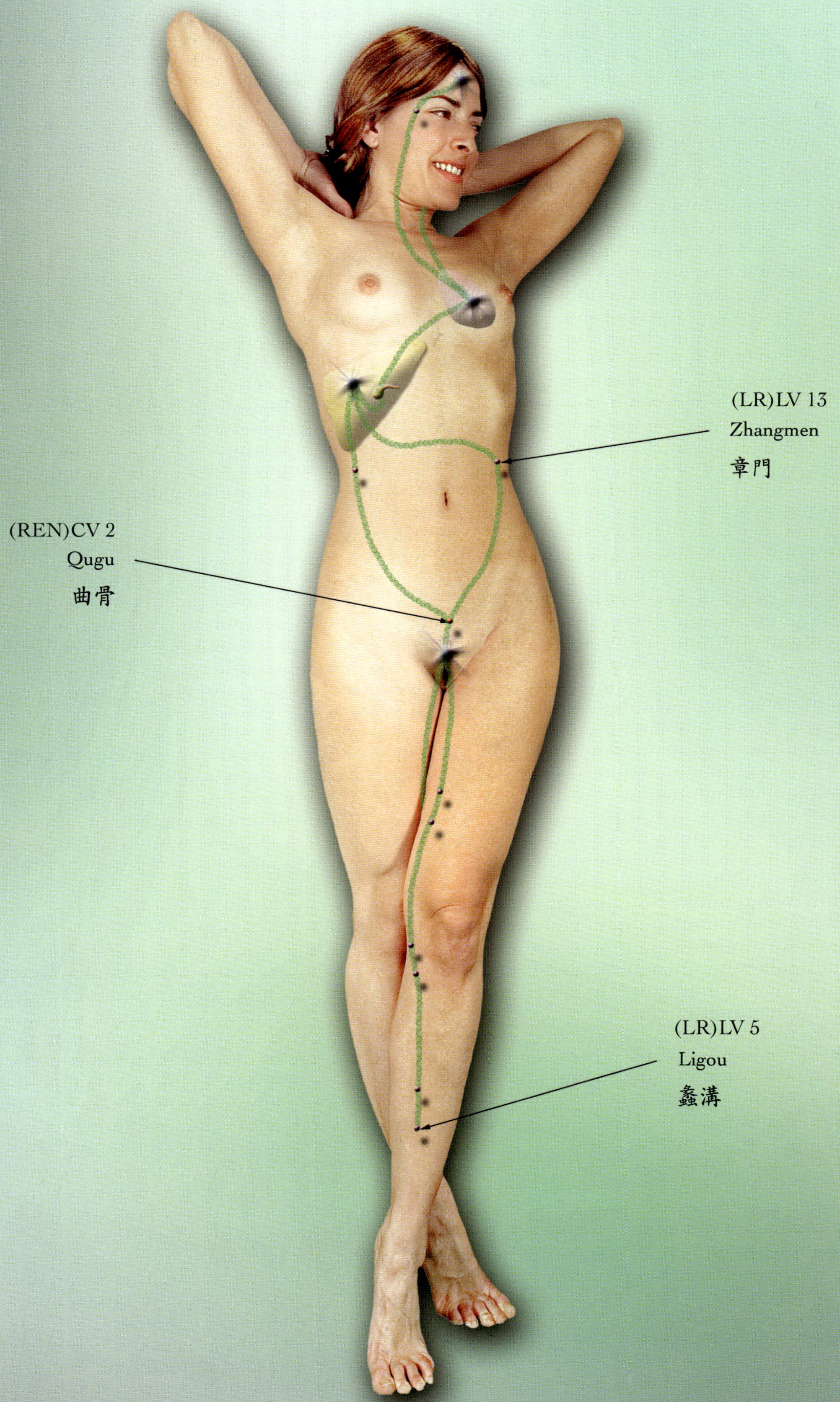

Distinct channel of the Zu Jue Yin (Liver)

Points of the Main channel Zu Jue Yin (Liver)

(LR)LV 1 Dadun
On the lateral side of the distal segment of the big toe, 0.1 Cun from the corner of the toenail.

(LR)LV 2 Xingjian
On the instep of the foot, between the 1st and 2nd toes, at the 'junction of the red and white skin', proximal to the margin of the web.

(LR)LV 3 Taichong
On the instep of the foot, in the depression of the posterior end of the 1st interosseous metatarsal space.

(LR)LV 4 Zhongfeng
On the instep of the foot, anterior to the medial malleolus, on the line connecting SP 5 Shangqiu and ST 41 Jiexi, in the depression medial to the tendon of the anterior tibial muscle.

(LR)LV 5 Ligou
On the medial side of the leg, 5 Cun above the tip of the medial malleolus, on the midline of the medial surface of the tibia.

(LR)LV 6 Zhongdu
On the medial side of the leg, 7 Cun above the tip of the medial malleolus, on the midline of the medial surface of the tibia.

(LR)LV 7 Xiguan
On the medial side of the leg, posterior and inferior to the medial epicondyle of the tibia, 1 Cun posterior to SP 9 Yinlingquan, at the upper end of the medial head of the gastrocnemius muscle.

(LR)LV 8 Ququan
On the medial side of the knee, at the medial end of the popliteal crease when the knee is flexed, posterior to the medial epicondyle of the tibia, in the depression of the anterior border of the insertions of the semi-membranous and semi-tendinous muscles.

(LR)LV 9 Yinbao
On the medial side of the thigh, 4 Cun above the medial epicondyle of the femur, between the medial vastus muscle and sartorius muscle.

(LR)LV 10 Zuwuli
On the medial side of the thigh, 3 Cun directly below ST 30 Qichong, at the proximal end of the thigh, below the pubic tubercle and on the lateral border of the long abductor muscle of the thigh.

(LR)LV 11 Yinlian
On the medial side of the thigh, 2 Cun directly below ST 30 Qichong, at the proximal end of the thigh, below the pubic tubercle and on the lateral border of the long abductor muscle of the thigh.

(LR)LV 12 Jimai
Lateral to the pubic tubercle, lateral and inferior to ST 30 Qichong, in the inguinal groove where the pulsation of the femoral artery is palpable, 2.5 Cun lateral to the anterior midline. (Do NOT needle directly into the artery!)

(LR)LV 13 Zhangmen
On the lateral side of the abdomen, below the free end of the 11th rib.

(LR)LV 14 Qimen
On the chest, directly below the nipple, in the 6th intercostal space, 4 Cun lateral to the anterior midline.

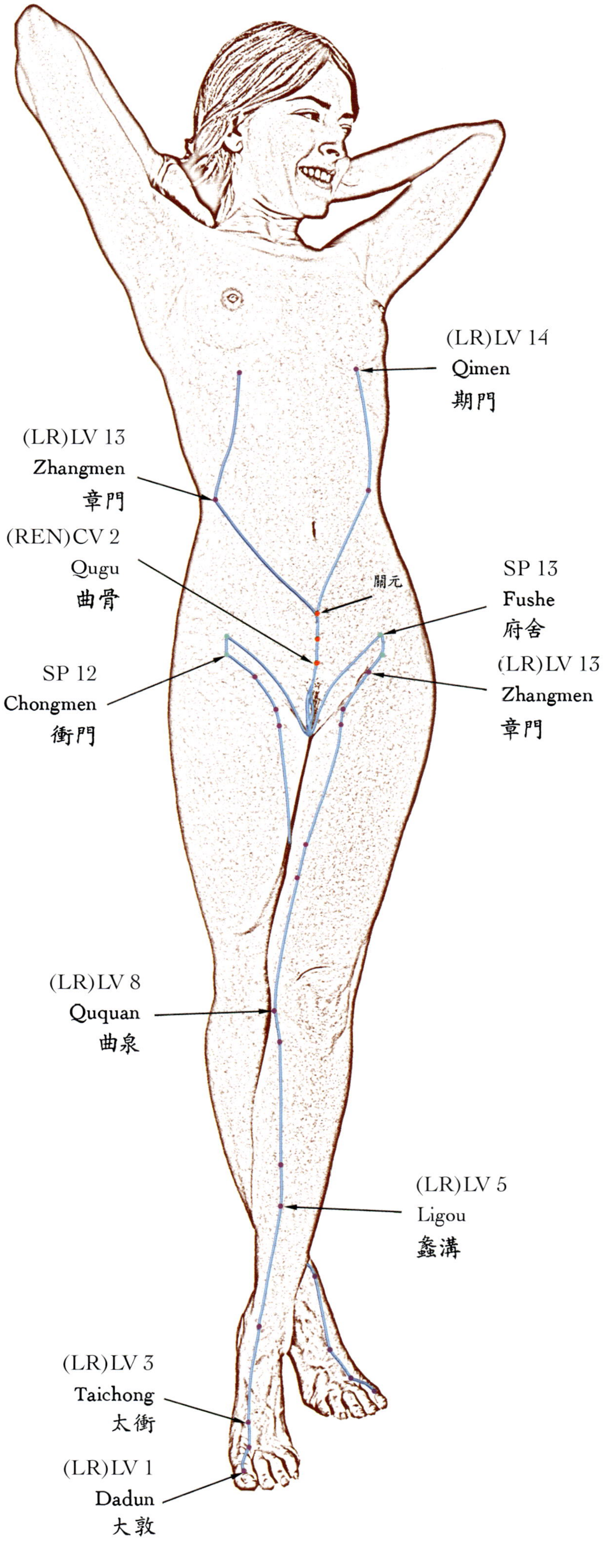

Points of the Main channel Zu Jue Yin (Liver)

Zu Shao Yin system (Kidney)

- Tendino-muscular channel
- Main channel
- Internal branches
- Luo vessels
- Distinct channel
- Points of the Main channel

Tendino-muscular of the Zu Shao Yin (Kidney)

***Ling Shu*, chapter IV, paragraph 13**

The tendino-muscular of the Kidney starts under the little toe.

It joins the tendino-muscular of the Spleen and passes obliquely under the internal malleolus.

It unites with the posterior part of the foot and joins the tendino-muscular of Tai Yang.

Going up, it attaches itself under the internal femoral condyle and joins the tendino-muscular of the Spleen.

It goes up by the internal side of the thigh and joins the genitalia.

It goes up along the backbone to the nape of the neck.

It unites with the occiput where it joins the tendino-muscular of the Urinary Bladder.

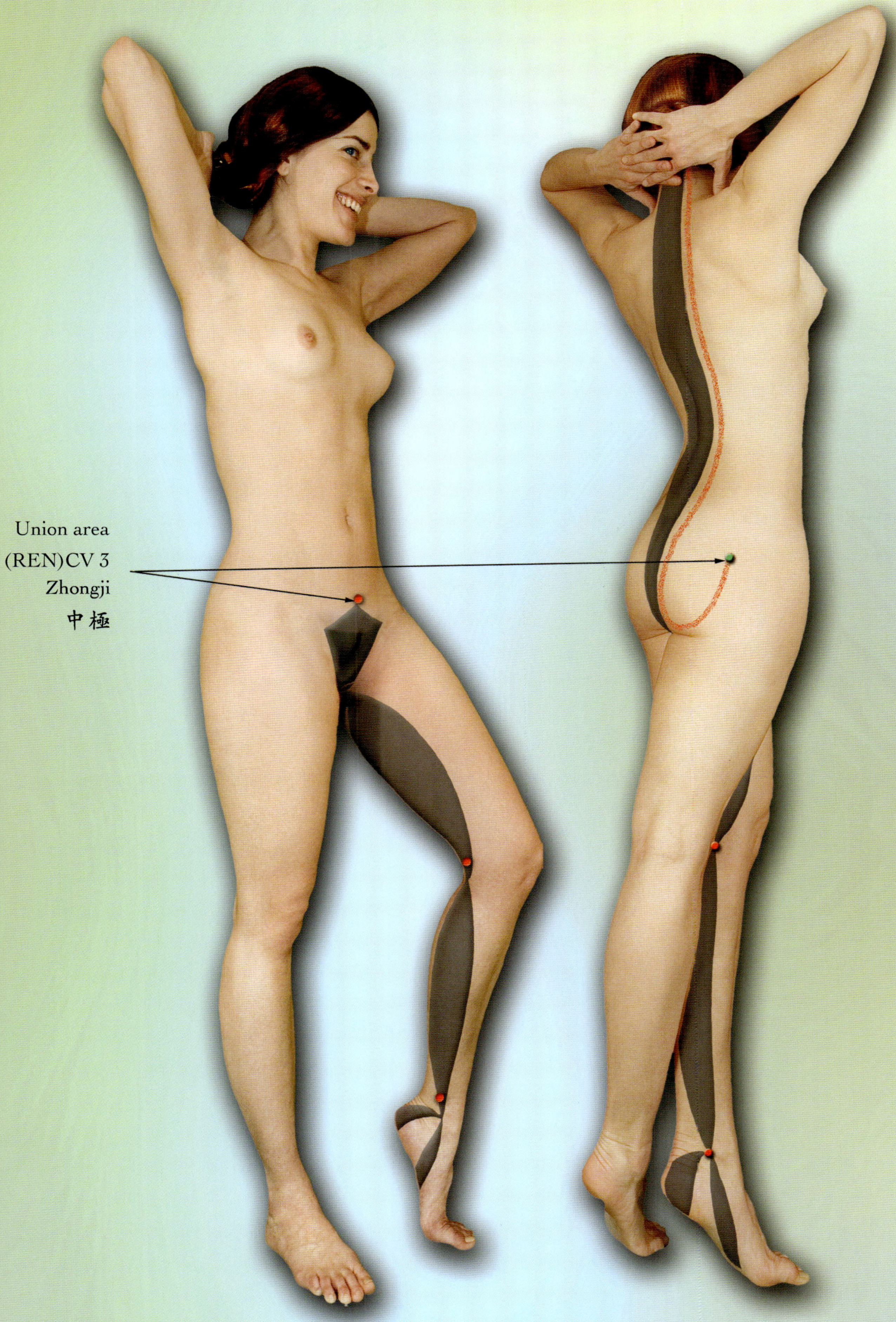

Tendino-muscular of the Zu Shao Yin (Kidney)

Main channel of the Zu Shao Yin (Kidney)

***Ling Shu*, chapter III, paragraph 10**

The channel of the Kidney, Zu Shao Yin of the foot, starts under the little toe.

It crosses obliquely the sole. It goes out at point KI 2 Rangu (under and before the internal malleolus), and runs along the posterior part of the internal malleolus.

This distinct branch goes inside at the midpoint of the heel and goes up by the internal side of the calf and goes out by the internal border of the popliteal crease.

It goes up by the posterior border of the internal side of the thigh.

It follows the vertebrae, joins the Kidneys and the Bladder.

A direct branch starts from the Kidneys and goes up by way of the Liver and the diaphragm.

It enters the Lungs, follows the throat and encircles the 'root of the tongue'.

The branch starting from the Lungs goes round the Pericardium and spreads at the midpoint of the chest.

Commentaries

The channel of the Kidney, Zu Shao Yin of the foot, starts under the little toe and travels on the sole of the foot where point KI 1 Yongquan is located.

It runs to the internal border of the foot and goes out under the scaphoid where point KI 2 Rangu is.

It then moves below the lower part of the internal malleolus and gets to point KI 3 Taixi.

It goes down again to the heel, crossing point KI 4 Dazhong.

It passes again below the internal malleolus where KI 6 Zhaohai is.

It goes up again by the postero-internal side of the leg, passing by point KI 7 Fuliu, and joins point SP 6 Sanyinjiao, the meeting place of the three lower Yin.

It keeps its ascending course and reaches the internal extremity of the popliteal crease, at point KI 10 Yingu.

It goes up by the internal side of the thigh, gets to the perineum, where its internal branch originates.

The superficial branch passes by point KI 11 Henggu, follows an abdominal and thoracic ascending course half a 0.5 Cun outside the Ren Mai channel.

It reaches the infra-clavicular fossa to end its course at point KI 27 Shufu.

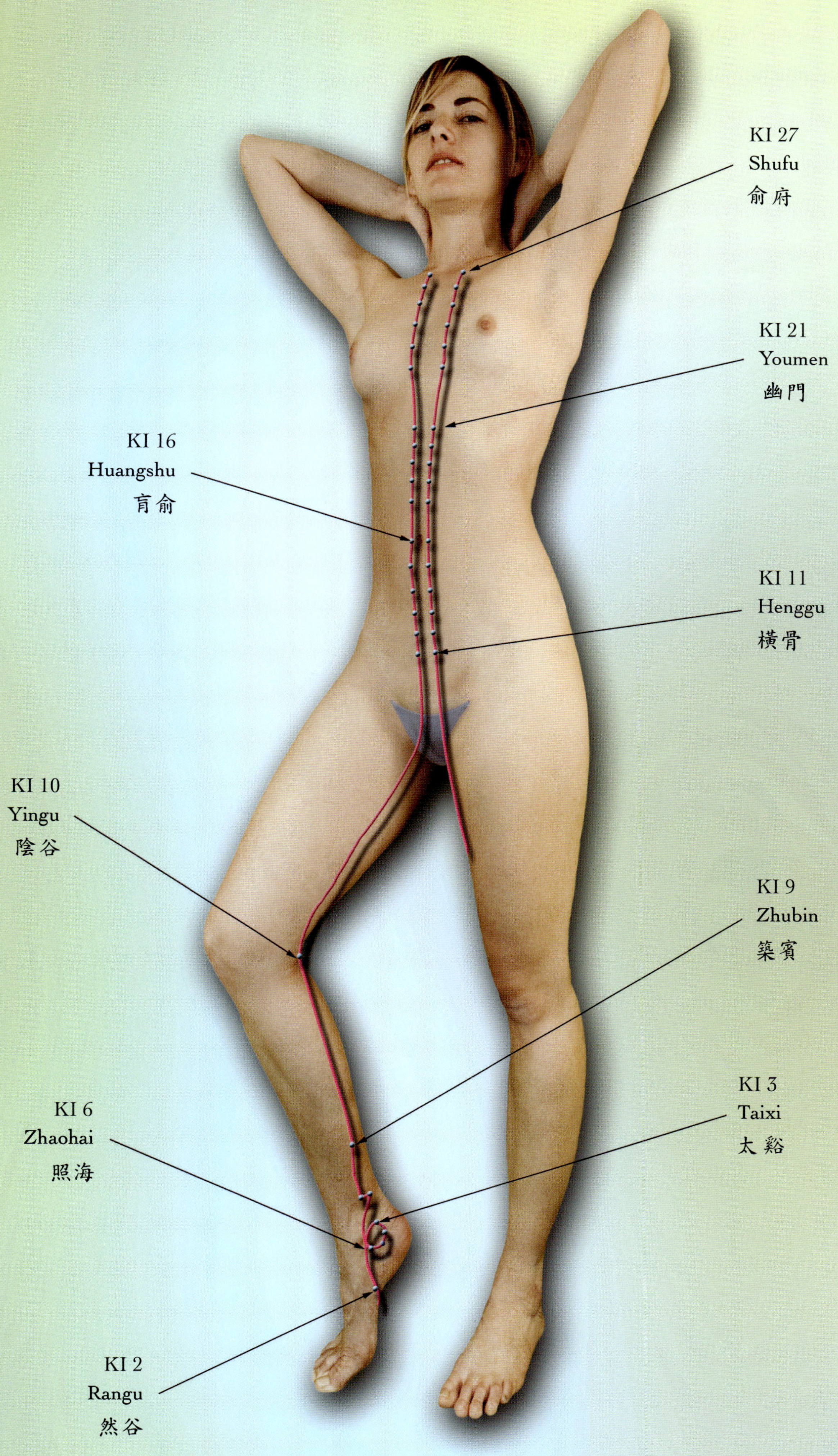

Main channel of the Zu Shao Yin (Kidney)

Internal branches of the Zu Shao Yin (Kidney)

***Ling Shu*, chapter III, paragraph 10**

The internal branch follows the vertebrae, belongs to the Kidneys and the Bladder.

A direct branch starts from the Kidneys and goes up, by way of the Liver and the diaphragm.

It enters the Lungs, goes along the throat and encircles the 'root of the tongue'.

The branch starting from the Kidneys goes round the Pericardium and spreads at the midpoint of the chest.

Commentaries

At the perineum starts the internal branch going to point (DU)GV 1 Changquiang, located between the anus and the tip of the coccyx.

It goes up inside the spinal column to reach the level of the 2nd lumbar vertebra and enter the Kidneys.

It then passes by the Urinary Bladder from which a small branch goes out and runs to points (REN)CV 4 Guanyuan and (REN)CV 3 Zhongji.

From the Kidneys, a branch reaches the Liver, crosses the diaphragm, enters the Lung where it spreads, goes up along the trachea and ends its course at the root of the tongue.

From the Lung, a branch runs to the Heart to join the channel of the Pericardium, passing by point (REN)CV 17 Danzhong, and ends its course at point (MH)PC 1 Tianchi.

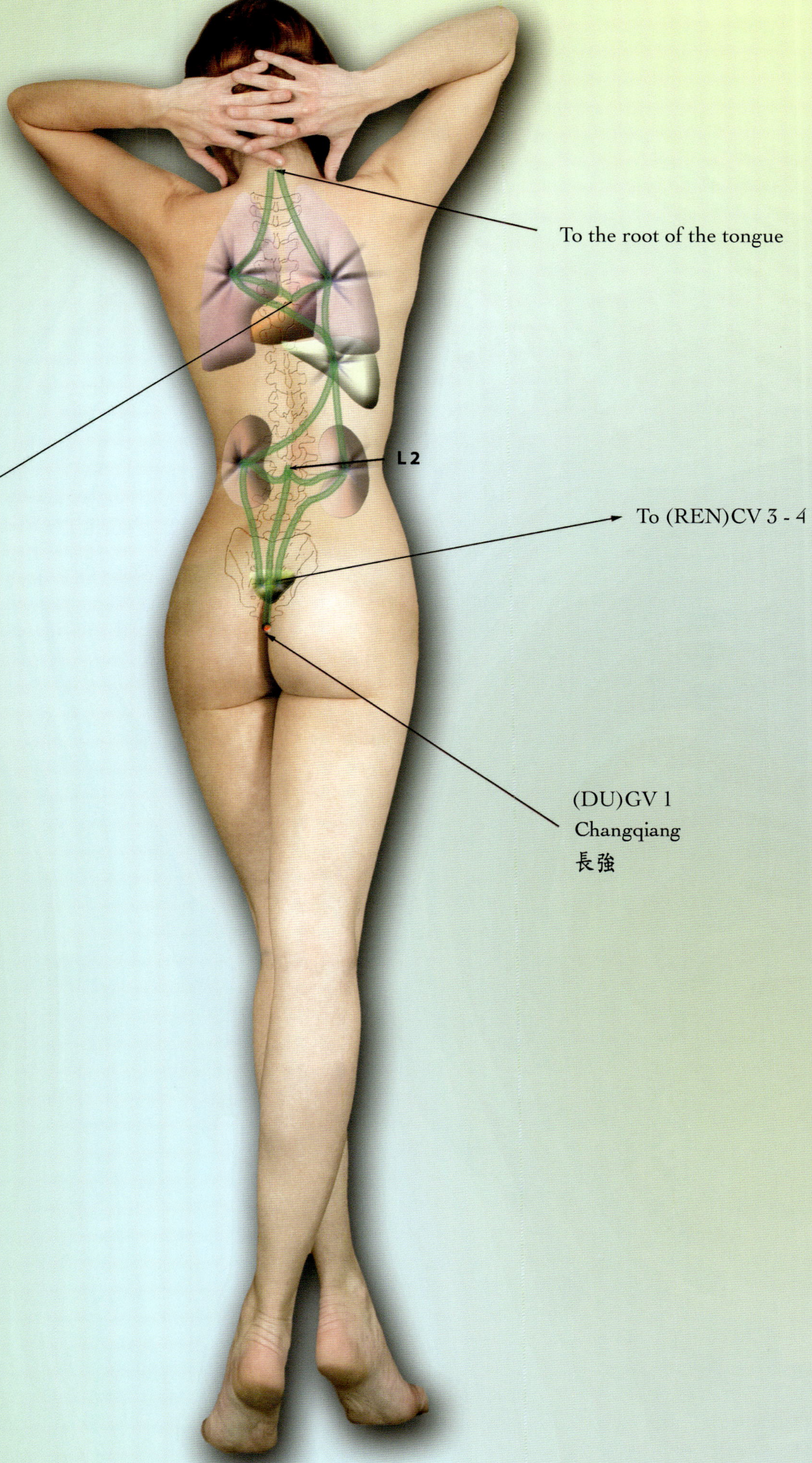

Internal branches of the Zu Shao Yin (Kidney)

Luo vessels of the Zu Shao Yin (Kidney)

Ling Shu, chapter III, paragraph 10

The Luo vessel of Zu Shao Yin of the foot is called Dazhong.

It goes round the heel from behind the internal malleolus and runs to Zu Tai Yang (Urinary Bladder).

A distinct branch joins with the channel of Zu Shao Yin of the foot, going up under the Pericardium.

It follows an outward course, crossing the lumbar vertebrae.

Commentaries

The Luo vessel goes out from point KI 4 Dazhong, situated in the angle between Achilles' heel and the roof of the calcaneum, and divides into two branches:

- The longitudinal branch follows the Main channel of the Kidney to the level of point (REN)CV 17 Danzhong, then gets inside to run to the spinal column in the zone of (DU) GV 11 Shendao, under the spinal process of the 5th thoracic vertebra.

- The transversal branch goes to the Urinary Bladder channel and joins with it at the level of point BL 64 Jinggu, before the head of the 5th metatarsal bone, at the 'junction of the red and white skin'.

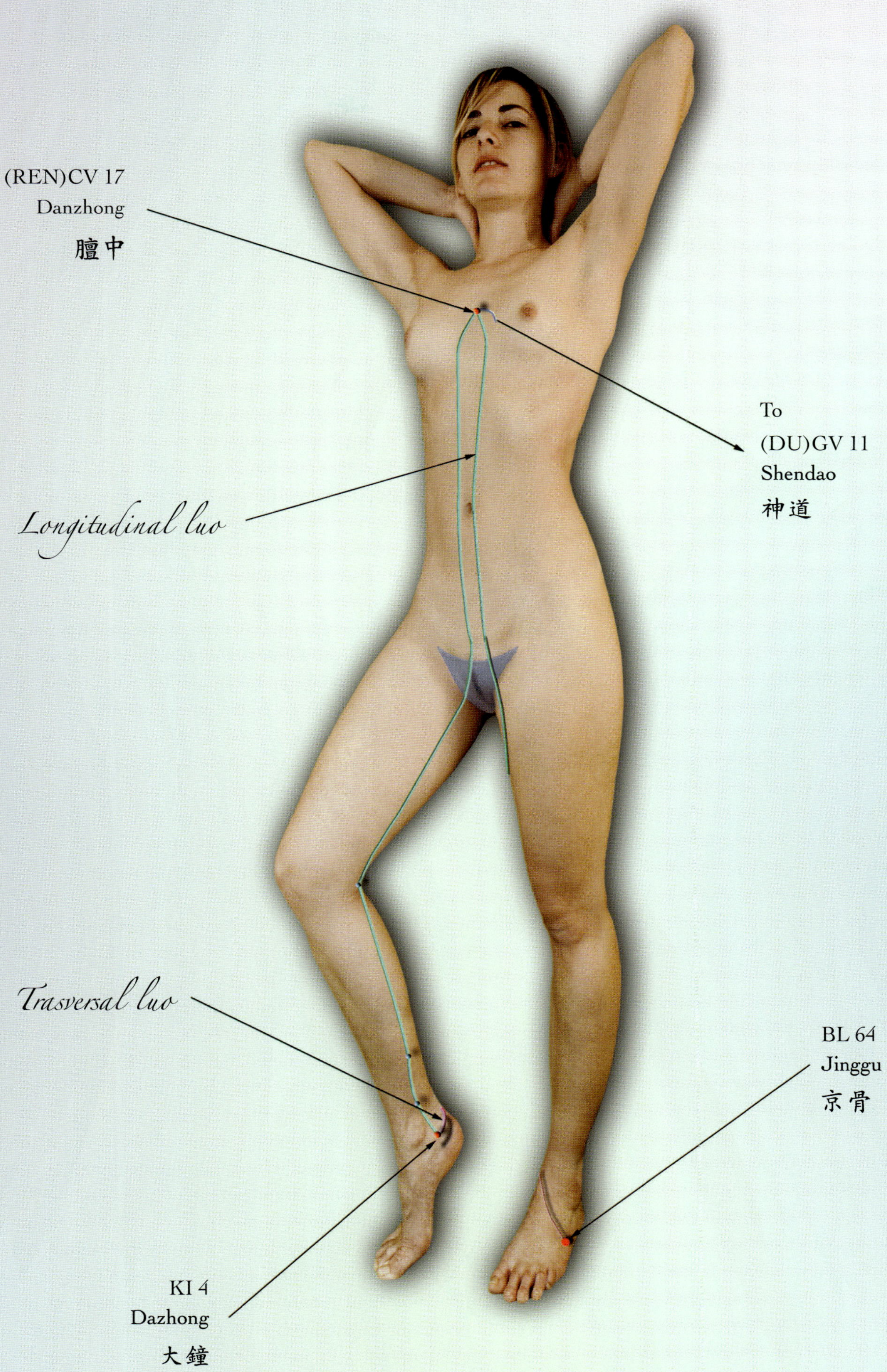

Luo vessels of the Zu Shao Yin (Kidney)

Distinct channel of the Zu Shao Yin (Kidney)

Ling Shu, chapter III, paragraph 11

From the regular vessel or Main channel Zu Shao Yin of the foot (Kidney), a distinct branch runs up to the popliteal fossa (its starting place is KI 1 Yongquan, 1 of the Kidneys, on the sole).

It joins Zu Tai Yang (channel of the Urinary Bladder) and goes up to the Kidneys.

It goes out on the level of the '14th thoracic vertebra', that is to say BL 23 Shenshu of the Urinary Bladder outside the tip of the spinal process of the 2nd lumbar vertebra.

It goes out and belongs to the Dai Mai (belt vessel), the direct branch runs to the root of the tongue and goes out by the nape of the neck.

It joins Zu Tai Yang and unites with it.

Commentaries

The distinct channel goes out of the Main channel at point KI 10 Yingu, and joins point BL 40 Weizhong, at the midpoint of the popliteal hollow. It goes up to the gluteal crease, to point BL 36 Chengfu.

From there, it enters the anus and unites with the Kidneys and the Urinary Bladder. It goes out again at point BL 23 Shenshu.

Following its course, it joins the Dai Mai vessel, goes up to the infra-clavicular fossa, passes by the throat, crossing point (REN)CV 23 Lianquan and it ramifies at the root of the tongue.

It goes out by the nape of the neck and joins the master and distinct channels of the Urinary Bladder, on the level of BL 10 Tianzhu, to form the first union.

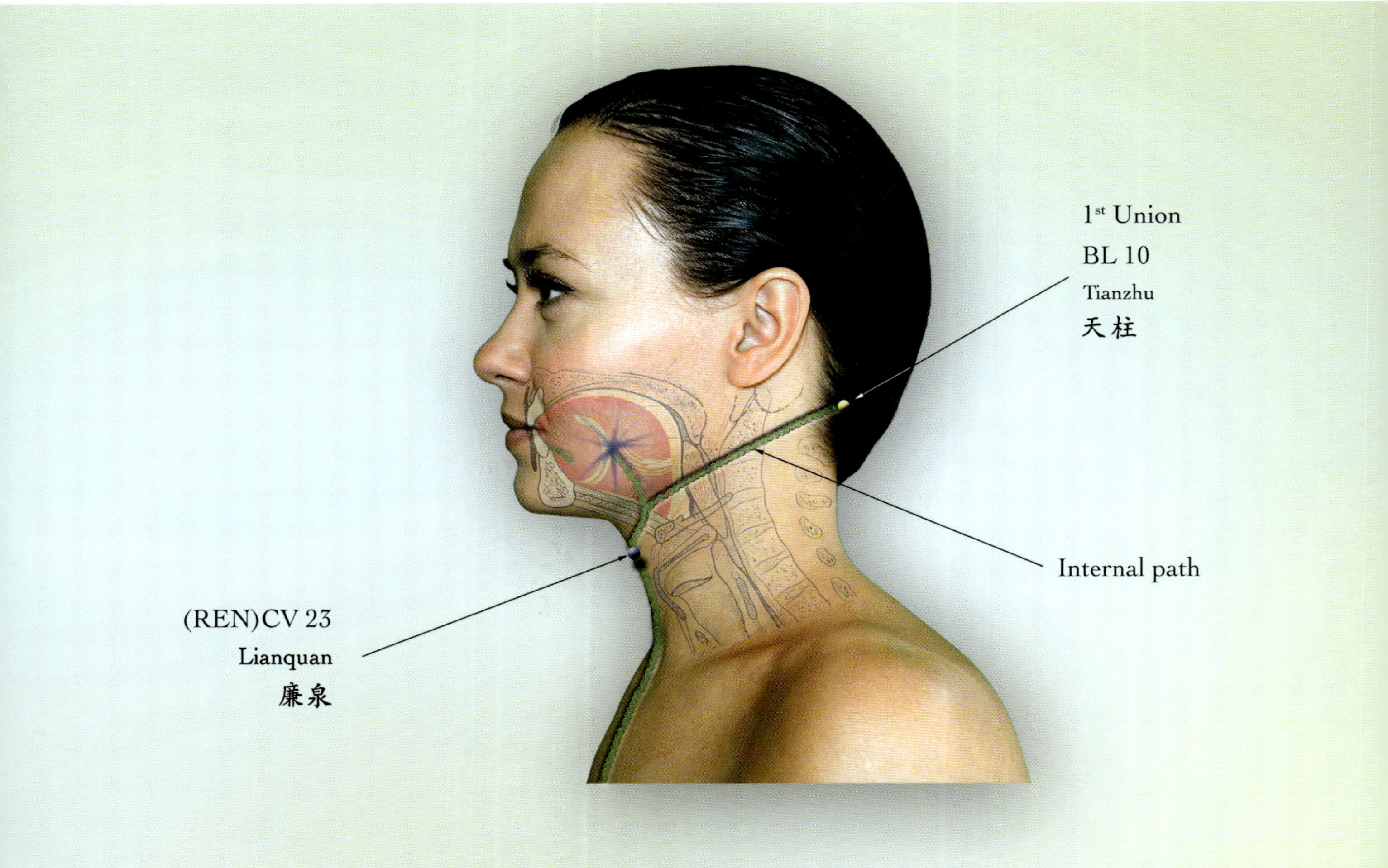

Distinct channel of the Zu Shao Yin (Kidney), cephalic course

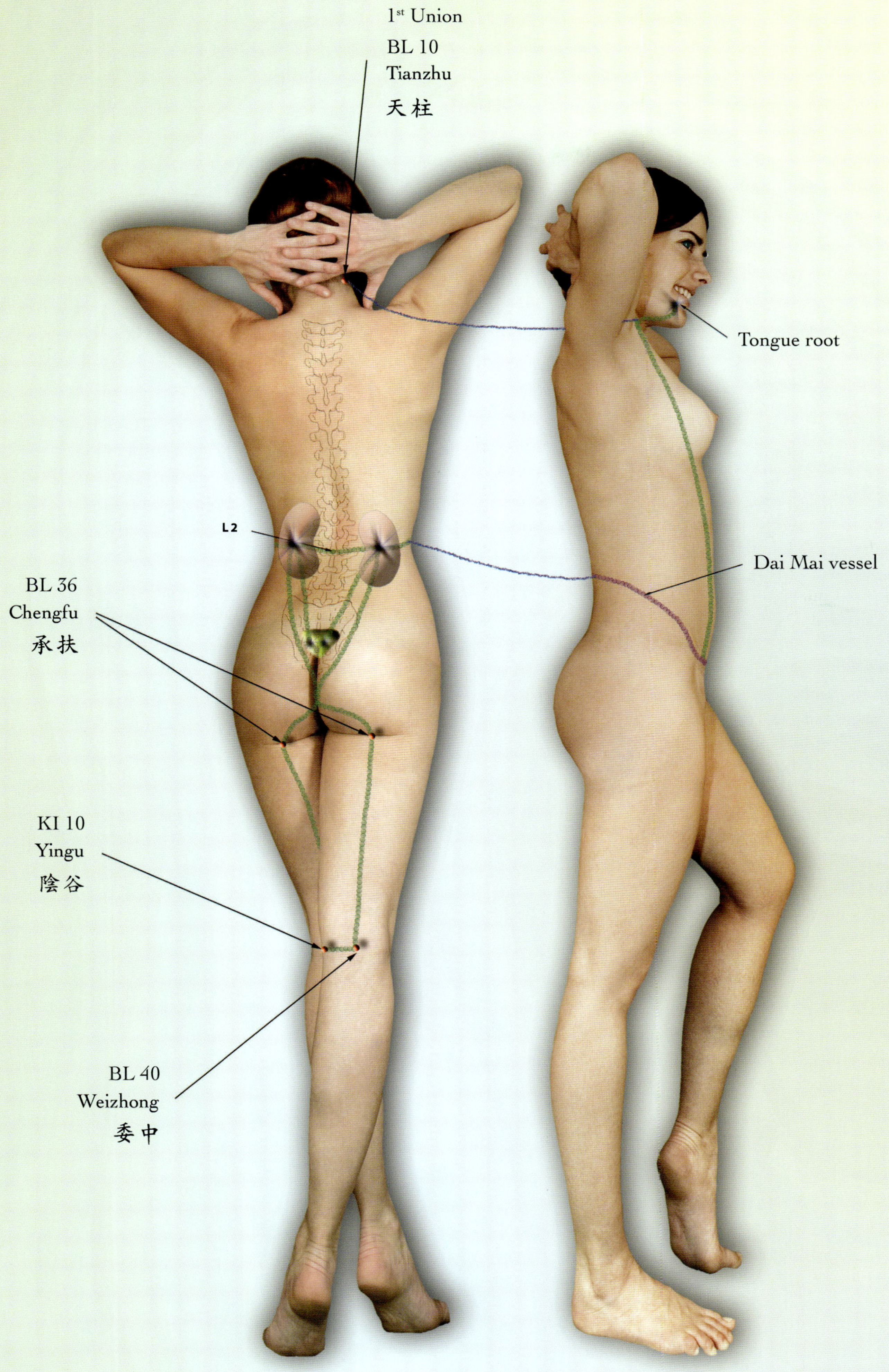

Distinct channel of the Zu Shao Yin (Kidney)

Main points of the Zu Shao Yin (Kidney)

KI 1 Yongquan
At the union of the anterior and middle two thirds of the foot, at the «V» formed by the muscle mass of the first two toes and the three remaining. It lies between the 2nd and 3rd metatarsal.

KI 2 Rangu
On the medial aspect of the foot, below the tuberosity of the navicular bone, 1 Cun behind SP 4 Gongsun.

KI 3 Taixi
On the posterior tibial artery, the posterior and inferior edge of the medial malleolus. In the depression between the medial malleolus and the calcaneal tendon, at the tip of the medial malleolus.

KI 4 Dazhong
In the angle formed by the Achilles tendon and the roof of the calcaneus.

KI 5 Shuiquan
1 Cun under KI 3 Taixi depression in the anterior superior medial side of the tuberosity of the calcaneus.

KI 6 Zhaohai
0.5 Cun in front of and 0.5 Cun under the tip of the medial malleolus. At 1 Cun under the medial malleolus itself.

KI 7 Fuliu
2 Cun directly above the protrusion of the medial malleolus, 0.5 Cun behind the posterior edge of the tibia.

KI 8 Jiaoxin
0.5 Cun in front of KI 7 Fuliu, at the posterior edge of the tibia.

KI 9 Zhubin
On the inside of the foot, at the union between the lower edge of the medial gastrocnemius and the Achilles tendon, 5 Cun above the medial malleolus, 1 Cun behind the postero-medial edge of the tibia.

KI 10 Yingu
On the medial side of the popliteal crease, between the tendons.

KI 11 Henggu
At 0.5 Cun from the Ren Mai, at the upper edge of the pubis, at 5 Cun below the umbilicus.

KI 12 Dahe
Horizontally from (REN)CV 3 Zhongji and 0.5 Cun outside the anterior midline.

KI 13 Qixue
Horizontally from (REN)CV 4 Guanyuan and 0.5 Cun outside the anterior midline.

KI 14 Siman
Horizontally from the (REN)CV 5 Shimen and 0.5 Cun outside the anterior midline.

KI 15 Zhongzhu
Horizontally from (REN)CV 7 Yinjiao and 0.5 Cun outside the anterior midline.

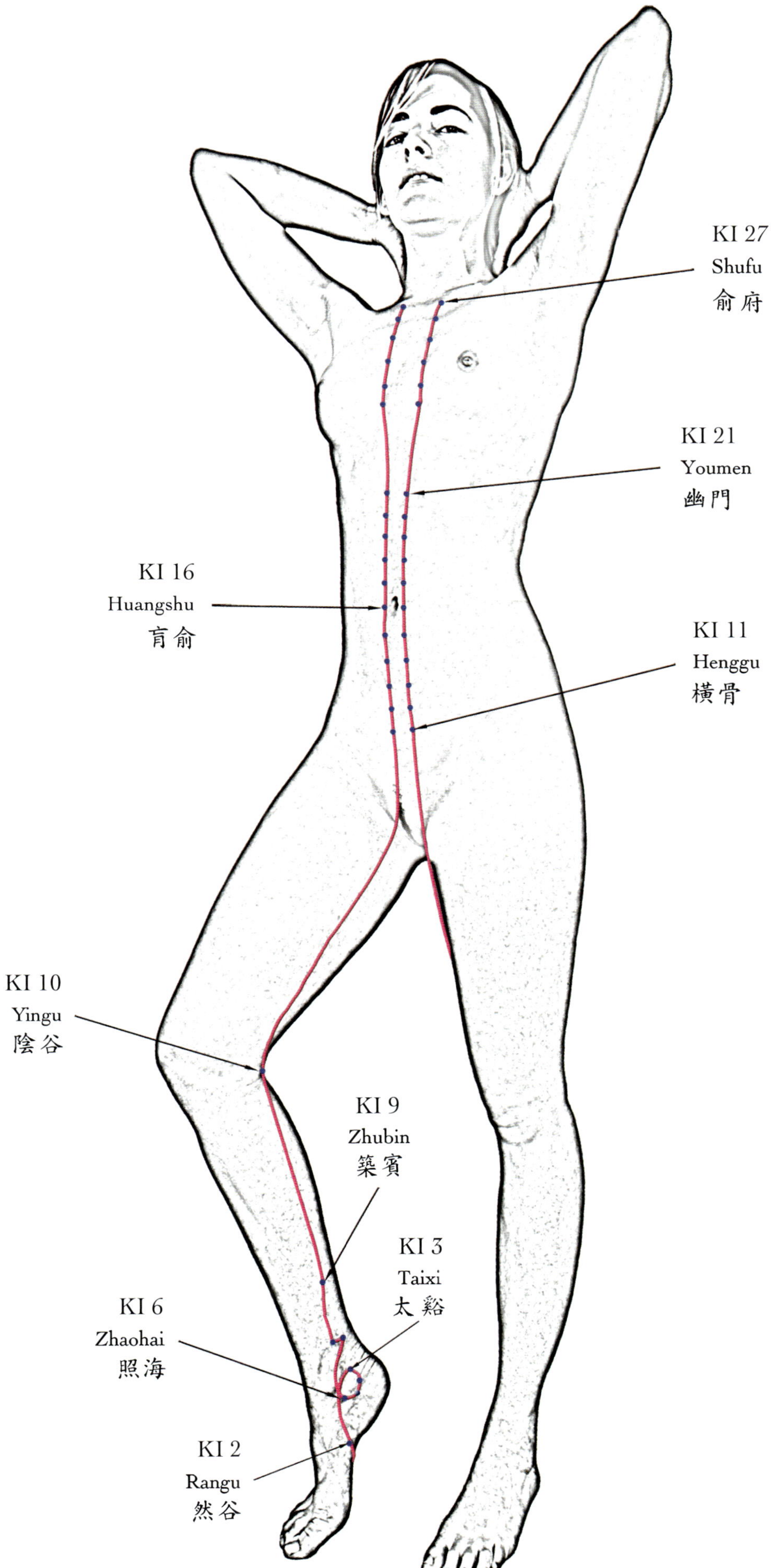

Main points of the Zu Shao Yin (Kidney)

KI 16 Huangshu
On the middle abdomen, 0.5 Cun lateral to the centre of the umbilicus.

KI 17 Shanqu
On the upper abdomen, 2 Cun above the centre of the umbilicus and 0.5 Cun lateral to the anterior midline.

KI 18 Shiguan
On the upper abdomen, 3 Cun above the umbilicus and 0.5 Cun lateral to the anterior midline.

KI 19 Yindu
On the upper abdomen, 4 Cun above the centre of the umbilicus and 0.5 Cun lateral to the anterior midline.

KI 20 Futonggu
On the upper abdomen, 5 Cun above the centre of the umbilicus and 0.5 Cun lateral to the anterior midline.

KI 21 Youmen
On the upper abdomen, 6 Cun above the umbilicus and 0.5 Cun lateral to the anterior midline.

KI 22 Bulang
On the chest, in the 5th intercostal space, 2 Cun lateral to the anterior midline.

KI 23 Shenfeng
On the chest, in the 4th intercostal space, 2 Cun lateral to the anterior midline.

KI 24 Lingxu
On the chest, in the 3rd intercostal space, 2 Cun lateral to the anterior midline.

KI 25 Shencang
On the chest, in the 2nd intercostal space, 2 Cun lateral to the anterior midline.

KI 26 Yuzhong
On the chest, in the 1st intercostal space, 2 Cun lateral to the anterior midline.

KI 27 Shufu
On the chest, below the lower border of the clavicle, 2 Cun lateral to the anterior midline.

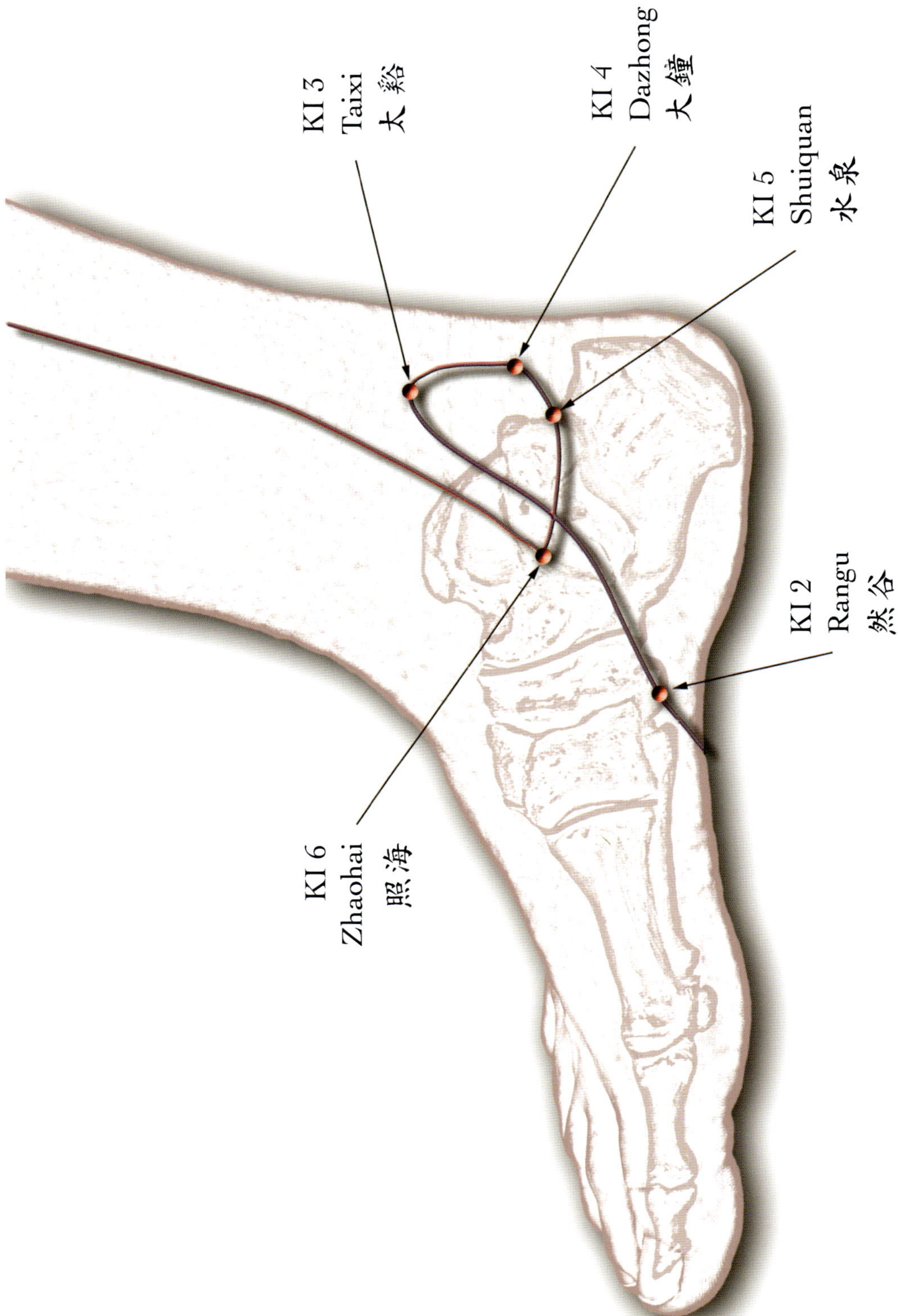

Main points of the Zu Shao Yin (Kidney) in the foot

The 3 lower Yang

- Zu Yang Ming (Stomach)
- Zu Shao Yang (Gall Bladder)
- Zu Tai Yang (Urinary Bladder)

Zu Yang Ming system (Stomach)

- Tendino-muscular channel
- Main channel
- Internal branches
- Luo vessels
- Distinct channel
- Points of the Main channel

Tendino-muscular of the Zu Yang Ming (Stomach)

Ling Shu, chapter IV, paragraph 13

The tendino-muscular of the Yang Ming of the foot starts at the 3 middle toes (2nd, 3rd and 4th).

It goes up on the dorsal side, connecting with the instep.

A branch goes up diagonally by the fibula, joining the external border of the knee.

An upward branch reaches the hip joint*, runs alongside the ribs and connects with the rachis.

The upward branch that follows the tibia connects with the knee.

The branch that connects with the fibula joins the Shao Yang (tendino-muscular of the foot).

The tendino-muscular of Yang Ming going up to Futu joins the biceps (antero-superior part of the thigh) and gathers in the genitalia**.

It fans out going up on the abdomen so as to reach the infra-clavicular fossa.

It goes up on the neck, encircles the mouth, joins the zygomatic, reaches the nose and then, still ascending, the Tai Yang (the tendino-muscular).

The Tai Yang belongs to the Mu Shang-Gang (free border of the upper eyelid); the Yang Ming to the free border of the lower eyelid; a branch forks out from the cheek.

It goes up and connects in front of the auricle.

* It sends out a branch in the area of GB 30 Huantiao, behind the great trochanter.

** It inserts itself on the middle part of the inguinal groove, and joins (REN)CV 2 Qugu (REN)CV 3 Zhongji.

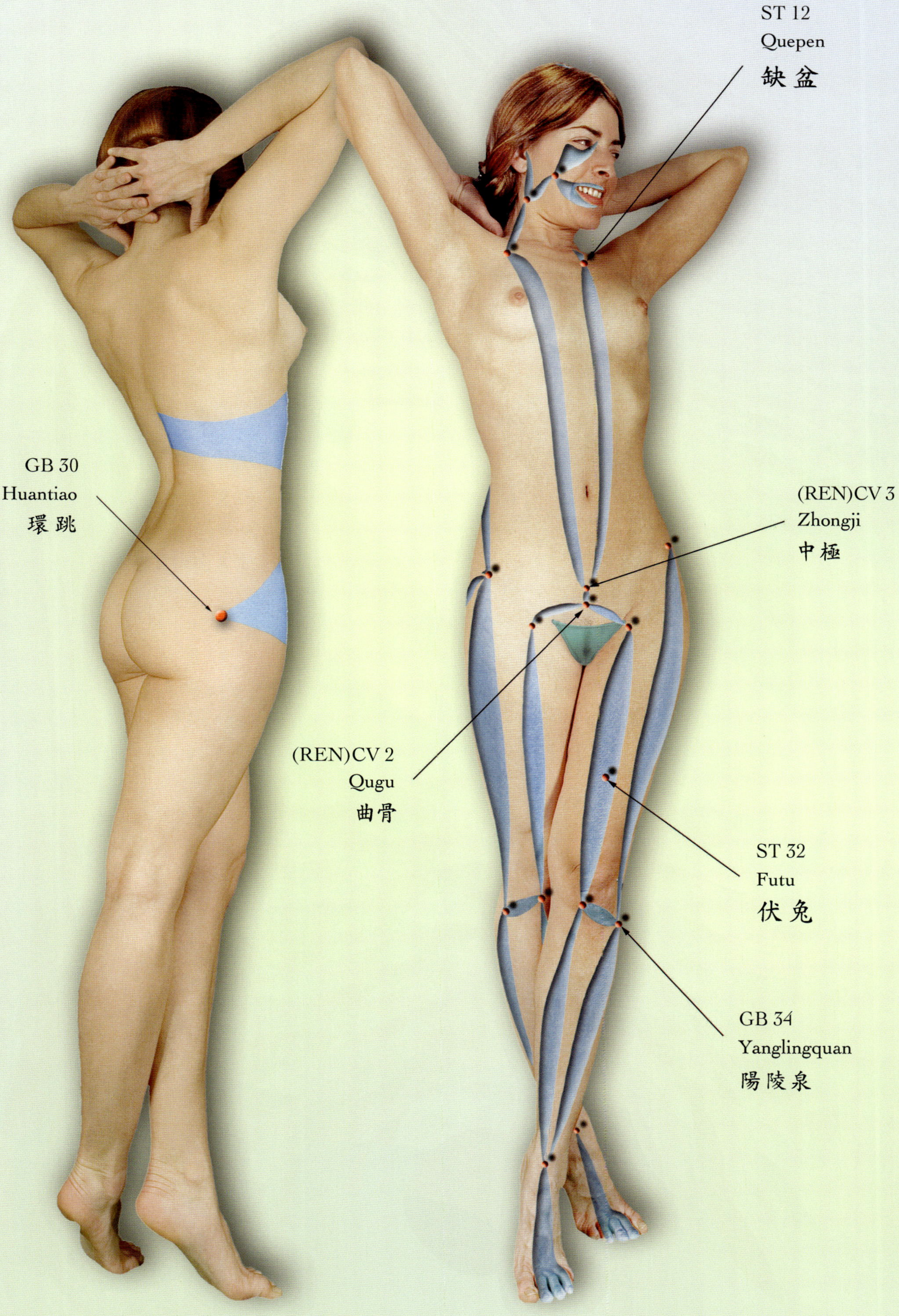
ST 12
Quepen
缺盆
GB 30
Huantiao
環跳
(REN)CV 3
Zhongji
中極
(REN)CV 2
Qugu
曲骨
ST 32
Futu
伏兔
GB 34
Yanglingquan
陽陵泉

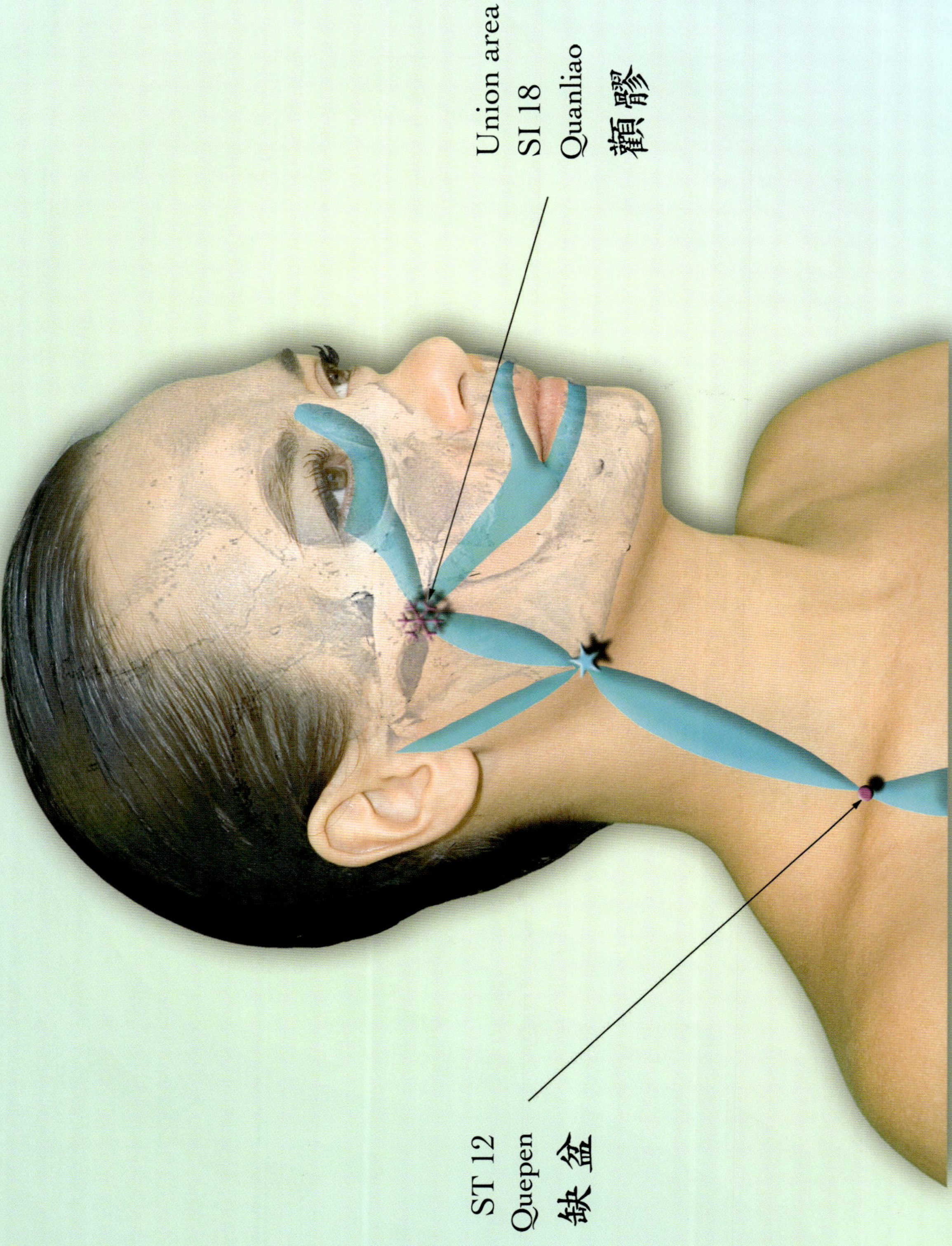

Tendino-muscular of the Zu Yang Ming (Stomach), cephalic course

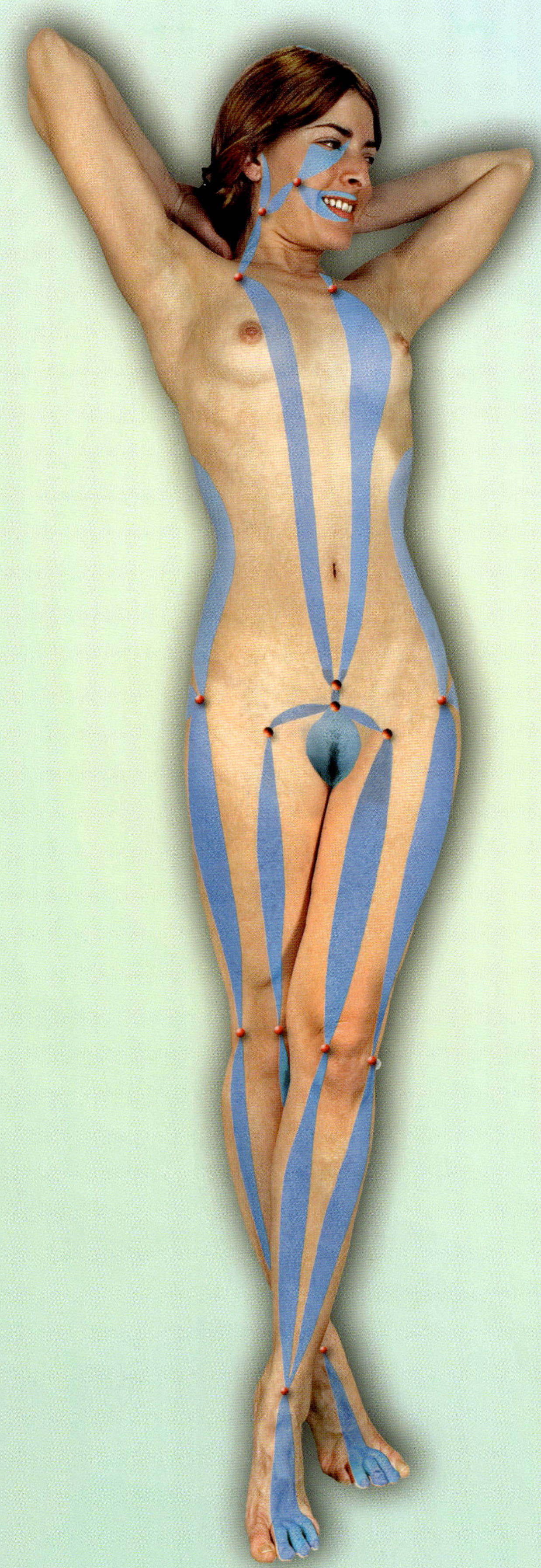

Another representation of the muscle-tendon Zu Yang Ming (Stomach)

Main channel of the Zu Yang Ming (Stomach)

Ling Shu, **chapter III, paragraph 10**

The vessel of the Stomach, Zu Yang Ming of the foot, starts on each side of the nasal ala (outside the root of the nasal ala).

It joins the side of the Zu Tai Yang vessel (internal border of the orbital cavity) and goes down (vertically on the face) along the external side of the nose.

It runs up penetrating in the middle of the teeth (gums) and goes out, encircling the lips.

It runs down crossing the Chengjiang (24 of the Conception vessel) below the lower lip and runs back along the lower maxilla, (it exits by Daying 5 of the Stomach, on the inferior border of the lower maxilla).

It follows the maxilla angle to Jiache (6 of the Stomach) and goes up in front of the auricle.

It passes by Kezhuren or Shangguan (3 of the Gall Bladder on the superior border of the zygomatic arch) and reaches the hairline on the forehead.

The branch originating in front of Daying (on the inferior border of the inferior maxilla) runs down on Renying (9 of the Stomach on the anterior border of the sterno-cleido-mastoid) by way of the throat.

It enters Quepen (12 of the Stomach, in the supra-clavicular fossa), runs down to the diaphragm, belongs to the Stomach and joins the Spleen.

The direct branch that goes down from Quepen (supra-clavicular fossa) on the internal side of the breast, surrounds the umbilicus and enters Qichong (30 of the Stomach, in the inguinal groove and 5 Cun above the umbilicus).

The branch that originates from the foramen of the Stomach goes down along the abdomen to reach Qichong, then runs down to Biguan (31 of the Stomach) and Futu (32 of the Stomach).

It reaches the midpoint of the patella. It runs down on the external aspect of the leg and on the dorsum of the foot.

It enters the interval between the second and third toes.

Another branch starts 3 cun below the patella and terminates on the external side of the third toe.

A (last) branch forks out of the dorsal side of the foot and ends at the tip of the big toe.

Commentaries

The Main channel of Zu Yang Ming (Stomach) starts at the edge of the nasal ala at point LI 20 Yingxiang.

It goes upward to the root of the nose and reaches the inner canthus where BL 1 Jingming is to be found.

It goes down to the cheek passing by the internal edge of the orbital cavity where ST 1 Chenqi is located.

It enters the gum of the upper maxilla, goes to the lower part of the nose, at point (DU)GV 26 Shuigou and reappears in the vicinity of the mouth, going around the lips.

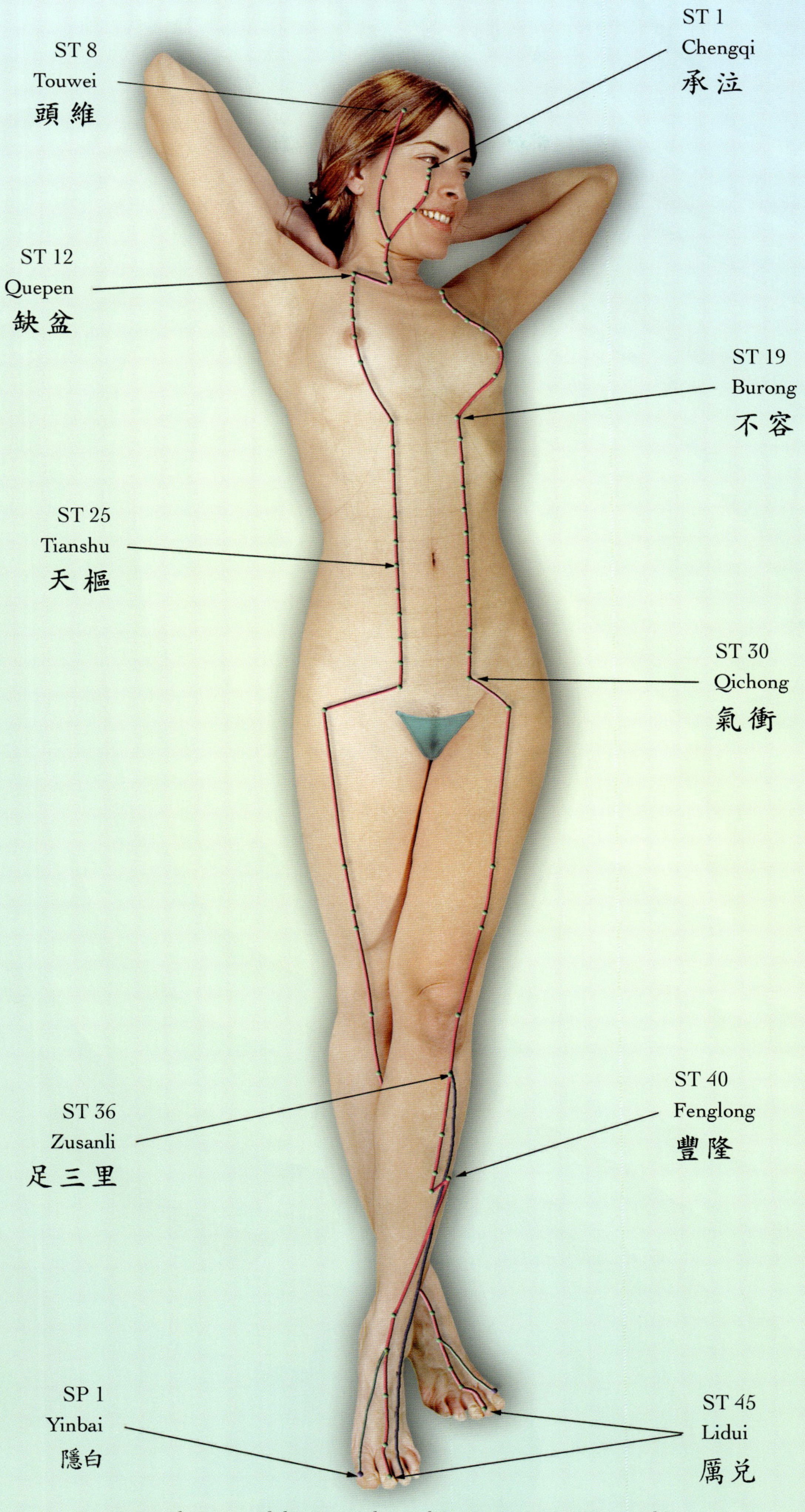

General course of the Main channel Zu Yang Ming (Stomach)

It goes to the mento-labial sulcus to join with its homologue at point (REN)CV 24 Chengjiang.

It follows the lower maxilla to point ST 5 Daying, in the hollow where the facial artery passes.

It reaches point ST 6 Jiache at the external border of the maxillary angle, under the ear.

It goes up in front of the ear at point ST 7 Xiaguan, located in front of the condyle of the upper maxilla and goes to point GB 3 Shangguan.

It keeps going up along the upper temple and joins points GB 6 Xuanli and GB 4 Hanyan.

It reaches the frontal angle at point ST 8 Touwei and runs toward the sagittal line to reach point (DU)GV 24 Shenting.

A branch leaves point ST 5 and goes to ST 9 Renying, runs downward to ST 11 Qishe, forks out to the shoulder and travels to the 7th cervical vertebra where (DU)GV 14 Dazhui is.

From that point, it goes back to the anterior part of the body to reach point ST 12 Quepen.

The internal branch of the channel originates at that point. The superficial branch goes down vertically under the nipple where point ST 18 Rugen is, and forks out internally to follow the abdomen along the intermammary line to the inguinal groove where ST 30 Quichong is.

It then goes downward by the anterior aspect of the thigh to the knee and passes along the external border of the tibia crest.

It passes by the dorsal side of the foot and ends in the external ungueal angle of the second toe at point ST 45 Lidui.

From point ST 36 Zusanli, situated 3 Cun below the knee, a branch goes downward and ends its course at the tip of the third toe.

Another branch goes out of point ST 42 Chongyang to unite with point SP 1 Yinbai at the internal ungueal angle of the big toe.

INTERNAL BRANCHES OF THE ZU YANG MING (STOMACH)

From point ST 12 Quepen, the channel enters the chest, goes through the diaphragm, passes by the Stomach and the Spleen.

It reaches the inguinal groove at point ST 30 Qichong.

From the cardia, a branch unites with point (REN)CV 12 Zhongwan, and, from the pylorus, another branch unites with point (REN)CV 10 Xiawan.

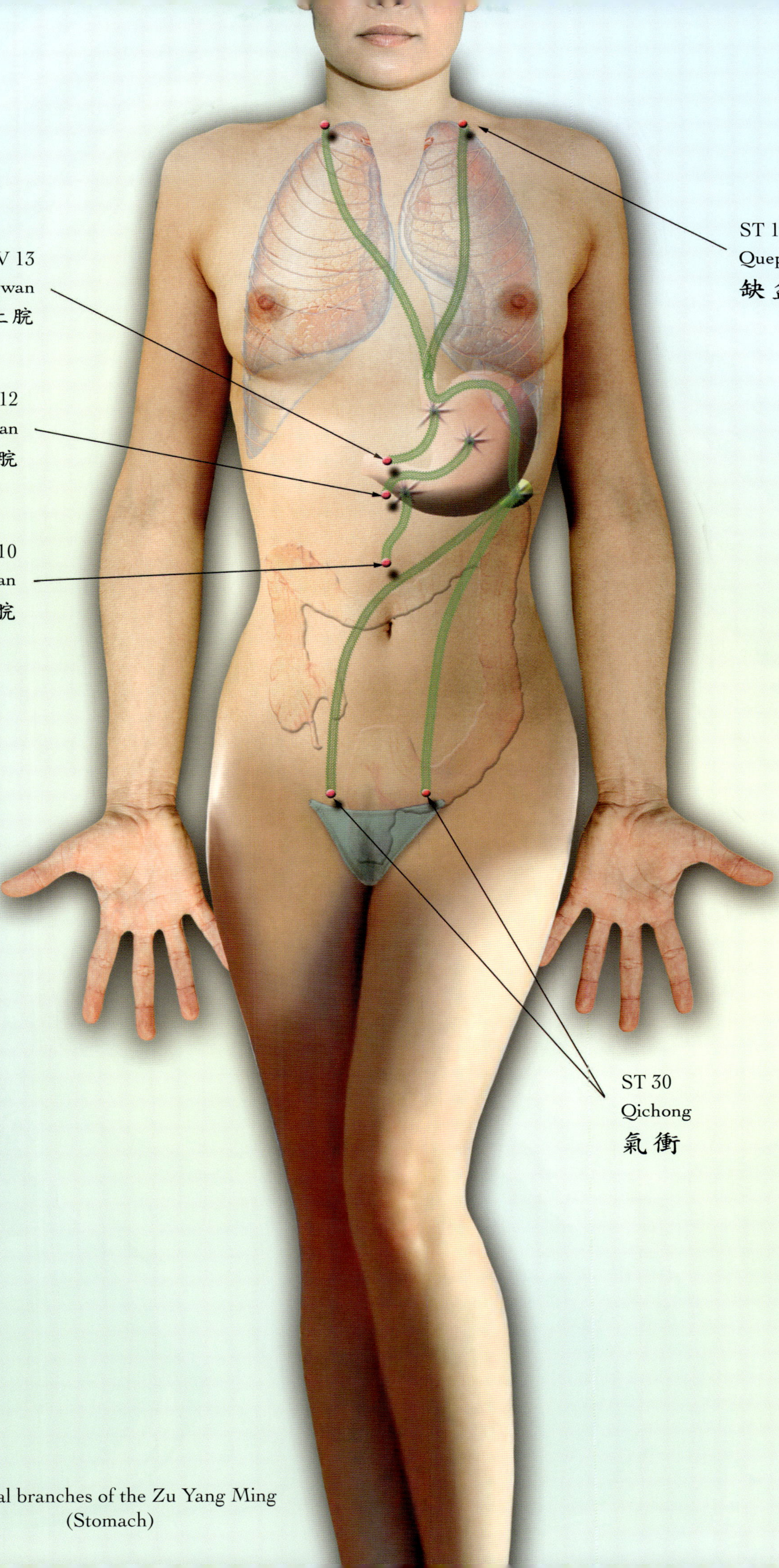

Internal branches of the Zu Yang Ming
(Stomach)

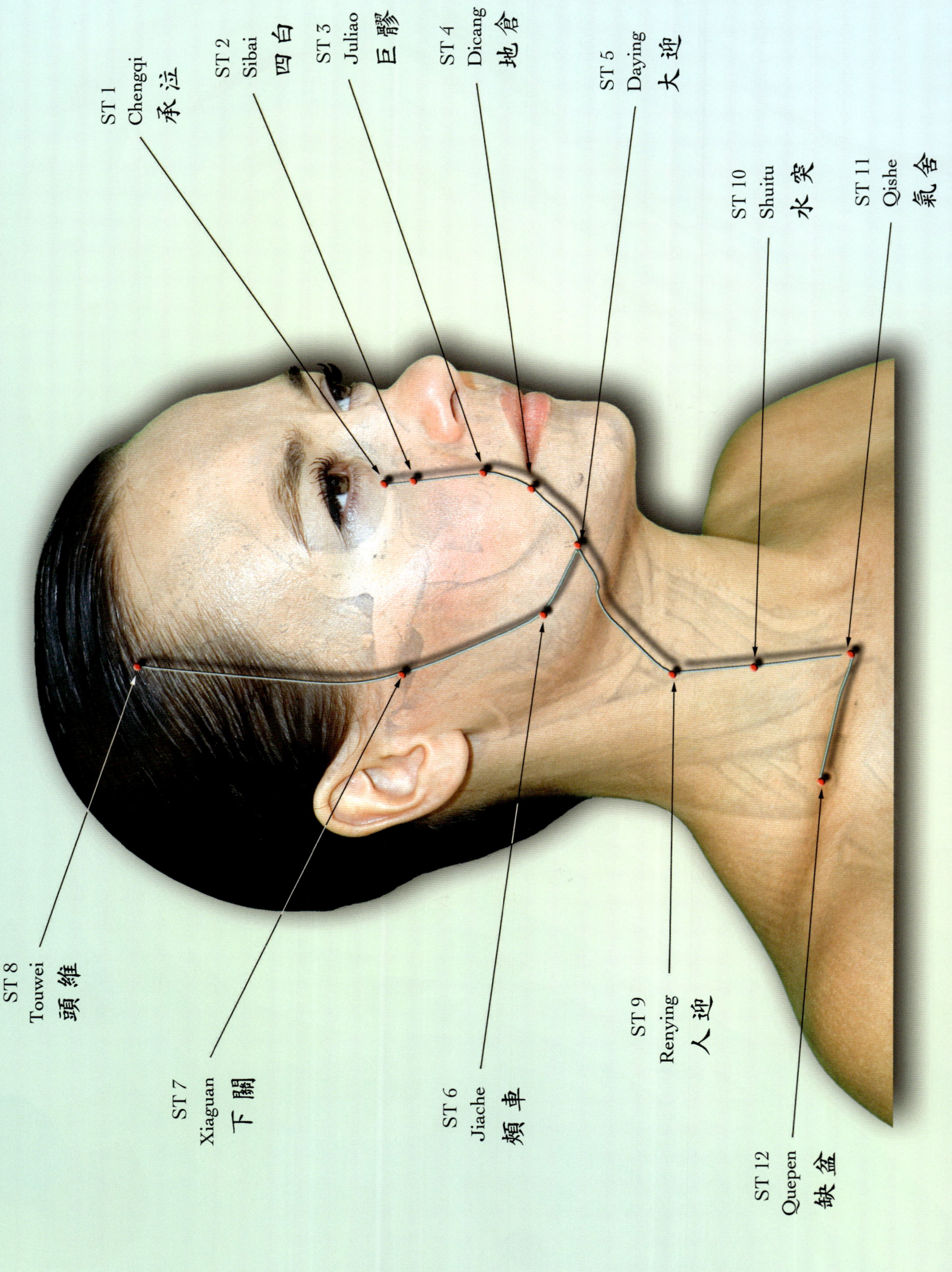

Traditional cephalic course of the Main Channel Zu Yang Ming (Stomach)

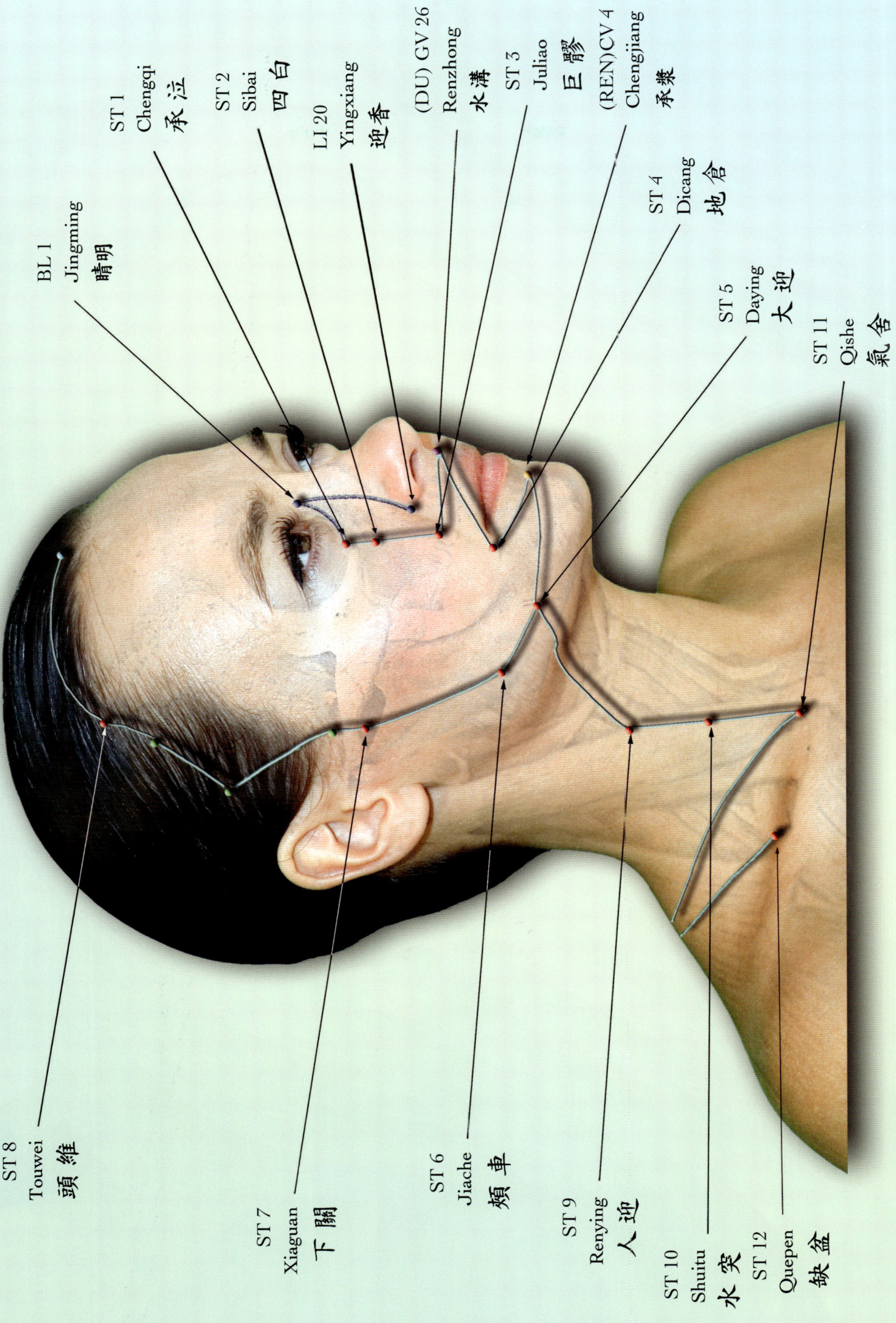

Complete cephalic course of the Main Channel Zu Yang Ming (Stomach)

Luo vessels of the Zu Yang Ming (Stomach)

***Ling Shu*, chapter III, paragraph 10**

The distinct branching of the Zu Yang Ming of the foot (channel of the Stomach) is called Fenglong.

It starts 8 Cun above the (external) malleolus.

It runs to the Zu Tai Yin (channel of the Spleen).

The distinct channel goes along the external border of the tibia (of the leg), then reaches the nape of the neck and the head.

It joins the Lie - where the vessels pour their energy - and goes down by way of the throat.

Commentaries

The Luo vessel starts at point ST 40 Fenlong, 7 Cun above the superior border of the external malleolus and one finger breadth outside ST 38 Tiaokou.

It divides into two branches:

- The first follows the anterio-external part of the leg and the anterio-external side of the thigh, goes up on the anterio-external side of the chest, and reaches the lateral side of the neck where it divides into two small branches:
  - the first one runs to the throat.
  - the second one goes up to the vertex to join the other Yang channels at point (DU)GV 20 Baihui.
- The second branch unites with the channel of the Zu Tai Yin (Spleen) at point SP 3 Taibai.

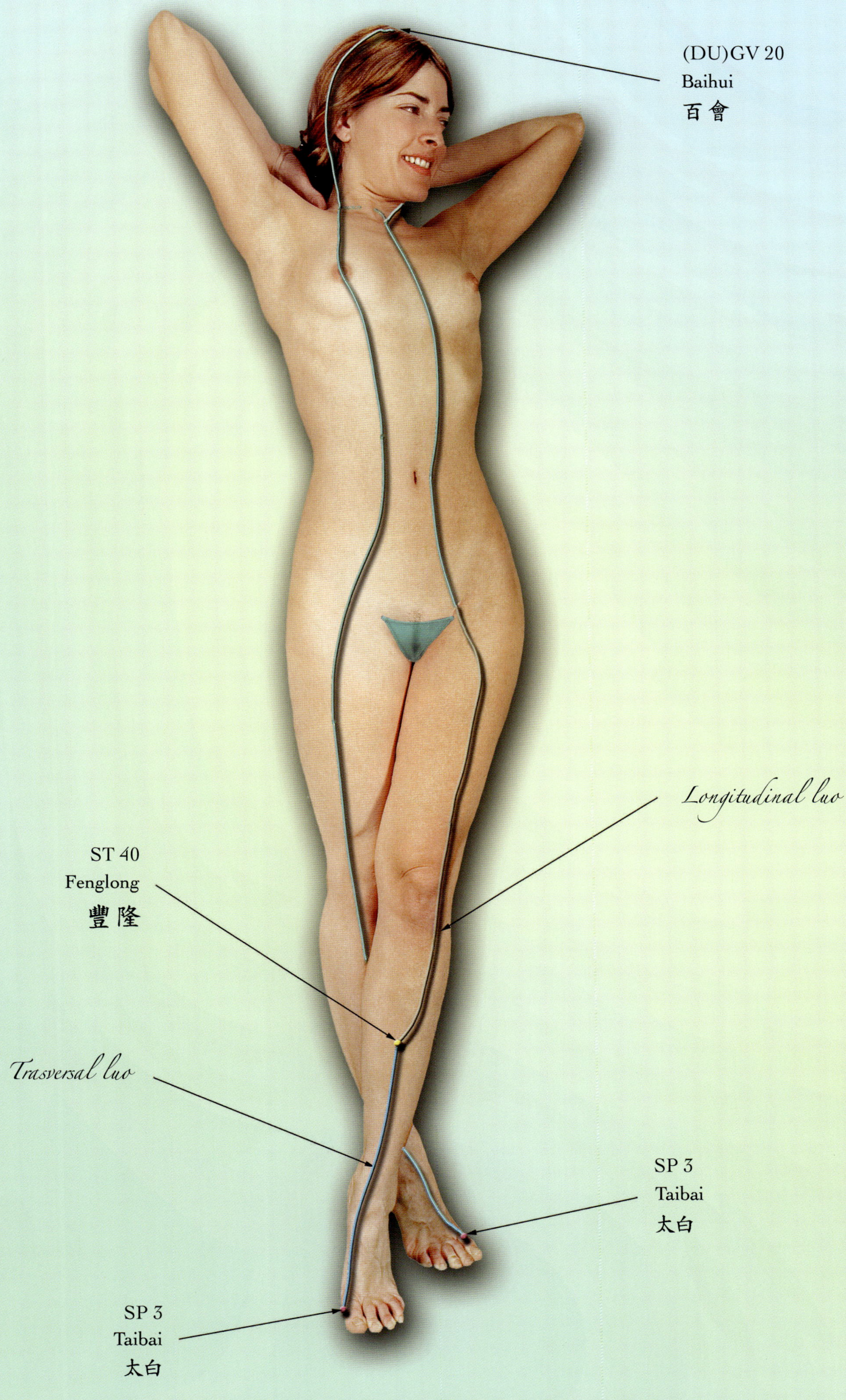

Course of the Luo vessels of the Main Channel Zu Yang Ming (Stomach)

Great Luo vessel of the Zu Yang Ming (Xu Li)

It leaves the Stomach, goes through the diaphragm and passes by point (REN)CV 17 Danzhong.

It reaches the Lung, passes by the Heart and goes out under the left breast at point ST 18 Rugen.

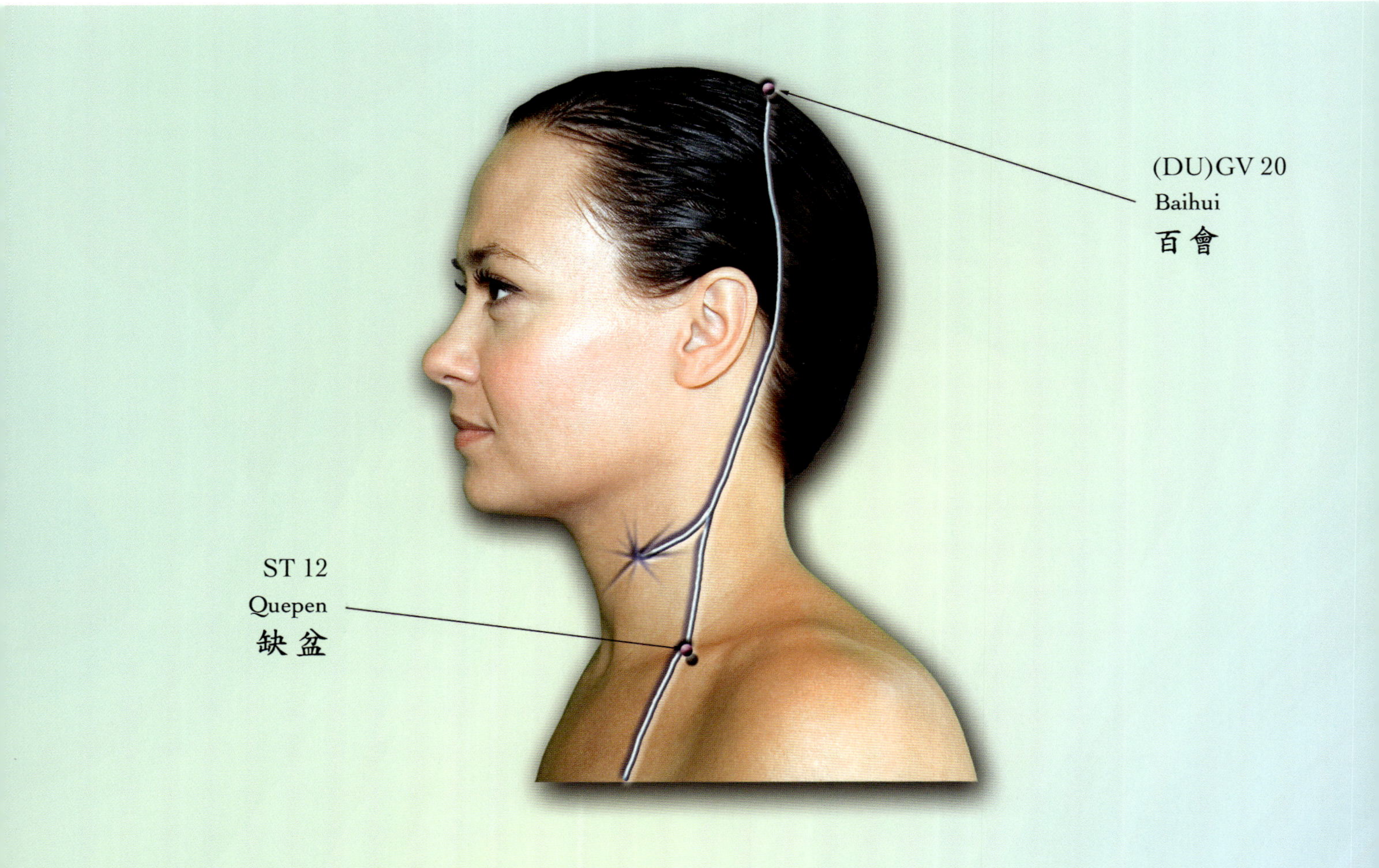

Cephalic course of a Luo vessel of the Zu Yang Ming (Stomach)

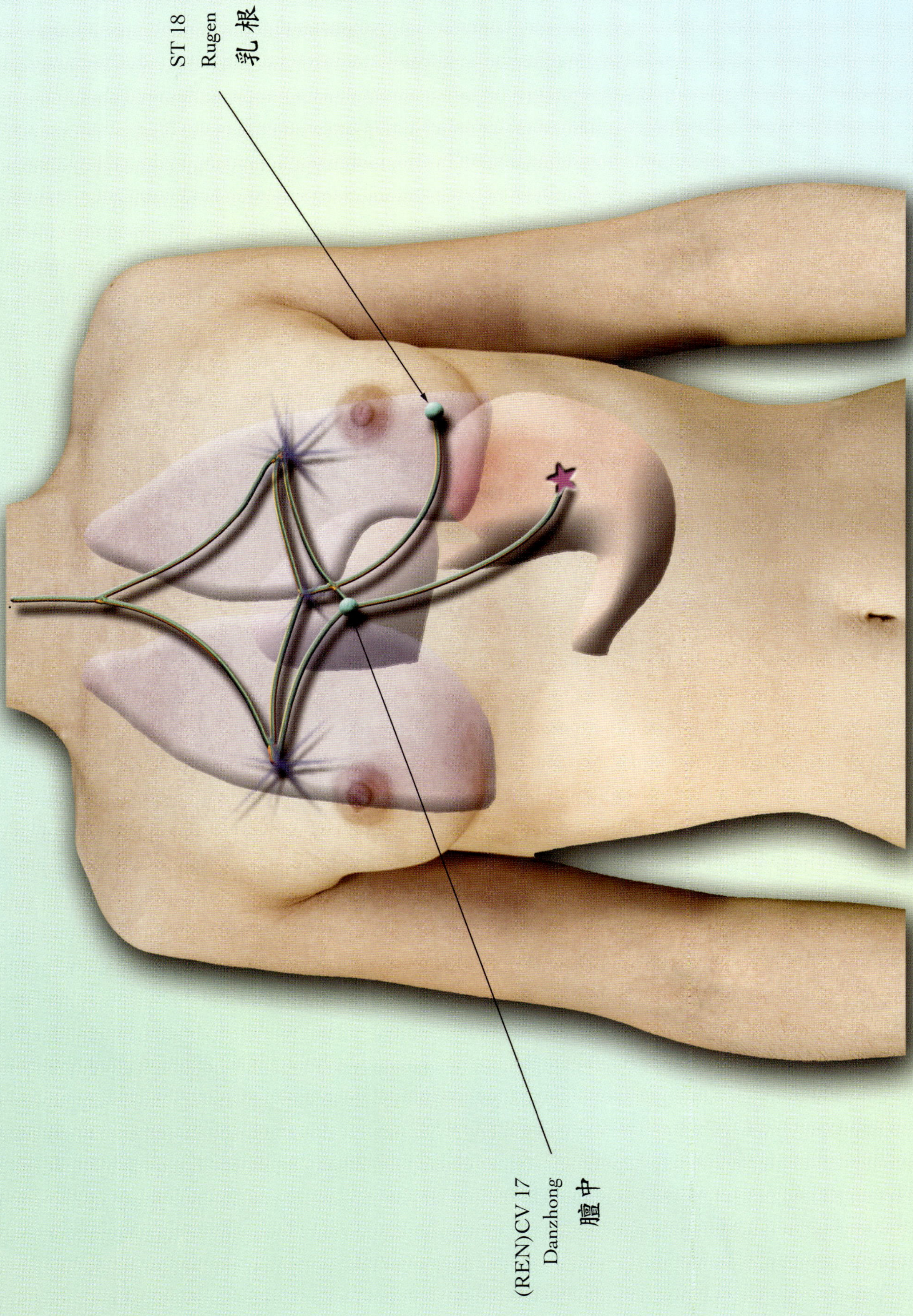

Course of the great Luo of Zu Yang Ming or Xu li

Distinct channel of the Zu Yang Ming (Stomach)

Ling Shu, chapter III, paragraph 11	Commentaries
The distinct channel of Zu Yang Ming goes up by the anterio-superior side of the thigh.	It goes out of the Main channel on the level of point ST 31 Biguan and runs to the abdomen, penetrates the area of point ST 30 Qichong, passes by the Stomach and the Spleen.
It enters the abdomen, unites with the Stomach and the Spleen.	
It goes up and connects with the Heart.	It goes through the Heart, passes by point ST 12 Quepen, goes up along the throat and unites with its Main channel at point ST 9 Renying to form the third union.
It passes by the throat and goes out by the mouth. It runs up by the bridge of the nose and joins the orbital cavity.	It goes up to the mouth and becomes superficial, passes by the nasal ala, reaches the inner cantus, unites with the eye, goes around the eyeball passing by the forehead and unites with its Main channel at point ST 1 Chengqi.
It unites with the 'eye connections' and joins the channel of Zu Yang Ming (Stomach).	

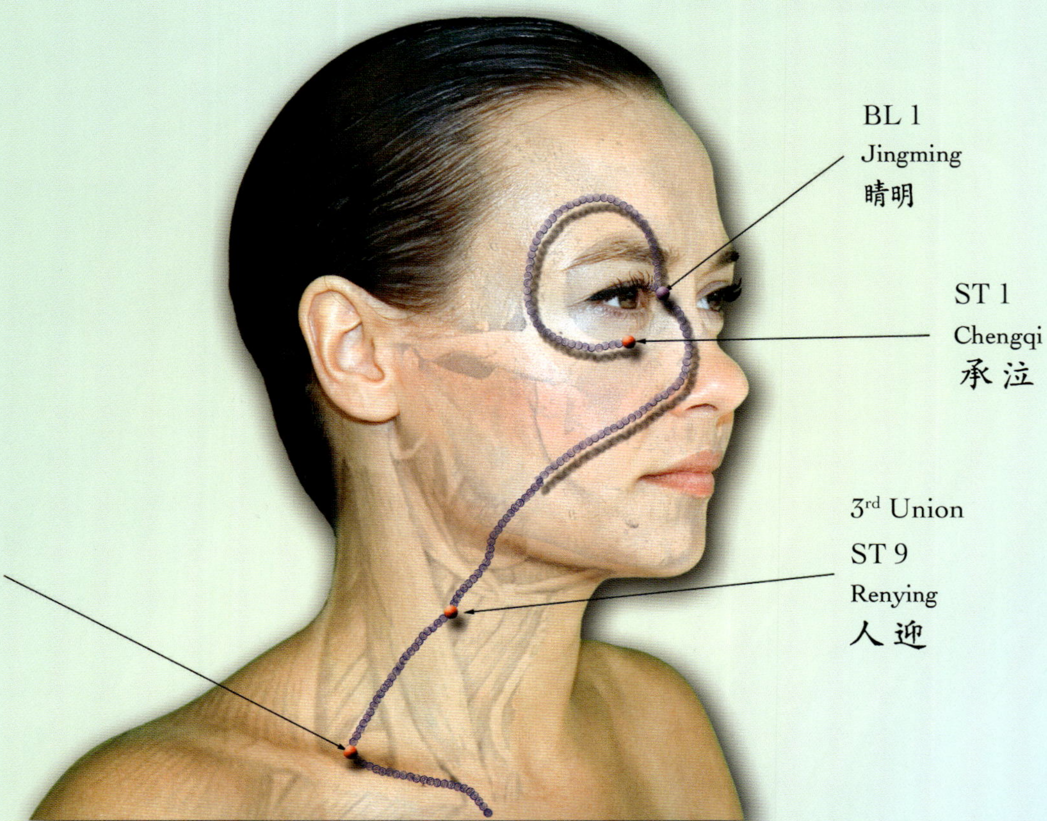

Cephalic course of the distinct channel Zu Yang Ming (Stomach)

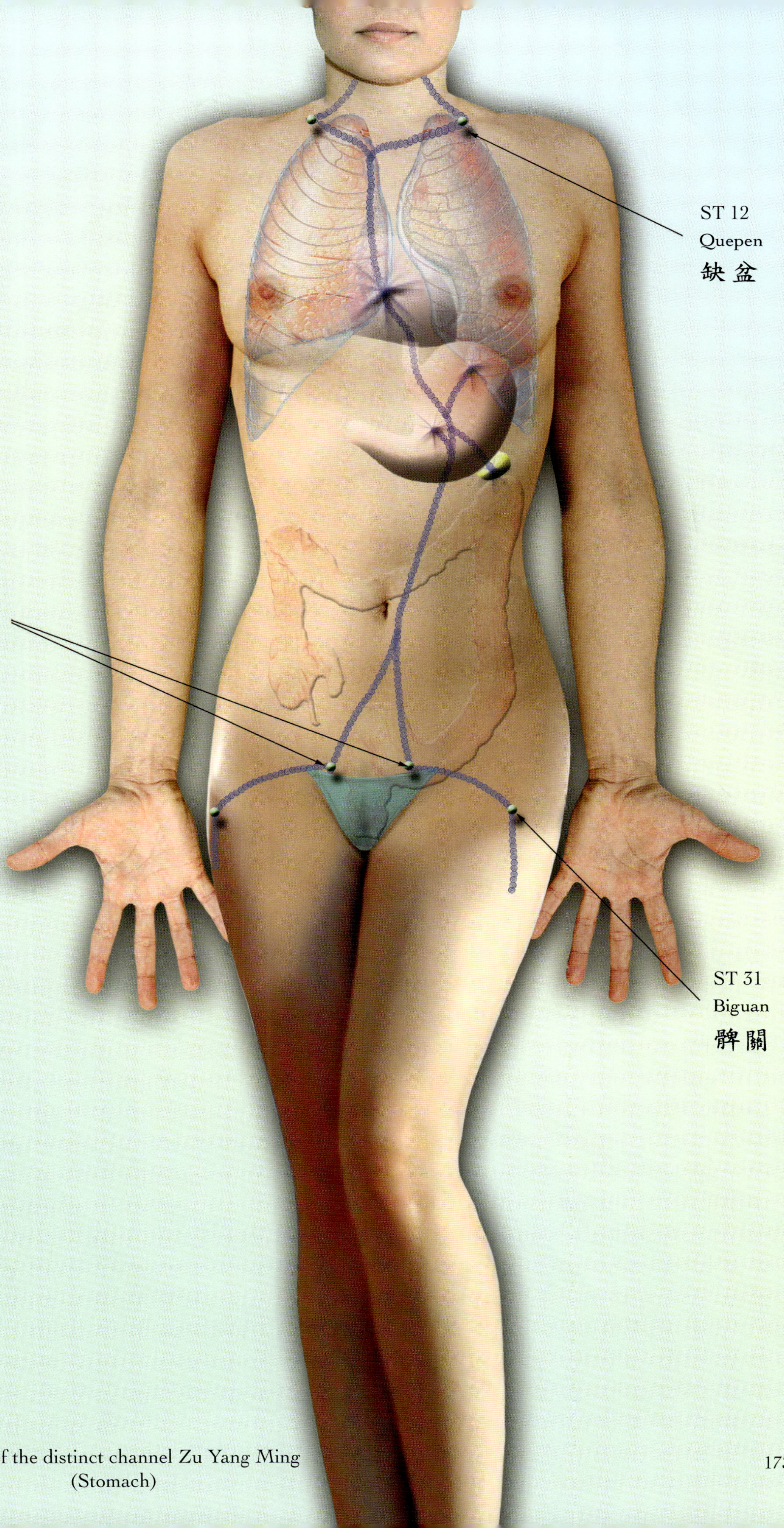

Course of the distinct channel Zu Yang Ming
(Stomach)

Points of the Main channel Yang Ming (Stomach)

ST 1 Chengqi
On the face, directly below the pupil, between the eyeball and the infra-orbital ridge

ST 2 Sibai
On the face, directly below the pupil, in the depression which is below the orbit, below the point ST 1 Chengqi.

ST 3 Juliao
On the face, directly below the pupil, on the level of the lower border of the nasal ala, beside the naso-labial groove.

ST 4 Dicang
On the face, directly below the pupil, beside the mouth angle.

ST 5 Daying
Anterior to the mandibular angle, on the anterior border of the masseter muscle, where the pulsation of the facial artery is palpable.

ST 6 Jiache
On the cheek, one finger width (middle finger) anterior and superior to the angle of the lower mandible, in the depression where the masseter bulges out when teeth are clenched.

ST 7 Xiaguan
On the face, anterior to the ear, in the depression between the zygomatic arch and mandibular notch.

ST 8 Touwei
On the lateral side of the head, 0.5 Cun above the anterior hairline at the corner of the forehead, and 4.5 Cun lateral to the midline of the head.

ST 9 Renying
On the neck, beside the laryngeal protuberance, and on the anterior border of the sterno-cleido-mastoid muscle, where the pulsation of the common carotid artery is palpable. (Do NOT needle directly into the artery!)

ST 10 Shuitu
On the neck and on the anterior border of the sterno-cleido-mastoid muscle, at the midpoint of the line connecting ST 9 Renying and ST 11 Qishe.

ST 11 Qishe
On the neck and on the upper border of the medial end of the clavicle, between the sternal and clavicular heads of the sterno-cleido-mastoid muscle.

ST 12 Quepen
At the centre of the supraclavicular fossa, 4 Cun lateral to the anterior midline.

ST 13 Qihu
On the chest, below the midpoint of the lower border of the clavicle, 4 Cun lateral to the anterior midline.

ST 14 Kufang
On the chest, in the 1st intercostal space, 4 Cun lateral to the anterior midline.

ST 15 Wuyi
On the chest, in the 2nd intercostal space, 4 Cun lateral to the anterior midline.

ST 16 Yingchuang
On the chest, in the 3rd intercostal space, 4 Cun lateral to the anterior midline.

ST 17 Ruzhong
On the chest, in the 4th intercostal space, at the centre of the nipple, 4 Cun lateral to the anterior midline.

ST 18 Rugen
On the chest, directly below the nipple, on the lower border of the breast, in the 5th intercostal space, 4 Cun lateral to the anterior midline.

ST 19 Burong
On the upper abdomen, 6 Cun above the umbilicus and 2 Cun lateral to the anterior midline.

ST 20 Chengman
On the upper abdomen, 5 Cun above the umbilicus and 2 Cun lateral to the anterior midline.

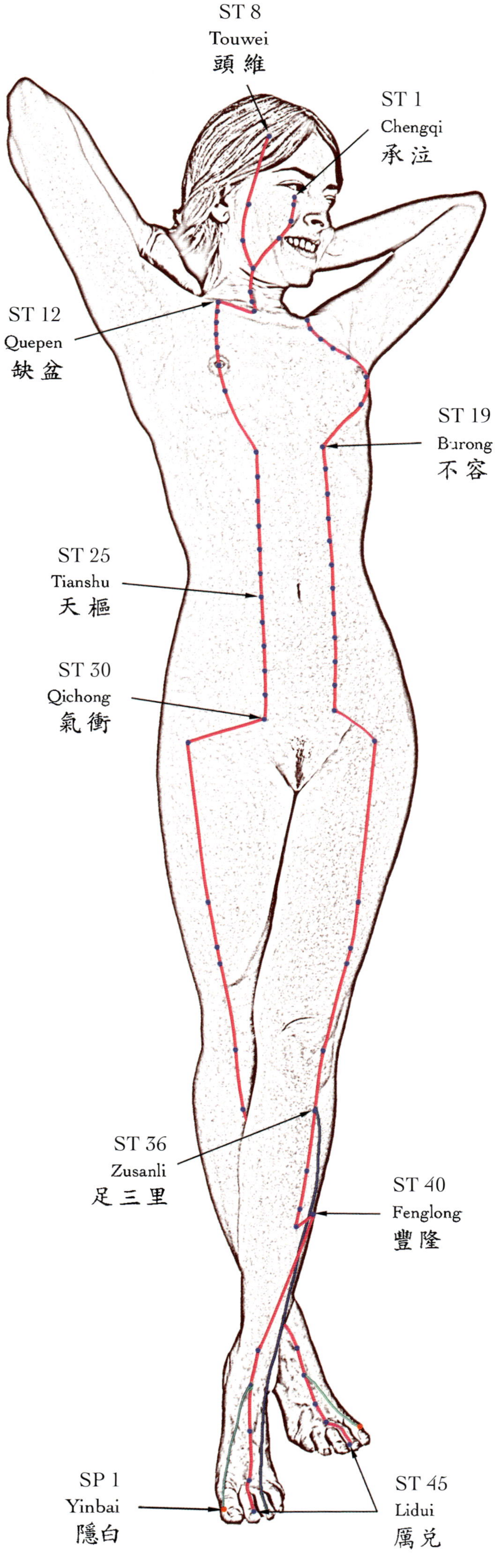

Points of the Main channel
Yang Ming (Stomach)

ST 21 Liangmen
On the upper abdomen, 4 Cun above the centre of the umbilicus and 2 Cun lateral to the anterior midline.

ST 22 Guanmen
On the upper abdomen, 3 Cun above the centre of the umbilicus and 2 Cun lateral to the anterior midline.

ST 23 Taiyi
On the upper abdomen, 2 Cun above the centre of the umbilicus and 2 Cun lateral to the anterior midline.

ST 24 Huaroumen
On the upper abdomen, 1 Cun above the centre of the umbilicus and 2 Cun lateral to the anterior midline.

ST 25 Tianshu
On the middle abdomen, 2 Cun lateral to the centre of the umbilicus.

ST 26 Wailing
On the lower abdomen, 1 Cun below the centre of the umbilicus and 2 Cun lateral to the anterior midline.

ST 27 Daju
On the lower abdomen, 2 Cun below the centre of the umbilicus and 2 Cun lateral to the anterior midline.

ST 28 Shuidao
On the lower abdomen, 3 Cun below the centre of the umbilicus and 2 Cun lateral to the anterior midline.

ST 29 Guilai
On the lower abdomen, 4 Cun below the centre of the umbilicus and 2 Cun lateral to the anterior midline.

ST 30 Qichong
Slightly above the inguinal groove, 5 Cun below the centre of the umbilicus and 2 Cun lateral to the anterior midline.

ST 31 Biguan
On the anterior side of the thigh and on the line connecting the anterio-superior iliac spine and the superiolateral corner of the patella, on the level of the perineum when the thigh is flexed, in the depression lateral to the sartorius muscle.

ST 32 Futu
On the anterior side of the thigh and on the line connecting the anteriosuperior iliac spine and the superiolateral corner of the patella, 6 Cun above this corner.

ST 33 Yinshi
On the anterior side of the thigh and on the line connecting the anterio-superior iliac spine and the superiolateral corner of the patella, 3 Cun above this corner.

ST 34 Liangqiu
With the knee flexed, on the anterior side of the thigh and on the line connecting the anteriosuperior iliac spine and the superiolateral corner of the patella, 2 Cun above this corner.

ST 35 Dubi
With the knee flexed, on the knee, in the depression lateral to the patella and its ligament.

ST 36 Zusanli
On the anterior lateral side of the leg, 3 Cun below ST 35 Dubi, one finger breadth (middle finger) from the anterior crest of the tibia.

ST 37 Shangjuxu
On the anterio-lateral side of the leg, 6 Cun below ST 35 Dubi, one finger breadth (middle finger) from the anterior crest of the tibia.

ST 38 Tiaokou
On the anteriolateral side of the leg, 8 Cun below ST 35 Dubi, one finger breadth (middle finger) from the anterior crest of the tibia.

ST 39 Xiajuxu
On the anteriolateral side of the leg, 9 Cun below ST 35 Dubi, one finger breadth (middle finger) from the anterior crest of the tibia.

ST 40 Fenglong
On the anteriolateral side of the leg, 8 Cun above the tip of the external malleolus, lateral to ST 38 Tiaokou and two finger breadths (middle finger) from the anterior crest of the tibia.

ST 41 Jiexi
In the central depression of the crease between the instep of the foot and leg, between the tendons of the long extensor muscle of the toes, where the pulsation of the dorsal artery of the foot is palpable.

ST 42 Chongyang
On the dome of the instep of the foot, between the tendons of the long extensor muscle of the big toe and the long extensor muscle of the toes, where the pulsation of the dorsal artery of the foot is palpable.

ST 43 Xiangu
On the instep of the foot, in the depression distal to the commissure of the 2nd and 3rd metatarsal bones.

ST 44 Neiting
On the instep of the foot, at the 'junction of the red and white skin', proximal to the margin of the web between the 2nd and 3rd toes.

ST 45 Lidui
On the lateral side of the distal segment of the 2nd toe, 0.1 Cun from the corner of the toenail.

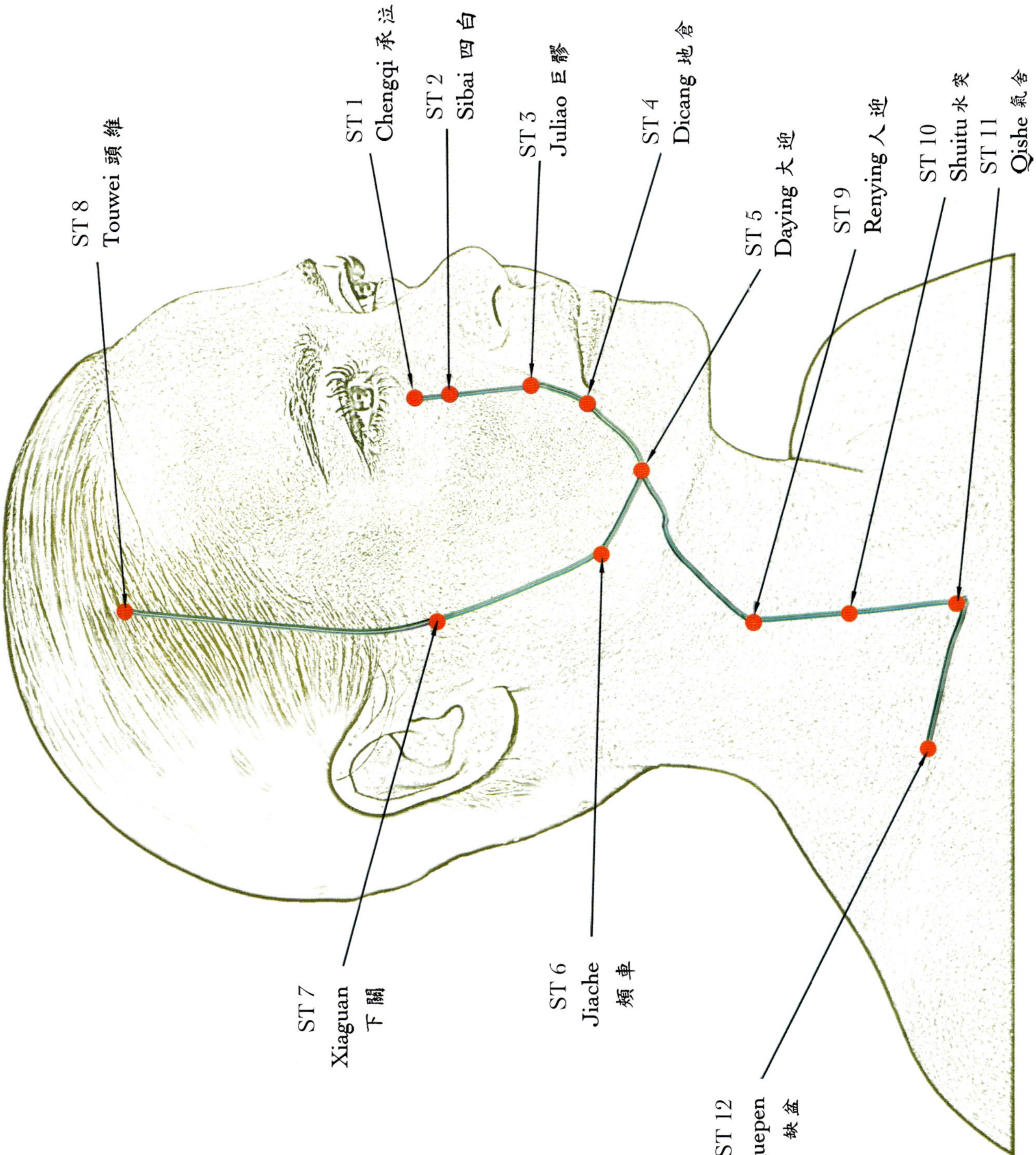

Points of the cephalic course of the Main channel Zu Yang Ming (Stomach)

Zu Shao Yang system (Gall Bladder)

- Tendino-muscular channel
- Main channel
- Internal branches
- Luo vessels
- Distinct channel
- Points of the Main channel

Tendino-muscular of the Zu Shao Yang (Gall Bladder)

***Ling Shu*, chapter IV, paragraph 13**

The tendino-muscular of the Zu Shao Yang of the foot starts at the 4th toe.

It ascends joining the external malleolus, running along the external border of the tibia and joining the external border of the knee.

The tendino-muscular starting from Waifugu (Fibula) goes up and runs on the Bi (thigh). From the front, it joins the top of Futu, (on the anterior part of the thigh) and from behind (a branch) joins Kao (buttock area).

It passes by the hypochondrium, reaches the costal area and the anterior part of the armpit.

On the side of the chest, a branch joins the breast on the way up to the supra-clavicular fossa. It joins the zygomatic. The tendino-muscular ascends, going out by the axilla and following the supraclavicular fossa. It emerges in front of the Zu Tai Yang (in front of the tendino-muscular of the Zu Tai Yang of the foot) while going round the ear.

It ascends to the angle of the forehead, crosses the apex of the cranium and descends on the malar region uniting with the zygomatic.

Another branch reaches the outer canthus.

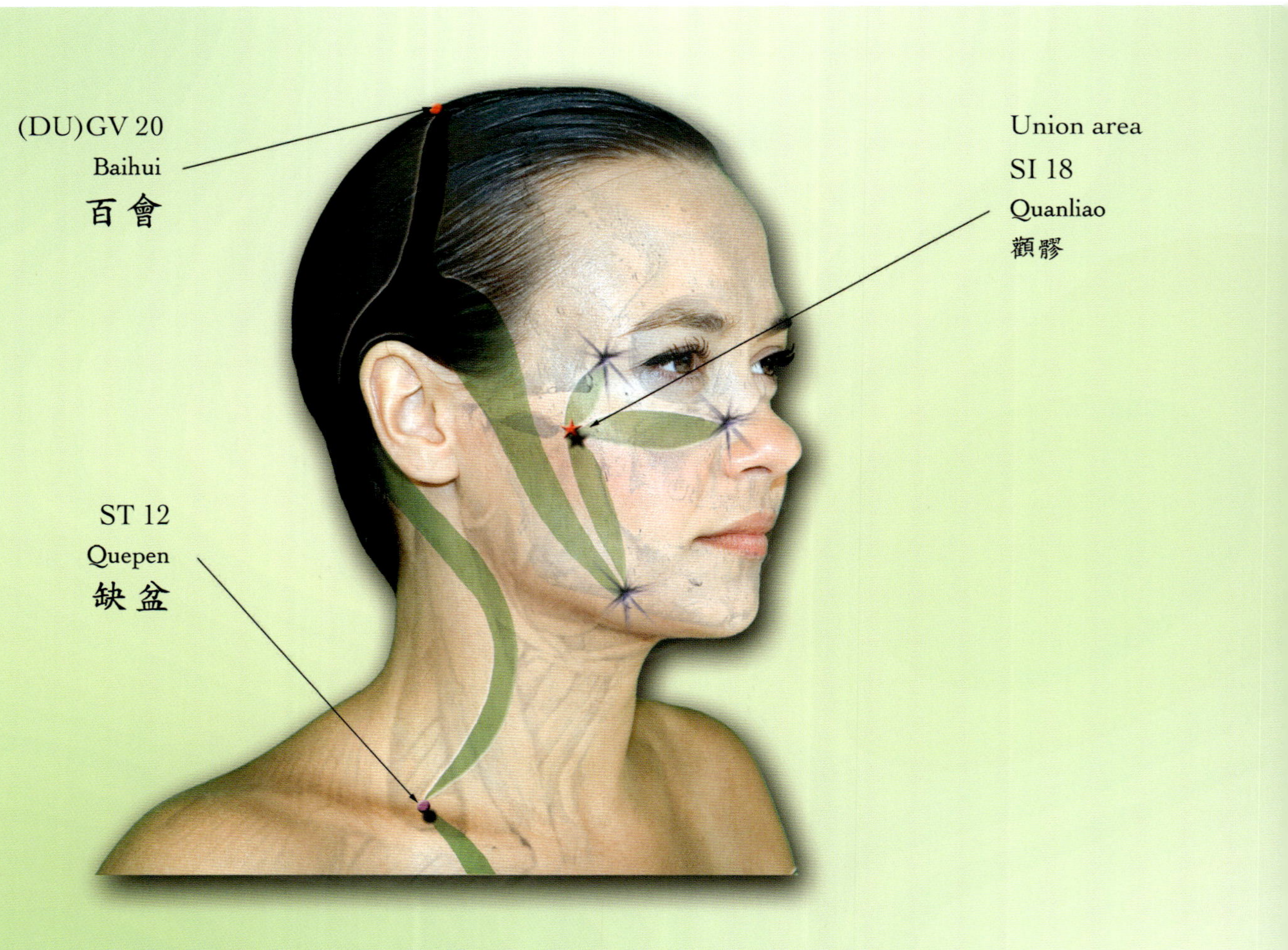

Cephalic course of the tendino-muscular of the Zu Shao Yang (Gall bladder)

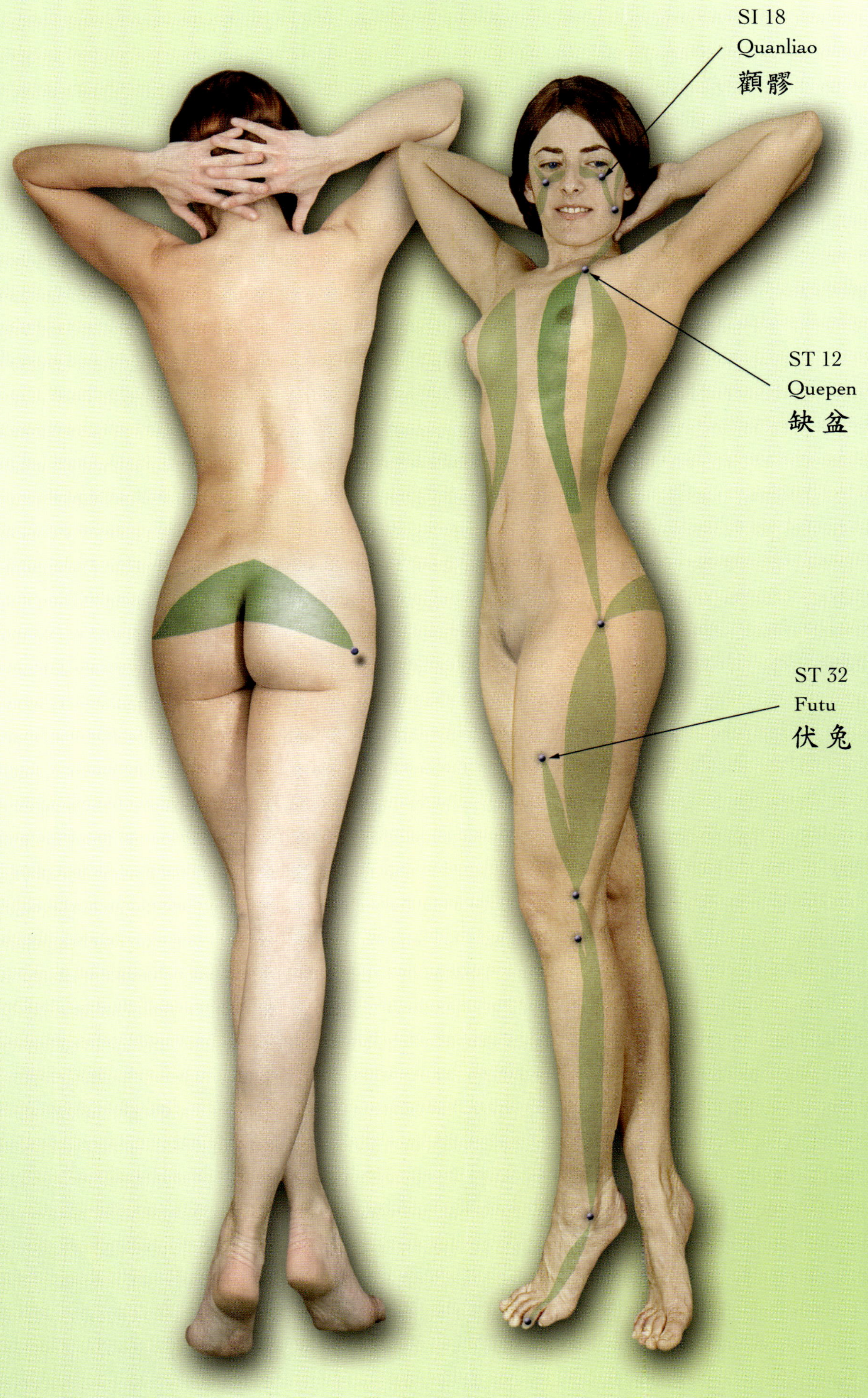

Course of tendino-muscular of Zu Shao Yang (Gall bladder)

Main channel of the Zu Shao Yang (Gall Bladder)

***Ling Shu*, chapter III paragraph 10**

The vessel of the Gall Bladder, Zu Shao Yang of the foot, starts at the outer canthus.

It goes up to the angle of the head (angle of the forehead) and goes down round the back of the ear (the auricle).

It runs along the neck and passes in front of the Shu Shao Yang of the hand (Triple Heater or Triple Burner channel).

It goes up on the shoulder and crosses again the Shu Shao Yang of the hand from behind.

It enters the supra clavicular fossa.
The branch that starts behind the auricle penetrates in the 2 PC Erzhong (midpoint of the ear), centre of the tragus and also travels on the area in front of the auricle.
It reaches the region behind the outer canthus.

A distinct branch descends from the outer canthus on the Daying (5 of the Stomach, on the inferior border of the inferior maxilla) and joins the Shu Shao Yang of the hand (channel of the Triple Heater or Triple Burner).

It reaches the area under the orbital cavity. It goes up by way of Jiache (6 of the Stomach, at the angle of the inferior maxilla).

It goes down again by the neck, joins the supra clavicular fossa so as to descend to the midpoint of the chest by crossing the diaphragm.

It joins the Liver and belongs to the Gall Bladder. It follows the inside of the ribs and goes out by ST 30 Qichong (point of the inguinal groove) going round the pubis. It penetrates obliquely in the area of the great trochanter.

The direct branch that leaves the supra clavicular fossa goes down under the armpits, follows the chest, passes by the floating ribs and, going down, it reaches the area of the great trochanter to travel down the antero superior part of the thigh and emerges at the external border of the knee.

It descends on the outside of the leg (by the external side of the Wai Fu Gu [fibula] and in the front).

It goes down vertically, reaching the lower extremity of Jue Gu (fibula) and, emerging in front of the external malleolus.

It crosses the dorsal side of the foot between the little toe and the 4th toe (at the angle of the 4th toe).

A distinct branch goes over the dorsal side of the foot and enters the area between the big toe and the 2nd toe, it follows the interdigital space, emerges at the tip of the big toe and goes back on the external nail (behind the ungual angle), in front of the metatarso-phalangeal joint where the hairs grow.

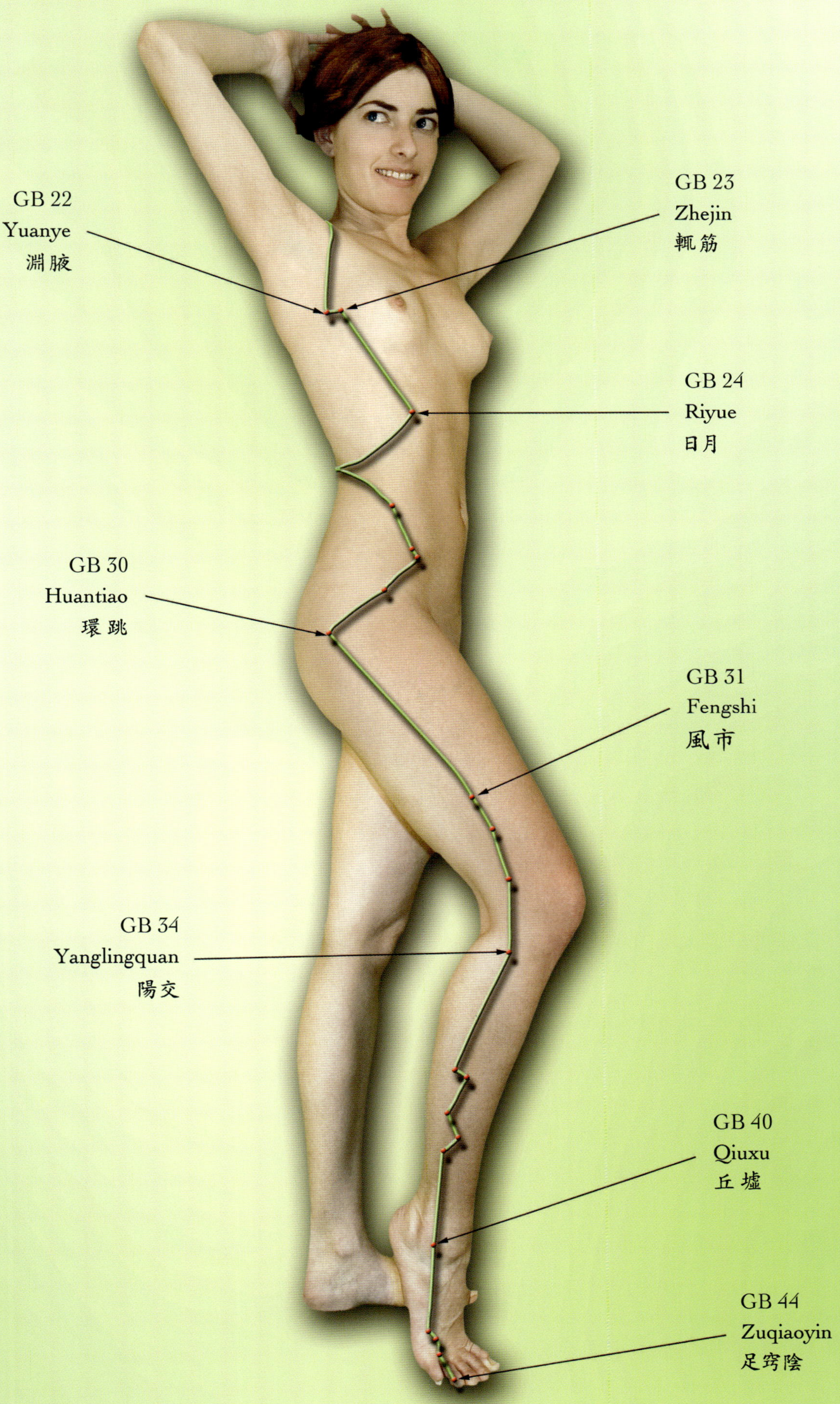

Course of Main channel Zu Shao Yang (Gall bladder)

Commentaries

The channel starts at the external end of the eyebrow at point GB 1 Tongziliao.

It passes in front of the lobule of the ear and ascends to the temporal area.

It goes down to the zygomatic arch where point (TH)TB 22 Erheliao is.

It passes behind the ear to reach the mastoid process by crossing point (TH)TB 20 Jiaosun.

It goes back to the forehead then goes back again to the nape of the neck.

It passes by the neck, crossing point SI 17 Tianrong and goes down to point GB 21 Jianjing.

It then travels to point (DU)GV 14 Dazhui, crosses point SI 12 Bingfeng and goes back in front, to the midpoint of the supra clavicular fossa at point ST 12 Quepen.

A branch goes out of GB 20 Fengchi, passes by point (TH)TB 17 Yifeng, runs across the ear, emerges in front of the central part of the tragus at point SI 19 Tinggong.

It passes then by point ST 7 Xiaguan and goes to GB 1 Tongziliao.

It descends again to point ST 5 Daying, goes up under the orbital cavity where it unites with the channel of the Triple Heater or Triple Burner, travels down again to point ST 6 Jiache and goes to point ST 12 Quepen, midpoint of the supra clavicular fossa.

From that point a deep branch and a superficial one go out.

The superficial route passes under the axilla in the 4th intercostal space, then it flows by the floating ribs, and crosses point (LR)LV 13 Zhangmen.

It then reaches the joint of the hip at GB 29 Juliao, passes by the sacrum and by the four sacral holes (from BL 31 Shangliao to BL 34 Xialiao) and it reaches point (DU)GV 1 Changquiang.

It goes back to point GB 30 Huantiao, goes down along the external side of the thigh, of the knee and of the foot.

It then passes by the hollow in front of the external malleolus, travels along the dorsal part of the foot between the 4th and the 5th metatarsal bones, and it ends in the external ungual angle of the big toe where it connects with the Liver channel at point (LR)LV 1 Dadun.

The route of the deeper branch will be considered with the internal branches.

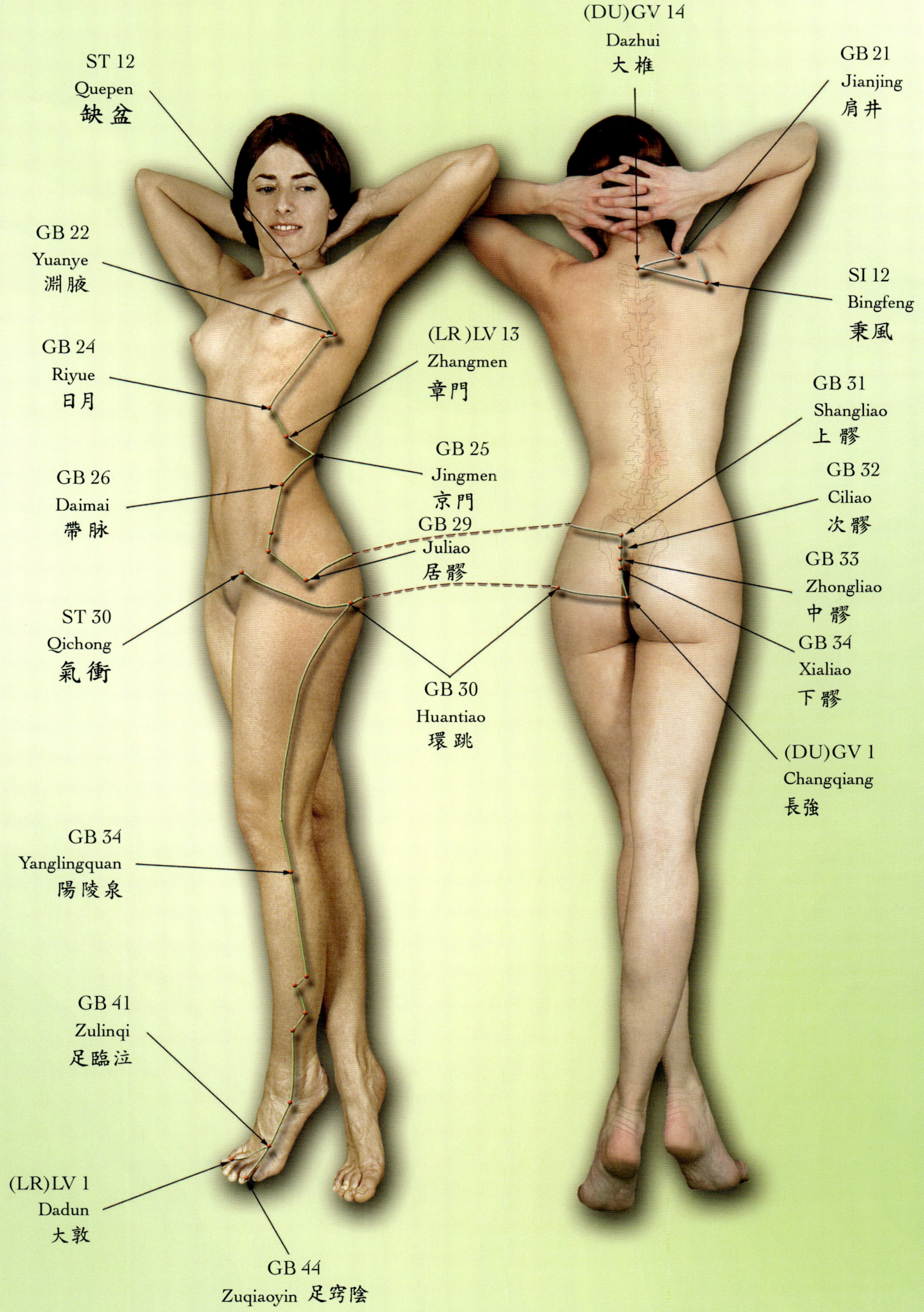

Complete course of Main channel Zu Shao Yang (Gall bladder)

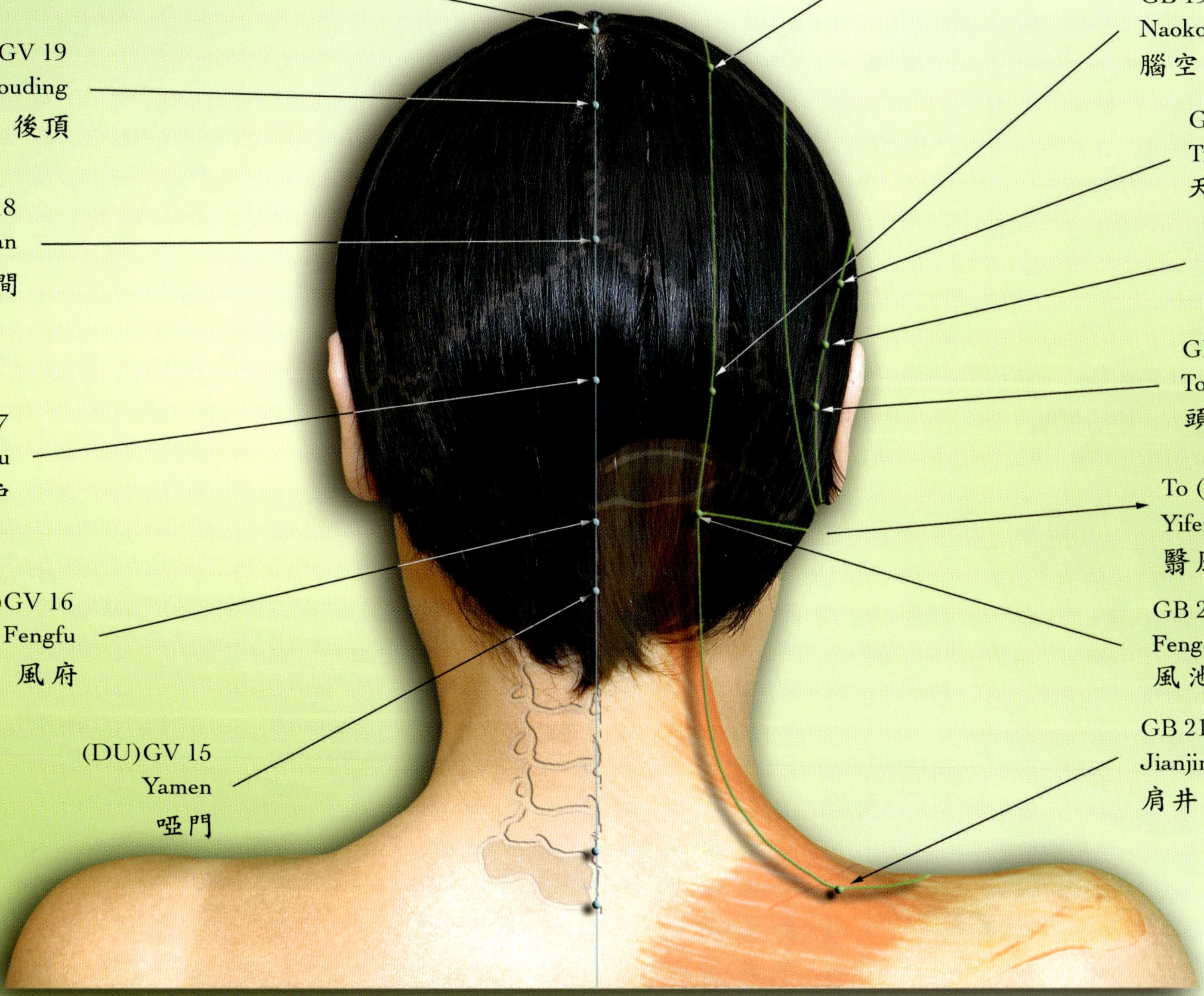

Posterior cephalic course of channel Zu Shao Yang (Gall bladder)

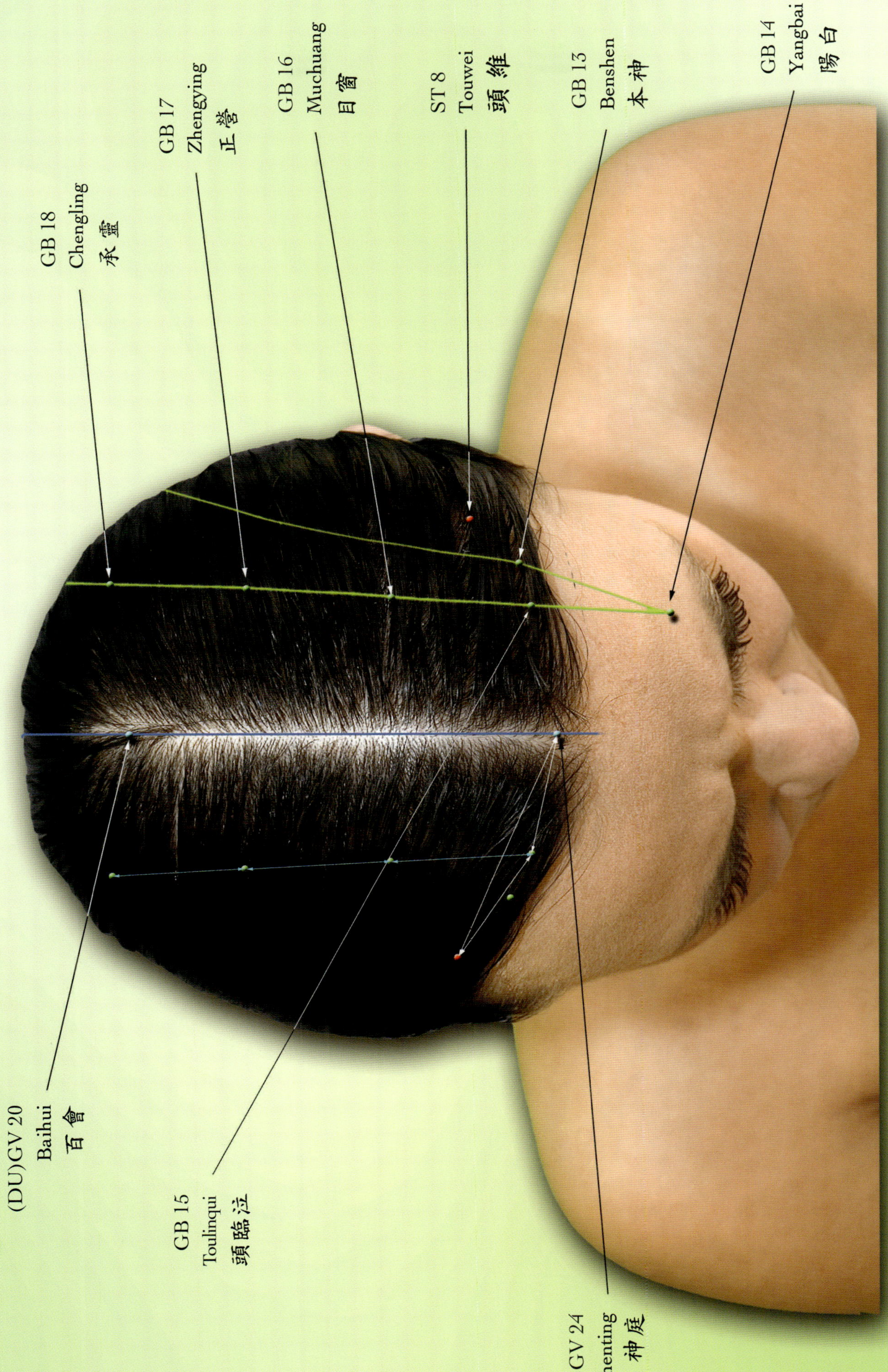

Anterior cephalic course of channel Zu Shao Yang (Gall Bladder)

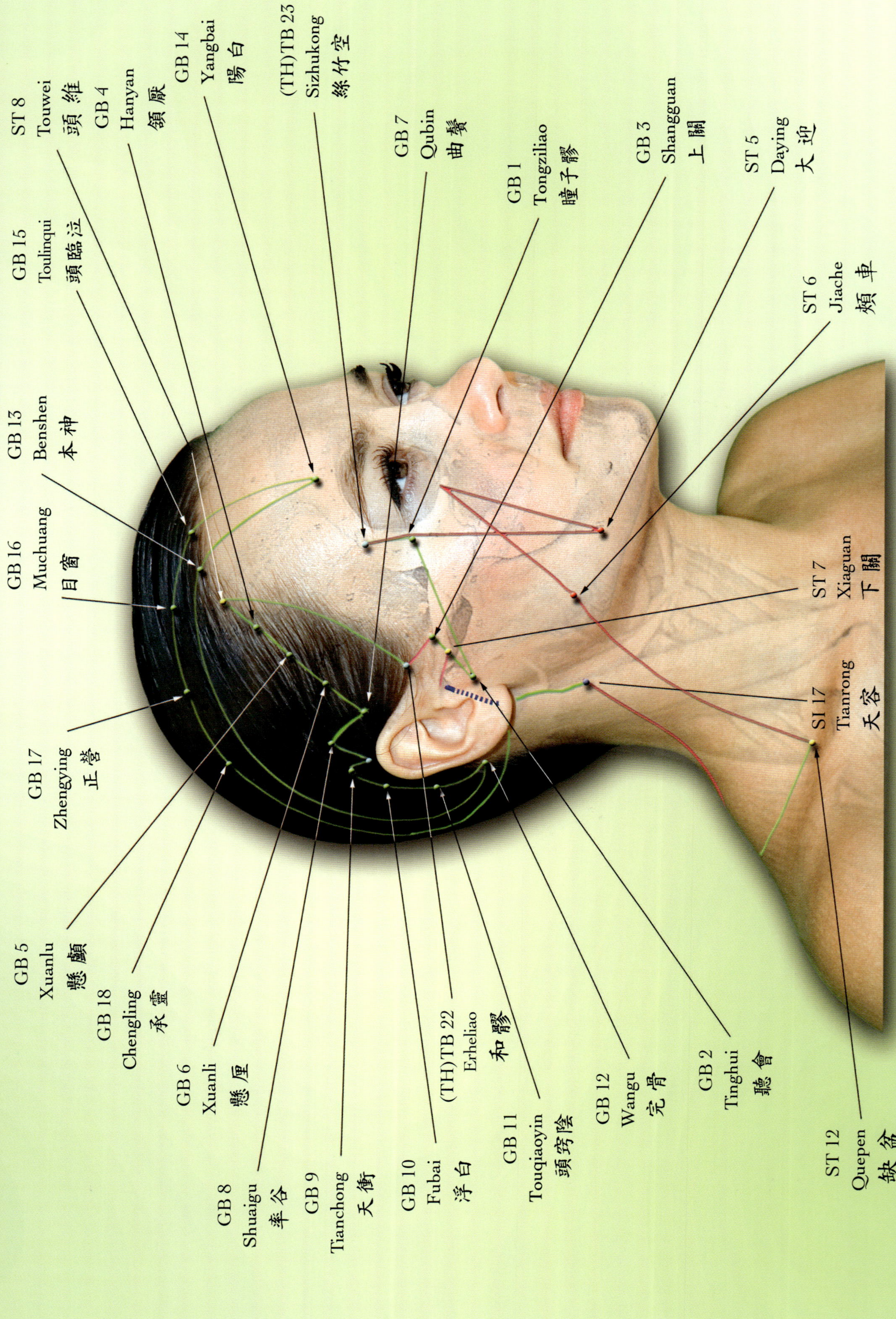

Complete lateral cephalic course of channel Zu Shao Yang (Gall bladder)

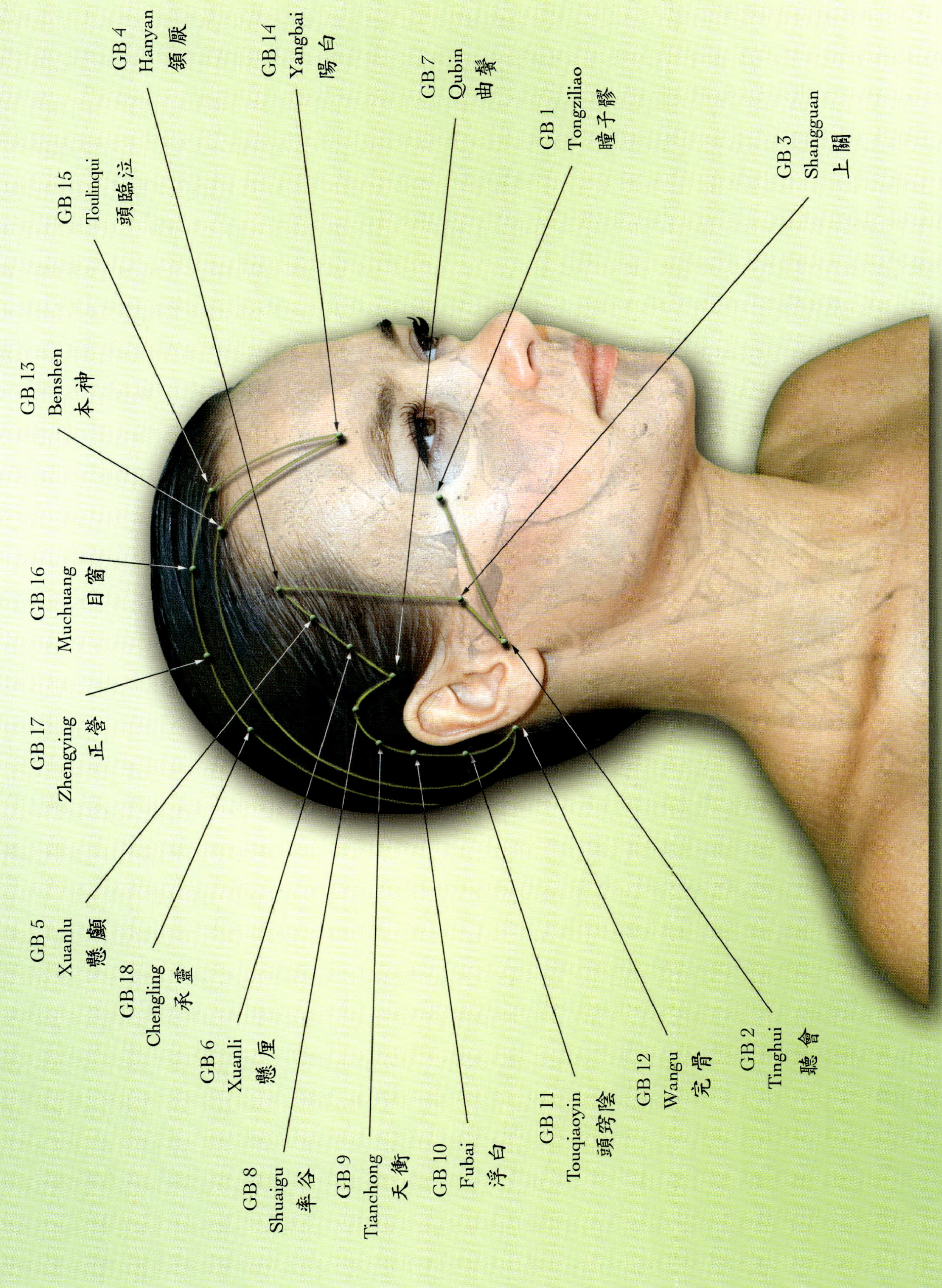

Traditional cephalic course of channel Zu Shao Yang (Gall bladder)

Internal branches of the Zu Shao Yang (Gall Bladder)

***Ling Shu*, chapter III, paragraph 10**

The internal branch leaves point ST 12 Quepen.

It enters the chest and connects deep inside with point (MH)PC 1 Tianchi.

It crosses the diaphragm, gets inside the Liver and ramifies in the Gall Bladder.

It goes out again at the groin at point ST 30 Qichong.

It goes to the joint of the hip at point GB 30 Huantiao.

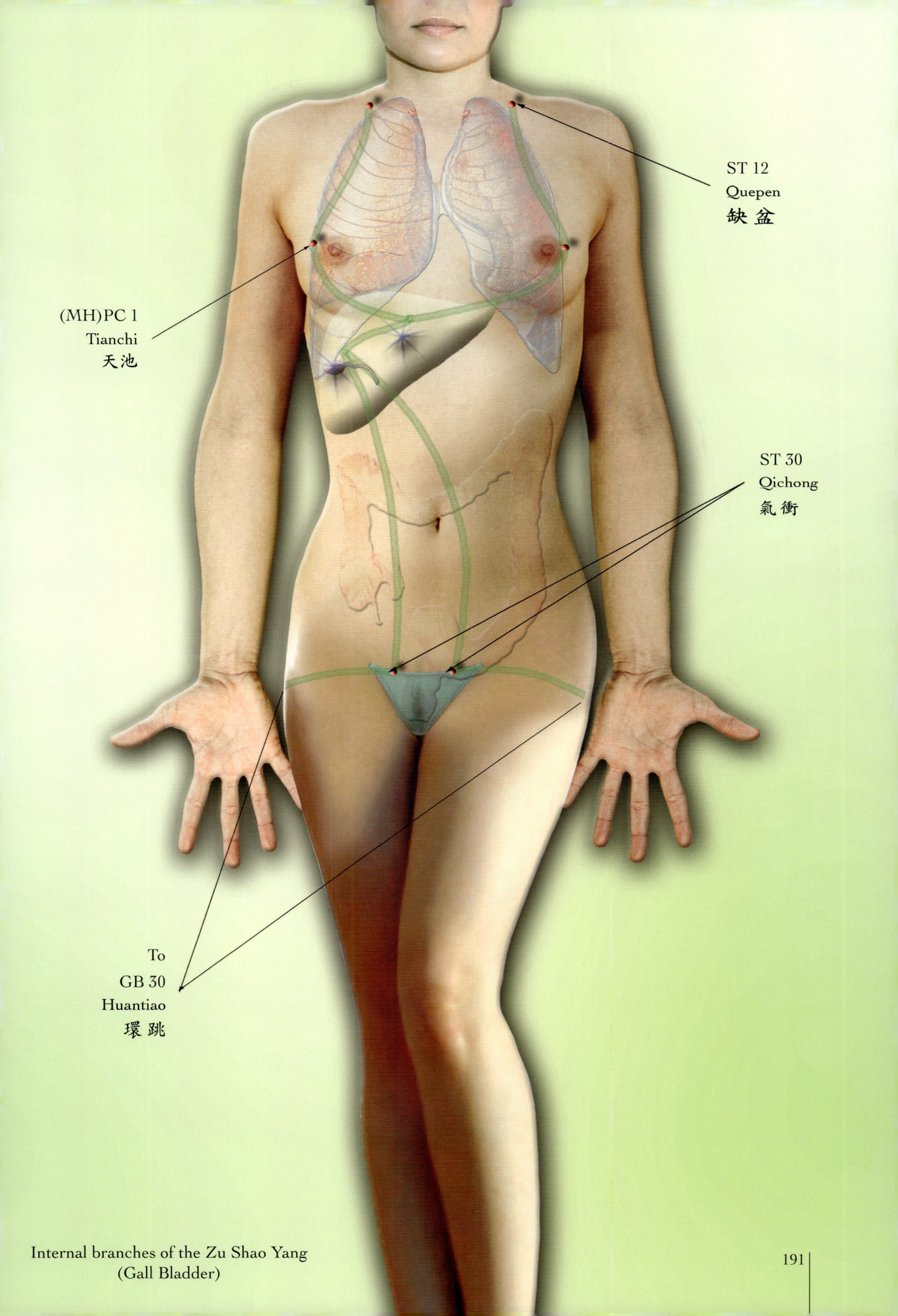

Internal branches of the Zu Shao Yang
(Gall Bladder)

LUO VESSELS OF THE ZU SHAO YANG (GALL BLADDER)

Ling Shu, chapter III, paragraph 10

The Luo vessel of the Zu Shao Yang of the foot (Gall Bladder) is called Guang Ming.

It starts 5 Cun above the external malleolus and travels to Zu Jue Yin (Liver).

It goes to the dorsal part of the foot.

Commentaries

It goes out of point GB 37 Guangming situated 5 Cun above the external malleolus and divides into two branches:

- The longitudinal branch that passes by the dorsal side of the foot and ramifies in the 3rd, 4th and 5th toes.

- The transversal branch that unites with the channel of Zu Jue Yin (Liver) at point (LR)LV 3 Taichong in the angle between the first two metatarsal bones.

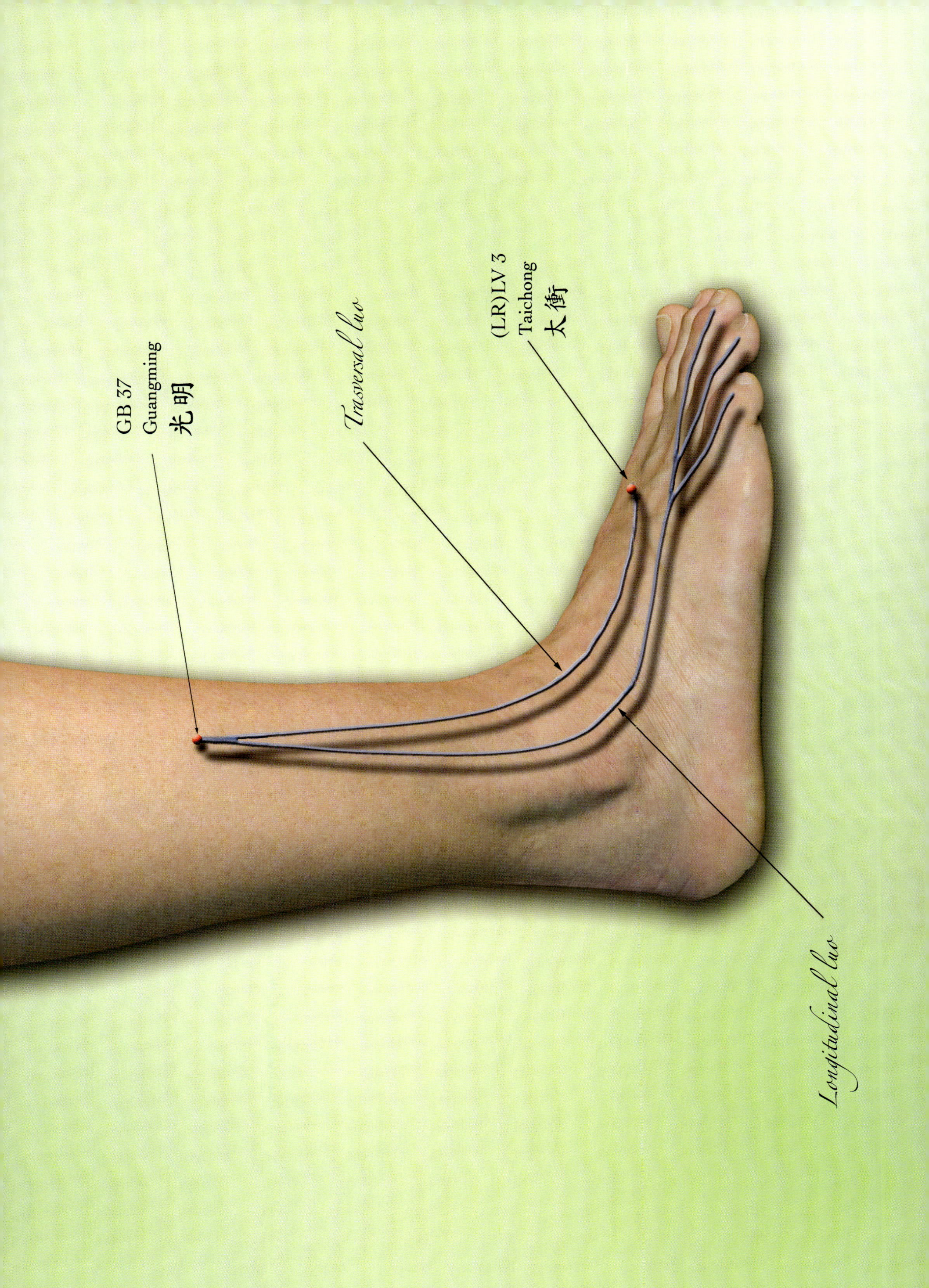

Course of Luo vessels of the Zu Shao Yang (Gall bladder)

Distinct channel of the Zu Shao Yang (Gall Bladder)

***Ling Shu*, chapter III, paragraph 11**

The distinct channel of the Main channel Zu Shao Yang of the foot[*] (channel of the Gall Bladder), a distinct branch, goes along the ribs and goes out by the inguinal groove at ST 30 Qichong, goes round the biceps (antero-superior part of the thigh) then reaches the Mao Fi (pubic hairs). It joins Zu Jue Yin[**] (Liver channel) and gets in between the ribs. (Ji Lei corresponds to the area situated between the 11th and 12th rib, that is to say the floating ribs[***]).

It goes along the inside of the rib cage and belongs to the Gall Bladder. It ramifies going up to the Liver. It passes by the area of the Heart and goes up encircling the throat.

It goes out by the Yi (lateral and superior area of the chin) and by the midpoint of the He (part of the maxilla situated under the ear). It spreads on the face and on 'the eye connections'.

It joins the Zu Shao Yang (Gall Bladder channel) at the outer canthus[****]

* Goes out of the GB 30 Huantiao area.

** Joins the Liver channel at point (REN)CV 2 Qugu.

*** Passing by point H 13 Zhangmen.

**** Joins at point GB 1 Tongziliao to form the 2nd union.

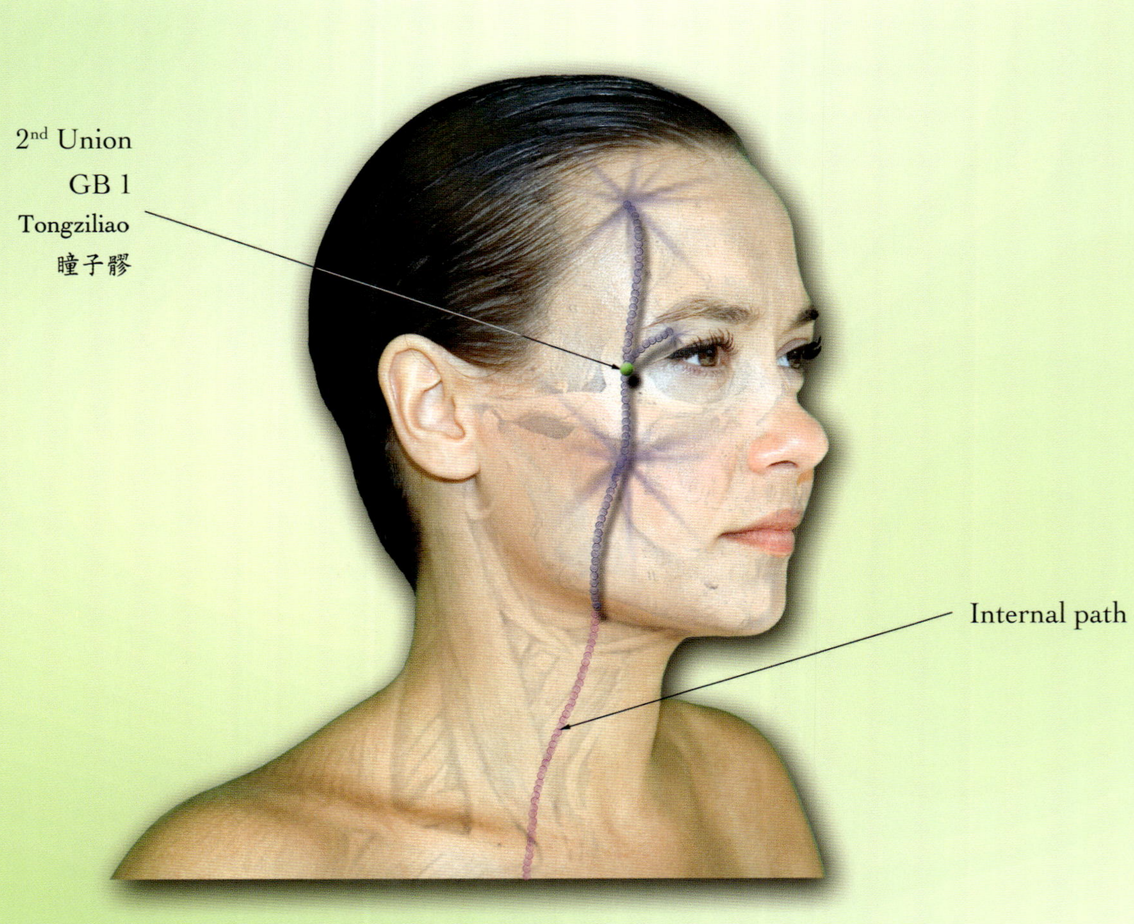

Cephalic course of the distinct channel of Zu Shao Yang (Gall bladder)

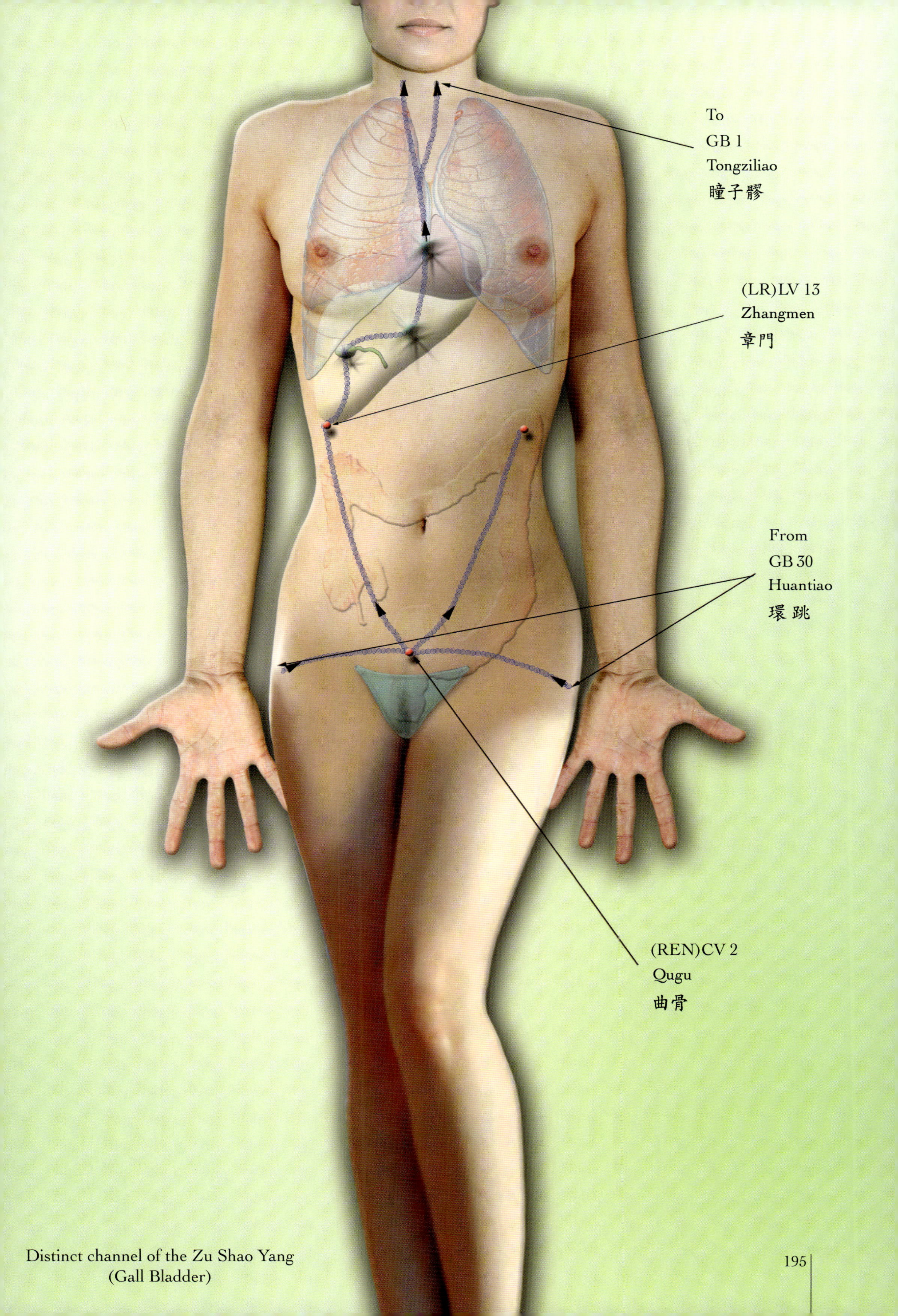

Distinct channel of the Zu Shao Yang (Gall Bladder)

Points of the Main channel Zu Shao Yang (Gall Bladder)

GB 1 Tongziliao
On the face, lateral to the outer canthus, on the lateral border of the orbit.

GB 2 Tinghui
On the face, anterior to the intertragic notch, in the depression posterior to the condyloid process of the mandible, when the mouth is open.

GB 3 Shangguan
Anterior to the ear, directly above ST 7 Xiaguan, in the depression above the upper border of the zygomatic arch.

GB 4 Hanyan
On the head, in the hair above the temples, at the junction of the upper fourth and lower three fourths of the curved line connecting ST 8 Touwei and GB 7 Qubin.

GB 5 Xuanlu
On the head, in the hair above the temples, at the midpoint of the curved line connecting ST 8 Touwei and GB 7 Qubin.

GB 6 Xuanli
On the head, in the hair above the temples, at the junction of the upper three fourths and lower fourth of the curved line connecting ST 8 Touwei and GB 7 Qubin.

GB 7 Qubin
On the head, at a crossing point of the vertical posterior border of the temples and horizontal line through the ear apex.

GB 8 Shuaigu
On the head, directly above the ear apex, 1.5 Cun above the hairline, directly above (TH)TB 20 Jiaosun.

GB 9 Tiangchong
On the head, directly above the posterior border of the ear root, 2 Cun above the hairline and 0.5 Cun posterior to GB 8 Shuaigu.

GB 10 Fubai
On the head, posterior and superior to the mastoid process, at the junction of the middle third and upper third of the curved line connecting GB 9 Tianchong and GB 12 Wangu.

GB 11 Touqiaoyin
On the head, posterior and superior to the mastoid process, at the junction of the middle third and lower third of the curved line connecting GB 9 Tianchong and GB 12 Wangu.

GB 12 Wangu
On the head, in the depression posterior and inferior to the mastoid process.

GB 13 Benshen
On the head, 0.5 Cun above the anterior hairline, 3 Cun lateral to (DU)GV 24 Shenting, at the junction of the medial two thirds and lateral third of the line connecting (DU)GV 24 Shenting and ST 8 Touwei.

GB 14 Yangbai
On the forehead, directly above the pupil, 1 Cun above the eyebrow.

GB 15 Toulinqi
On the head, directly above the pupil and 0.5 Cun above the anterior hairline, at the midpoint of the line connecting (DU)GV 24 Shenting and ST 8 Touwei.

GB 16 Muchuang
On the head, 1.5 Cun above the anterior hairline and 2.25 Cun lateral to the midline of the head.

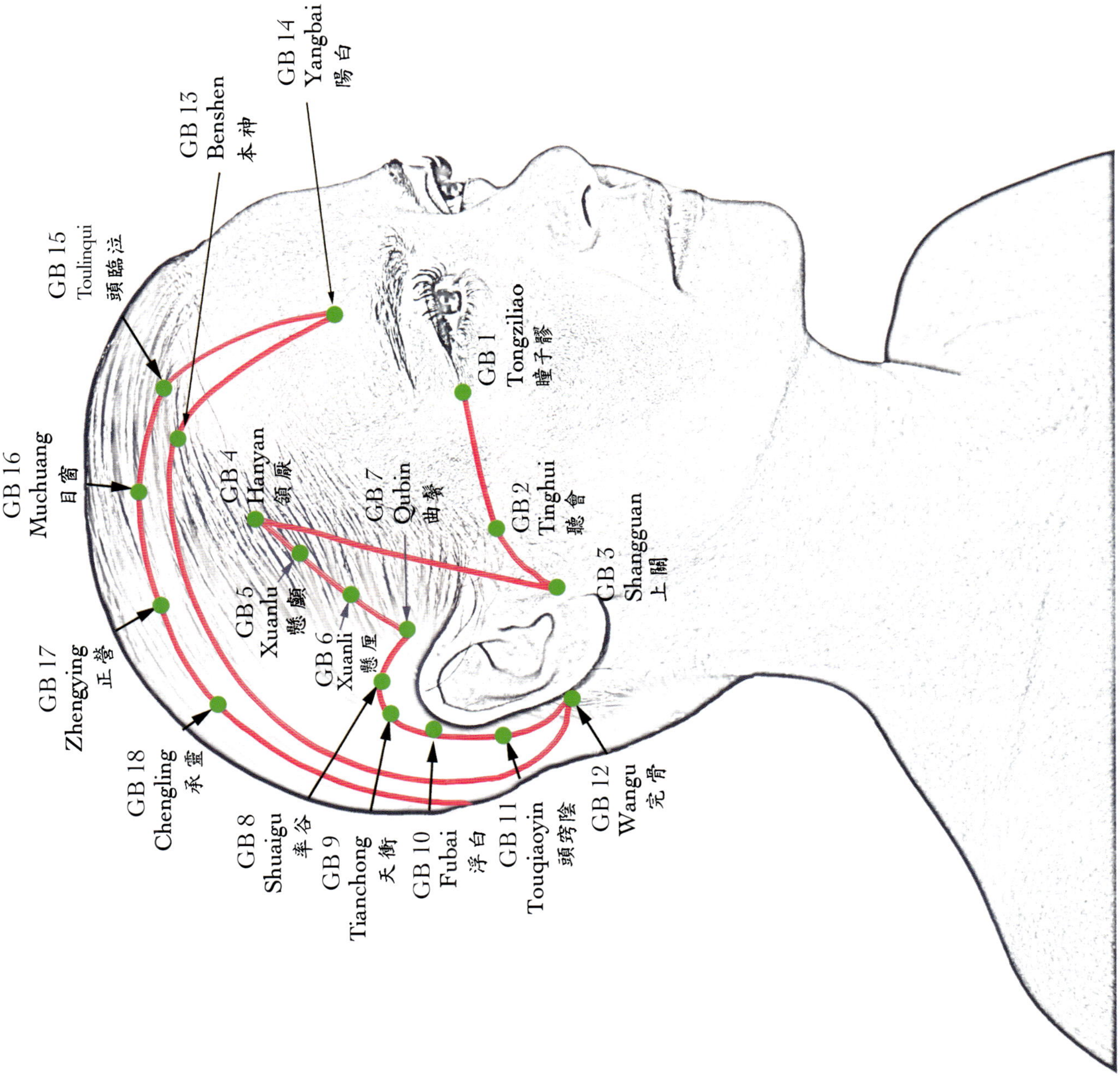
GB 1
Tongziliao
瞳子髎
GB 2
Tinghui
聽會
GB 3
Shangguan
上關
GB 4
Hanyan
頷厭
GB 5
Xuanlu
懸顱
GB 6
Xuanli
懸厘
GB 7
Qubin
曲鬢
GB 8
Shuaigu
率谷
GB 9
Tianchong
天衝
GB 10
Fubai
浮白
GB 11
Touqiaoyin
頭竅陰
GB 12
Wangu
完骨
GB 13
Benshen
本神
GB 14
Yangbai
陽白
GB 15
Toulinqui
頭臨泣
GB 16
Muchuang
目窗
GB 17
Zhengying
正營
GB 18
Chengling
承靈

GB 17 Zhengying
On the head, 2.5 Cun above the anterior hairline and 2.25 Cun lateral to the midline of the head.

GB 18 Chengling
On the head, 4 Cun above the anterior hairline and 2.25 Cun lateral to the midline of the head.

GB 19 Naokong
On the head and on the level of the upper border of the external occipital protuberance of (DU)GV 17 Naohu, 2.25 Cun lateral to the midline of the head.

GB 20 Fengchi
On the nape, below the occipital bone, on the level of (DU)GV 16 Fengfu, in the depression between the upper ends of the sterno-cleido-mastoid and trapezius muscles.

GB 21 Jianjing
On the shoulder, directly above the nipple, at the midpoint of the line connecting (DU)GV 14 Dazhui and the acromion.

GB 22 Yuanye
On the lateral side of the chest, on the midaxillary line when the arm is raised, 3 Cun below the axilla, in the 4th intercostal space.

GB 23 Zhejin
On the lateral side of the chest, 1 Cun anterior to GB 22 Yuanye, on the level of the nipple, and in the 4th intercostal space.

GB 24 Riyue
On the upper abdomen, directly below the nipple, in the 7th intercostal space, 4 Cun lateral to the anterior midline.

GB 25 Jingmen
On the lateral side of the waist, 1.8 Cun posterior to (LR)LV 13 Zhangmen, below the free end of the 12th rib.

GB 26 Daimai
On the lateral side of the abdomen, 1.8 Cun below (LR)LV 13 Zhangmen, at the crossing point of a vertical line through the free end of the 11th rib and a horizontal line through the umbilicus.

GB 27 Wushu
On the lateral side of the abdomen, anterior to the anteriosuperior iliac spine, 3 Cun below the horizontal level of the umbilicus.

GB 28 Weidao
On the lateral side of the abdomen, anterior and inferior to GB 27 Wushu.

GB 29 Juliao
On the hip, at the midpoint of the line connecting the anterio-superior iliac spine and the prominence of the great trochanter.

GB 30 Huantiao
On the lateral side of the thigh, at the junction of the middle third and lateral third of the line connecting the prominence of the great trochanter and the sacral hiatus, when the patient is in a lateral recumbent position with the thigh flexed.

GB 31 Fengshi
On the lateral midline of the thigh, 7 Cun above the popliteal crease, or at the place touching the tip of the middle finger when the patient stands erect with the arms hanging down freely.

GB 32 Zhongdu
On the lateral side of the thigh, 2 Cun below GB 31 Fengshi, or 5 Cun above the popliteal crease, between the lateral vastus muscle and biceps muscle of the thigh.

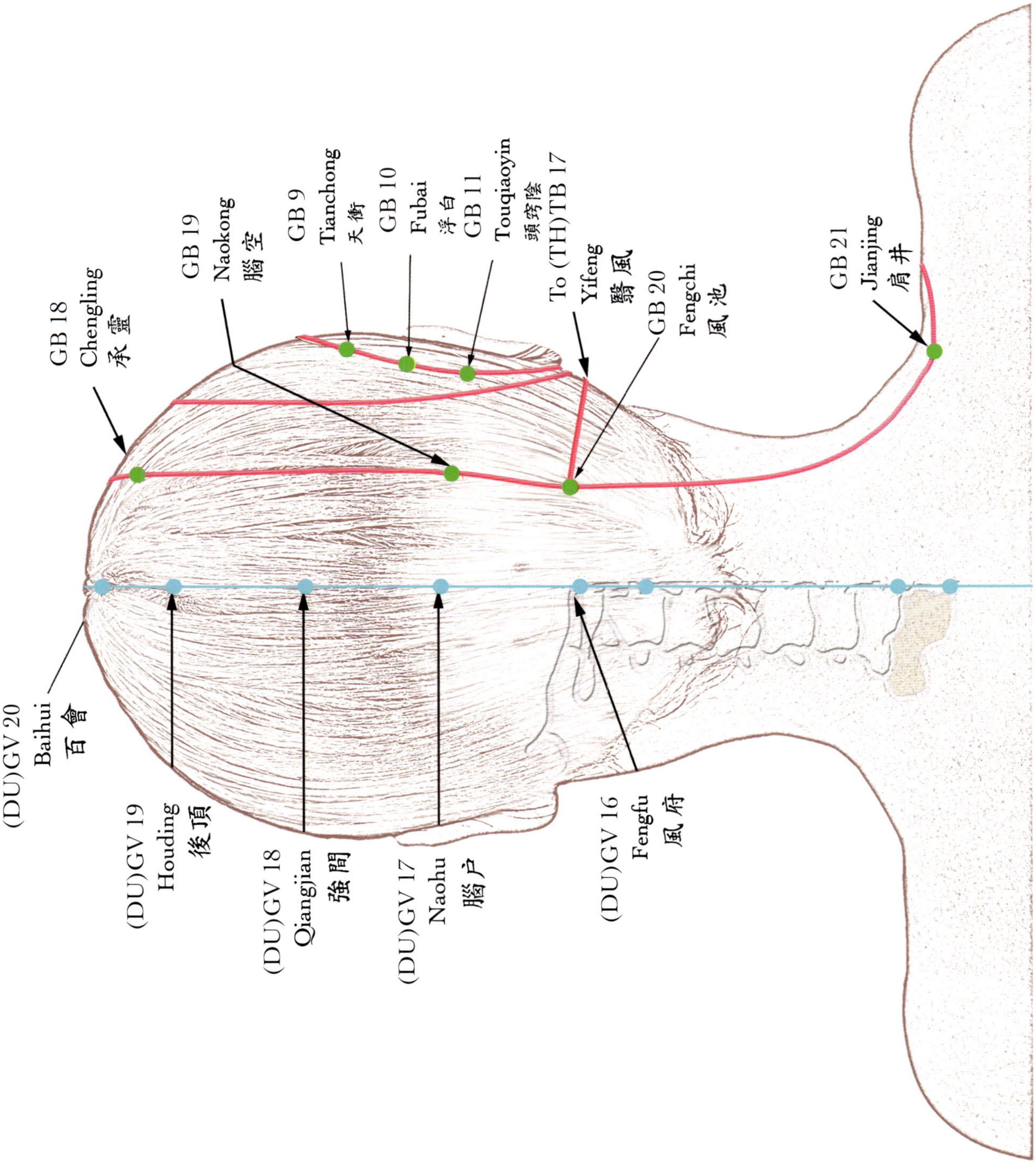

Posterior head points of Zu Shao Yang (Gall bladder)

GB 33 Xiyangguan
On the lateral side of the knee, 3 Cun above GB 34 Yanglingquan, in the depression above the external epicondyle of the femur.

GB 34 Yanglingquan
On the lateral side of the leg, in the depression anterior and inferior to the head of the fibula.

GB 35 Yangjiao
On the lateral side of the leg, 7 Cun above the tip of the external malleolus, on the posterior border of the fibula.

GB 36 Waiqiu
On the lateral side of the leg, 7 Cun above the tip of the external malleolus, on the anterior border of the fibula and on the level of GB 35 Yangjiao.

GB 37 Guangming
On the lateral side of the leg, 5 Cun above the tip of the external malleolus, on the anterior border of the fibula.

GB 38 Yangfu
On the lateral side of the leg, 4 Cun above the tip of the external malleolus, slightly anterior to the anterior border of the fibula.

GB 39 Xuanzhong
On the lateral side of the leg, 3 Cun above the tip of the external malleolus, on the anterior border of the fibula.(*)

GB 40 Qiuxu
Anterior and inferior to the external malleolus, in the depression lateral to the tendon of the long extensor muscle of the toes.

GB 41 Zulinqi
On the lateral side of the instep of the foot, posterior to the 4th metatarso-phalangeal joint, in the depression lateral to the tendon of the extensor muscle of the little toe.

GB 42 Diwuhui
On the lateral side of the instep of the foot, posterior to the 4th metatarsophalangeal joint, between the 4th and 5th metatarsal bones, medial to the tendon of the extensor muscle of the little toe.

GB 43 Xiaxi
On the lateral side of the instep of the foot, between the 4th and 5th toes, at the 'junction of the red and white skin', proximal to the margin of the web.

GB 44 Zuqiaoyin
On the lateral side of the distal segment of the 4th toe, 0.1 Cun from the corner of the toenail.

* 50 % of the writers also position it on the posterior border of the fibula.

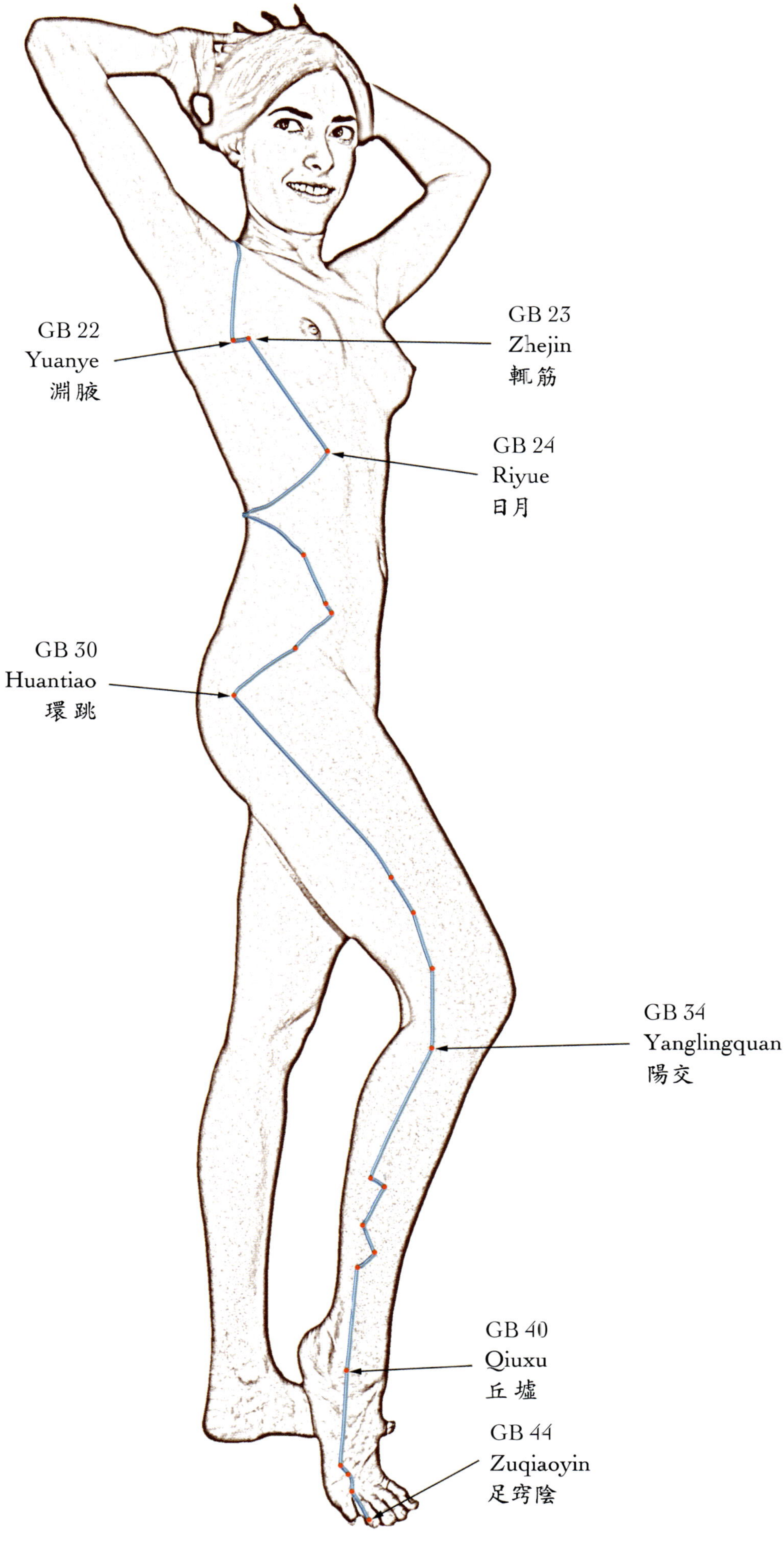

Points of the Main channel Zu Shao Yang (Gall Bladder)

Zu Tai Yang system (Urinary Bladder)

- Tendino-muscular channel
- Main channel
- Internal branches
- Luo vessels
- Distinct channel
- Points of the Main channel

Tendino-muscular of the Zu Tai Yang (Urinary Bladder)

Ling Shu, chapter IV, paragraph 13

The tendino-muscular of the Zu Tai Yang of the foot starts at the fifth toe.

It goes up joining the external malleolus°.

It goes up obliquely and connects with the knee.

It goes down by the external malleolus and unites with the Zhong (posterior purchase of the foot).

It goes up on the heel and joins the popliteal crease. A distinct branch joins the external part of the Zhuan (Gastrocnemius) and goes up on the internal part of the popliteal crease°°.

From the popliteal crease, the tendino-muscular joins the buttock while going up.

It runs up to the nape along the rachis°°°. A distinct branch goes inside and joins the 'root of the tongue' (base of the tong).

The tendino-muscular attaches itself to the occipital bone.

It goes up on the head (vertex) and goes down on the face uniting with the nose.

A branch goes up like a 'net over the eye' (on the upper eyelid).

It goes down uniting with the Jiu (zygomatic)°°°°.

A branch starts behind the axilla, from the external border. It joins point Jianyu at the acromioclavicular joint.

A branch penetrates under the armpit and, going up, goes out by the supraclavicular fossa.

It goes up uniting with Wanggu (behind the posterior border of the mastoid).

A branch goes out by the supra-clavicular fossa. It goes up obliquely and out by the Jiu (zygomatic).

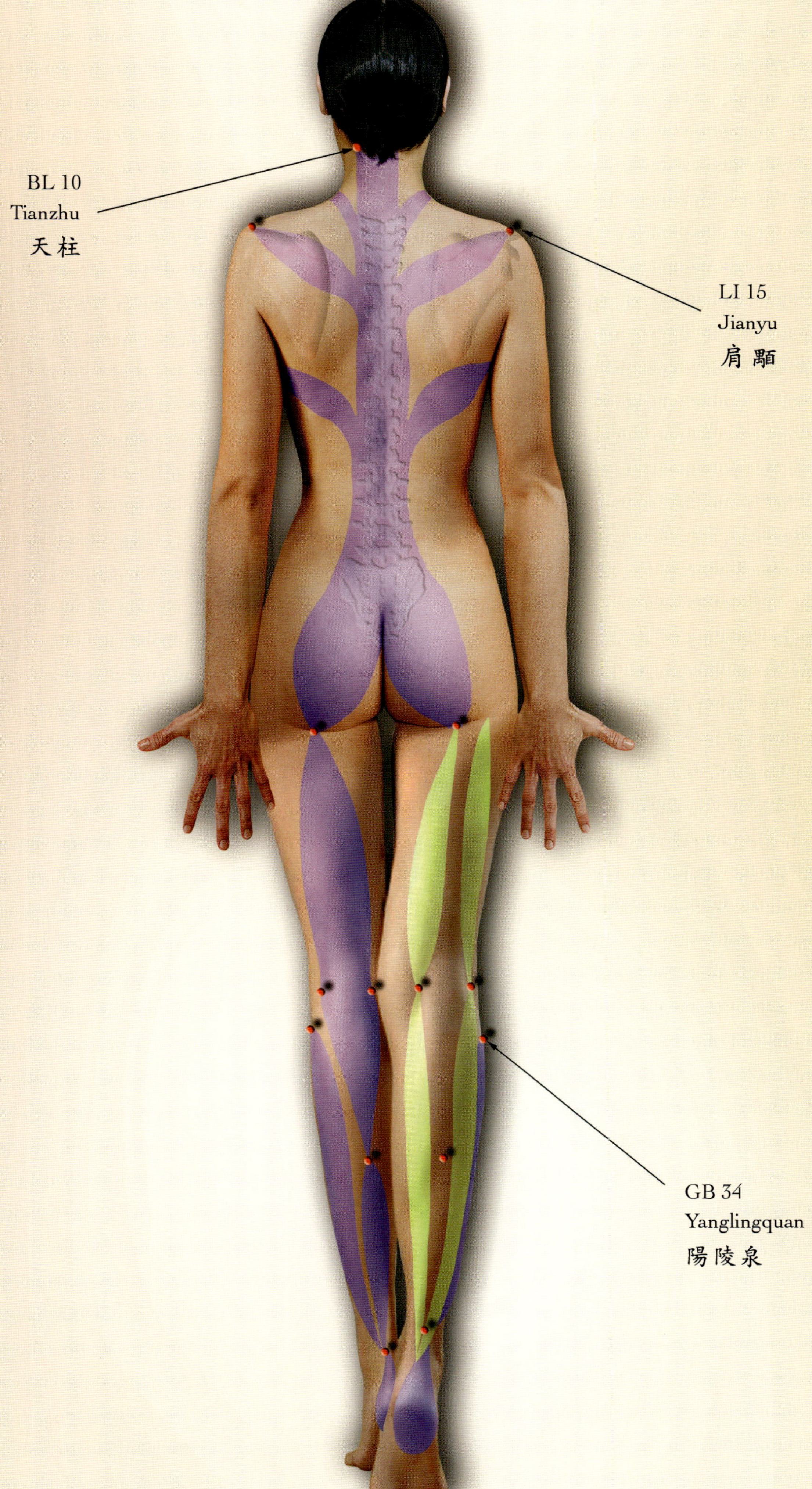

Tendino-muscular of the Zu Tai Yang (Urinary Bladder)

Commentaries

* The channel is divided into three branches:

- The first one joins the external part of the calcaneum and goes to the heel.
- The second one follows the external side of the leg and goes to point GB 34 Yanglingquan.
- The third one unites with Achilles' heel and goes up following the gemellus muscles.

** Some writers consider the third branch to be double. (cf right-hand side path)

*** It joins the external border of the trapezius in the area of BL 10 Tianzhu.
From that point, two branches start:

- The first one joins point GB 20 Fengchi, then goes deep inside to unite with the root of the tongue.
- The other one passes by the cranium and reaches the inner canthus, from where several ramifications go out:
  - a branch joins the side of the nose and unites with the cheek, the other one passes by the upper eyelid and by the deep part of the orbital cavity to finally end in the palate.

**** Along its dorsal course, two branches start:

- The first one goes to the shoulder and reaches point LI 15 Jianyu.
- The second one enters the axilla and goes up in the supra-clavicular fossa, passing by the chest.

From the supraclavicular fossa, at point ST 12 Quepen, two branches are born:

- The first one joins the mastoid at point GB 12 Wangu.
- The second one joins the other inferior tendino-muscular channels at point SI 18 Quanliao.

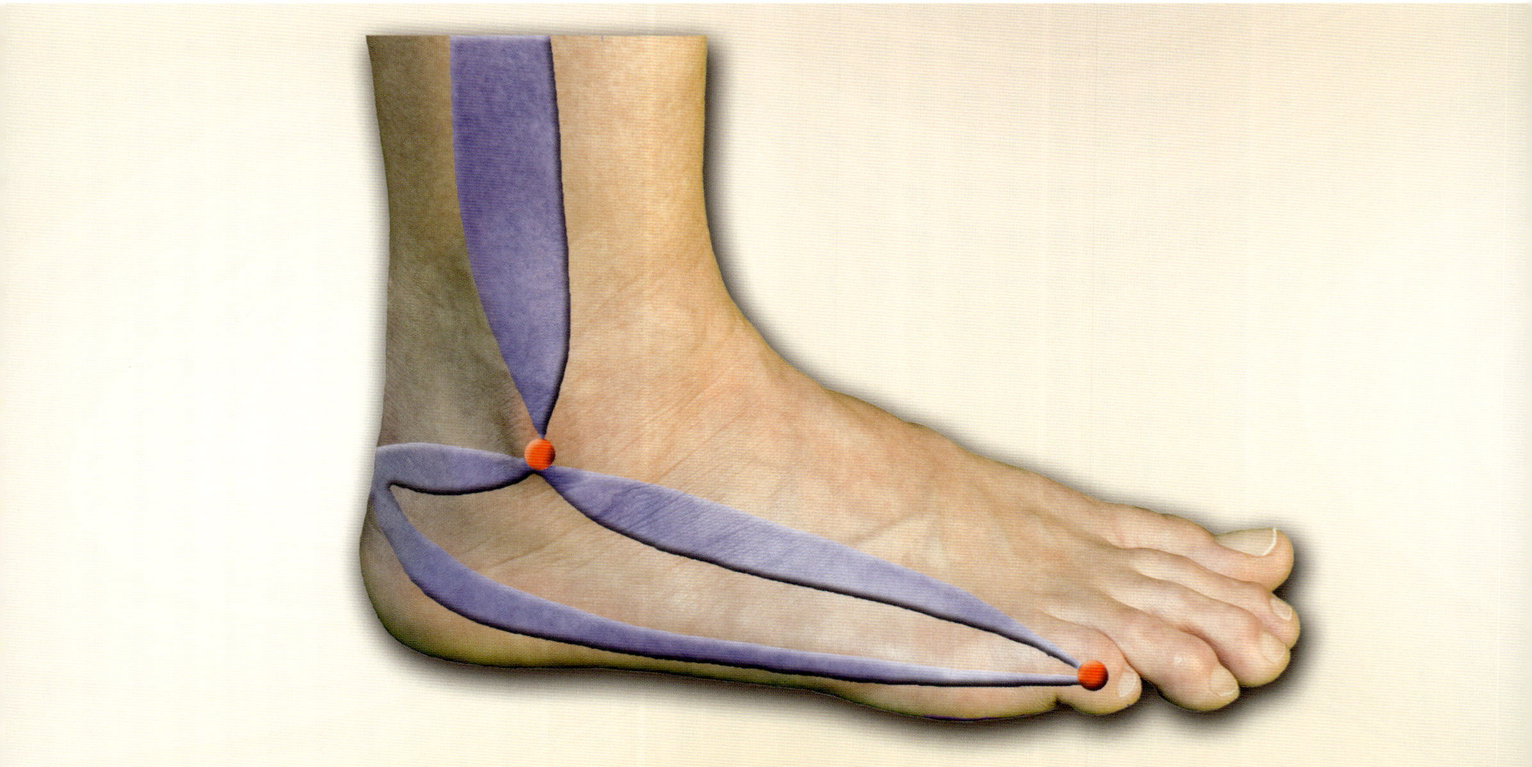

Course of tendino-muscular of Zu Tai Yang (Urinary Bladder) with various possibilities

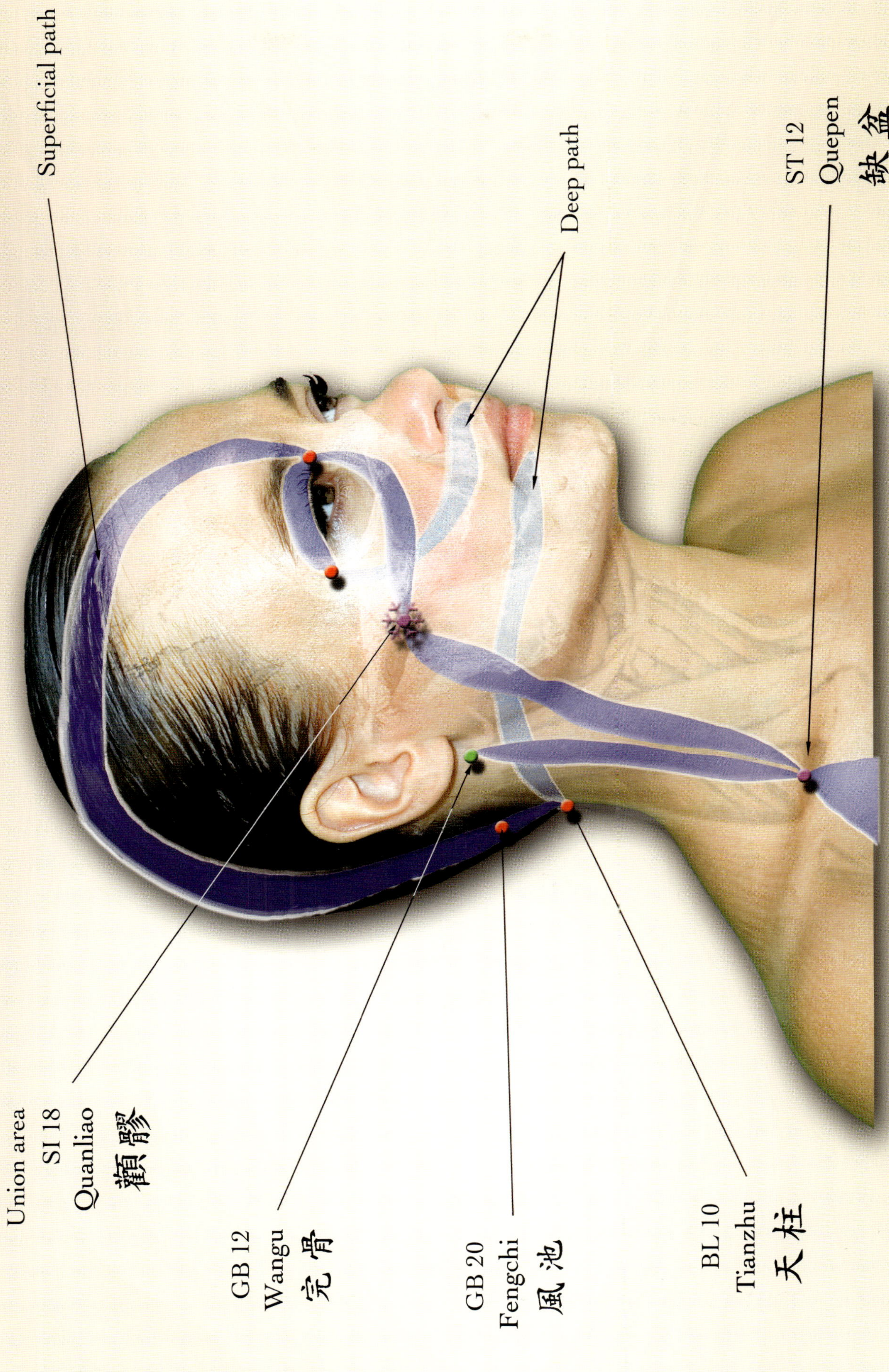

Cephalic course of tendino-muscular of Zu Tai Yang (Urinary bladder)

Main channel of the Zu Tai Yang (Urinary Bladder)

Ling Shu, Chapter III, paragraph 10

The Urinary Bladder vessel, Zu Tai Yang of the foot, starts at the inner canthus.

It goes up on the forehead and crosses the top of the head. A branch runs from the top of the head to the apex of the ear.

The branch that starts at the top of the head joins the brain, goes out again and differentiates descending on the nape. It follows the internal part of the shoulder and encircles the backbone (goes down along the rachis) and reaches the middle part of the lumbar area. It goes on and enters the paravertebral muscles, joins the Kidneys and belongs to the Bladder.

The branch that starts from the middle region of the lumbar area goes down encircling the vertebrae (along the transverse processes), it runs on the buttock area and goes in at the midpoint of the popliteal crease.

The branch that starts at the internal part of the shoulder-blade differentiates while going down along the internal side of the shoulder-blade.

It encompasses the internal side of the vertebrae (goes down along the rachis).

It passes by the joint of the hip, that is to say by the Bi Shu (great trochanter), follows the external side of the thigh and starting at the posterior border, it goes down and reaches the midpoint of the popliteal crease to run down along the internal side of the leg (calf).

It goes out behind the external malleolus. It follows the Jing Gu (capital bone), behind the 5th metatarsal bone, on the external border of the foot and reaches the external border of the little toe.

This distinct branch penetrates at the midpoint of the heel so as to go up on the internal side of the calf and go out by the internal border of the popliteal crease.

It goes up by the posterior border of the internal side of the thigh. It follows the vertebrae (vertebral axis), belongs to the Kidneys and joins the Urinary Bladder.

A direct branch starts at the Kidneys, goes up by way of the Liver and the diaphragm. It penetrates the Lungs, goes along the throat and encompasses the 'root of the tongue'.

The branch starting at the Lungs goes round the Pericardium and spreads at the midpoint of the thorax.

Commentaries

The Main channel starts at point BL 1 Jingming, on the superio-internal border of the orbital cavity.

It goes up on the forehead and joins points (DU)GV 24 Shenting and GB 15 Toulinqi.

It keeps ascending outside Du Mai and unites with point (DU)GV 20 Baihui.

On that level, it divides into 2 branches:

- The first one runs to the upper part of the auricle, crossing points GB 8 Shuaigu, GB 7 Qubin, GB 10 Fubai, GB 11 Touqiaoyin and GB 12 Wangu.

- The second one goes into the brain then goes out at point (DU)GV 17 Naohu.

From (DU)GV 20 Baihui, the channel follows its course passing by points BL 8 Luoque, BL 9 Yuzhen and unites with (DU)GV 17 Naohu, then descends by the nape where it unites with point BL 10 Tianzhu.

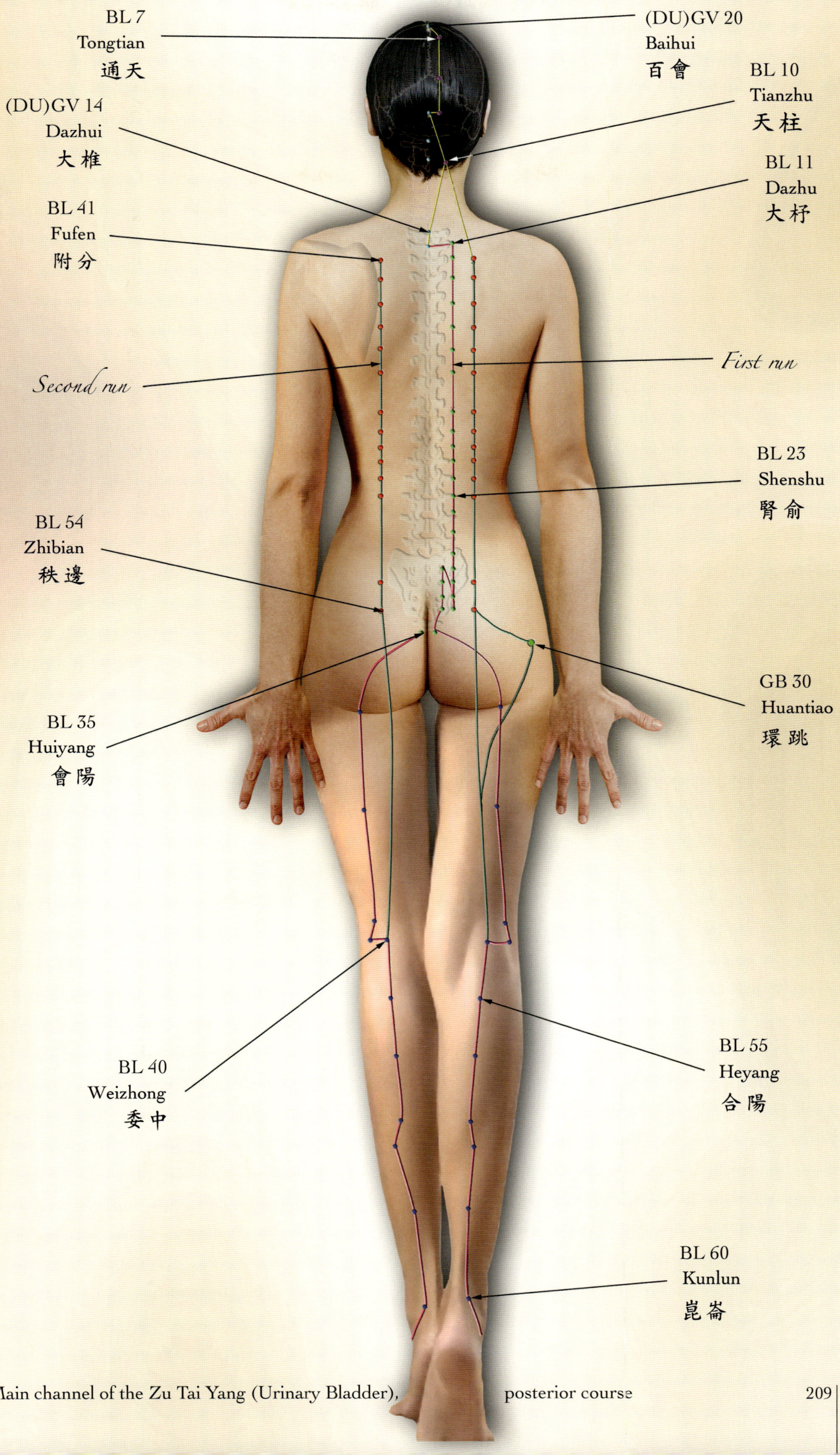

Main channel of the Zu Tai Yang (Urinary Bladder), posterior course

From point BL 10 Tianzhu emerge 2 superficial paravertebral branches, one internal, the other external.

- The internal branch passes by points (DU)GV 13 Taodao and (DU)GV 14 Dazhui to then follow a lateral course and reach the first point of the first chain of the Urinary Bladder at point BL 11 Dazhu, 1.5 Cun from the spinous process of T 1.

It goes down vertically, still 1.5 Cun from the posterior midline, reaches point BL 23 Shenshu, between L 2 and L 3, where a deep branch originates that goes inside to join Kidneys and Urinary Bladder.

From point BL 23, the channel continues its course to point BL 30, still 1.5 Cun from the posteriomedian line, on the level of the fourth sacral hole.

It goes up and passes by the first sacral hole, descends vertically, passes by each of the sacral holes, and goes to point BL 35 Huiyang situated horizontally, that passes by the upper part of the coccyx, 0.5 Cun outside the midline.

It keeps its course running obliquely to the outside and goes over the buttock.

It crosses the posterior part of the thigh and gets to the external extremity of the flex crease of the knee where point BL 39 Weiyang is and goes to the midpoint of the popliteal crease at point BL 40 Weichong.

- The external branch goes, from BL 10, to point BL 41 Fufen located 3 Cun outside the posteriomedian line, to form the second Urinary Bladder chain.

It goes down vertically along the external border of the scapula, still 3 Cun outside the posteriomedian line, to reach the level of the 4th sacral hole, where, according to some authors, it swerves to join point GB 30 Huantiao.

It keeps goins down along the external border of the posterior side of the thigh and joins the other internal branch at point BL 40 Weizhong.

From that point, the channel passes by the posterior side of the leg under the gemellus muscles.

It keeps descending and passes behind the external malleolus, travels along the external border of the foot and ends its course at the external ungueal angle of the little toe at point BL 67 Zhiyin.

Antero-superior cephalic course of channel Zu Tai Yang (Urinary Bladder)

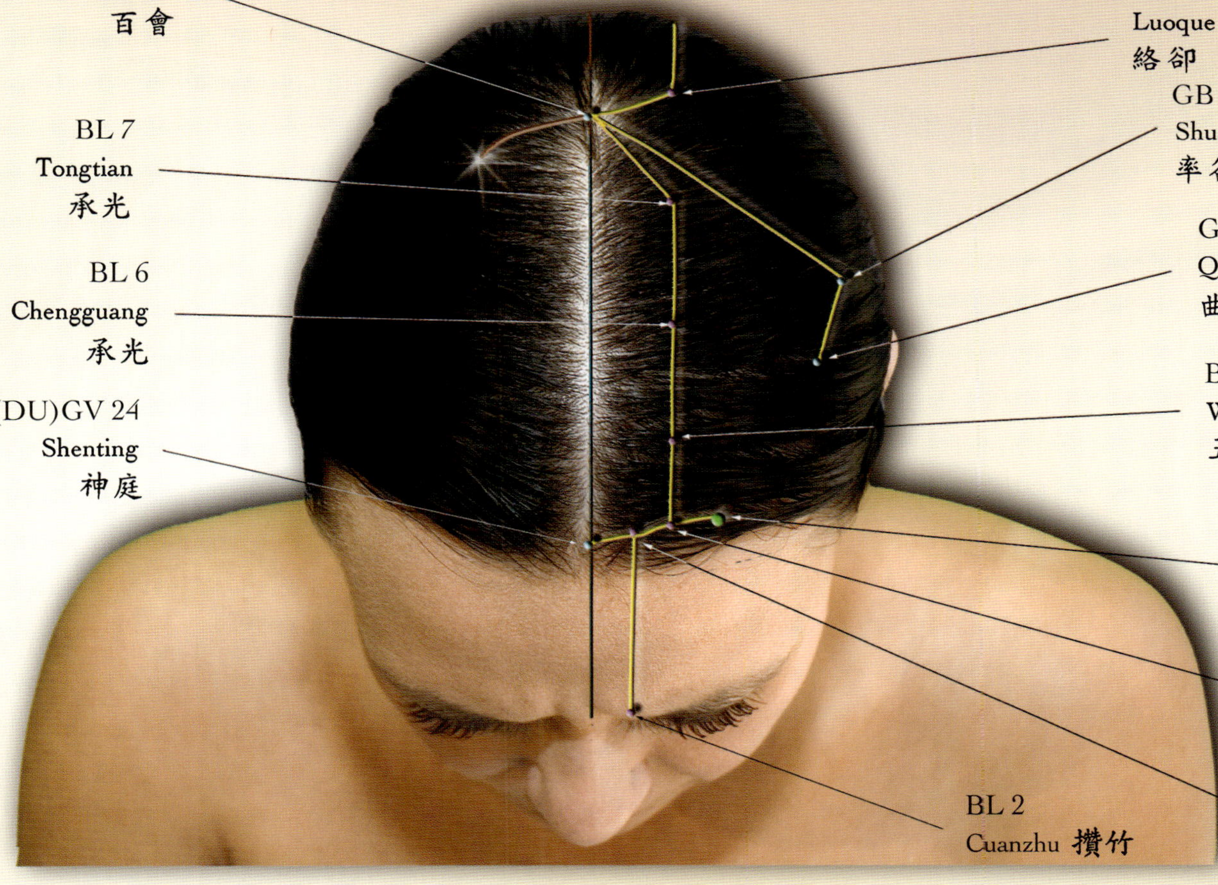

Complete antero-superior cephalic course of channel Zu Tai Yang (Urinary Bladder)

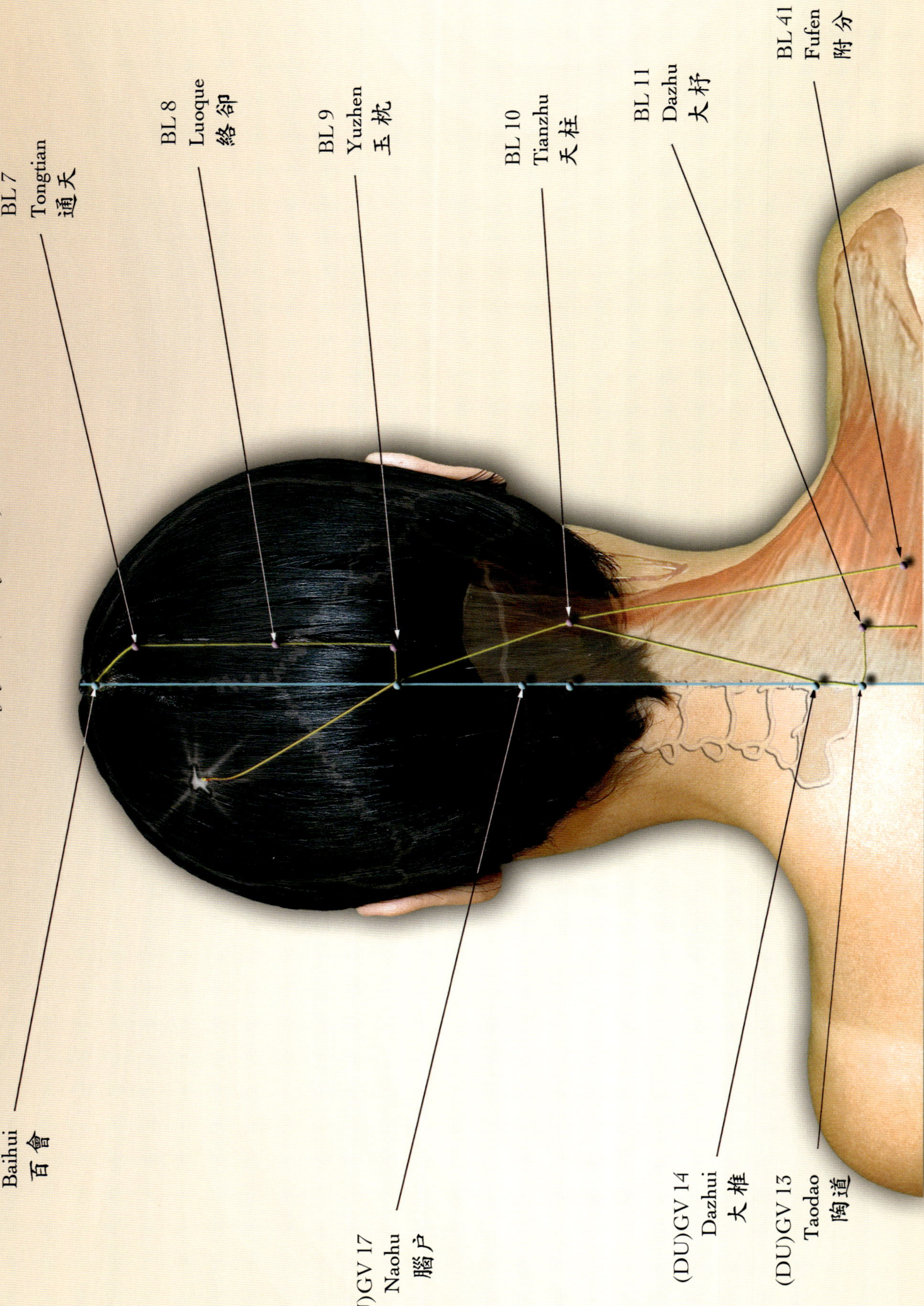

 Complete posterior cephalic course of channel Zu Tai Yang (Urinary Bladder)

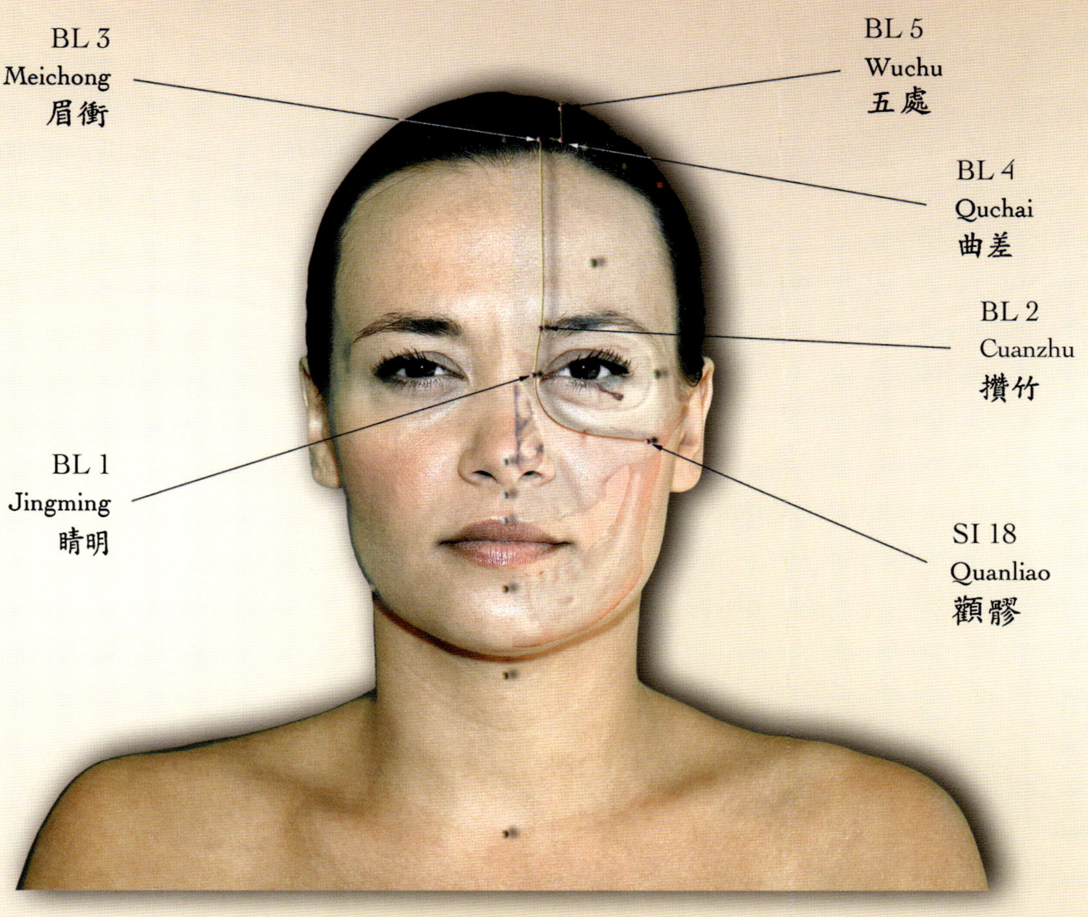

Facial cephalic course of channel Zu Tai Yang (Urinary Bladder)

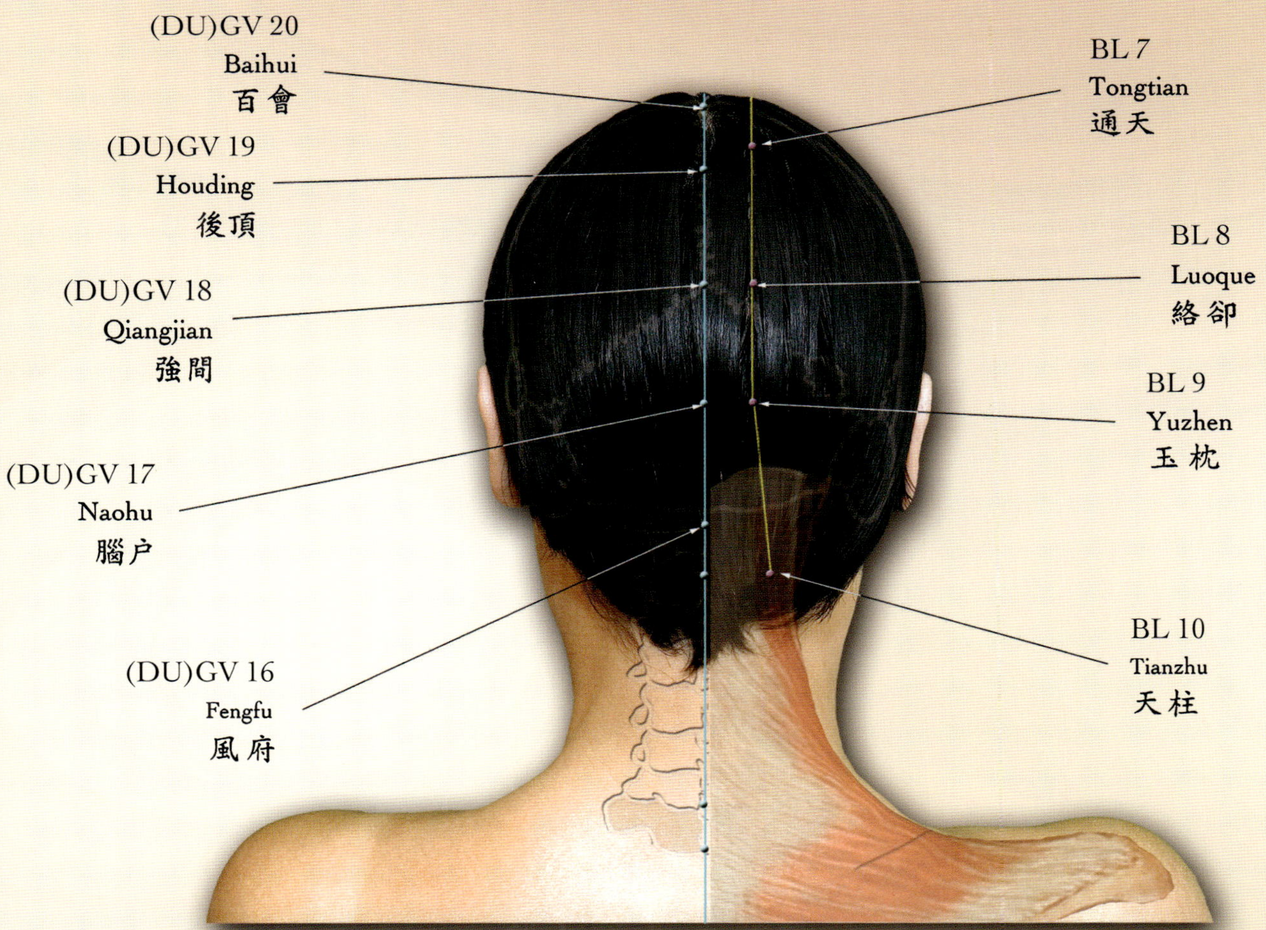

Posterior cephalic course of channel Zu Tai Yang (Urinary Bladder)

Internal branches of the Zu Tai Yang (Urinary Bladder)

***Ling Shu*, Chapter III, paragraph 10**

The internal branches are divided into 2 categories:

- the first one starts at the upper part of the cranium, at point (DU)GV 20 Baihui, where the channel penetrates the brain. The branch then goes out at point (DU)GV 17 Naohu.

- The second one leaves the first Bladder chain on the level of point BL 23 Shenshu, where it gets inside, to join Kidneys and Bladder.

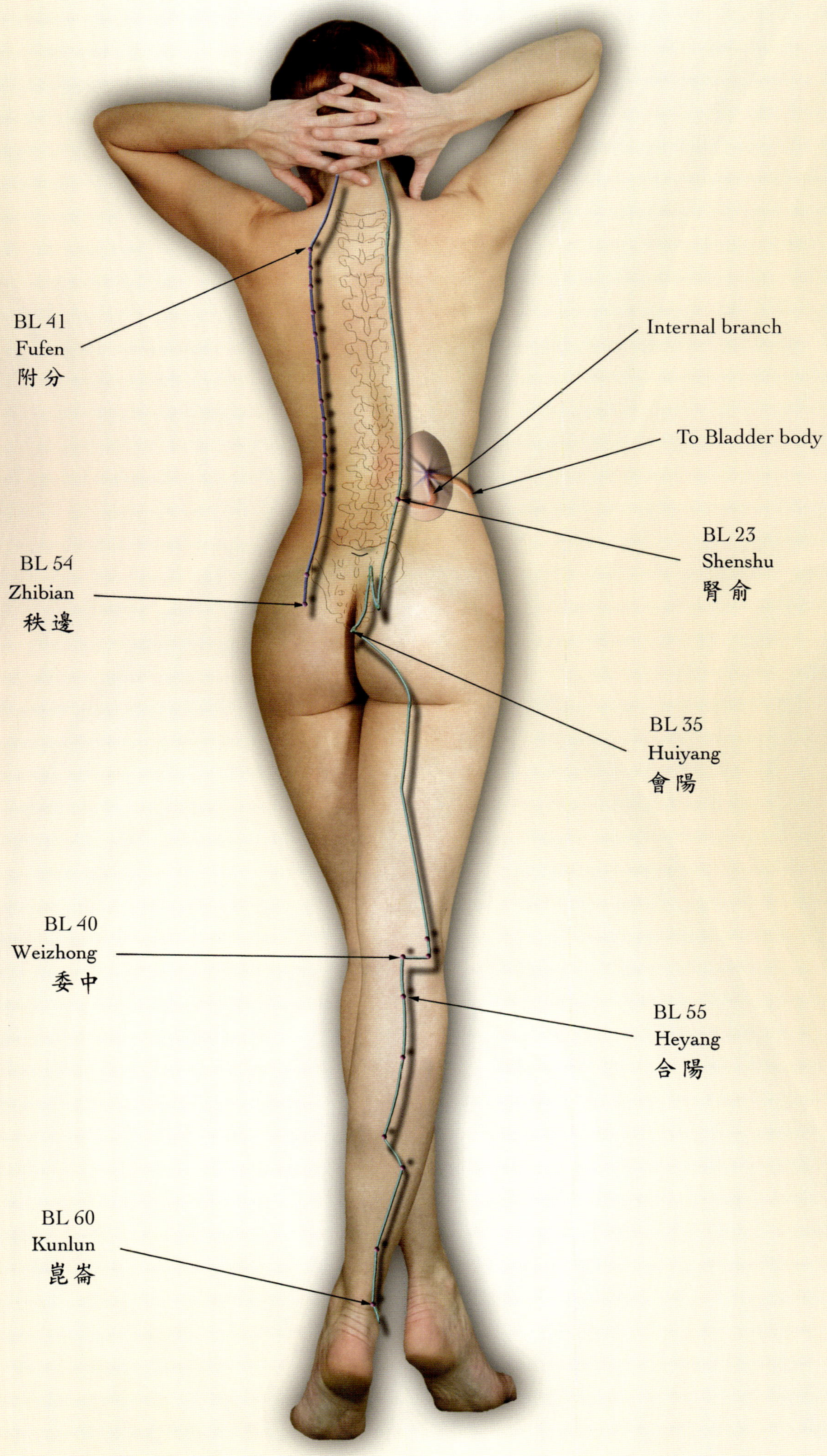

Internal branches of the Zu Tai Yang (Urinary Bladder)

Luo vessels of the Zu Tai Yang (Urinary Bladder)

Ling Shu, **chapter III, paragraph 10**

The Luo vessel of the Zu Tai Yang of the foot (Urinary Bladder) is called Fei Yang and goes out 7 Cun above the external malleolus.

It enters Zu Shao Yin (Kidney channel).

Commentaries

The Luo of the Urinary Bladder emerges from point BL 58 Fei Yang, located 7 Cun above point BL 60 Kunlun, on the posterior border of the fibula.

It is divided into 2 branches:

- The longitudinal Luo follows the Main channel of the Urinary Bladder to the head where it penetrates into the nose and the mouth.
- The transversal Luo connects with Zu Shao Yin channel (Kidney) and joins point KI 3 Taixi, situated in the depression between the internal malleolus and the tendon of the calcaneum, on the level of the tip of the internal malleolus.

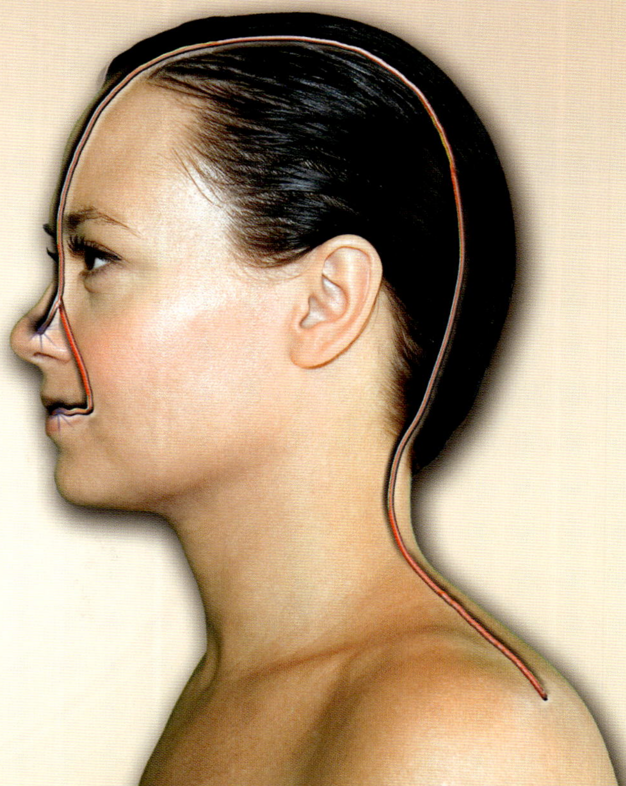

Cephalic course of longitudinal Luo of Zu Tai Yang (Urinary Bladder)

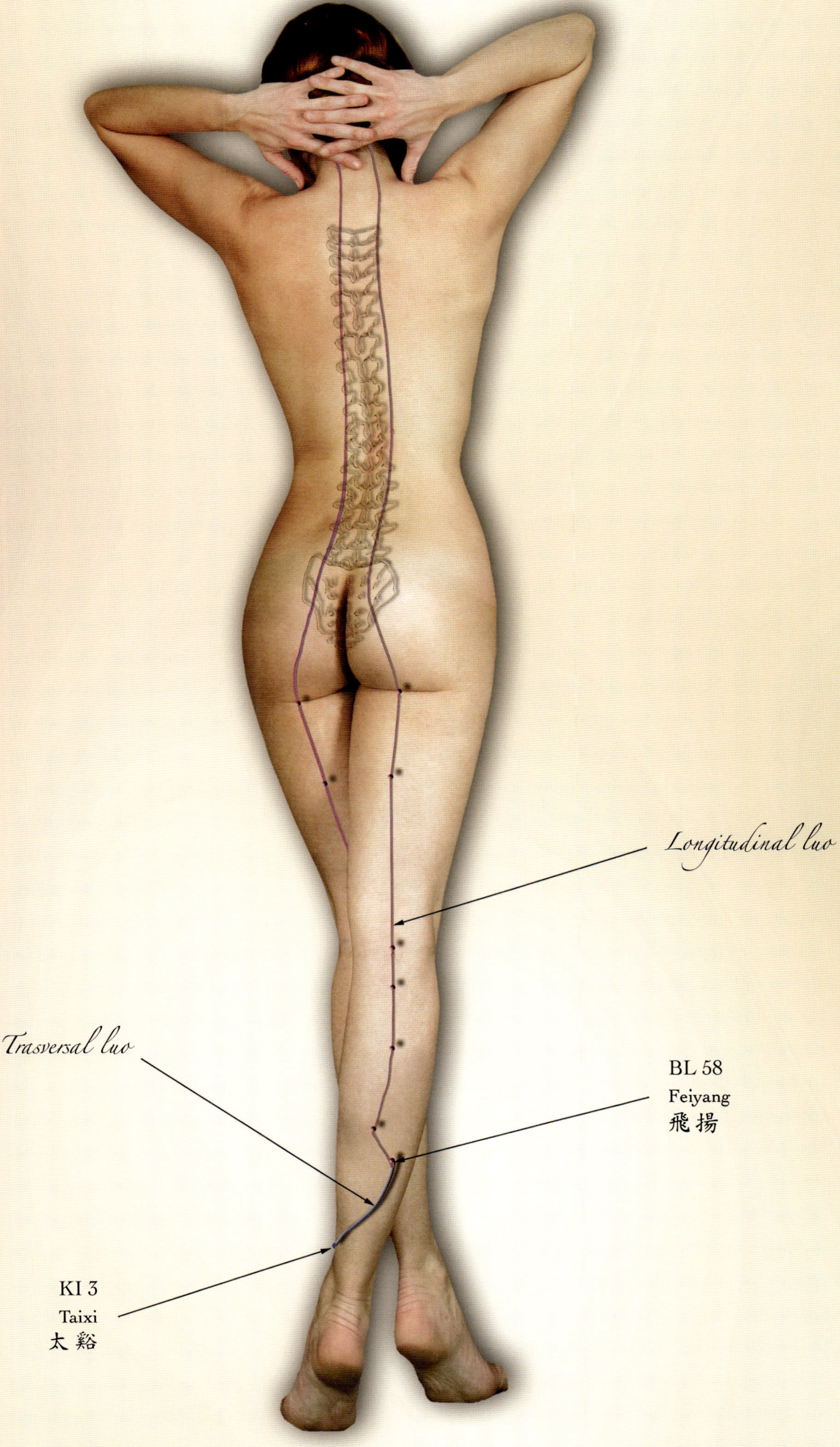

Course of Luo vessels of Zu Tai Yang (Urinary Bladder)

Distinct channel of the Zu Tai Yang (Urinary Bladder)

Ling Shu*, Chapter III, paragraph 11

From the regular vessel of the Main channel Zu Tai Yang of the foot (Bladder channel), a distinct branch gets inside at the midpoint of the popliteal crease (at point Weizhong, 40 of the Bladder).

It alters its course (goes up) 5 Cun below the sacrum.

It joins the anus (at point Chengfu, 36 of the Bladder, midpoint of the gluteal crease.

It belongs to the Urinary Bladder, spreads in the Kidneys while going along the rachis (the Lu, paravertebral muscles of the lumbar region.)

The direct branch of the rachis goes up to the nape.

It goes in and spreads in the cardial area.

It also belongs to the Zu Tai Yang (channel of the Urinary Bladder) forming with it a single vessel.

Commentaries

The distinct channel of Zu Tai Yang (Urinary Bladder) starts at point BL 40 Weizhong.

It goes up to arrive at the midpoint of the gluteal crease at point BL 36 Chengfu.

It gets to the anus where it gets inside, and joins Urinary Bladder and Kidneys.

It ascends along the backbone and reaches the Heart where it ramifies.

It keeps ascending to the neck where it emerges on the surface to join its own Main channel and the distinct channel of Zu Shao Yin (Kidney) at point BL 10 Tianzhu, thus forming the first union.

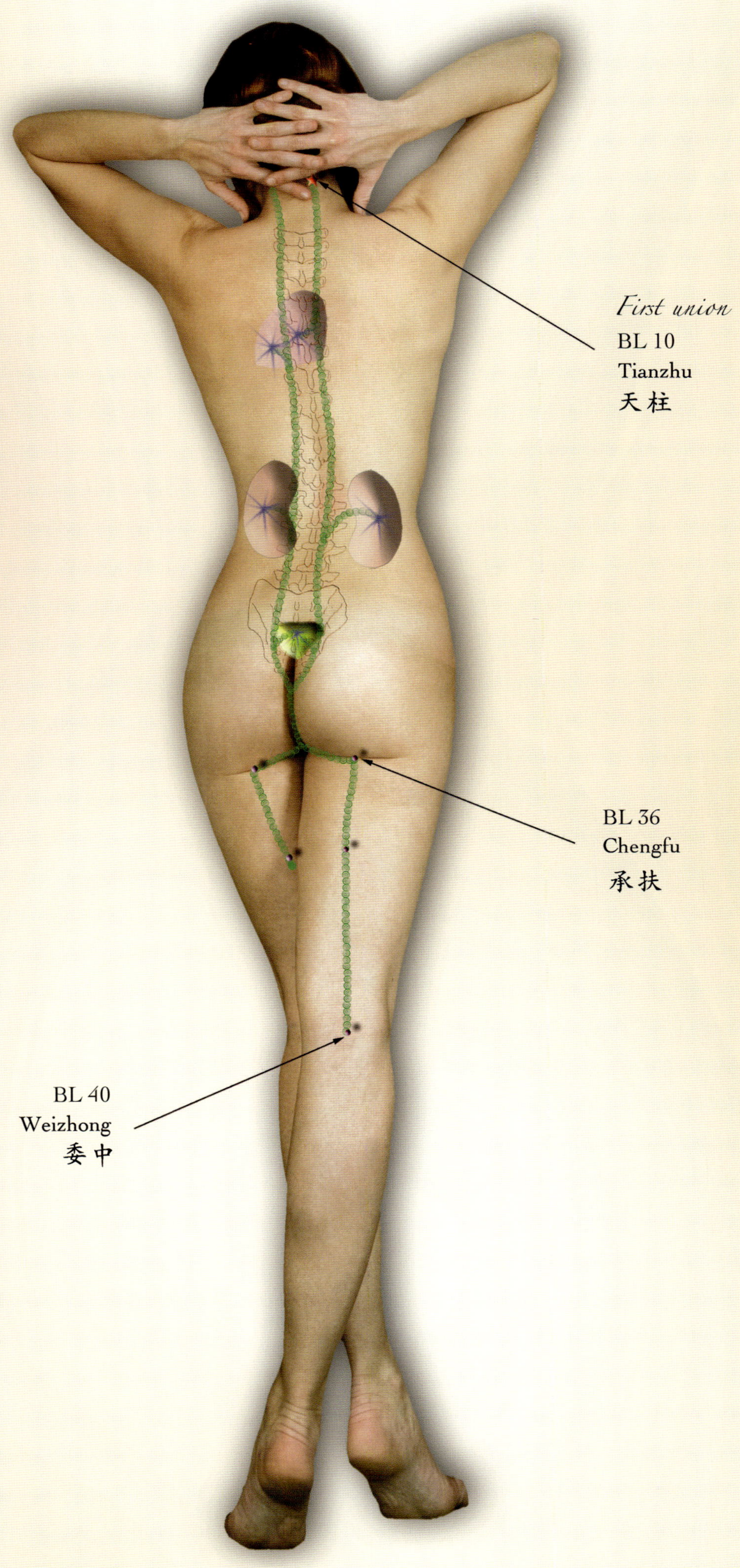

Course of distinct channel of Zu Tai Yang (Urinary Bladder)

Points of the Main channel Zu Tai Yang (Urinary Bladder)

BL 1 Jingming
On the face, in the depression slightly above the inner canthus.

BL 2 Cuanzhu
On the face, in the depression of the medial end of the eyebrow, at the supraorbital notch.

BL 3 Meichong
On the head, directly above BL 2 Cuanzhu, 0.5 Cun above the anterior hairline, on the line connecting (DU)GV 24 Shenting and BL 4 Qucha.

BL 4 Qucha
On the head, 0.5 Cun directly above the midpoint of the anterior hairline and 1.5 Cun lateral to the midline, at the junction of the medial third and middle third of the line connecting (DU)GV 24 Shenting and ST 8 Touwei.

BL 5 Wuchu
On the head, 1 Cun directly above the midpoint of the anterior hairline and 1.5 Cun lateral to the midline.

BL 6 Chengguang
On the head, 2.5 Cun directly above the midpoint of the anterior hairline and 1.5 Cun lateral to the midline.

BL 7 Tongtian
On the head, 4 Cun directly above the midpoint of the anterior hairline and 1.5 Cun lateral to the midline.

BL 8 Luoque
On the head, 5.5 Cun directly above the midpoint of the anterior hairline and 1.5 Cun lateral to the midline.

BL 9 Yuzhen
On the occiput, 2.5 Cun directly above the midpoint of the posterior hairline and 1.3 Cun lateral to the midline, in the depression on the level of the upper border of the external occipital protuberance.

BL 10 Tianzhu
On the nape, in the depression of the lateral border of the trapezius muscle and 1.3 Cun lateral to the midpoint of the posterior hairline.

BL 11 Dazhu
On the back, below the spinous process of the 1st thoracic vertebra, 1.5 Cun lateral to the posterior midline.

BL 12 Fengmen
On the back, below the spinous process of the 2nd thoracic vertebra, 1.5 Cun lateral to the posterior midline.

BL 13 Feishu
On the back, below the spinous process of the 3rd thoracic vertebra, 1.5 Cun lateral to the posterior midline.

BL 14 Jueyinshu
On the back, below the spinous process of the 4th thoracic vertebra, 1.5 Cun lateral to the posterior midline.

BL 15 Xinshu
On the back, below the spinous process of the 5th thoracic vertebra, 1.5 Cun lateral to the posterior midline.

BL 16 Dushu
On the back, below the spinous process of the 6th thoracic vertebra, 1.5 Cun lateral to the posterior midline.

BL 17 Geshu
On the back, below the spinous process of the 7th thoracic vertebra, 1.5 Cun lateral to the posterior midline.

BL 18 Ganshu
On the back, below the spinous process of the 9th thoracic vertebra, 1.5 Cun lateral to the posterior midline.

BL 19 Danshu
On the back, below the spinous process of the 10th thoracic vertebra, 1.5 Cun lateral to the posterior midline.

BL 20 Pishu
On the back, below the spinous process of the 11th thoracic vertebra, 1.5 Cun lateral to the posterior midline.

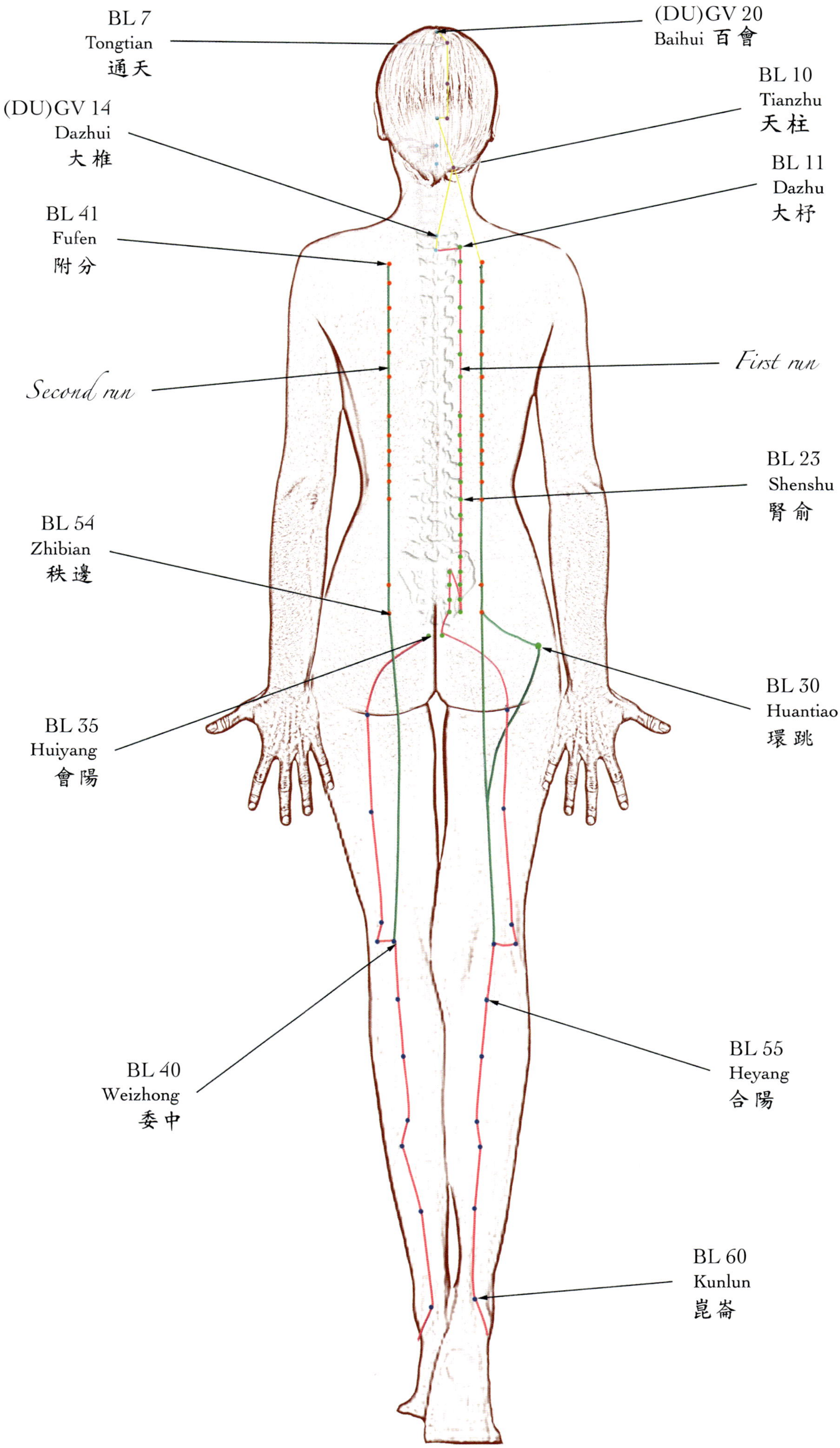

Points the Main channel Zu Tai Yang (Urinary Bladder)

BL 21 Weishu
On the back, below the spinous process of the 12th thoracic vertebra, 1.5 Cun lateral to the posterior midline.

BL 22 Sanjiaoshu
On the low back, below the spinous process of the 1st lumbar vertebra, 1.5 Cun lateral to the posterior midline.

BL 23 Shenshu
On the low back, below the spinous process of the 2nd lumbar vertebra, 1.5 Cun lateral to the posterior midline.

BL 24 Qihaishu
On the low back, below the spinous process of the 3rd lumbar vertebra, 1.5 Cun lateral to the posterior midline.

BL 25 Dachangshu
On the low back, below the spinous process of the 4th lumbar vertebra, 1.5 Cun lateral to the posterior midline.

BL 26 Guanyuanshu
On the low back, below the spinous process of the 5th lumbar vertebra, 1.5 Cun lateral to the posterior midline.

BL 27 Xiaochangshu
On the sacrum and on the level of the 1st posterior sacral foramen, 1.5 Cun lateral to the median sacral crest.

BL 28 Pangguangshu
On the sacrum and on the level of the 2nd posterior sacral foramen, 1.5 Cun lateral to the median sacral crest.

BL 29 Zhongliishu
On the sacrum and on the level of the 3rd posterior sacral foramen, 1.5 Cun lateral to the median sacral crest.

BL 30 Baihuanshu
On the sacrum and on the level of the 4th posterior sacral foramen, 1.5 Cun lateral to the median sacral crest.

BL 31 Shangliao
On the sacrum, at the midpoint between the posteriosuperior iliac spine and the posterior midline, just at the 1st posterior sacral foramen.

BL 32 Ciliao
On the sacrum, medial and inferior to the posterio-superior iliac spine, just at the 2nd posterior sacral foramen.

BL 33 Zhongliao
On the sacrum, medial and inferior to BL 32 Ciliao, just at the 3rd posterior sacral foramen.

BL 34 Xialiao
On the sacrum, medial and inferior to BL 33 Zhongliao, just at the 4th posterior sacral foramen.

BL 35 Huiyang
On the sacrum, 0.5 Cun lateral to the tip of the coccyx.

BL 36 Chengfu
On the posterior side of the thigh, at the midpoint of the inferior gluteal crease.

BL 37 Yinmen
On the posterior side of the thigh and on the line connecting BL 36 Chengfu and BL 40 Weizhong, 6 Cun below BL 36 Chengfu.

BL 38 Fuxi
At the lateral end of the popliteal crease, 1 Cun above BL 39 Weiyang, medial to the tendon of the biceps muscle of the thigh.

BL 39 Weiyang
At the lateral end of the popliteal crease, medial to the tendon of the biceps muscle of the thigh.

BL 40 Weizhong
At the midpoint of the popliteal crease, between the tendons of the biceps muscle of the thigh and the semitendinous muscle.

BL 41 Fufen
On the back, below the spinous process of the 2nd thoracic vertebra, 3 Cun lateral to the posterior midline.

BL 42 Pohu
On the back, below the spinous process of the 3rd thoracic vertebra, 3 Cun lateral to the posterior midline.

BL 43 Gaohuang
On the back, below the spinous process of the 4th thoracic vertebra, 3 Cun lateral to the posterior midline.

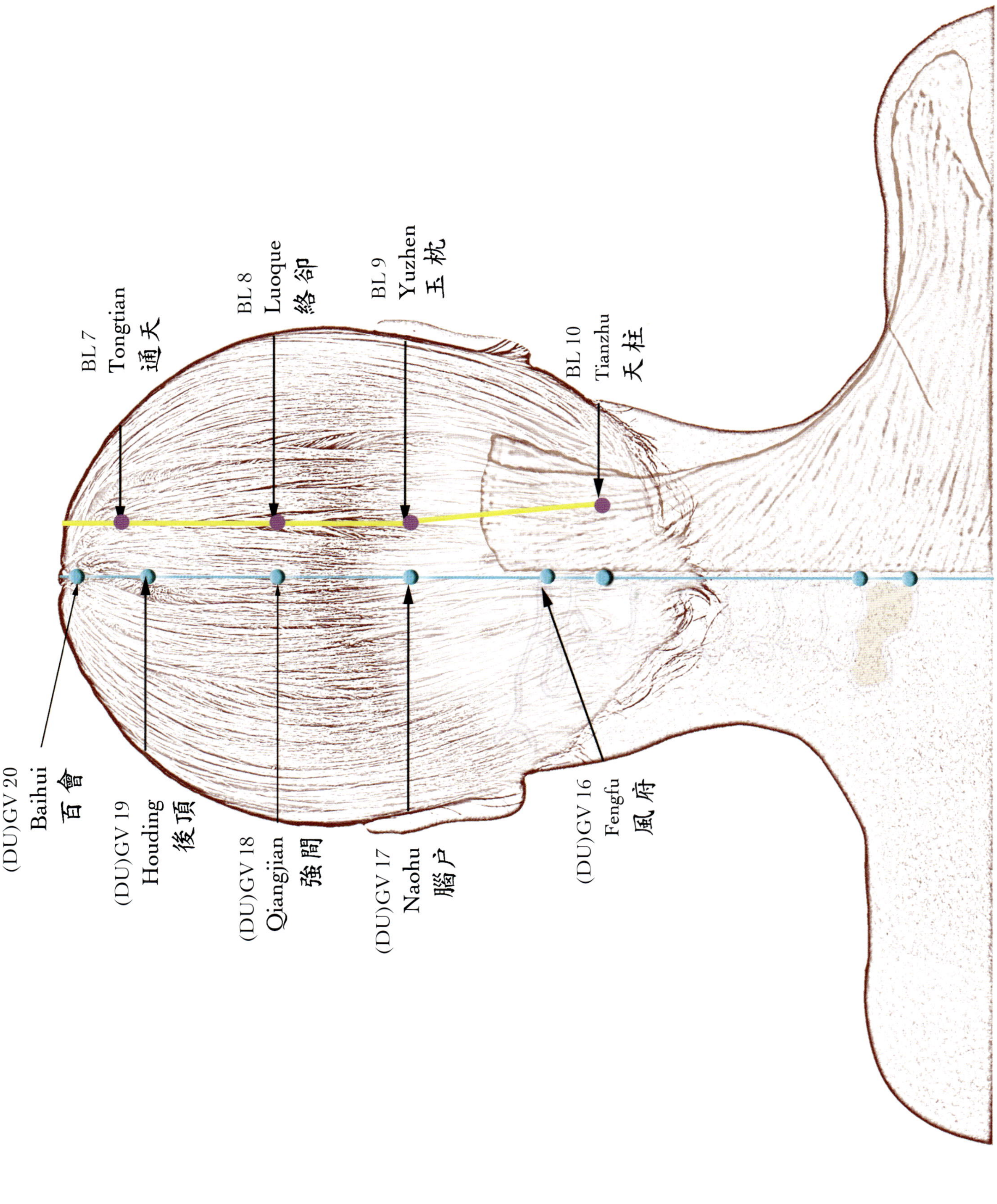

Posterior cephalic points of channel Zu Tai Yang (Urinary Bladder)

BL 44 Shentang
On the back, below the spinous process of the 5th thoracic vertebra, 3 Cun lateral to the posterior midline.

BL 45 Yixi
On the back, below the spinous process of the 6th thoracic vertebra, 3 Cun lateral to the posterior midline.

BL 46 Geguan
On the back, below the spinous process of the 7th thoracic vertebra, 3 Cun lateral to the posterior midline.

BL 47 Hunmen
On the back, below the spinous process of the 9th thoracic vertebra, 3 Cun lateral to the posterior midline.

BL 48 Yanggang
On the back, below the spinous process of the 10th thoracic vertebra, 3 Cun lateral to the posterior midline.

BL 49 Yishe
On the back, below the spinous process of the 11th thoracic vertebra, 3 Cun lateral to the posterior midline.

BL 50 Weicang
On the back, below the spinous process of the 12th thoracic vertebra, 3 Cun lateral to the posterior midline.

BL 51 Huangmen
On the low back, below the spinous process of the 1st lumbar vertebra, 3 Cun lateral to the posterior midline.

BL 52 Zhishi
On the low back, below the spinous process of the 2nd lumbar vertebra, 3 Cun lateral to the posterior midline.

BL 53 Baohuang
On the buttock and on the level of the 2nd posterior sacral foramen, 3 Cun lateral to the median sacral crest.

BL 54 Zhibian
On the buttock and on the level of the 4th posterior sacral foramen, 3 Cun lateral to the median sacral crest.

BL 55 Heyang
On the posterior side of the leg and on the line connecting BL 40 Weizhong and BL 57 Chengshan, 2 Cun below BL 40 Weizhong.

BL 56 Chengjin
On the posterior side of the leg and on the line connecting BL 40 Weizhong and BL 57 Chengshan, at the centre of the gastrocnemius muscle belly, 5 Cun below BL 40 Weizhong.

BL 57 Chengshan
On the posterior midline of the leg, between BL 40 Weizhong and BL 60 Kunlun, in a pointed depression formed below the gastrocnemius muscle belly when the leg is stretched or the heel is lifted.

BL 58 Fei Yang
On the posterior side of the leg, 7 Cun directly above BL 60 Kunlun and 1 Cun lateral and inferior to BL 57 Chengshan.

BL 59 Fu Yang
On the posterior side of the leg, posterior to the lateral malleolus, 3 Cun directly above BL 60 Kunlun.

BL 60 Kunlun
Posterior to the lateral malleolus, in the depression between the tip of the external malleolus and Achilles tendon.

BL 61 Pucan
On the lateral side of the foot, posterior and inferior to the external malleolus, directly below BL 60 Kunlun, lateral to the calcaneum, at the 'junction of the red and white skin'.

BL 62 Shenmai
On the lateral side of the foot, in the depression directly below the external malleolus.

BL 63 Jinmen
On the lateral side of the foot, directly below the anterior border of the external malleolus, on the lower border of the cuboid bone.

BL 64 Jinggu
On the lateral side of the foot, below the tuberosity of the 5th metatarsal bone, at the 'junction of the red and white skin'.

BL 65 Shugu
On the lateral side of the foot, posterior to the 5th metatarso-phalangeal joint, at the 'junction of the red and white skin'.

BL 66 Zutonggu
On the lateral side of the foot, anterior to the 5th metatarso-phalangeal joint, at the 'junction of the red and white skin'.

BL 67 Zhiyin
On the lateral side of the distal segment of the little toe, 0.1 Cun from the corner of the toenail.

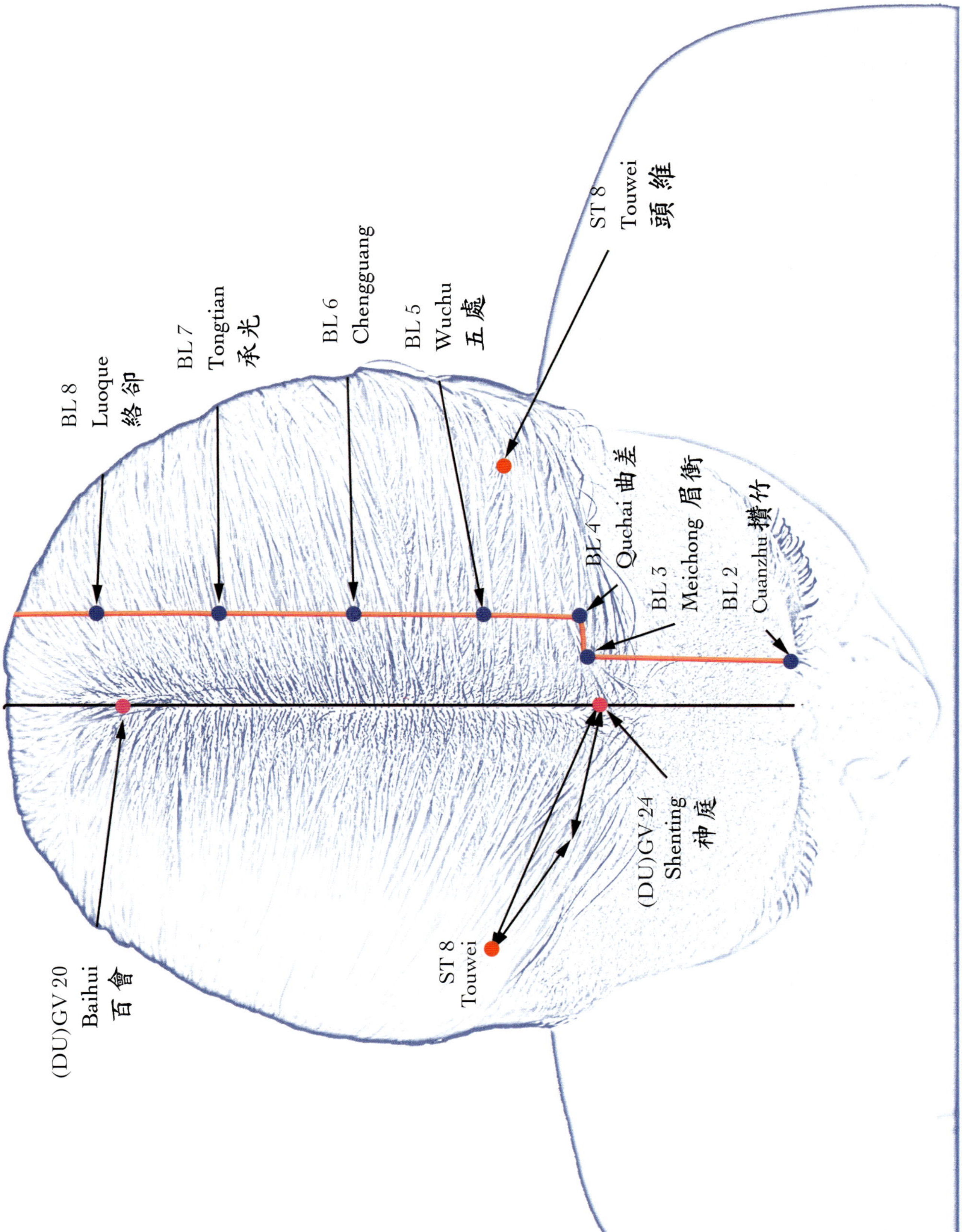

Anterior cephalic points of channel Zu Tai Yang (Urinary Bladder)

BL 10 Tianzhu
On the nape, in the depression of the lateral border of the trapezius muscle and 1.3 Cun lateral to the midpoint of the posterior hairline.

BL 11 Dazhu
On the back, below the spinous process of the 1st thoracic vertebra, 1.5 Cun lateral to the posterior midline.

BL 23 Shenshu
On the low back, below the spinous process of the 2nd lumbar vertebra, 1.5 Cun lateral to the posterior midline.

BL 35 Huiyang
On the sacrum, 0.5 Cun lateral to the tip of the coccyx.

BL 40 Weizhong
At the midpoint of the popliteal crease, between the tendons of the biceps muscle of the thigh and the semi-tendinous muscle.

BL 41 Fufen
On the back, below the spinous process of the 2nd thoracic vertebra, 3 Cun lateral to the posterior midline.

BL 54 Zhibian
On the buttock and on the level of the 4th posterior sacral foramen, 3 Cun lateral to the median sacral crest.

BL 55 Heyang
On the posterior side of the leg and on the line connecting BL 40 Weizhong and BL 57 Chengshan, 2 Cun below BL 40 Weizhong.

BL 60 Kunlun
Posterior to the lateral malleolus, in the depression between the tip of the external malleolus and Achilles tendon.

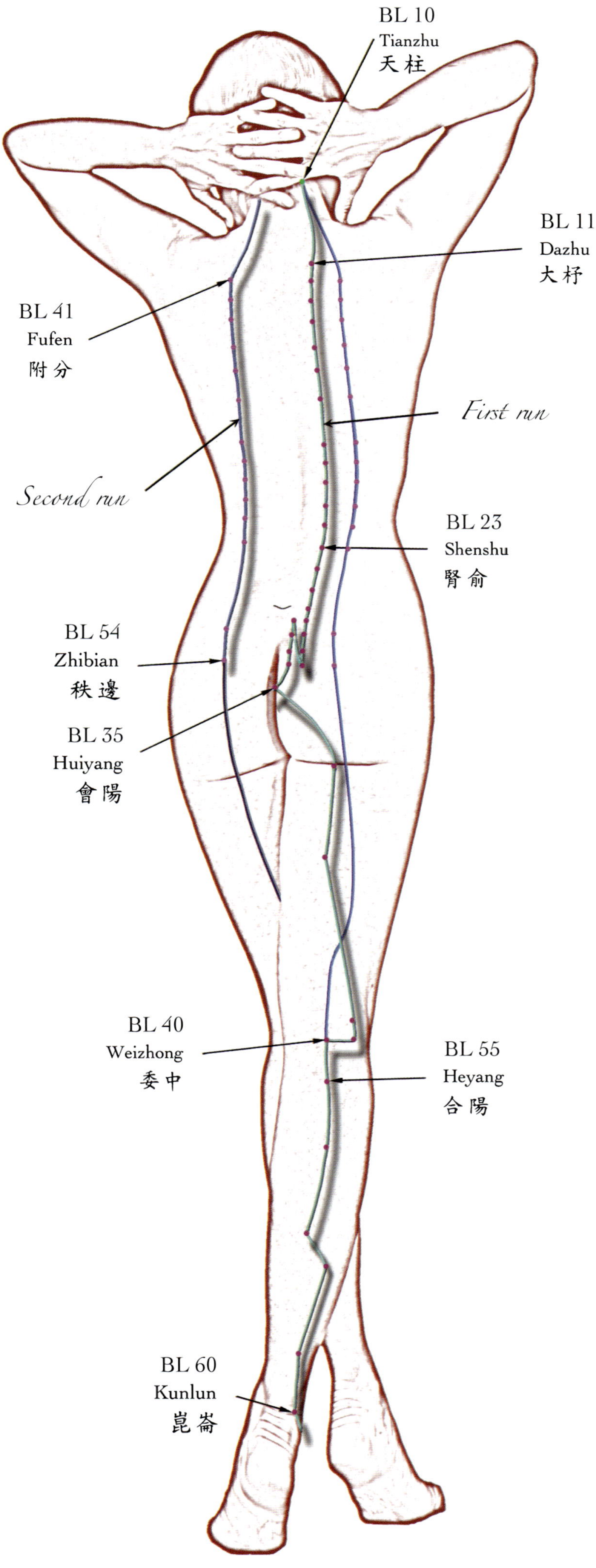
BL 10
Tianzhu
天柱
BL 11
Dazhu
大杼
BL 41
Fufen
附分
First run
Second run
BL 23
Shenshu
腎俞
BL 54
Zhibian
秩邊
BL 35
Huiyang
會陽
BL 40
Weizhong
委中
BL 55
Heyang
合陽
BL 60
Kunlun
崑崙

A summary of the channels

- **The tendino-muscular channels**
  - The 3 upper Yin
  - The 3 upper Yang
  - The 3 lower Yin
  - The 3 lower Yang
- **Main channels**
  - The 3 upper Yin
  - The 3 upper Yang (head)
  - The 3 lower Yin
  - The 3 lower Yang (head)
- **Cutaneous territories**
  - Tai Yang and Shao Yang
  - Yang Ming and Tai Yin
  - Jue Yin and Shao Yin

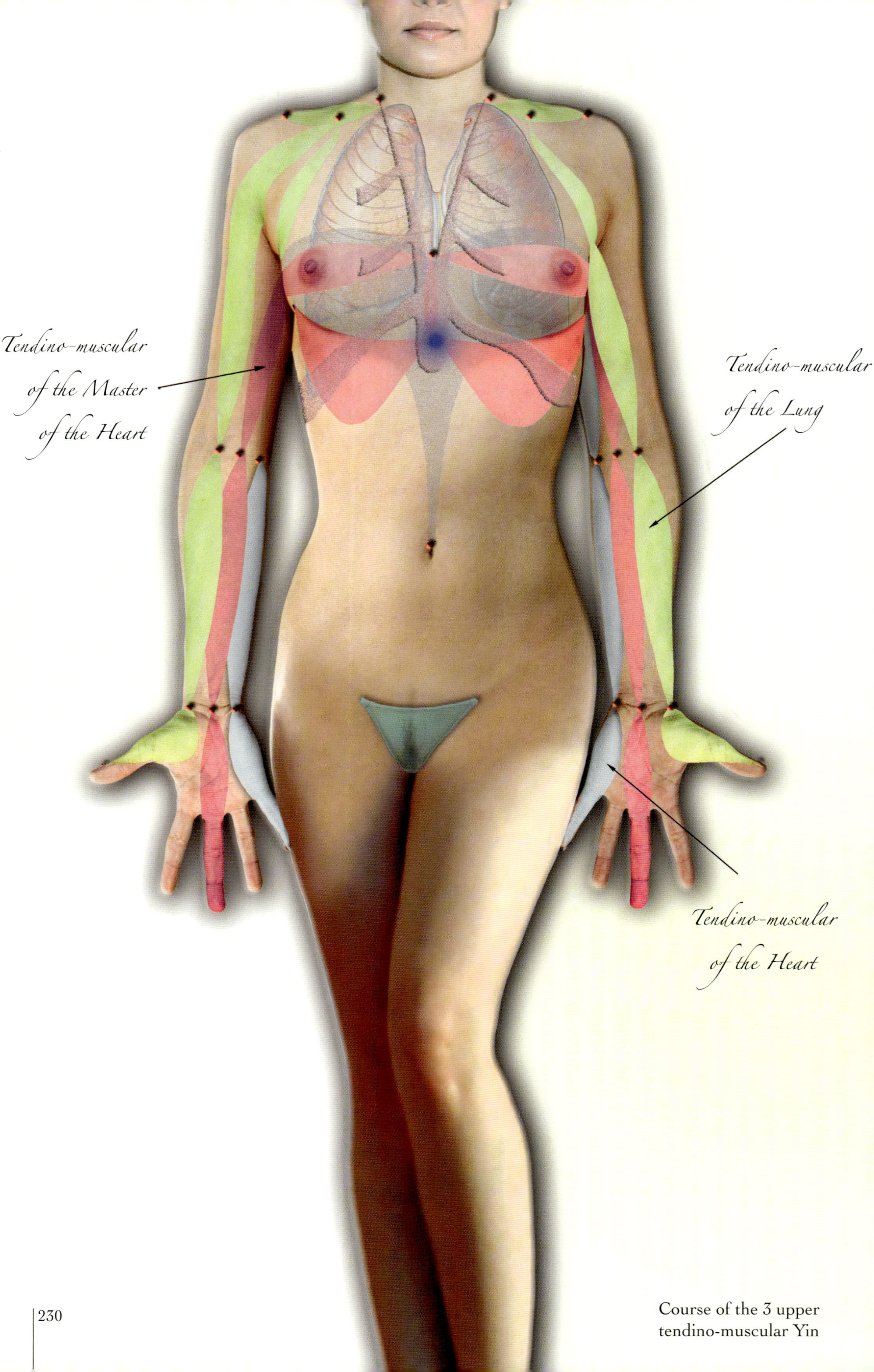

Course of the 3 upper tendino-muscular Yin

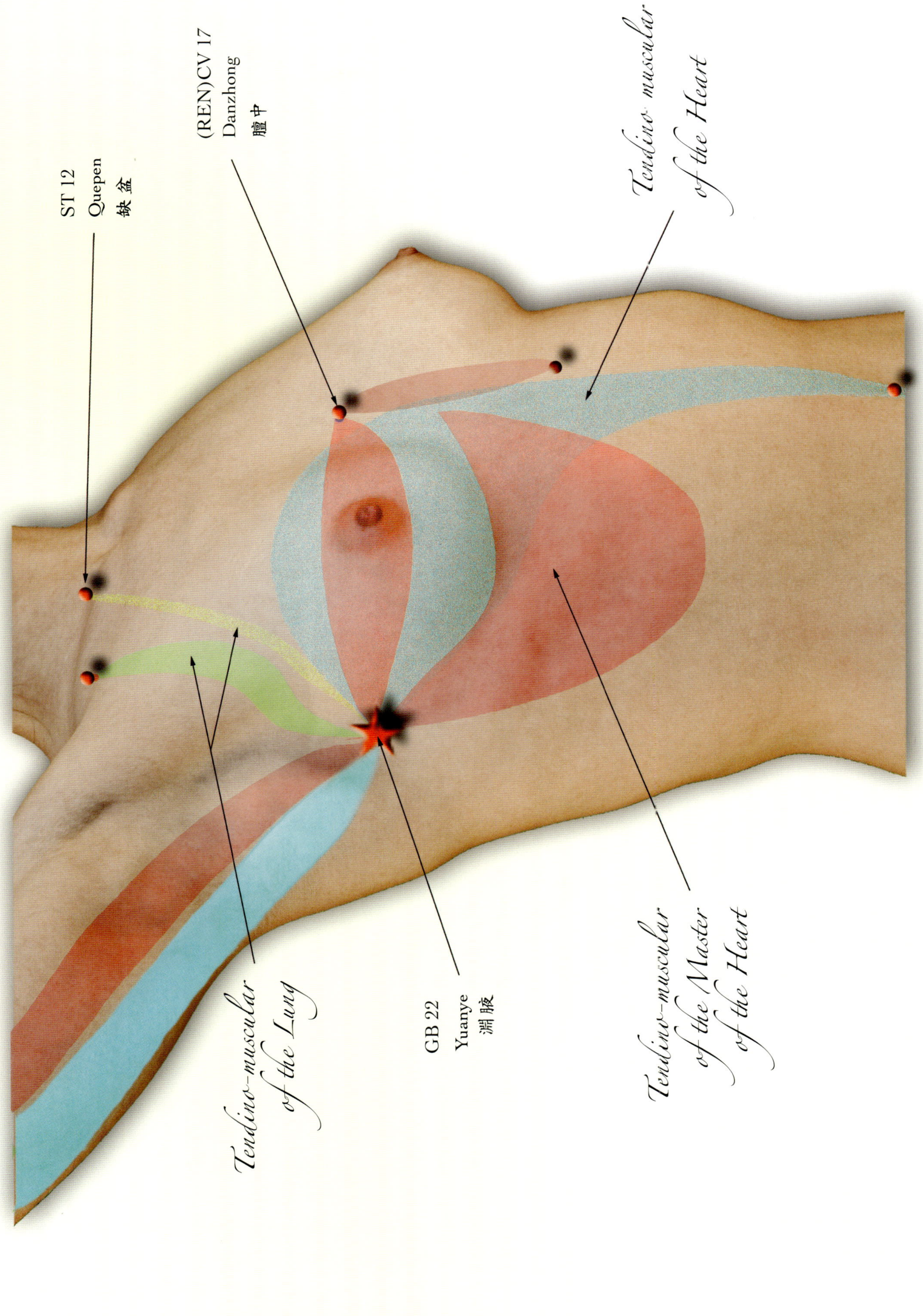

Junction of the 3 upper tendino-muscular Yin

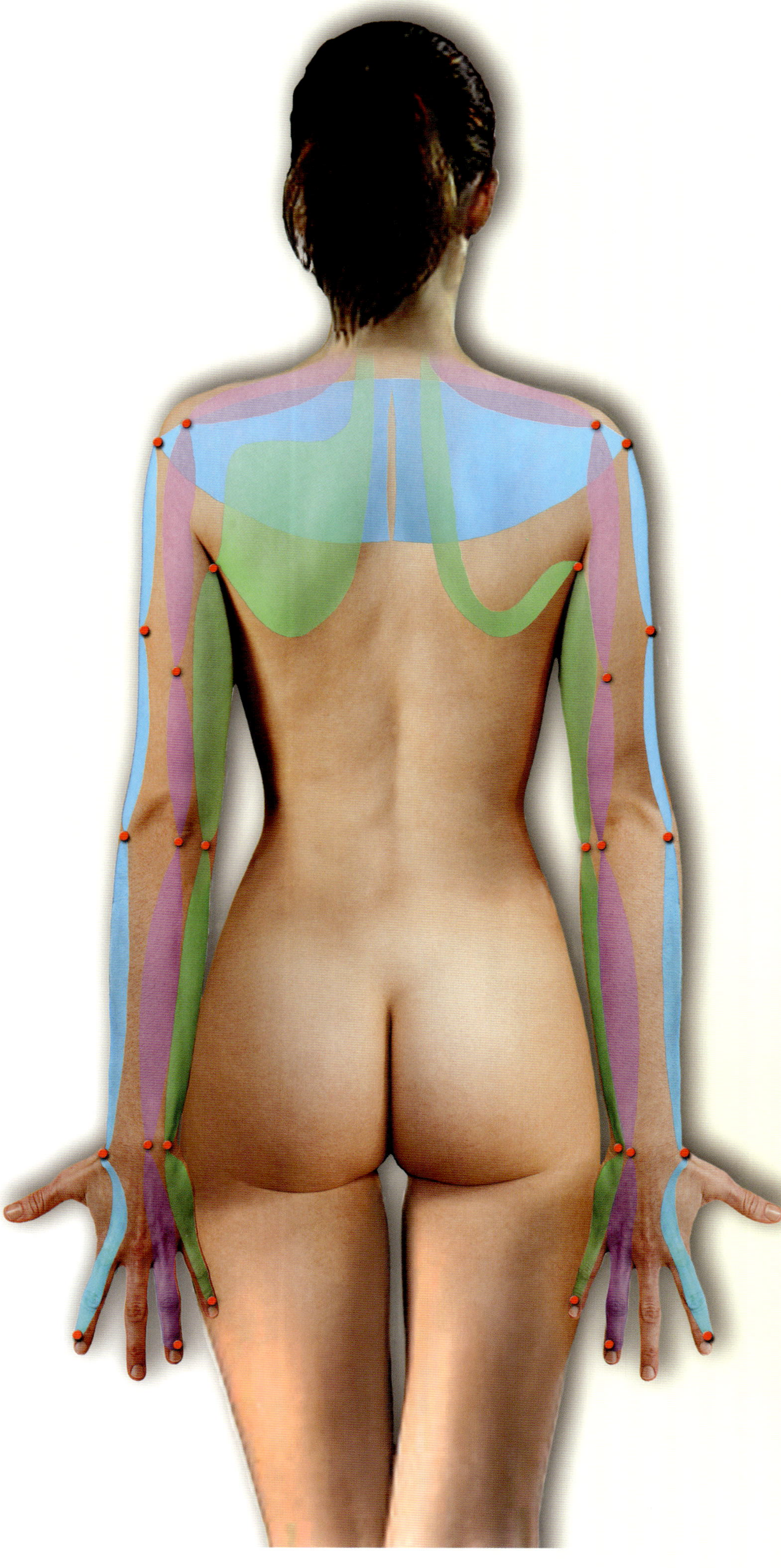

Course of 3 upper tendino-muscular Yang

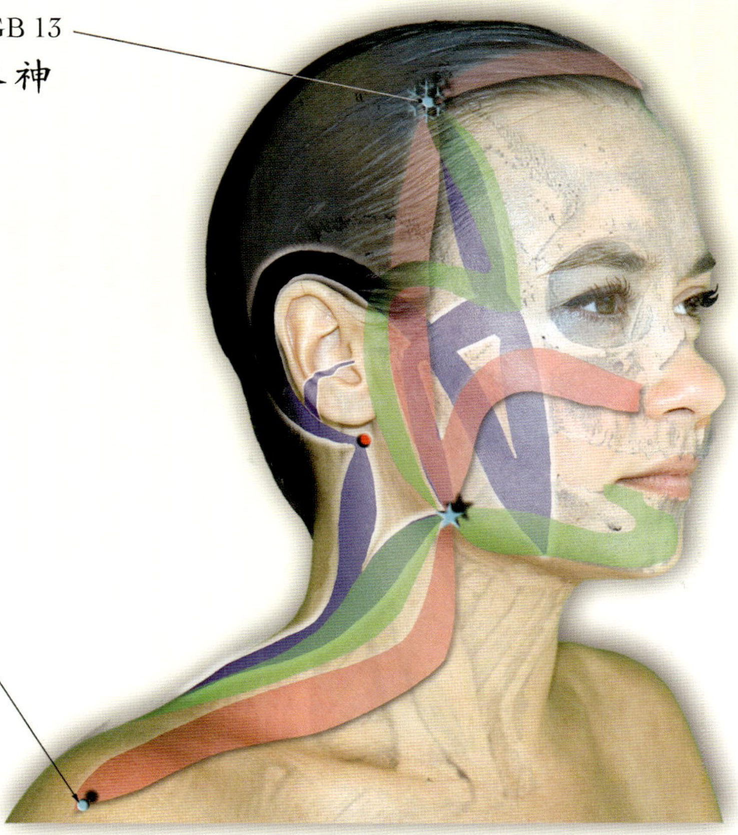

Cephalic course of 3 upper tendino-muscular Yang

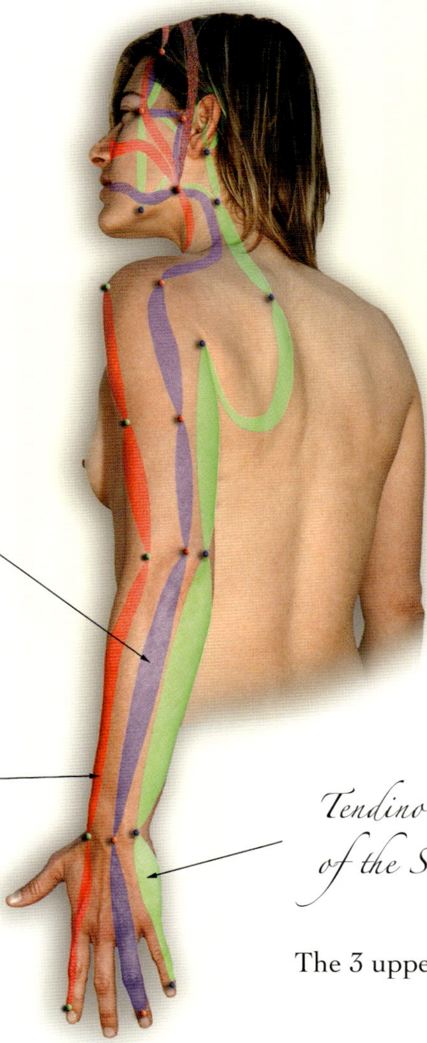

The 3 upper tendino-muscular Yang

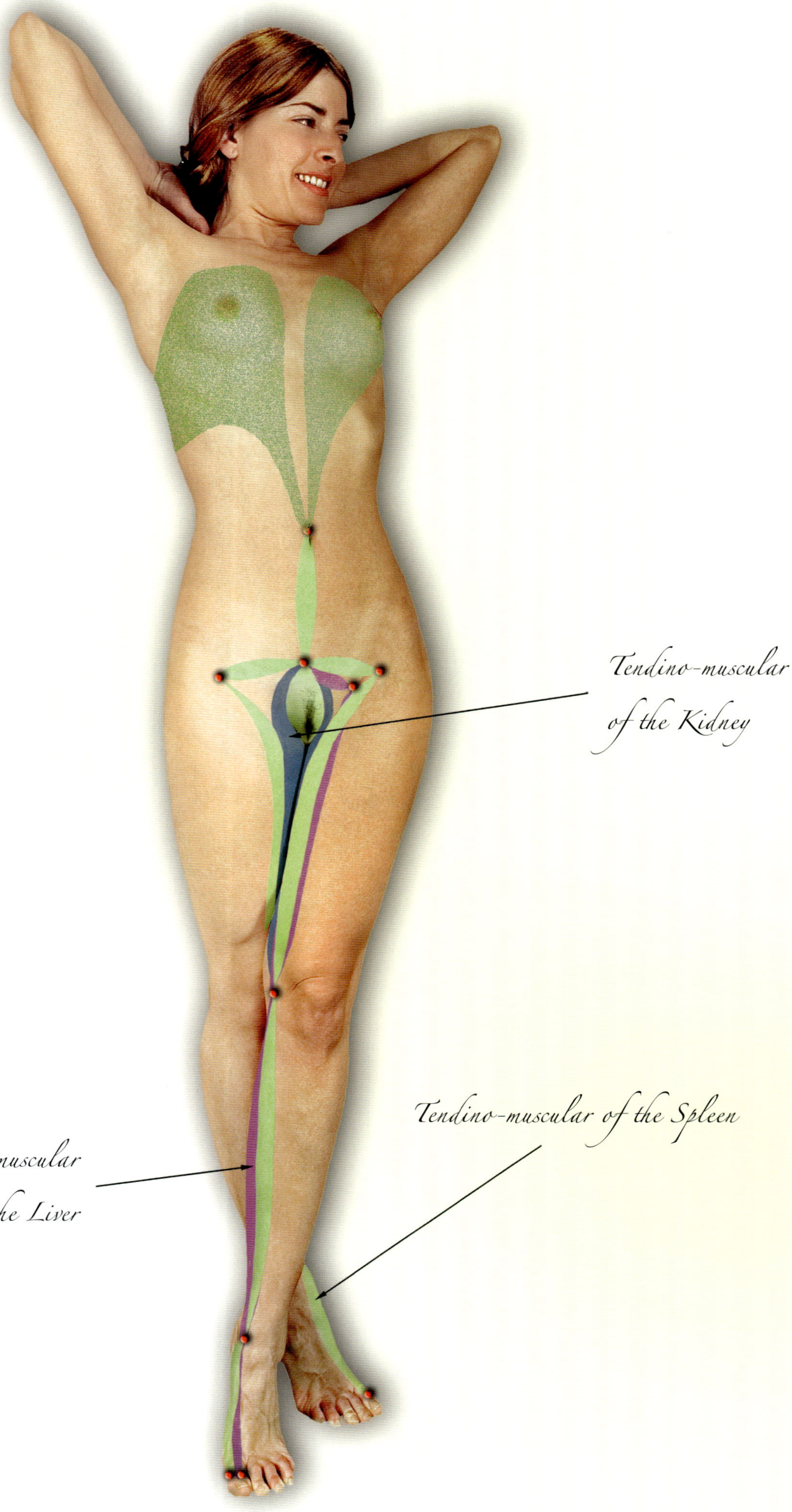

Course of 3 lower tendino-muscular Yin

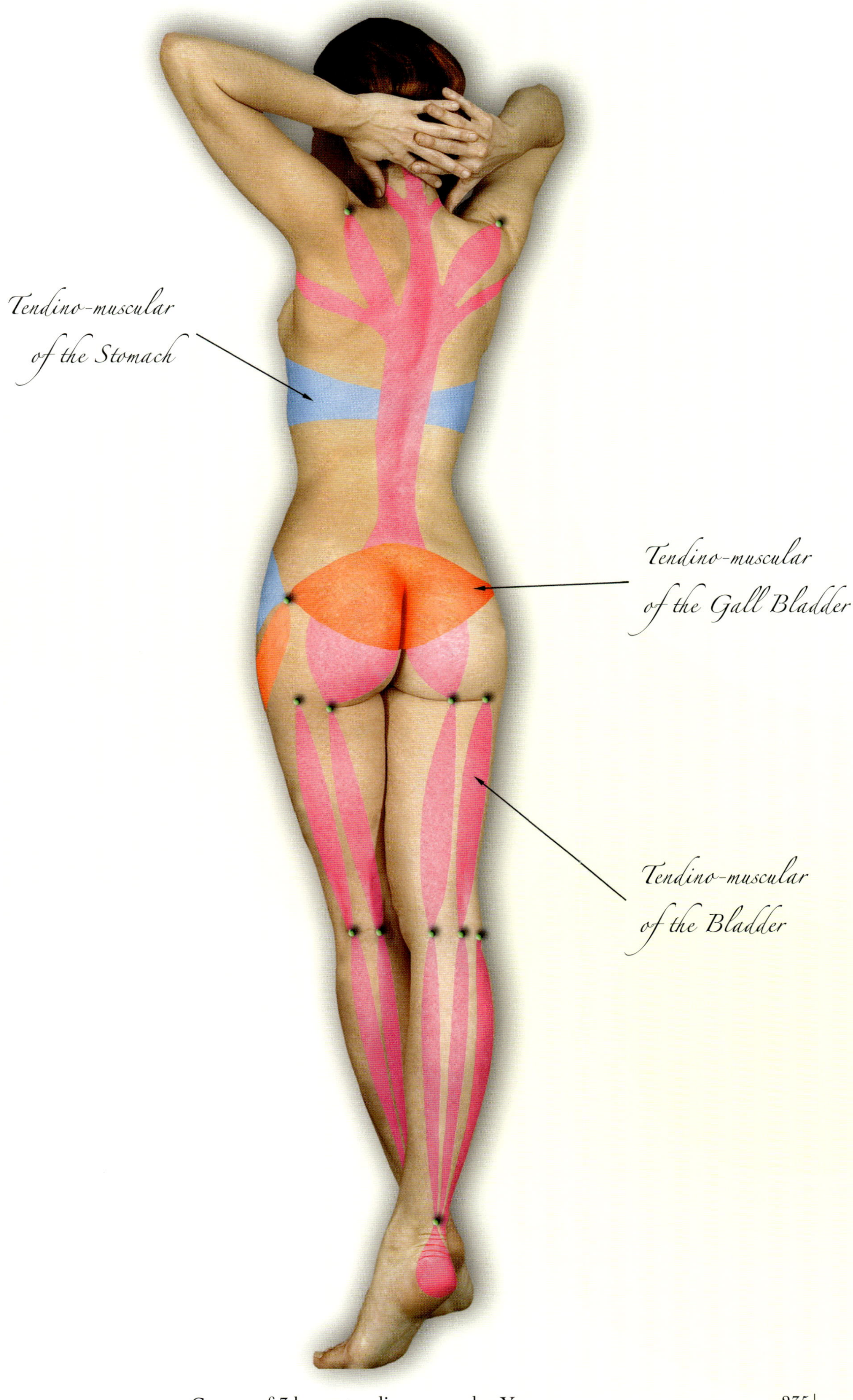

Course of 3 lower tendino-muscular Yang

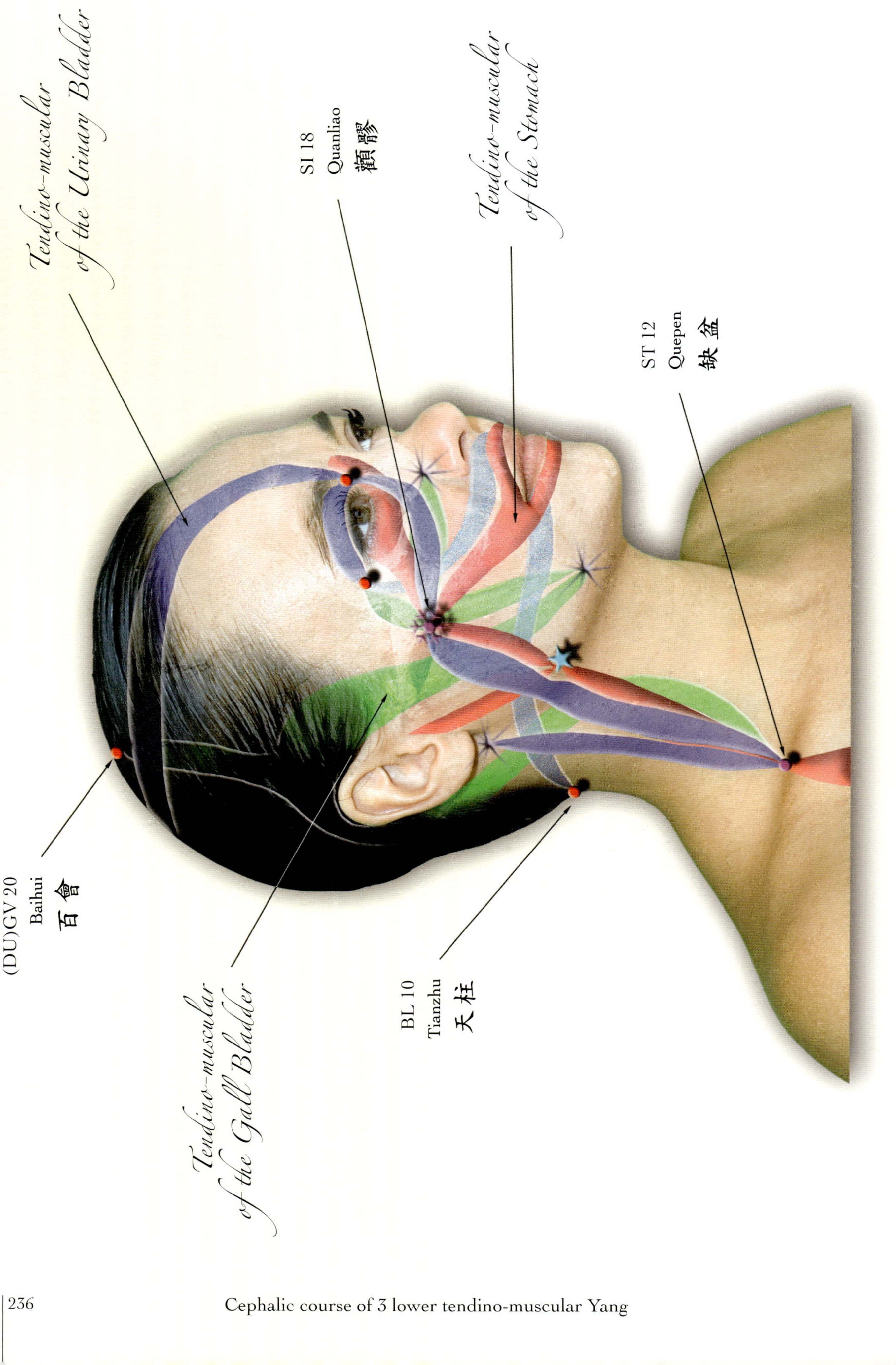

Cephalic course of 3 lower tendino-muscular Yang

THE MAIN CHANNELS

- The 3 upper Yin
- The 3 upper Yang (head)
- The 3 lower Yin
- The 3 lower Yang (head)

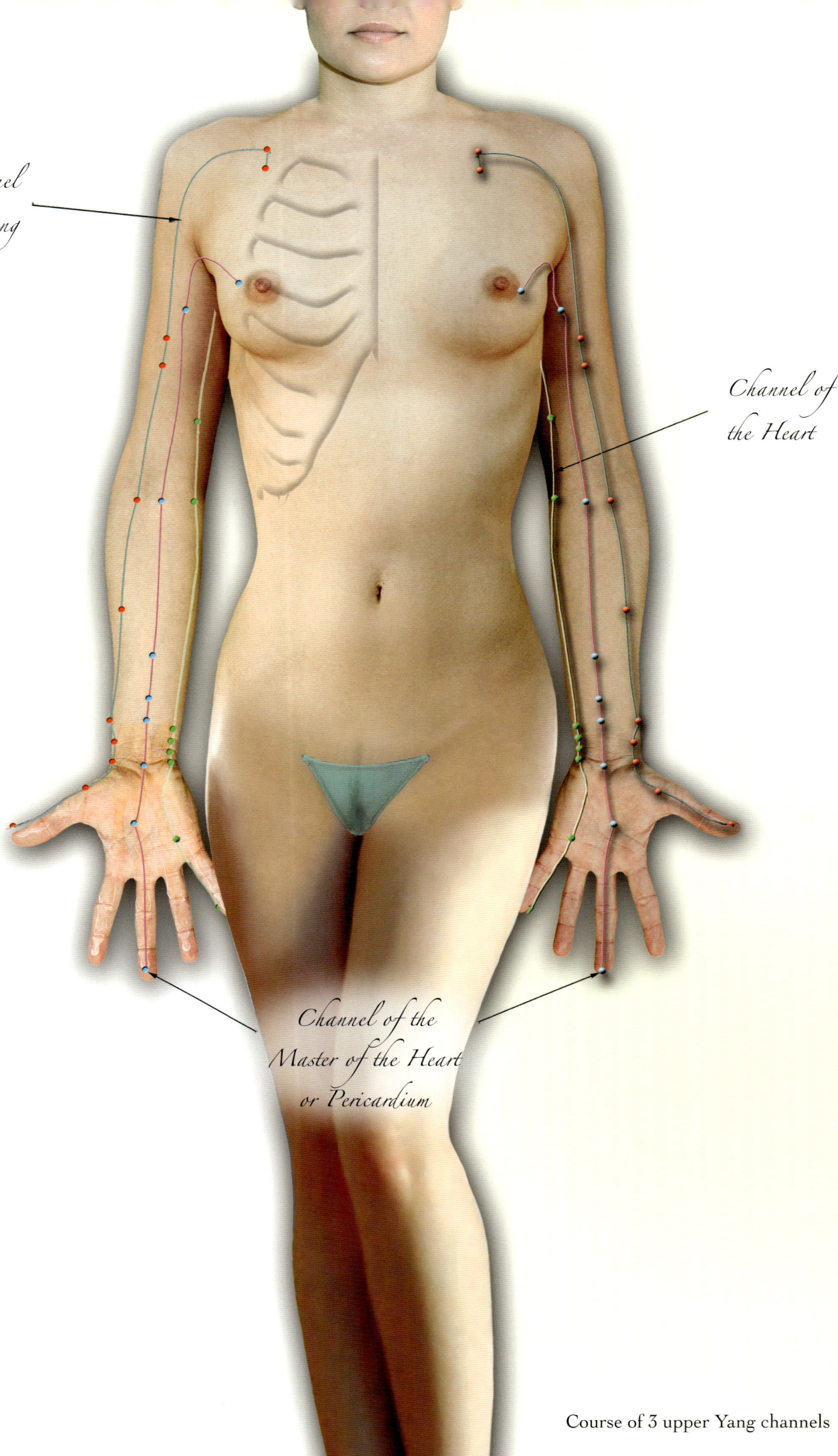

Course of 3 upper Yang channels

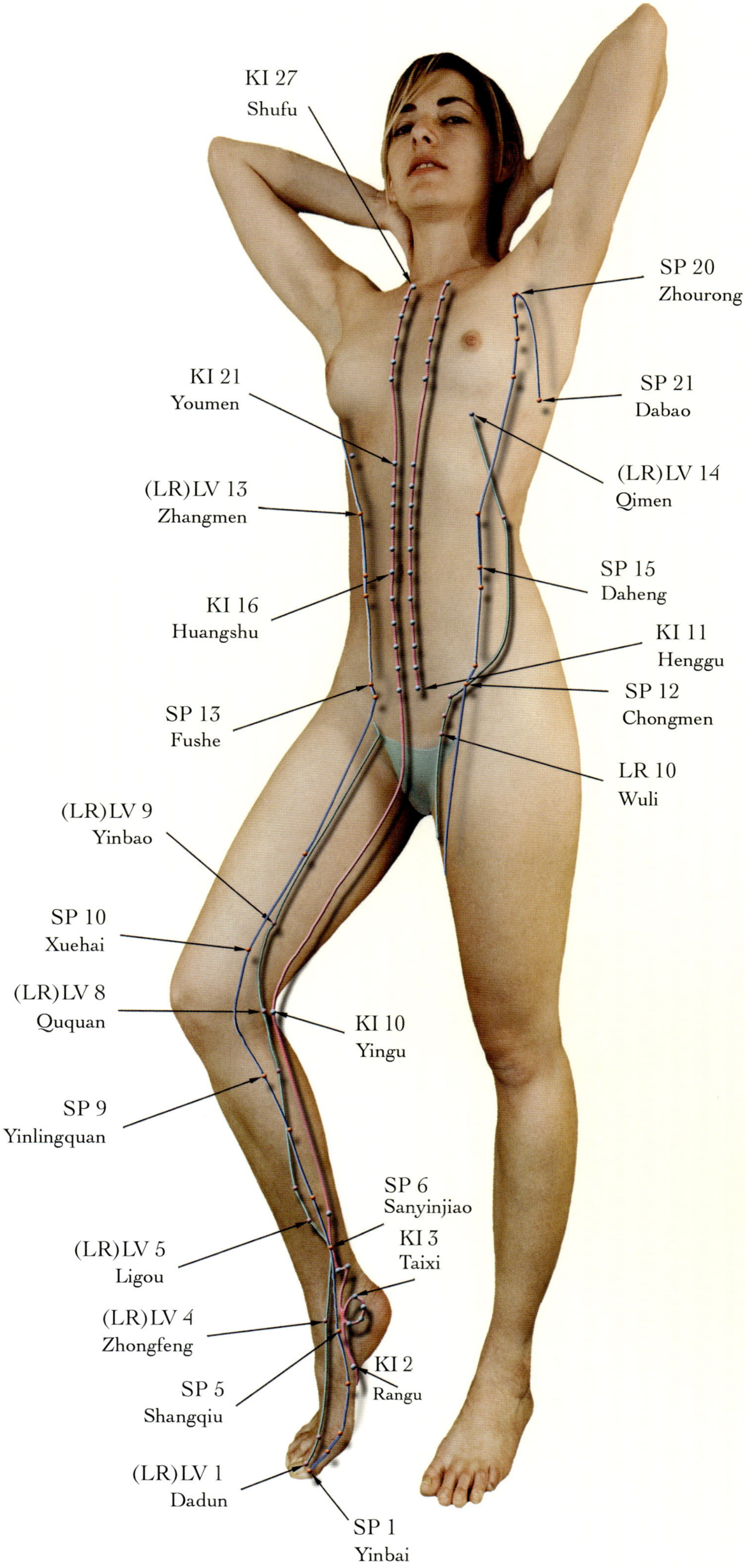

Course of 3 lower Yin channels

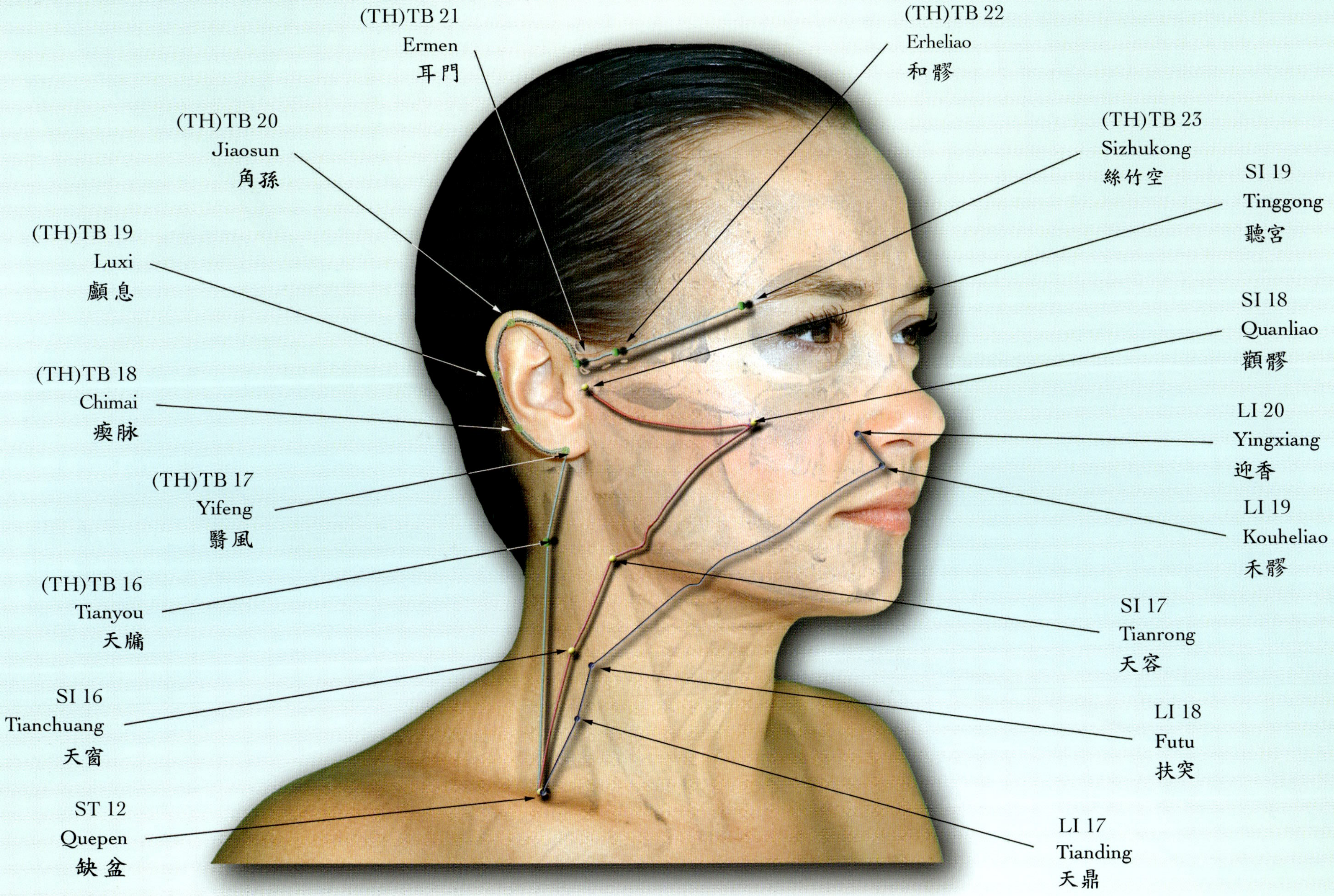

Cephalic course of 3 upper Yang channels

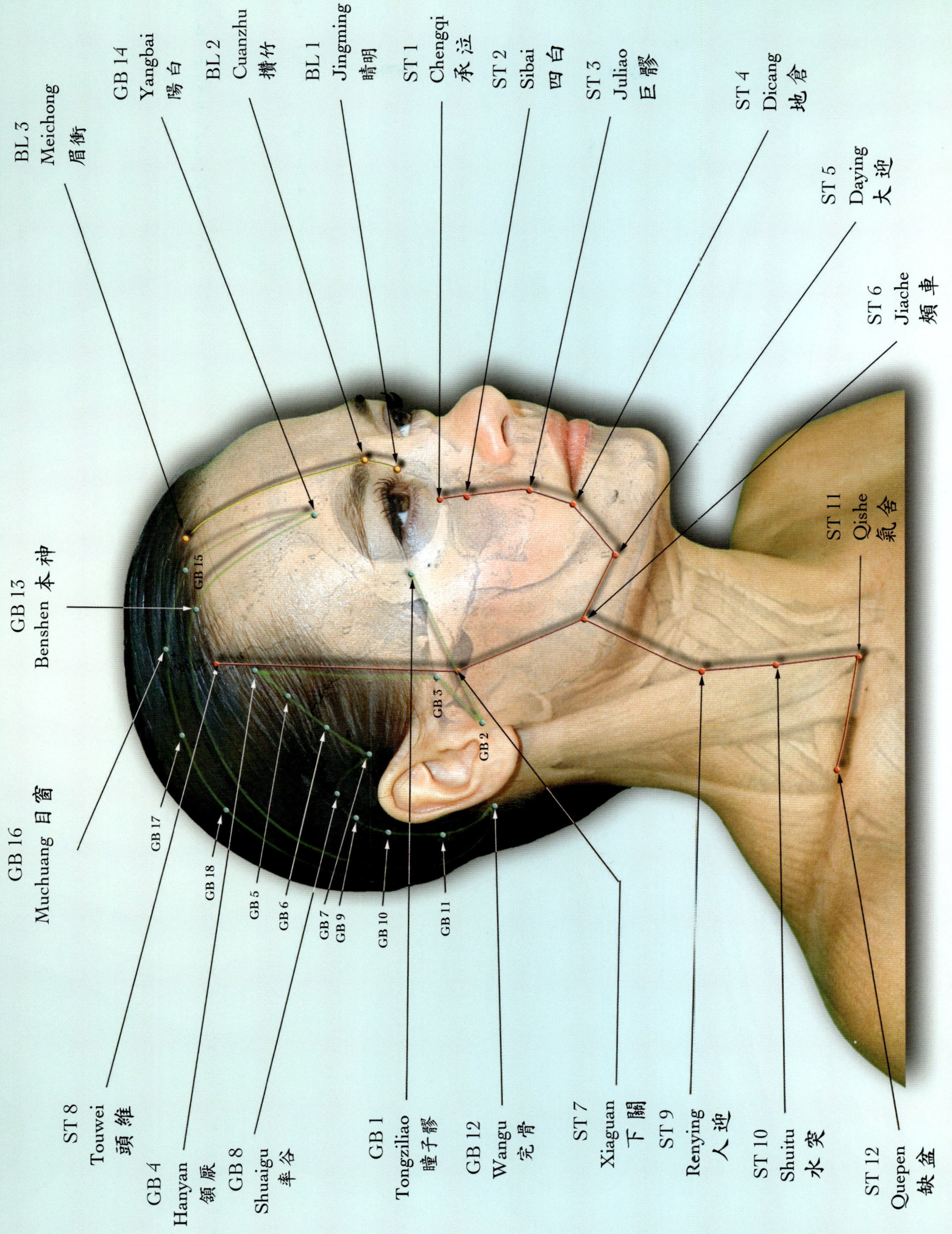

Cephalic course of 3 lower Yin channels

Cutaneous territories

- Tai Yang
  (Urinary Bladder-Small Intestine)
- Shao Yang
  (Gall Bladder-Triple Heater or Triple Burner)
- Yang Ming
  (Stomach-Large Intestine)
- Tai Yin
  (Spleen-Lung)
- Jue Yin
  (Liver-Master of the Heart or Pericardium)
- Shao Yin
  (Kidney-Heart)

Cutaneous territories

1. Tai Yang

Includes a part of the median axis of the skull and the whole posterior part of the body with the Urinary Bladder channel and the posterior part of the upper member with the Small Intestine channel.

2. Shao Yang

Includes the posterior and middle part of the upper member with the course of the Triple Heater or Triple Burner, and the lateral part of the body with the channel of the Gall Bladder.

3. Yang Ming

Represents the posterio-external side of the upper member with the Large Intestine channel and nearly the whole anterior part of the body with the Stomach channel.

4. Tai Yin

Includes the anterio-external side of the upper member with the Lung channel and the lateral parts of the abdomen, as well as the latero-external side of the lower members with the Spleen channel.

5. Jue Yin

Located on the anterior and middle part of the upper member with the Pericardium channel, and the anterio-inferior and latero-external side of the chest as well as the internal side of the lower member with the Liver channel.

6. Shao Yin

Represents the anterior and latero-internal side of the upper member with the Pericardium channel and the middle part of the chest and of the abdomen, including the internal side of the lower member with the Kidney channel.

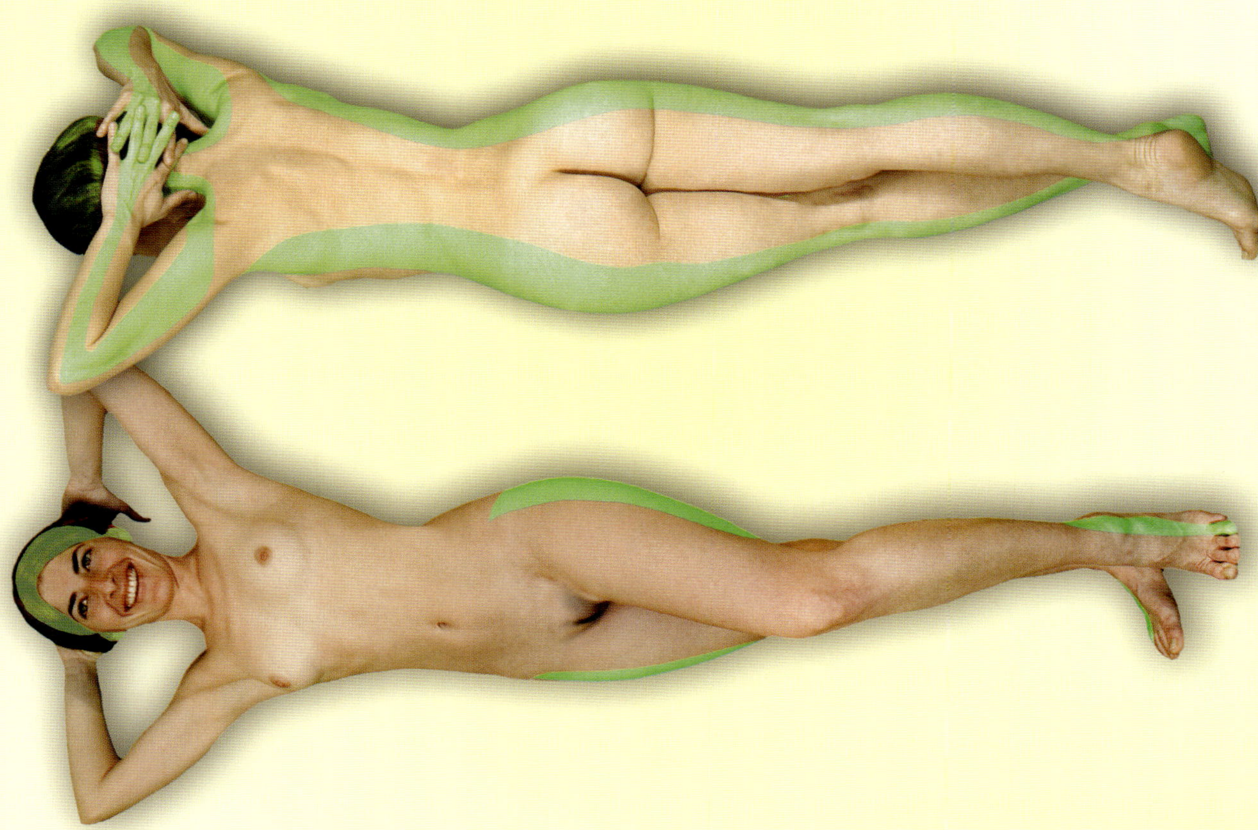

Shao Yang

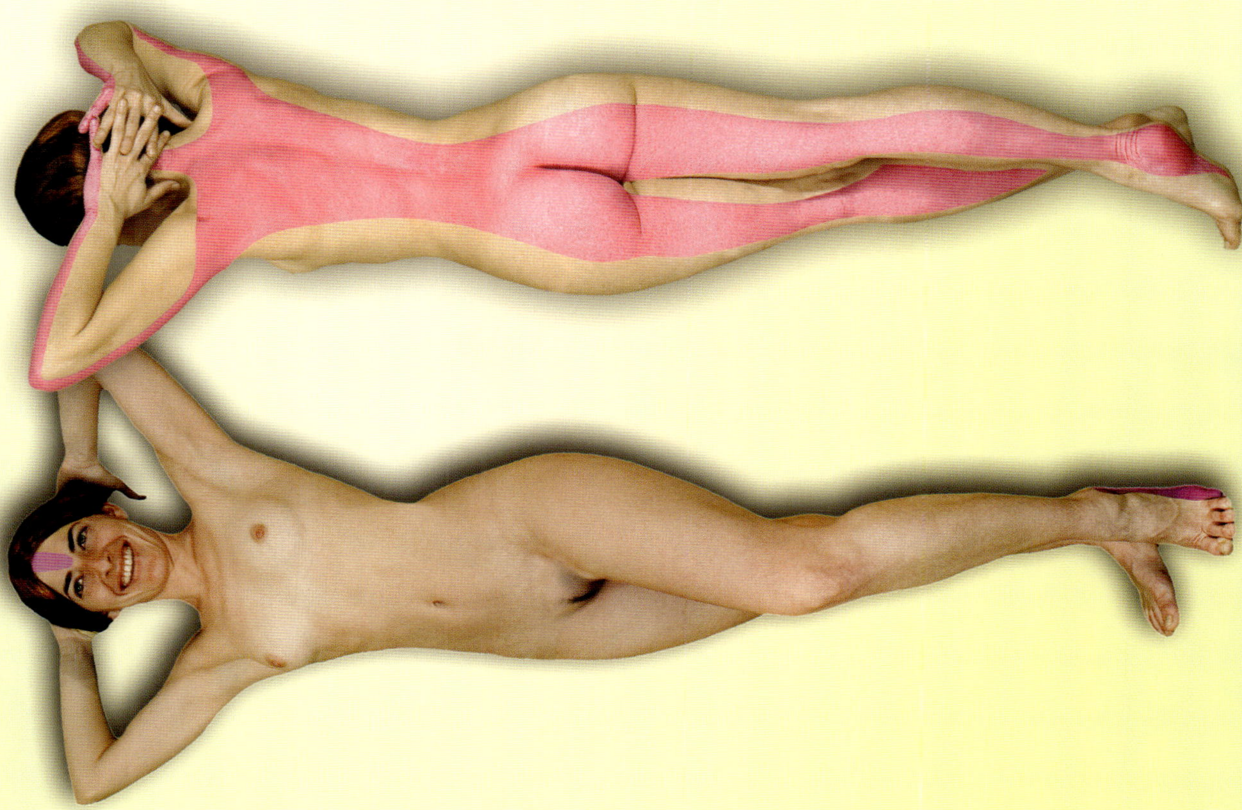

Tai Yang

Yang Ming

Tai Yin

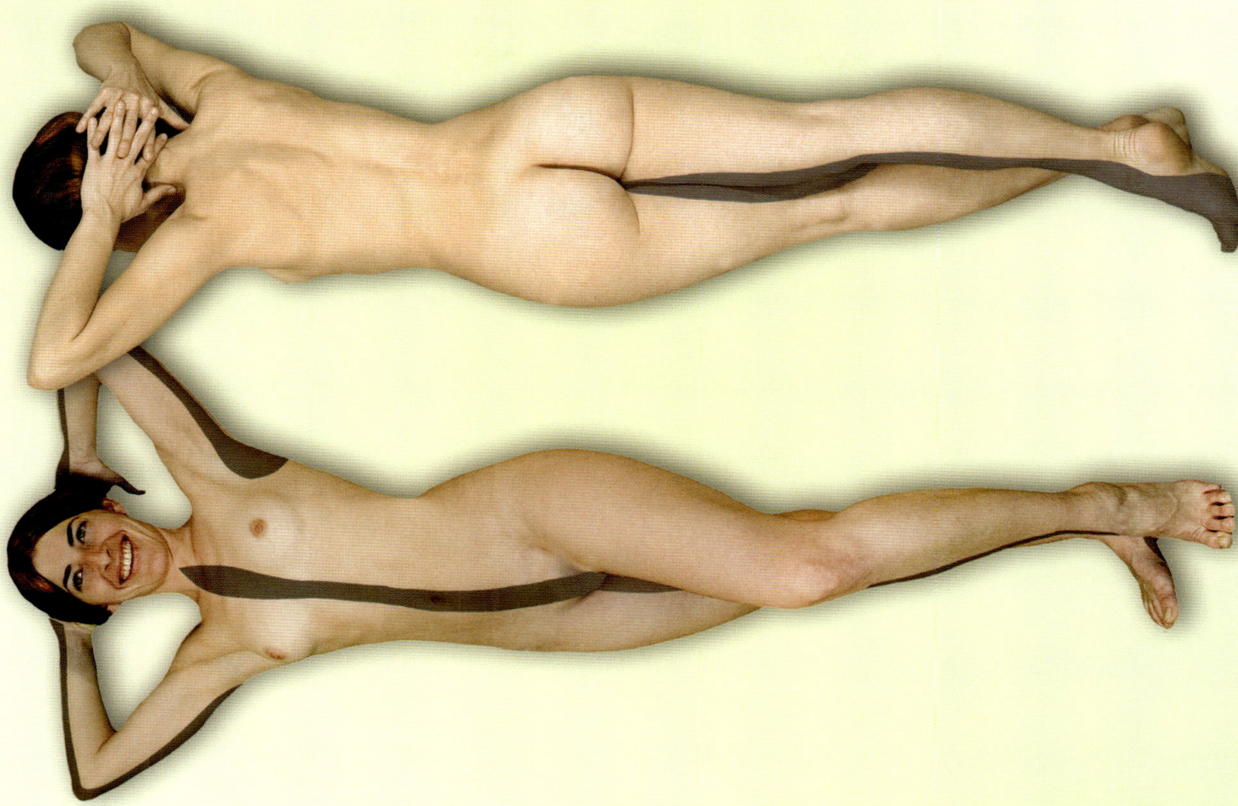

Shao Yin

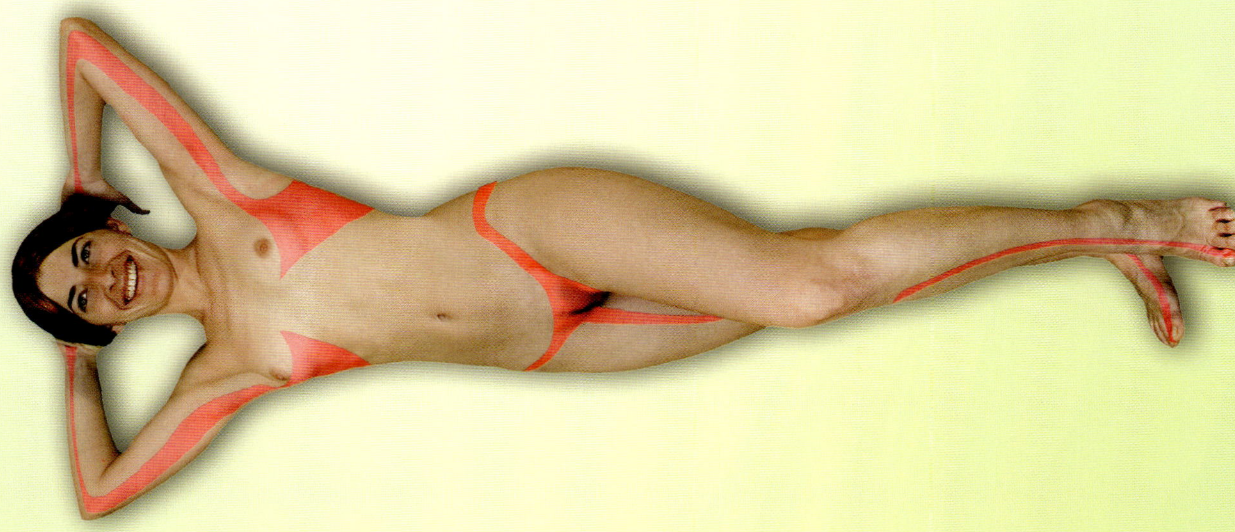

Jue Yin

Marvellous vessels or Extraordinary vessels

- Qi Jing Ba Mai

Chong Mai vessel (Thorongfare vessel)
Yin Wei Mai vessel (Yin Link vessel)
Yin Qiao Mai vessel (Yin Heel vessel)
Ren Mai vessel (Conception vessel)
Du Mai vessel (Governor vessel)
Yang Qiao Mai vessel (Yang Heel vessel)
Yang Wei Mai vessel (Yang Link vessel)
Dai Mai vessel (Belt vessel)

Chong Mai vessel (Thorongfare vessel)

Master point: SP 4 Gongsun

On the edge of the foot and under the base of the first metatarsal.

Course

Starts in the pelvian cavity (Baozhong), gets to the perineum at point (REN)CV 1 Huiyin where it separates into 2 branches, one posterior and one anterior.

The posterior branch, or deep course, passes by the deep and posterior side of the pelvic cavity and goes up along the backbone (in its lower part, according to some authors), and it ends on the level of T 1-T 2, or at point BL 11 Dazhu for some.

The anterior branch goes to the anterior abdominal wall and, on the level of CV 4 Guanyan or on the level of ST 30 Qichong, divides into two branches.

An ascending branch that passes by point KI 11 Hengu and follows the Kidney channel to point KI 15 Zhongzhu, where it alters its course to pass by point CV 7 Yinjiao.

It starts following the course of the Kidney channel at point KI 16 Huangyu and goes up to point KI 21 Youmen.

From there, it passes in the middle of the chest and in the intercostal spaces, to reach point KI 27 Shufu.

It runs up to the throat to encircle the mouth and join on each side with point BL 1 Jingming. Some authors consider that after flowing round the mouth, the channel spreads on the upper part of the pharynx and on the upper nasal fossae.

The second branch that leaves ST 30 Qichong, joins with the Main channel of the Kidney, passing by the internal side of the thigh to reach the internal retromalleolar zone where KI 4 Dazhong is.

On that level, it divides into 2 small branches:

- The first passes by the internal border of the dorsal side of the foot to reach the big toe. According to Nguyen Van Nghi, the branch goes round the big toe and back to the internal malleolus.

- The second runs to the sole of the foot. According to the authors, these two small branches might join again on the level of the 1st and 2nd toe.

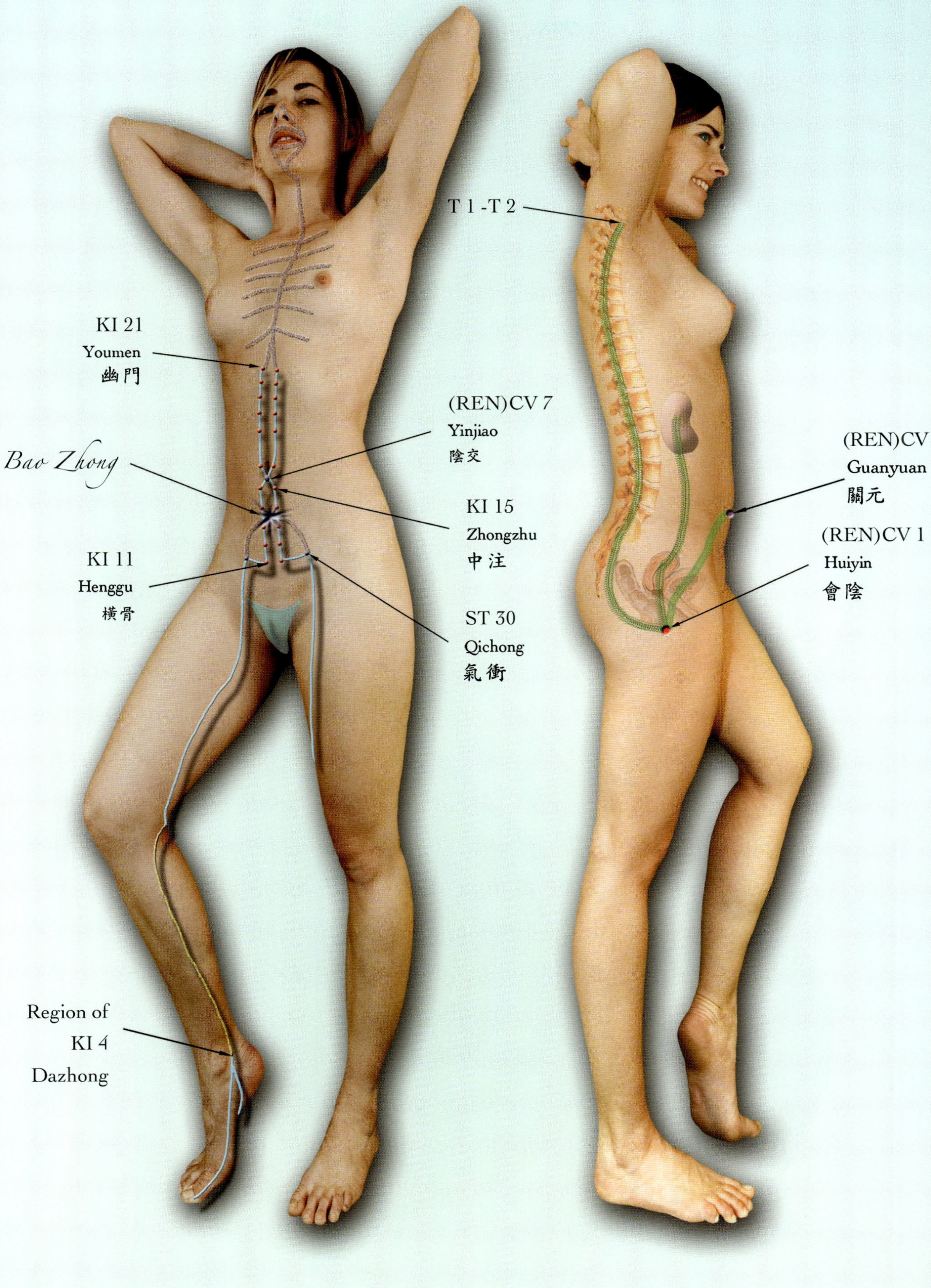

Superficial and deep course of Chong Mai vessel

Points of the Chong Mai vessel (Thorongfare vessel)

KI 11 Henggu
On the lower abdomen, 5 Cun the umbilicus and 0.5 Cun to the anterior midline.

KI 12 Dahe
On the lower abdomen, 4 Cun below the umbilicus and 0.5 Cun to the anterior midline.

KI 13 Qixue
On the lower abdomen, 3 Cun below the umbilicus and 0.5 Cun to the anterior midline.

KI 14 Siman
On the lower abdomen, 2 Cun below the umbilicus and 0.5 lateral to the anterior midline.

KI 15 Zhongzhu
On the lower abdomen, 1 Cun below the umbilicus and 0.5 Cun lateral to the anterior midline.

KI 16 Huangshu
On the middle abdomen, 0.5 Cun lateral to the umbilicus.

KI 17 Shanqu
On the upper abdomen, 2 Cun above the umbilicus, and 0.5 Cun to the anterior midline.

KI 18 Shiguan
On the upper abdomen, 3 Cun above the centre of the umbilicus and 0.5 Cun to the anterior midline.

KI 19 Yindu
On the upper abdomen, 4 Cun above the umbilicus and 0.5 Cun to the anterior midline.

KI 20 Futonggu
On the upper abdomen, 5 Cun above the umbilicus and 0.5 Cun to the anterior midline.

KI 21 Youmen
On the upper abdomen, 6 Cun above the umbilicus and 0.5 Cun to the anterior midline.

KI 22 Bulang
On the chest, in the 5th intercostal space, 2 Cun lateral to the anterior midline.

KI 23 Shenfeng
On the chest, in the 4th intercostal space, 2 Cun lateral to the anterior midline.

KI 24 Lingxu
On the chest, in the 3rd intercostal space, 2 Cun to the anterior midline.

KI 25 Shencang
On the chest, in the 2nd intercostal space, 2 Cun to the anterior midline.

KI 26 Yuzhong
On the chest, in the 1st intercostal space, 2 Cun to the anterior midline.

KI 27 Shufu
On the chest, below the lower border of the clavicle, 2 Cun to the midline.

CV 7 Yinjiao
On the lower abdomen and on the anterior midline, 1 Cun below the umbilicus.

SP 4 Gongsun
On the medial border of the foot, anterior and inferior to the proximal end of the 1st metatarsal bone.

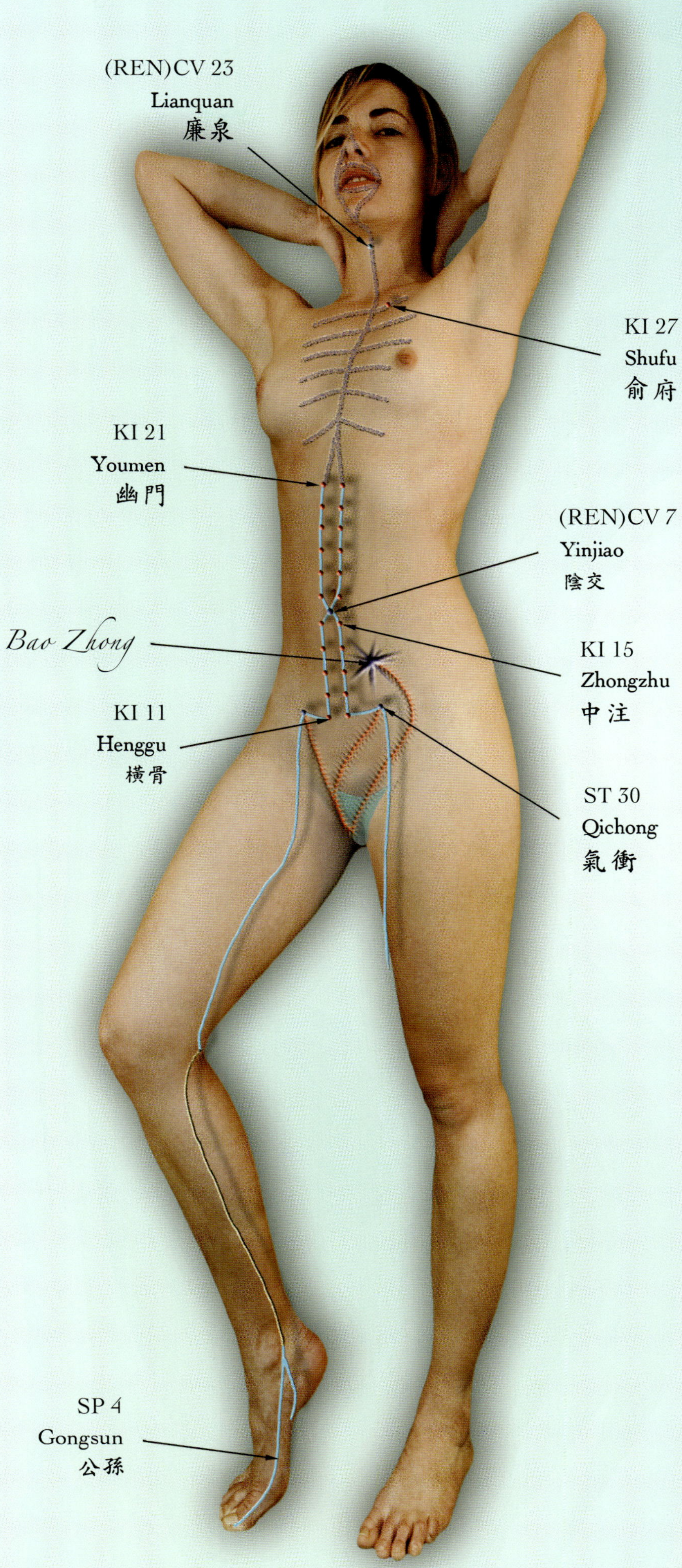

Course variations of Chong Mai vessel

Yin Wei Mai vessel (Yin Link vessel)

Master point: (MH)PC 6 Neiguan

On the palmar side of the forearm and on the line connecting (MH)PC 3 Quze and (MH)PC 7 Daling, 2 Cun above the crease of the wrist, between the tendons of the long palmar and radial flexor muscle of the wrist.

Course

It starts on the internal aspect of the leg at point KI 9 Zhubin, 1 Cun behind the postero-internal border of the tibia and 5 Cun above the most prominent part of the internal malleolus.

It goes up along the internal side of the leg and the thigh, passes by the anterior side of the chest, joining the mammary line.

On the upper part of the chest, it follows a curved course to join the midline, above the sternal manubrium.

It joins Ren Mai vessel on the level of the anterior side of the neck, and with points (REN)CV 22 Tiantu and (REN)CV 23 Lianquan.

Variations

For some other authors (Nguyen Van Nghi and Darras), the channel goes out by the same point KI 9 Zhubin, it follows the same course on the level of the lower members, but it divides into 3 different branches:

- The first passes by points SP 15 Daheng and SP 16 Fuai, following the Spleen channel.
- The second passes by points SP 13 Fushe and SP 14 Fujie, using the Liver channel.
- The third uses Conception vessel and passes by points (REN)CV 22 Tiantu and (REN)CV 23 Lianquan.

For De La Fuÿe, there is an anastomotic branch that joins (MH)PC 1 Tianchi with (REN)CV 17 Shanzhong and crosses Yin Wei vessel in the area of point KI 23 Taixi.

It might be thanks to this branch that the connection with master point (MH)PC 6 Neiguan is made.

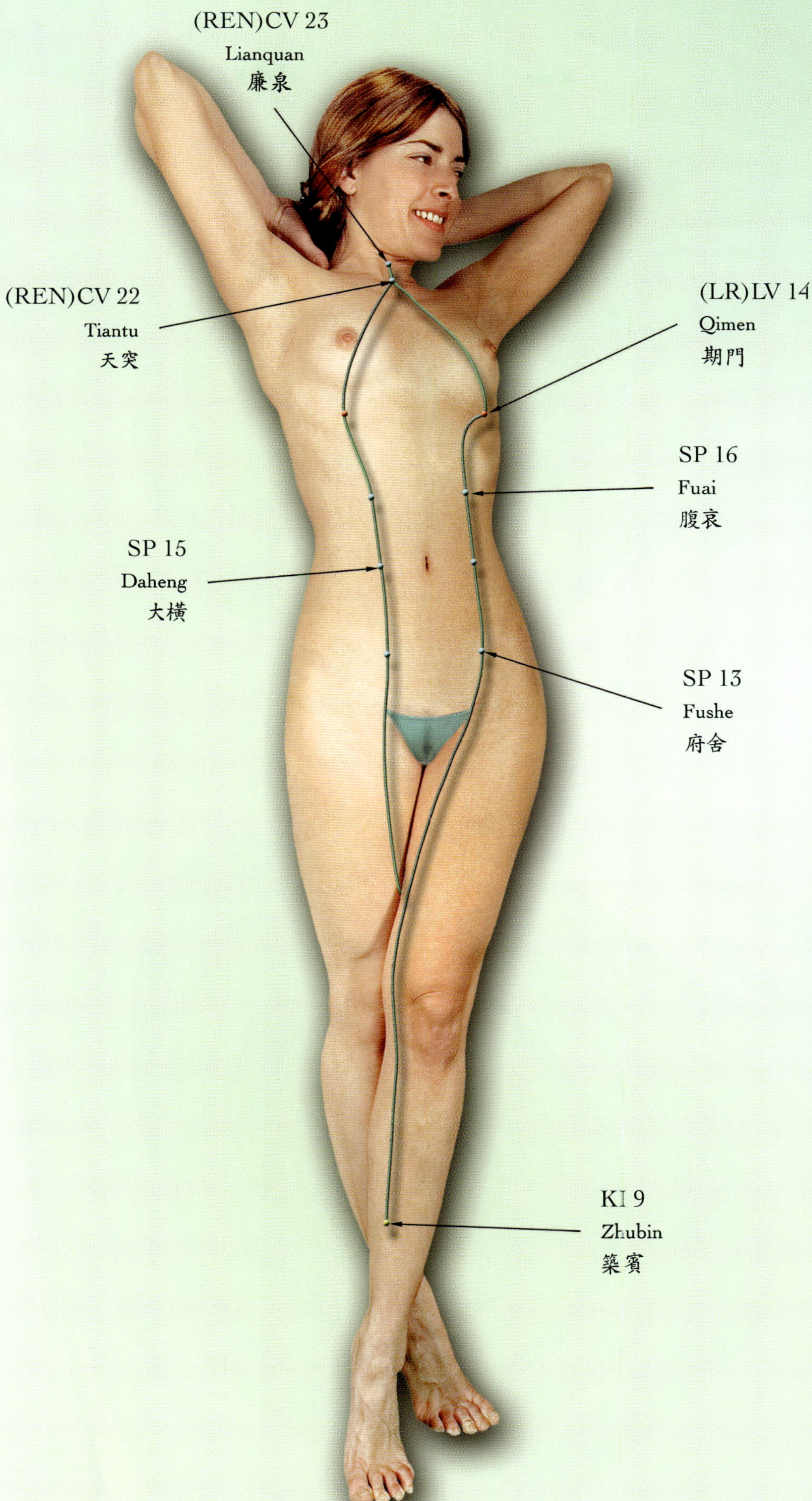

Course of the Yin Wei Mai vessel

Points of the Yin Wei Mai vessel (Yin Link vessel)

KI 9 Zhubin
On the medial side of the leg and on the line connecting KI 3 Taixi and KI 10 Yingu, 5 Cun above KI 3 Taixi, medial and inferior to the gastrocnemius muscle.

SP 13 Fushe
On the lower abdomen, 4 Cun below the centre of the umbilicus, 0.7 Cun above SP 12 Chongmen and 4 Cun lateral to the anterior midline.

SP 15 Daheng
On the middle abdomen, 4 Cun to the umbilicus.

SP 16 Fu'ai
On the upper abdomen, 3 Cun above the centre of the umbilicus, and 4 Cun lateral to the anterior midline.

LR 14 Qimen
On the chest, directly below the nipple, in the 6th intercostal space, 4 Cun lateral to the anterior midline.

CV 22 Tiantu
On the neck and on the anterior midline, at the centre of the supra-sternal fossa.

CV 23 Lianquan
On the neck and on the anterior midline, above the laryngeal protuberance, in the depression above the upper border of the hyoid bone.

MH 6 Neiguan
On the palmar side of the forearm and on the line connecting (MH)PC 3 Quze and (MH)PC 7 Daling, 2 Cun above the crease of the wrist, between the tendons of the long palmar and radial flexor muscle of the wrist.

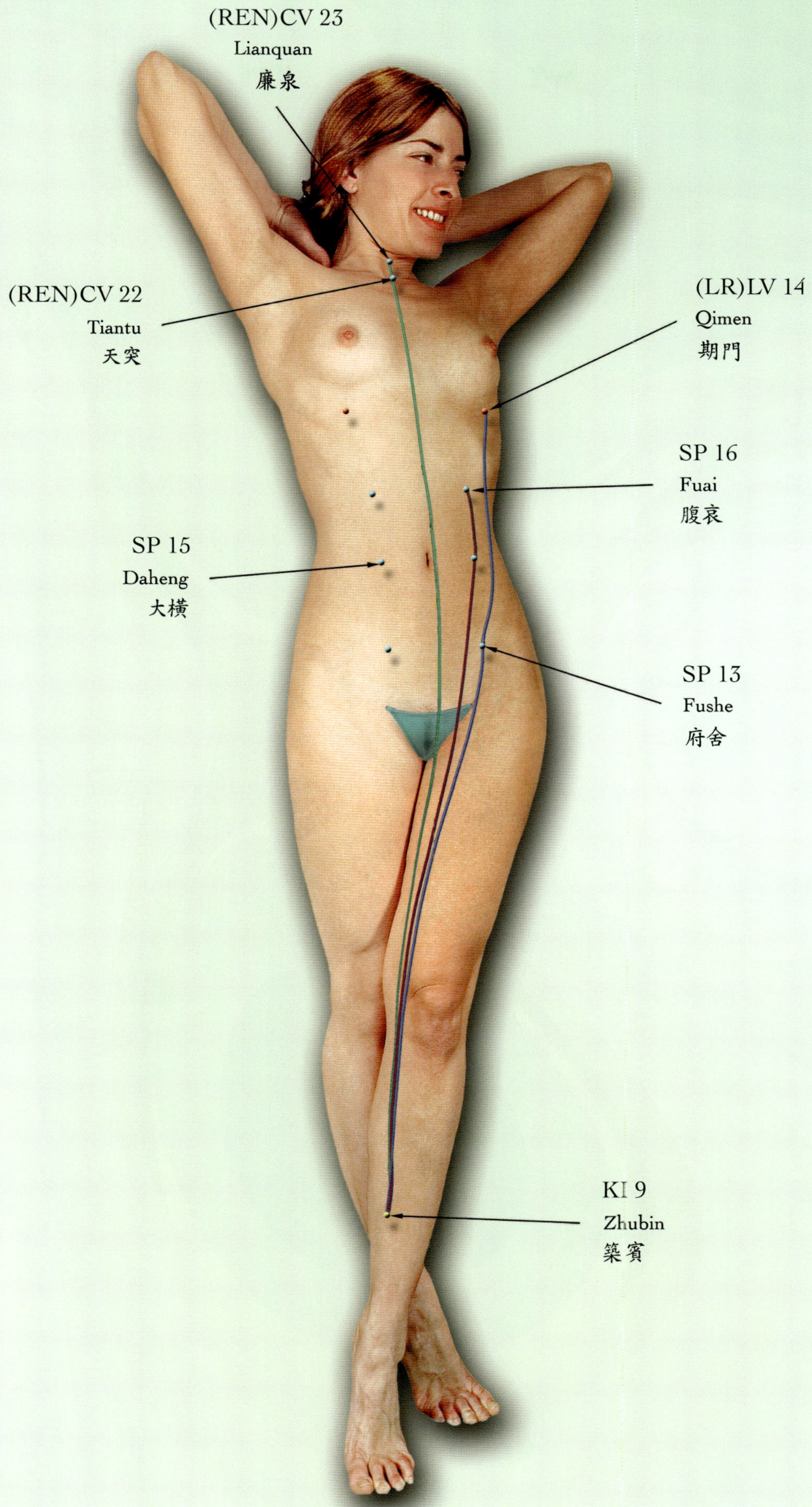

Course variations of the Yin Wei Mai vessel

Yin Qiao Mai vessel (Yin Heel vessel)

Master point: KI 6 Zhaohai

On the edge of the foot, in the depression formed below the tip of the medial malleolus.

Course

It starts at point KI 2 Rangu, on the internal side of the foot, in a depression under the tubercle of the scaphoid, 1 Cun behind SP 4 Gongsun.

It passes successively by point KI 6 Zhaohai, under the internal malleolus, and by point KI 8 Jiaoxin halfway in front of KI 7 Fuliu, on the posterior border of the tibia.

It follows the internal side of the leg and of the thigh and enters the genitalia.

It then goes again in the abdomen, follows its internal wall and reaches the chest.

It goes out in the supra-clavicular area, at point ST 12 Quepen.

It passes by point ST 9 Renying, goes up to the face, attaches itself to the zygomatic arch and terminates in the inner canthus at point BL 1 Jingming. (According to certain writers, it penetrates in the brain where it ends its course.)

Points of the Yin Qiao Mai channel (Yin Heel vessel)

KI 2 Rangu
On the medial border of the foot, below the tuberosity of the scaphoid bone, and at the 'junction of the red and white skin'.

KI 6 Zhaohai
On the medial side of the foot, in the depression below the tip of the medial malleolus.

KI 8 Jiaoxin
On the medial side of the leg, directly 2 Cun above KI 3 Taixi and 0.5 Cun anterior to KI 7 Fuliu, posterior to the medial border of the tibia.

ST 9 Renying
On the neck, beside the laryngeal protuberance, and on the anterior border of the sterno-cleido-mastoid muscle where the pulsation of the common carotid artery is palpable. (Do NOT needle directly into the artery!)

ST 12 Quepen
At the centre of the supraclavicular fossa, 4 Cun lateral to the anterior midline.

BL 1 Jingming
On the face, in the depression slightly above the inner canthus.

Course of Yin Qiao Mai vessel

Ren Mai vessel (Conception vessel)

Master point: LU 7 Lieque

On the edge of the radial forearm, on the styloid process of the radius, 1.5 Cun proximal to the fold of the wrist, between the brachioradialis muscle and the tendon of the adductor muscle of the thumb.

Course

It also starts in the pelvic cavity (Baozhong). Its Main course starts from (REN)CV 1 Huiyin, then goes up by the middle of the pubic symphysis.

It follows the medial abdominal line, then thoracic and reaches the throat.

It passes by the mento-labial sulcus where point (REN)CV 24 Chengjiang can be found, to unite with the Stomach channel and goes deep inside.

It runs inside and round the lips.

It reaches point (DU)GV 28 Yinjiao and divides into 2 branches that flow up to the cheeks.

It reaches point ST 1 Chengqi, where it joins the Stomach channel and marvellous vessel Yang Qiao Mai.

It enters the eyes.

Another branch leaves the perineum, passes inside the spinal column to reach the upper part of the shoulder.

Location of the points

One point is perineal: (REN)CV 1 Huiyin.

Seven points are between the symphysis and the umbilicus:

- (REN)CV 2, (REN)CV 3, (REN)CV 4, (REN)CV 5, (REN)CV 6, (REN)CV 7, (REN)CV 8.

Seven points are supra-umbilical:

- (REN)CV 9, (REN)CV 10, (REN)CV 11, (REN)CV 12, (REN)CV 13, (REN)CV 14, (REN)CV 15.

Six points are thoracic:

- (REN)CV 16, (REN)CV 17, (REN)CV 18, (REN)CV 19, (REN)CV 20, (REN)CV 21.

Two points are cervical:

- (REN)CV 22, (REN)CV 23.

The last point is situated on the level of the chin:

- (REN)CV 24.

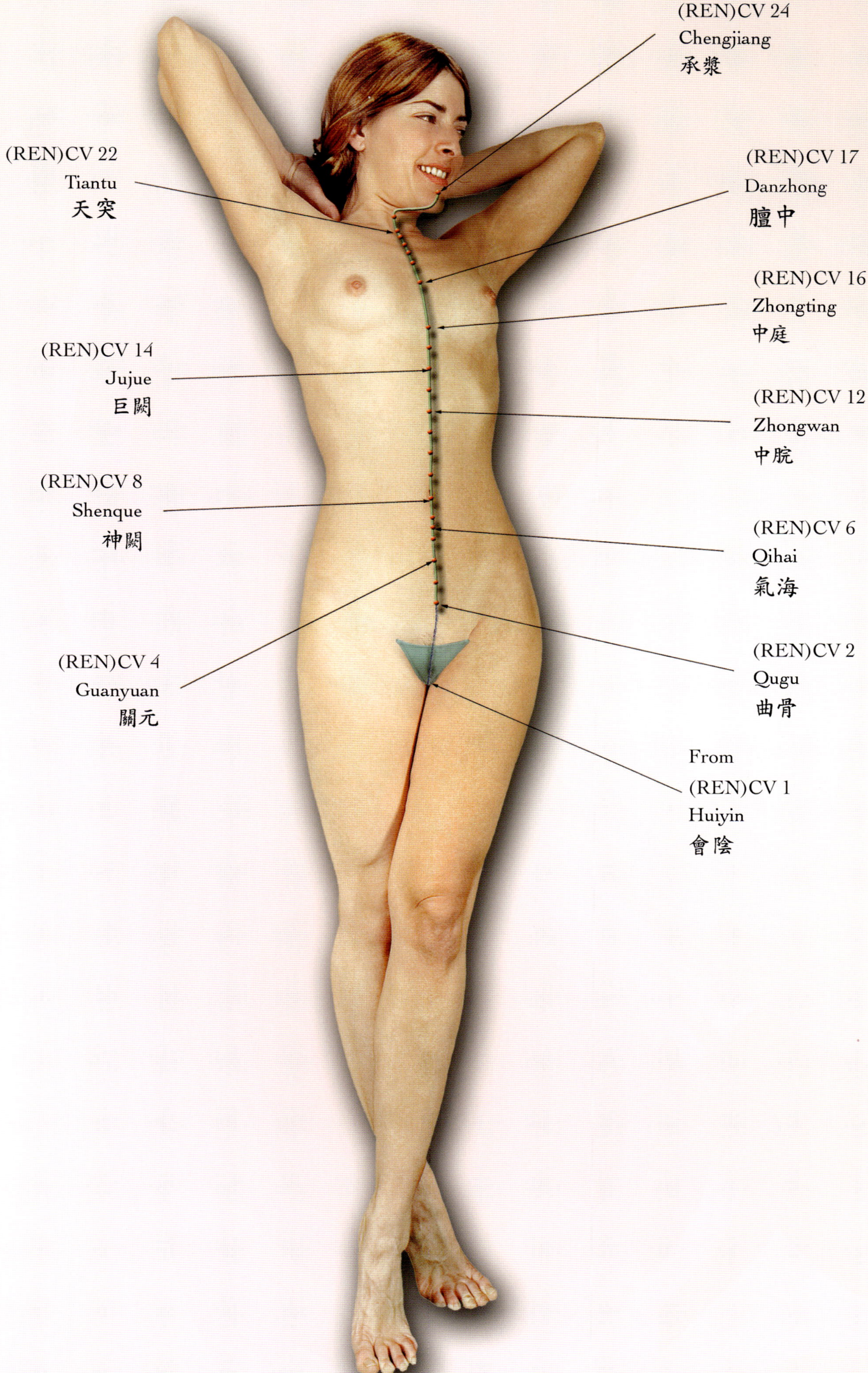

Course of Ren Mai vessel

Points of the Ren Mai vessel (Conception vessel)

(REN)CV 1 Huiyin
On the perineum, at the midpoint between the posterior border of the scrotum and anus in male, and between the posterior commissure of the large labia and anus in female.

(REN)CV 2 Qugu
On the lower abdomen and on the anterior midline, at the midpoint of the upper border of the pubic symphysis.

(REN)CV 3 Zhongji
On the lower abdomen and on the anterior midline, 4 Cun below the umbilicus.

(REN)CV 4 Guanyuan
On the lower abdomen and on the anterior midline, 3 Cun below the umbilicus.

(REN)CV 5 Shimen
On the lower abdomen and on the anterior midline, 2 Cun below the umbilicus.

(REN)CV 6 Qihai
On the lower abdomen and on the anterior midline, 1.5 Cun below the umbilicus.

(REN)CV 7 Yinjiao
On the lower abdomen and on the anterior midline, 1 Cun below the umbilicus.

(REN)CV 8 Shenque
On the middle abdomen and at the centre of the umbilicus.

(REN)CV 9 Shuifen
On the upper abdomen and on the anterior midline, 1 Cun above the umbilicus.

(REN)CV 10 Xiawan
On the upper abdomen and on the anterior midline, 2 Cun above the umbilicus.

(REN)CV 11 Jianli
On the upper abdomen and on the anterior midline, 3 Cun above the umbilicus.

(REN)CV 12 Zhongwan
On the upper abdomen and on the anterior midline, 4 Cun above the umbilicus.

(REN)CV 13 Shangwan
On the upper abdomen and on the anterior midline, 5 Cun above the umbilicus.

(REN)CV 14 Juque
On the upper abdomen and on the anterior midline, 6 Cun above the umbilicus.

(REN)CV 15 Jiuwei
On the upper abdomen and on the anterior midline, 1 Cun below the xiphi-sternal synchondrosis.

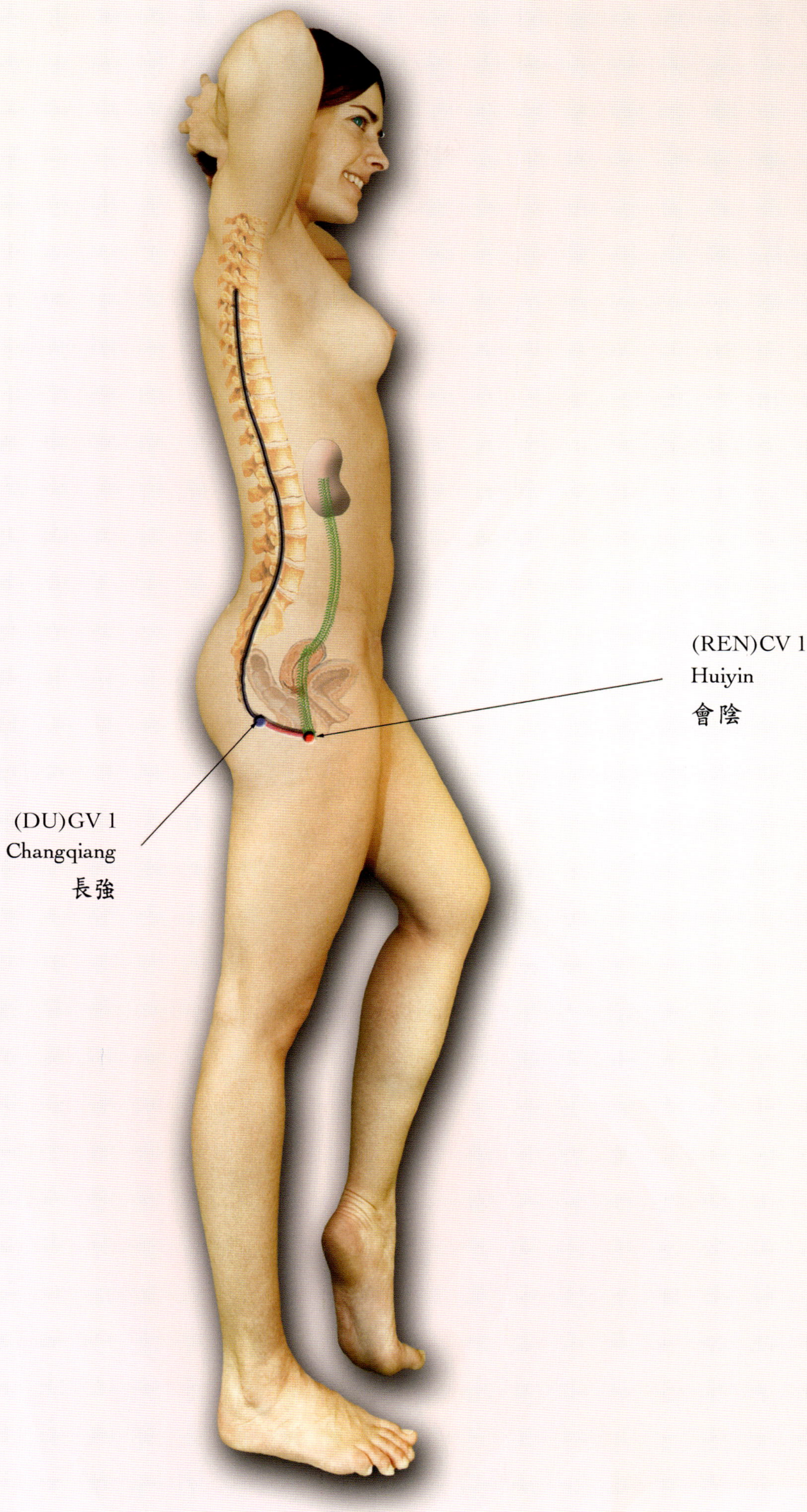

Internal course of Ren Mai vessel

(REN)CV 16 Zhongting
On the chest and on the anterior midline, on the level of the 5th intercostal space, on the xiphi-sternal synchondrosis.

(REN)CV 17 Danzhong
On the chest and on the anterior midline, on the level of the 4th intercostal space, at the midpoint of the line connecting both nipples.

(REN)CV 18 Yutang
On the chest and on the anterior midline, on the level of the 3rd intercostal space.

(REN)CV 19 Zigong
On the chest and on the anterior midline, on the level of the 2nd intercostal space.

(REN)CV 20 Huagai
On the chest and on the anterior midline, on the level of the 1st intercostal space.

(REN)CV 21 Xuanji
On the chest and on the anterior midline, 1 Cun below (REN)CV 22 Tiantu.

(REN)CV 22 Tiantu
On the neck and on the anterior midline, at the centre of the suprasternal fossa.

(REN)CV 23 Lianquan
On the neck and on the anterior midline, above the laryngeal protuberance, in the depression above the upper border of the hyoid bone.

(REN)CV 24 Chengjiang
On the face, in the depression at the midpoint of the mento-labial sulcus.

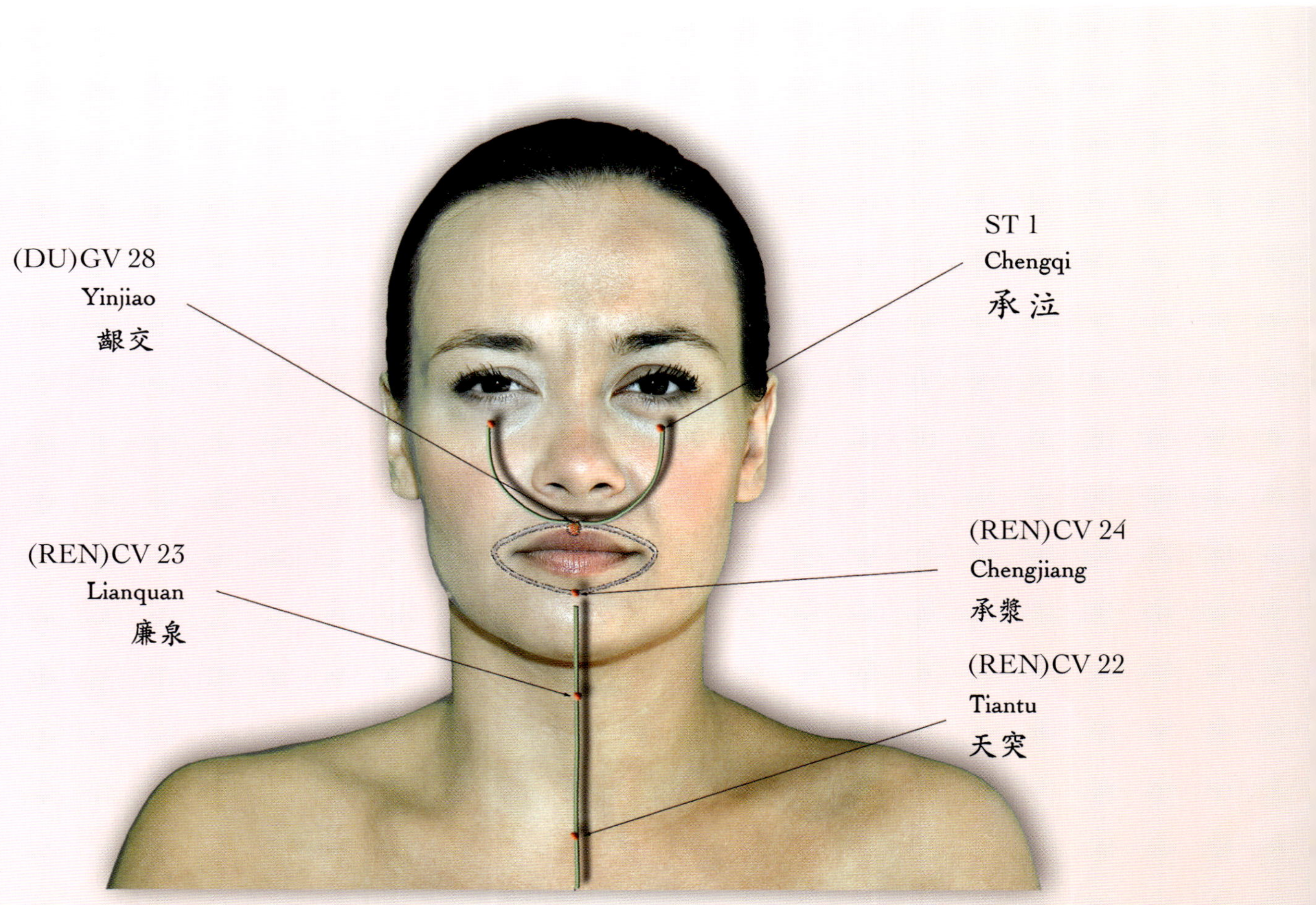

Cephalic course of Ren Mai vessel

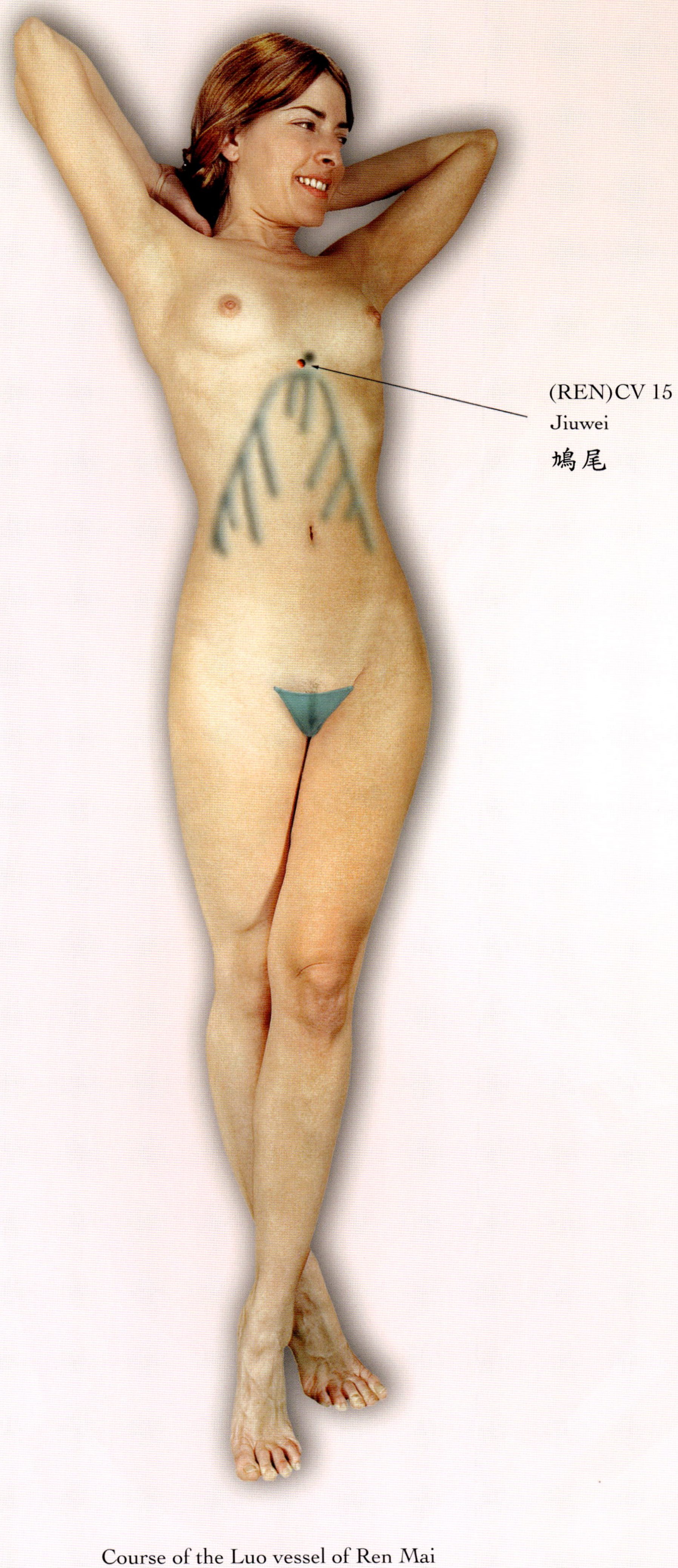

Course of the Luo vessel of Ren Mai

Du Mai vessel (Governor vessel)

Master point: SI 3 Houxi

On the ulnar side of the hand, at the boundary between the dorsal skin and the palmar skin, at the end of the distal palmar crease and behind the metacarpal-phalangeal joint of the little finger when the fist is closed.

Course

The Du Mai vessel originates in the Kidneys (or in the pelvic cavity Baozhong).

It reaches the uro-genital apparatus, and joins the perineum at point (REN)CV 1 Huiyin.

It then passes by point (DU)GV 1 Changqiang, at the tip of the coccyx.

It rises along the spinal column following the tip of the vertebral processes*.

It reaches the nape of the neck at point (DU) GV 16 Fengfu, where a branch enters the brain and another joins point (DU)GV 20 Baihui.

It follows its course passing by the forehead and goes down to the tip of the nose at point (DU) GV 25 Suliao.

It terminates at point (DU)GV 28 Yinjiao where it joins Conception vessel and Stomach.

From the genitalia starts a branch that goes out at point (REN)CV 2** and divides into two branches:

- An ascending branch with an abdominal course that follows the tendinomuscular channel of the Spleen, goes out at the umbilicus, rises along the internal wall of the abdomen and crosses the Heart.
- A descending branch, that passes by the rectum, goes round the buttock and joins the tendino muscular of the Urinary Bladder, then goes up to the head to point BL 1 Jingming. This branch penetrates the brain and follows the Main channel of the Urinary Bladder, descends on the neck down to the Kidney region at point BL 23 Shenshu and gets inside the Kidneys.

There is also a description of a vertebral course that goes out of point (DU)GV 1 Huiyin, and crossing the Kidney channel, passes by the coccyx, goes up in the spinal column and ends in the Kidneys.

Luo vessel of Du Mai

It goes out of Du Mai at point (DU)GV 1 Changqiang, ascends to the neck, ramifies in the occiput and joins the Main channel of the Urinary Bladder.
It goes down to the shoulders and goes deep inside to ramify with the paravertebral muscles on the level of the spine of the shoulder-blade.

* A branch that forks out on the level of point (DU)GV 12 Shenzhu, goes to point BL 12 Fengmen, and passes by point (DU)GV 13 Taodao is also described.

** For certain writers, the branches start from (REN)CV 1 Huiyin.

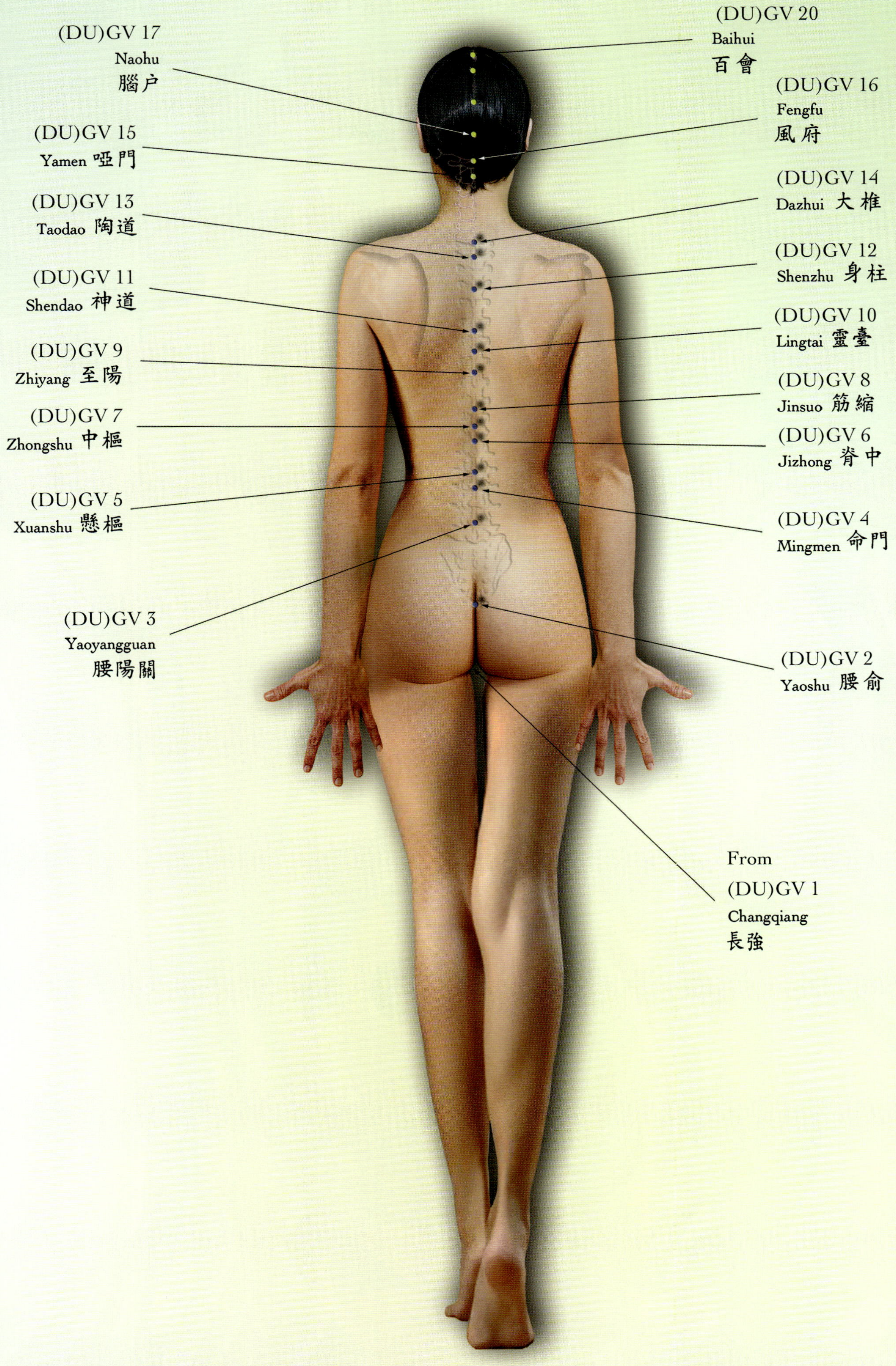

Course of Du Mai vessel

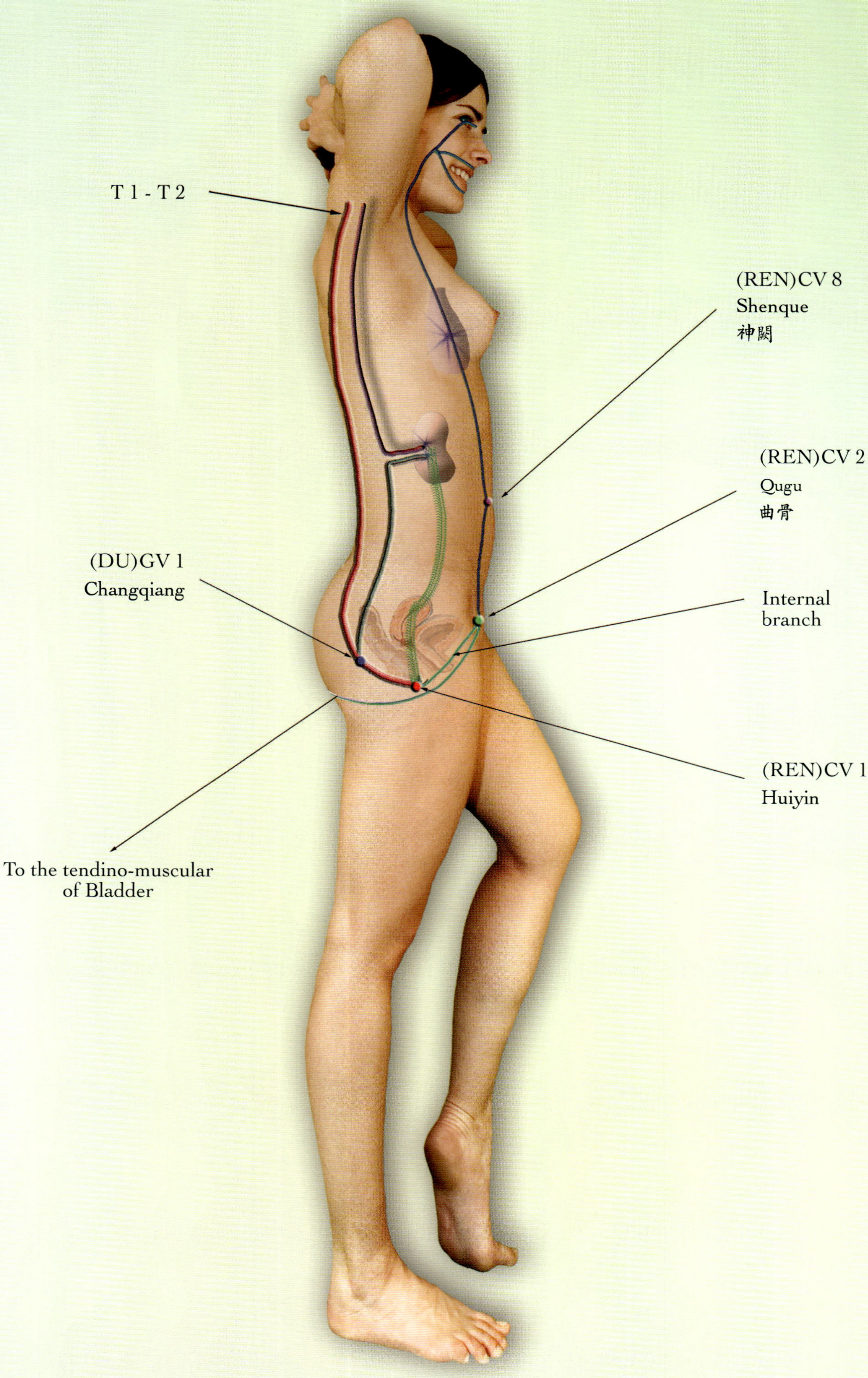

Internal branches of Du Mai vessel

Course of Du Mai vessel

Points of the Du Mai vessel (Governor vessel)

(DU)GV 1 Changqiang
Below the tip of the coccyx, at the midpoint of the line connecting the tip of the coccyx and anus.

(DU)GV 2 Yaoshu
On the sacrum and on the posterior midline, just at the sacral hiatus.

(DU)GV 3 Yaoyangguan
On the low back and on the posterior midline, in the depression below the spinous process of the 4th lumbar vertebra.

(DU)GV 4 Mingmen
On the low back and on the posterior midline, in the depression below the spinous process of the 2nd lumbar vertebra.

(DU)GV 5 Xuanshu
On the low back and on the posterior midline, in the depression below the spinous process of the 1st lumbar vertebra.

(DU)GV 6 Jizhong
On the back and on the posterior midline, in the depression below the spinous process of the 11th thoracic vertebra.

(DU)GV 7 Zhongshu
On the back and on the posterior midline, in the depression below the spinous process of the 10th thoracic vertebra.

(DU)GV 8 Jinsuo
On the back and on the posterior midline, in the depression below the spinous process of the 9th thoracic vertebra.

(DU)GV 9 Zhiyang
On the back and on the posterior midline, in the depression below the spinous process of the 7th thoracic vertebra.

(DU)GV 10 Lingtai
On the back and on the posterior midline, in the depression below the spinous process of the 6th thoracic vertebra.

(DU)GV 11 Shendao
On the back and on the posterior midline, in the depression below the spinous process of the 5th thoracic vertebra.

(DU)GV 12 Shenzhu
On the back and on the posterior midline, in the depression below the spinous process of the 3rd thoracic vertebra.

(DU)GV 13 Taodao
On the back and on the posterior midline, in the depression below the spinous process of the 1st thoracic vertebra.

(DU)GV 14 Dazhui
On the posterior midline, in the depression below the 7th cervical vertebra.

(DU)GV 15 Yamen
On the nape, 0.5 Cun directly above the midpoint of the posterior hairline, below the 1st cervical vertebra.

(DU)GV 16 Fengfu
On the nape, 1 Cun directly above the midpoint of the posterior hairline, directly below the external occipital protuberance, in the depression between the trapezius muscle of both sides.

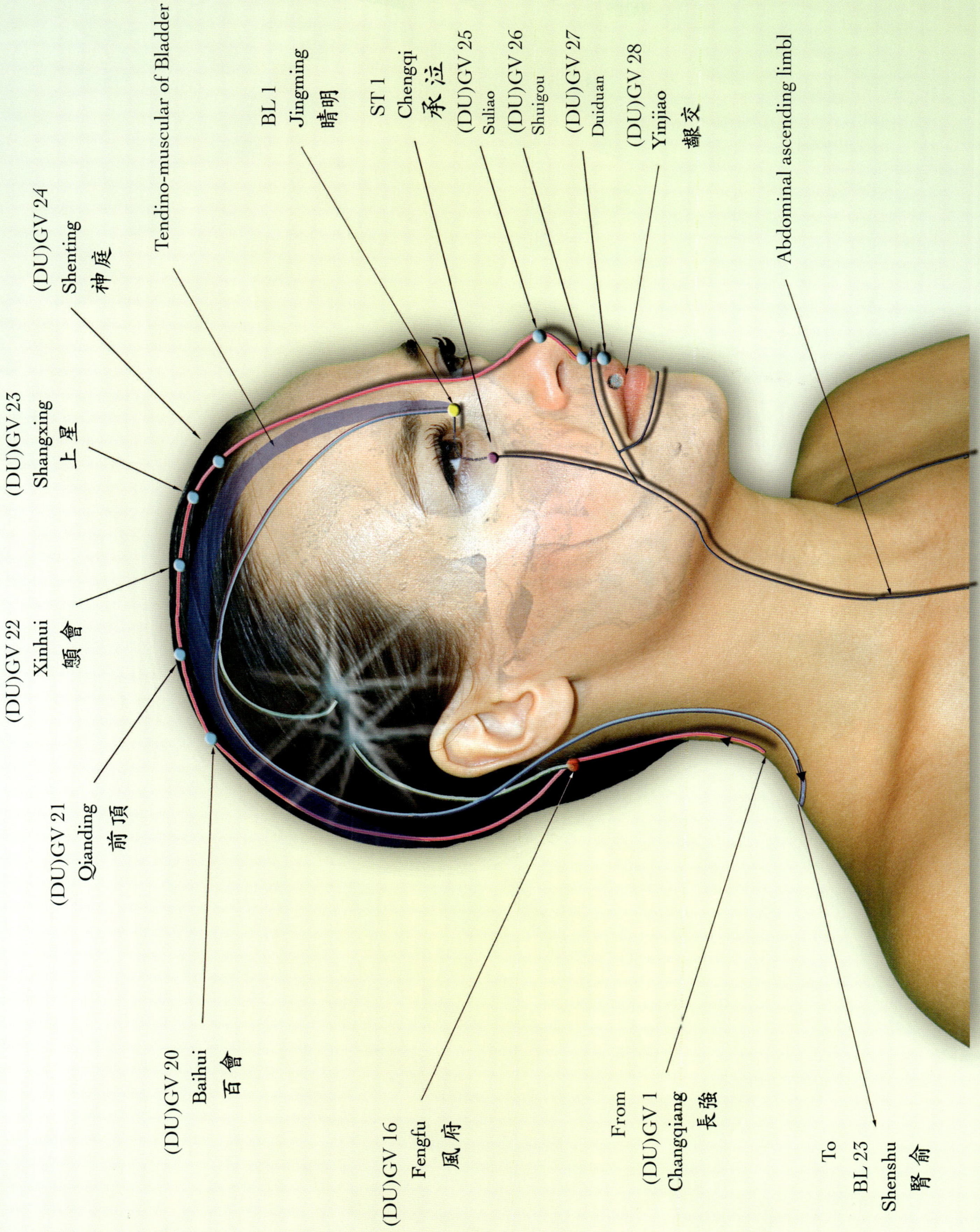

Cephalic course of Du Mai vessel

(DU)GV 17 Naohu
On the head, directly 2.5 Cun above the midpoint of the posterior hairline, 1.5 Cun above (DU)GV 16 Fengfu, in the depression on the upper border of the external occipital protuberance.

(DU)GV 18 Qiangjian
On the head, directly 4 Cun above the midpoint of the posterior hairline and 1.5 Cun above (DU)GV 17 Naohu

(DU)GV 19 Houding
On the head, directly 5.5 Cun above the midpoint of the posterior hairline and 3 Cun above (DU)GV 17 Naohu.

(DU)GV 20 Baihui
On the head, directly 5 Cun above the midpoint of the anterior hairline, at the midpoint of the line connecting the apexes of both ears.

(DU)GV 21 Qianding
On the head, directly 3.5 Cun above the midpoint of the anterior hairline and 1.5 Cun anterior to (DU)GV 20 Baihui.

(DU)GV 22 Xinhui
On the head, directly 2 Cun above he midpoint of the anterior hairline and 3 Cun anterior to (DU)GV 20 Baihui between the philtrum and upper lip.

(DU)GV 23 Shangxing
On the head, directly 1 Cun above the midpoint of the anterior hairline.

(DU)GV 24 Shenting
On the head, directly 0.5 Cun above the midpoint of the anterior hairline.

(DU)GV 25 Suliao
On the face, at the centre of the nose apex.

(DU)GV 26 Shuigou
On the face, at the junction of the upper third and middle third of the philtrum.

(DU)GV 27 Duiduan
On the face, on the labial tubercle of the upper lip, on the vermilion border.

(DU)GV 28 Yinjiao
Inside of the upper lip, at the junction of the labial frenum and upper gum.

SI 3 Houxi
At the 'junction of the red and white skin' along the ulnar border of the hand, at the ulnar end of the distal palmar crease, proximal to the 5th metacarpophalangeal joint when a hollow fist is made.

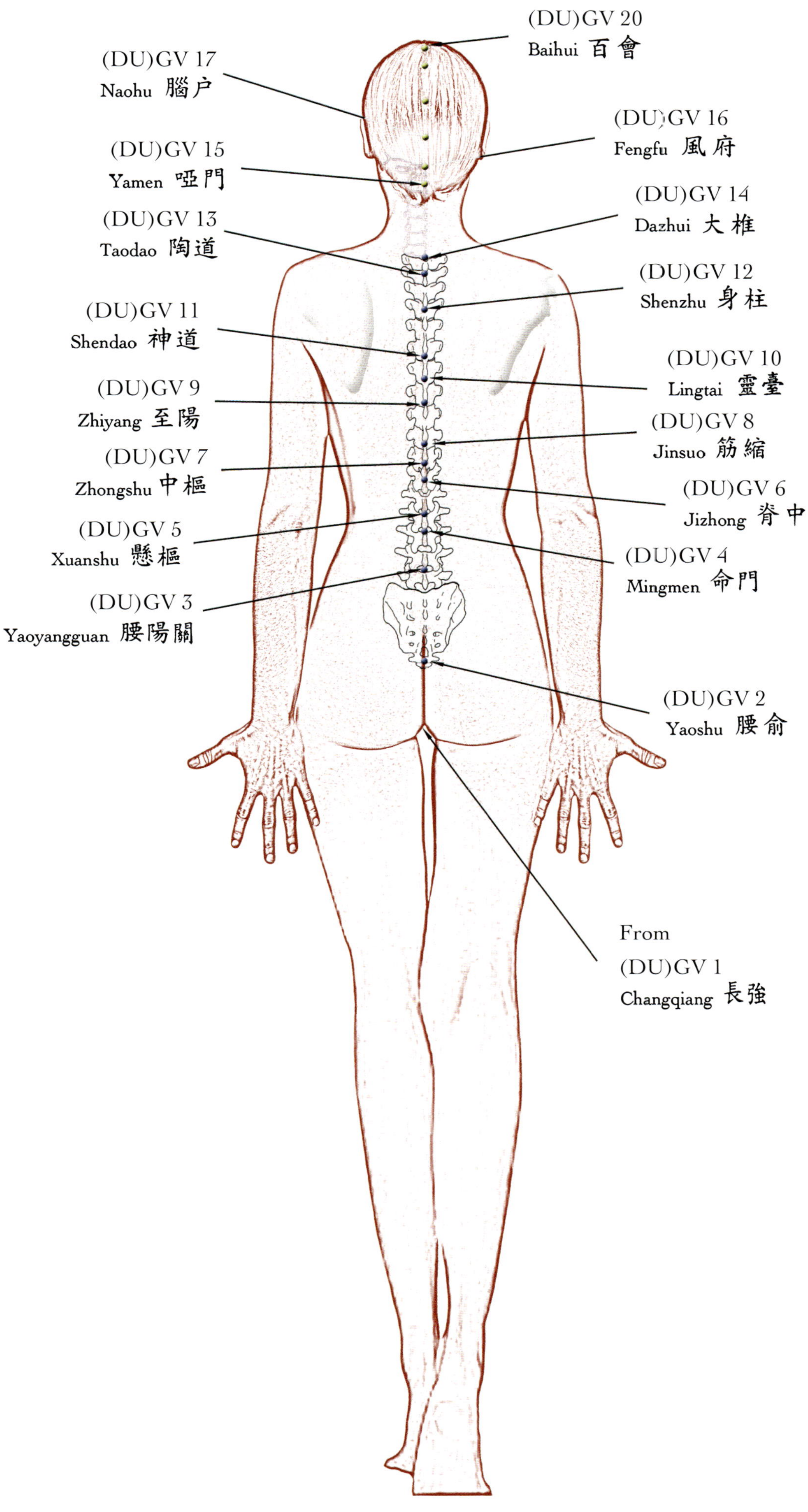

Points of the Du Mai (Governor vessel)

Yang Qiao Mai vessel (Yang Heel vessel)

Master point: BL 62 Shenmai

On the outer edge of the foot, in depression that is located under the lateral malleolus.

Course

It starts at point BL 63 Jinmen or at point BL 62 Shenmai according to the authors.

It then passes by point BL 61 Pushen, ascends by the external side of the leg and joins the Gall Bladder channel at point GB 35 Yangjiao. (Certain writers have it pass by BL 59 Fuyang).

It passes by the external aspect of the thigh at point GB 29 Juliao.

It follows the lateral side of the body to the shoulder, there it joins Tai Yang of the foot and hand (Urinary Bladder + Small Intestine) and vessel Yang Wei at point SI 10 Naoshu.

It crosses the Large Intestine channel at points LI 15 Jianyu and LI 16 Jugu.

It ascends to the face to join Yang Ming of the foot and hand (Stomach + Large Intestine) at points ST 4 Dicang and ST 3 Juliao.

It unites with Stomach and Ren Mai at point ST 1 Chengqi and runs to the inner canthus at point BL 1 Jingming where it joins Marvellous vessel Yin Wei Mai.

It goes up to the forehead and passes under the tip of the mastoid process at point GB 20 Fengchi, where it ends its course according to certain writers. For others, it goes on, passes by point (DU)GV 16 Fengfu and enters the brain.

Variants

Some authors say that this vessel starts at BL 1 Jingming and follows a descending course to BL 62 Shenmai and an ascending course to GB 20 Fengchi.

Niboyet thinks that point BL 59 Fuyang doesn't exist, the channel passing by GB 38 Yangfu.

According to De La Fuÿe, there are 4 more points:

- BL 63 Jinmen
- ST 11 Qishe
- ST 9 Renying
- ST 5 Daying

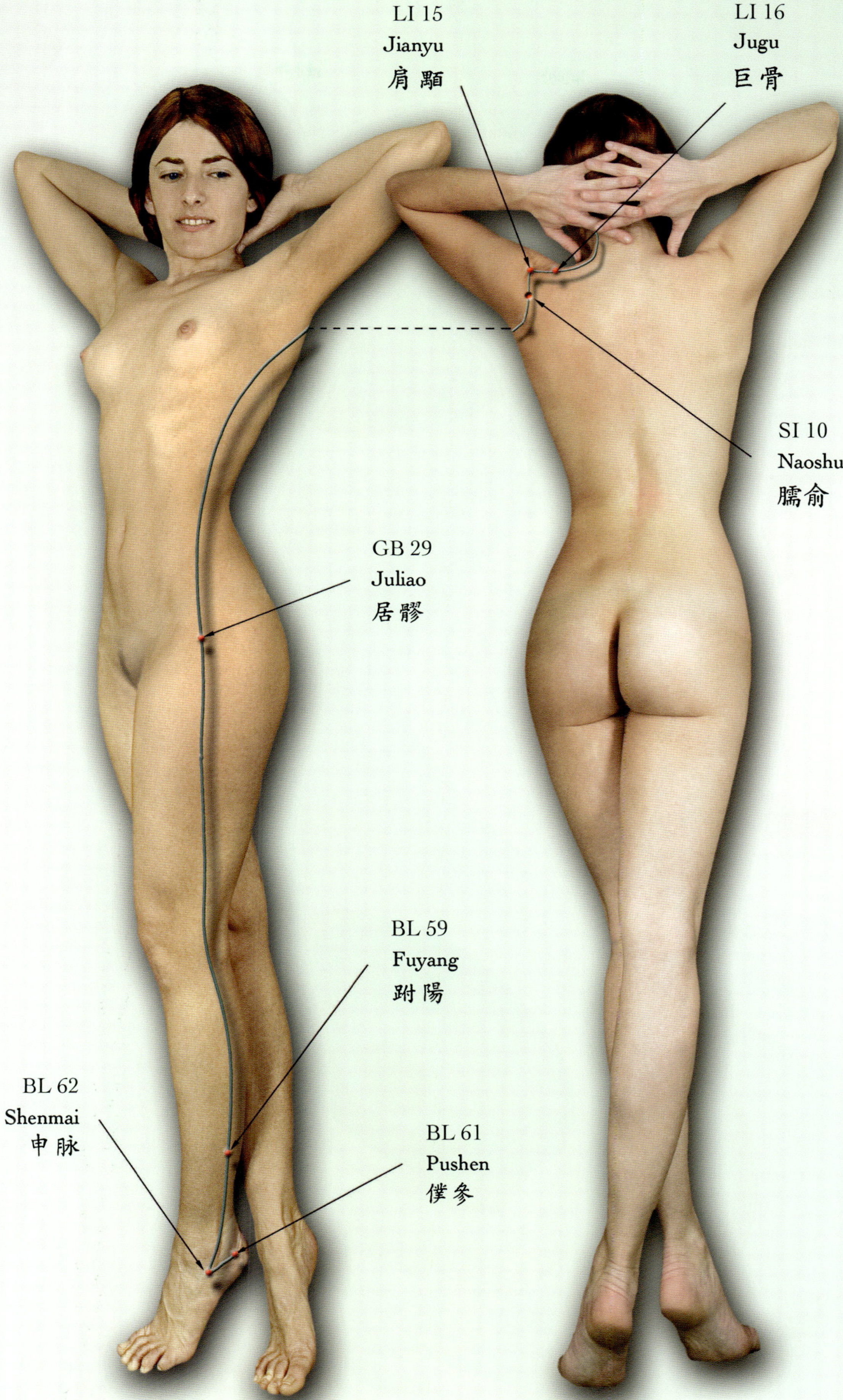

Course of Yang Qiao Mai vessel

Points of the Yang Qiao Mai vessel (Yang Heel vessel)

BL 61 Pucan
On the lateral side of the foot, posterior and inferior to the external malleolus, directly below BL 60 Kunlun, lateral to the calcaneum, at the 'junction of the red and white skin'.

BL 62 Shenmai
On the lateral side of the foot, in the depression directly below the external malleolus.

BL 59 Fuyang
On the posterior side of the leg, posterior to the lateral malleolus, 3 Cun directly above BL 60 Kunlun.

GB 29 Juliao
On the hip, at the midpoint of the line connecting the anterio-superior iliac spine and the prominence of the great trochanter.

GB 35 Yangjiao
On the lateral side of the leg, 7 Cun above the tip of the external malleolus, on the posterior border of the fibula.

SI 10 Naoshu
On the shoulder, above the posterior end of the axillary fold, in the depression below the lower border of the scapular spine.

LI 15 Jianyu
On the shoulder, superior to the deltoid muscle, in the depression anterior and inferior to the acromion when the arm is abducted or raised on the level of the shoulder.

LI 16 Jugu
On the shoulder, in the depression between the acromial extremity of the clavicle and the scapular spine.

ST 1 Chengqi
On the face, directly below the pupil, between the eyeball and the infraorbital ridge.

ST 2 Sibai
On the face, directly below the pupil, in the depression of the infraorbital foramen.

ST 3 Juliao
On the face, directly below the pupil, on the level of the lower border of the nasal ala, beside the nasolabial groove.

ST 4 Dicang
On the face, directly below the pupil, beside the mouth angle.

BL 1 Jingming
On the face, in the depression slightly above the inner canthus.

GB 20 Fengchi
On the nape, below the occipital bone, on the level of (DU)GV 16 Fengfu, in the depression between the upper ends of the sternocleidomastoid and trapezius muscles.

(DU)GV 16 Fengfu
On the nape, 1 Cun directly above the midpoint of the posterior hairline, directly below the external occipital protuberance, in the depression between the trapezius muscle of both sides.

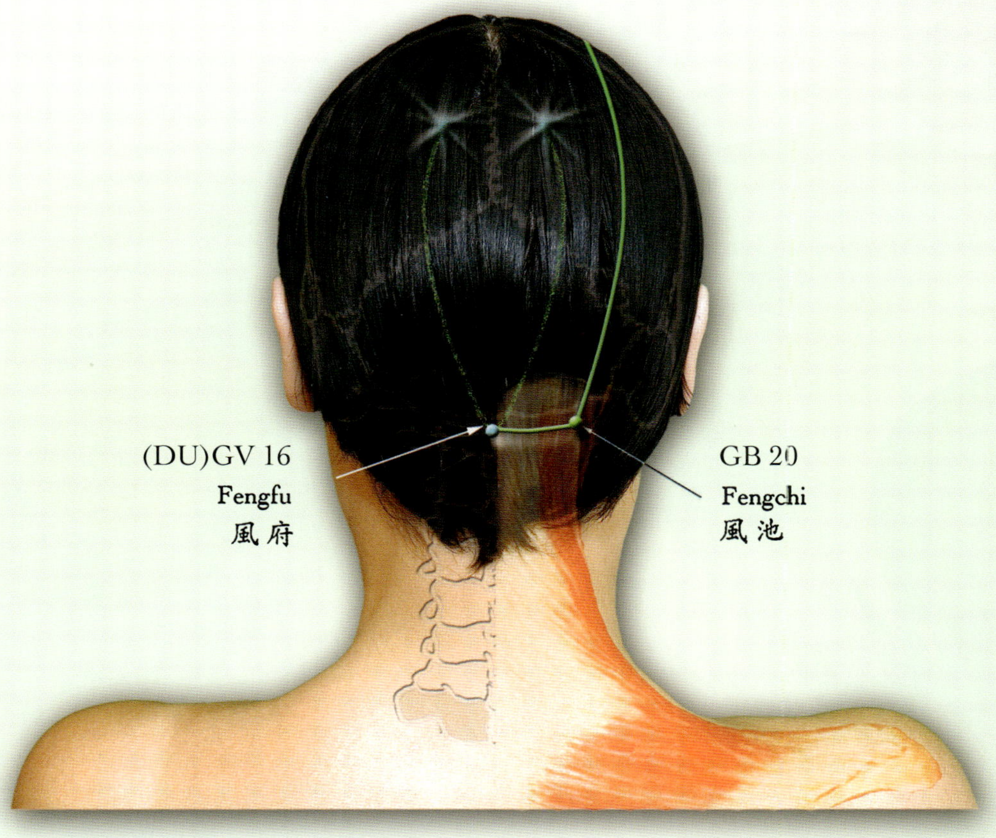

Course of Yang Wei Mai vessel

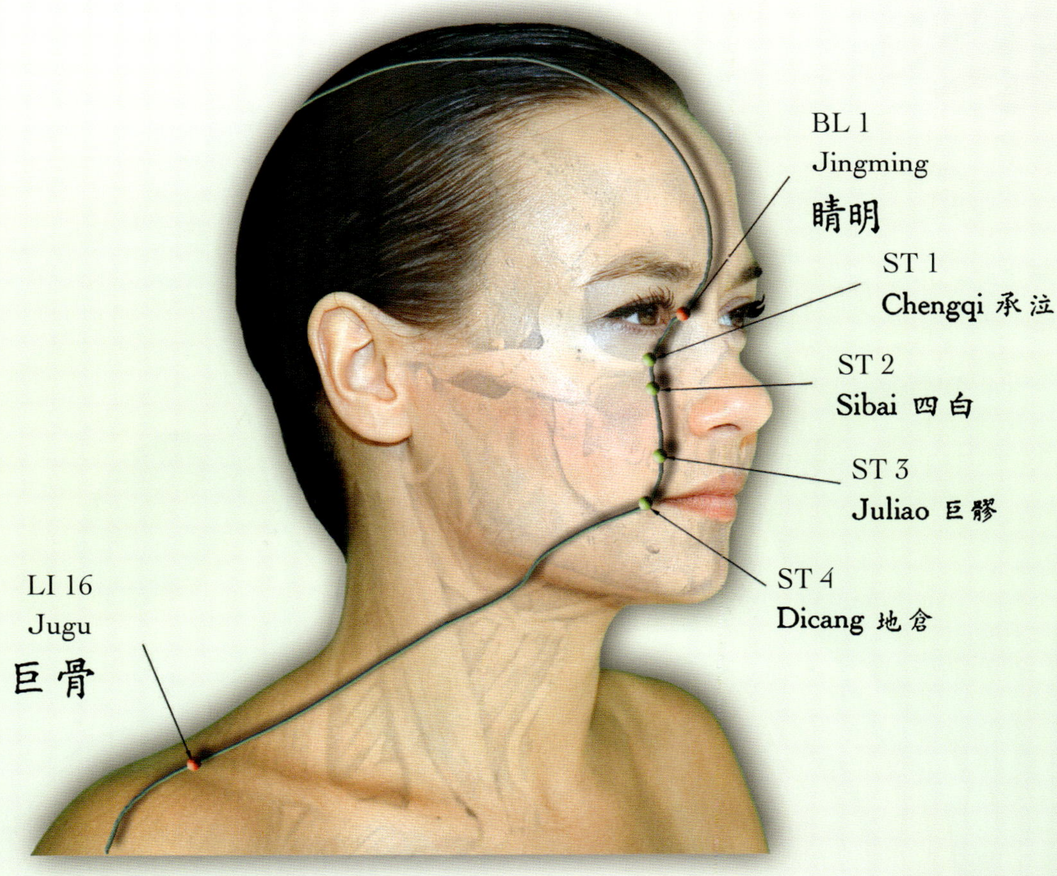

Course of Yang Wei Mai vessel

Yang Wei Mai vessel (Yang Link vessel)

Master point: (TH)TB 5 Waiguan

On the back of the forearm, on the line which connects (TH)TB 4 Yangchi and the tip of the elbow, 2 cun proximal to TH 4, between the radius and the ulna.

Course

It starts at point BL 63 Jinmen on the external border of the calcaneum in front of the external malleolus.

Its course is composed of three parts:

– the lower limb:
  It flows round the external malleolus by the front (Niboyet) leaving the Urinary Bladder channel and the Yang Qiao Mai behind to join the Gall Bladder channel at point GB 35 Yangjiao. It follows this channel along the external part of the leg, the thigh and the hip.

– on the level of the chest:
  its course is posteriolateral and runs to the posterior axillary fold at point SI 10 Naoshu.

– the head: it travels on the face and the forehead to go to the lateral part of the cranium at point ST 8 Touwei, following the Gall Bladder channel, passing successively by points GB 13, 14, 15, 16, 17, 18, 19, 20, getting progressively nearer to the posterior midline.

– It ends its course on the nape on the level of Du Mai at points (DU)GV 16 Fengfu and (DU)GV 15 Yamen.

Variants

For certain writers, the course alters from point GB 21 Jianjing. The channel then passes by the points of Du Mai, (DU)GV 15 Yamen and (DU) GV 16 Fengfu to join point GB 20 Fengchi and follow the course of Gall Bladder, passing by GB 19, 17, 15, 14, to end at GB 13 Benshen.

Nguyen Van Nghi and Darras think the channel must be composed of 4 segments:

– Lower limb: BL 63, GB 35, GB 29.
– Shoulder: BL 63, SI 10, (TH)TB 15, GB 21.
– Cranium: BL 63, GB 20 to GB 13.
– Nape: BL 63, (DU)GV 15, (DU)GV 16.

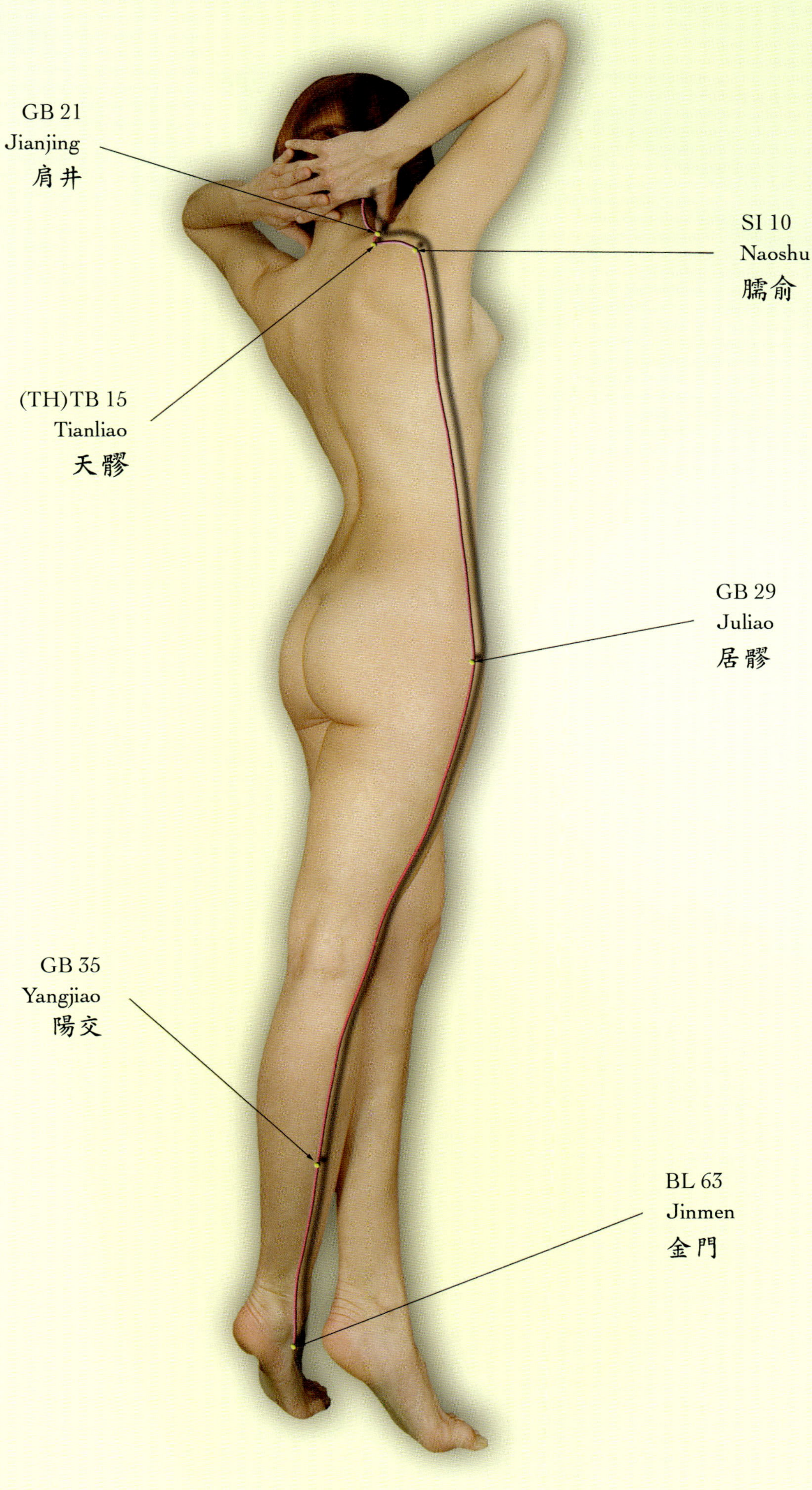

Course of Yang Wei Mai vessel

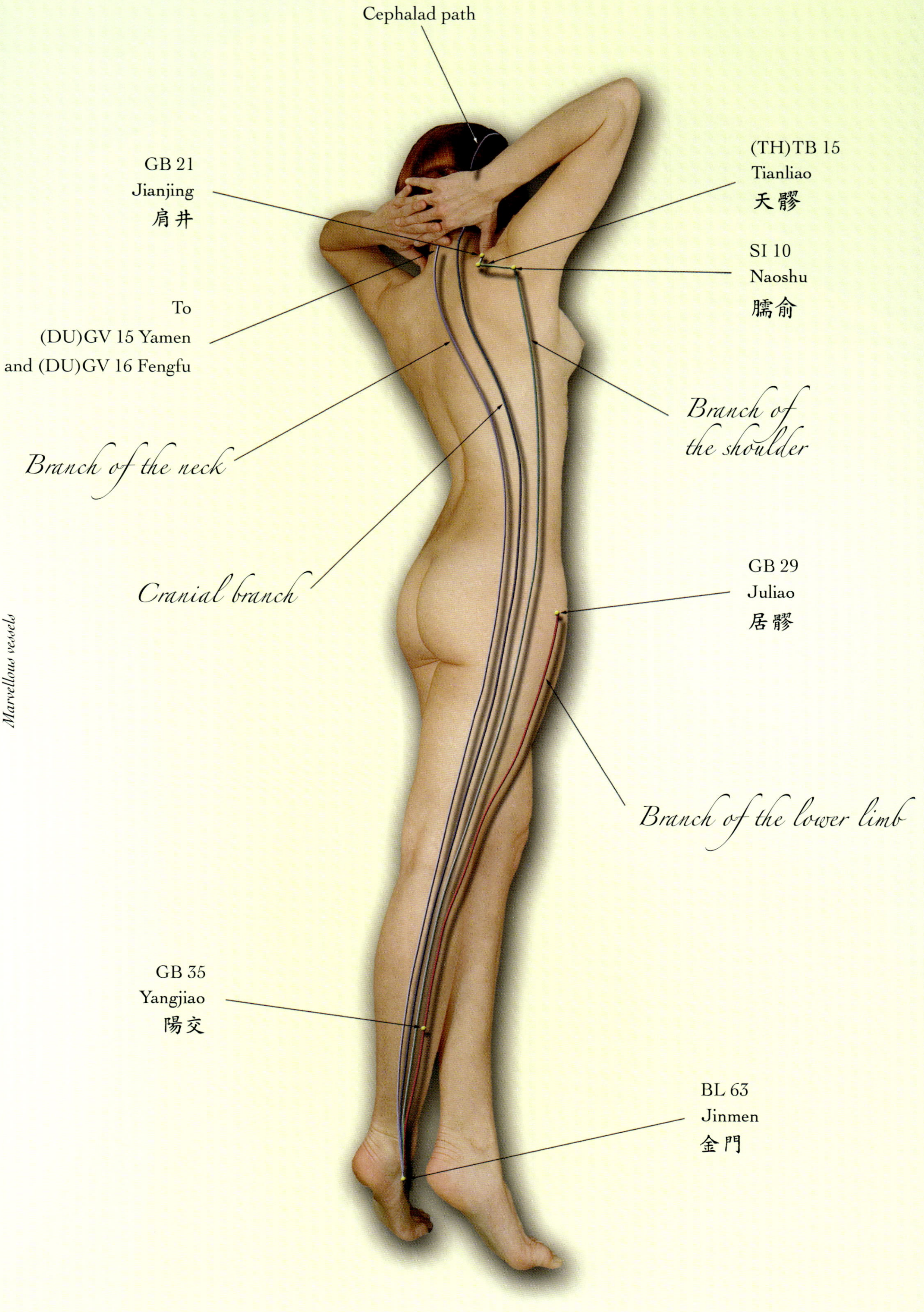

Course variations of the Yang Wei Mai vessel

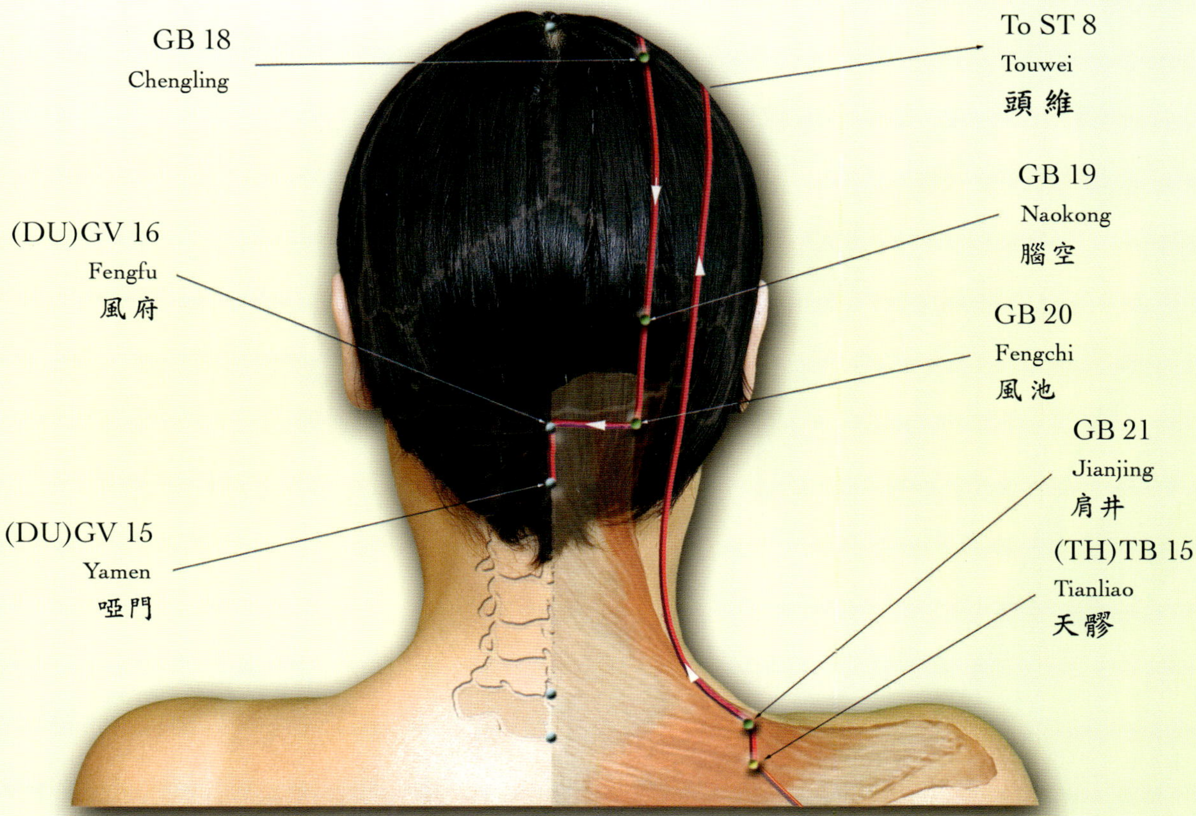

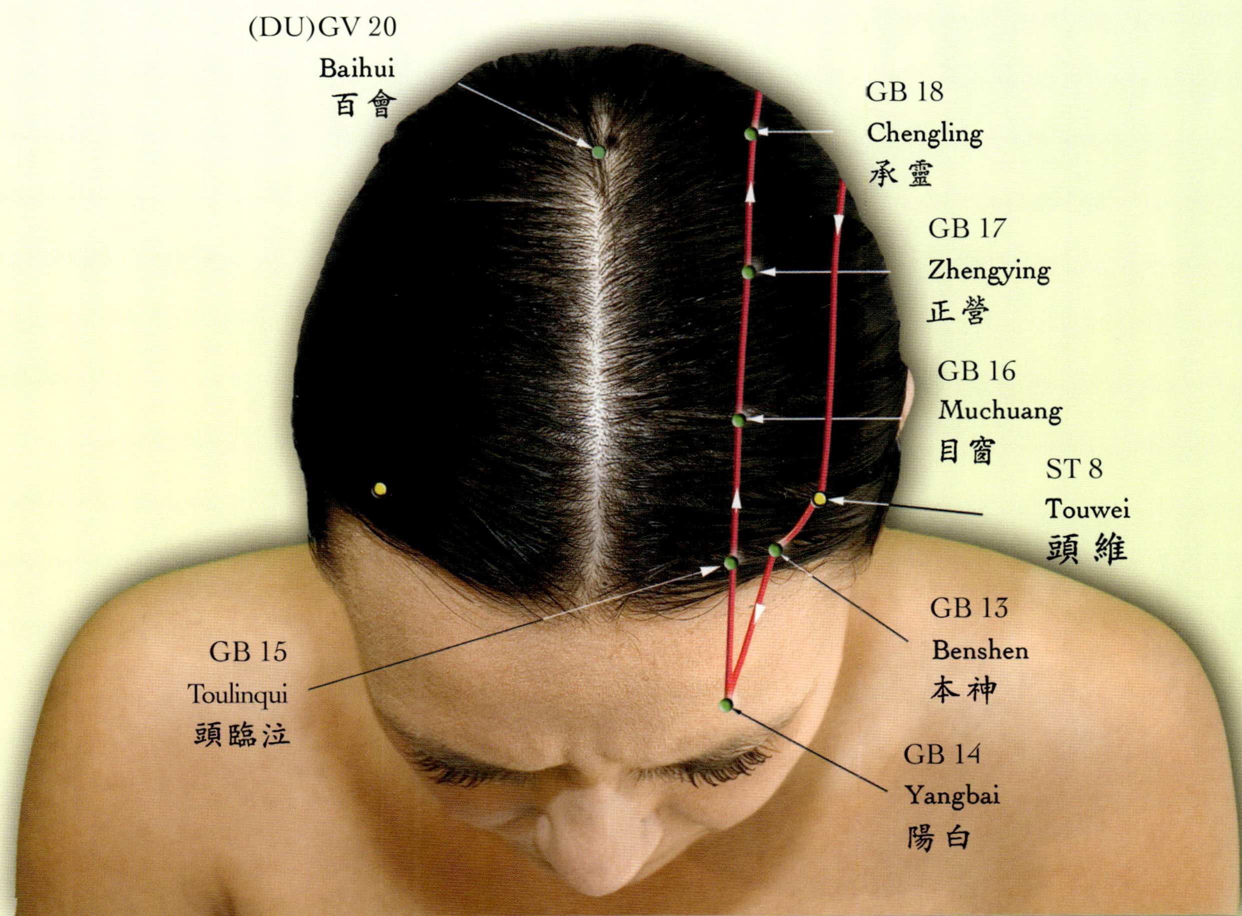

Cephalic course of Yang Wei Mai vessel

Points of the Yang Wei Mai vessel (Yang Link vessel)

BL 23 Jinmen
On the low back, below the spinous process of the 2nd lumbar vertebra, 1.5 Cun lateral to the posterior midline.

SI 10 Naoshu
On the shoulder, above the posterior end of the axillary fold, in the depression below the lower border of the scapular spine.

(TH)TB 15 Tianliao
On the scapula, between GB 21 Jianjing and SI 13 Quyuan, at the superior angle of the scapula.

ST 8 Touwei
On the lateral side of the head, 0.5 Cun above the anterior hairline at the corner of the forehead, and 4.5 Cun lateral to the midline of the head.

GB 13 Benshen
On the head, 0.5 Cun above the anterior hairline, 3 Cun lateral to (DU)GV 24 Shenting, at the junction of the medial two thirds and lateral third of the line connecting (DU)GV 24 Shenting and ST 8 Touwei.

GB 14 Yangbai
On the forehead, directly above the pupil, 1 Cun above the eyebrow.

GB 15 Toulinqi
On the head, directly above the pupil and 0.5 Cun above the anterior hairline, at the midpoint of the line connecting (DU)GV 24 Shenting and ST 8 Touwei.

GB 16 Muchuang
On the head, 1.5 Cun above the anterior hairline and 2.25 Cun lateral to the midline of the head.

GB 17 Zhengying
On the head, 2.5 Cun above the anterior hairline and 2.25 Cun lateral to the midline of the head.

GB 18 Chengling
On the head, 4 Cun above the anterior hairline and 2.25 Cun lateral to the midline of the head.

GB 19 Naokong
On the head and on the level of the upper border of the external occipital protuberance of (DU)GV 17 Naohu, 2.25 Cun lateral to the midline of the head.

GB 20 Fengchi
On the nape, below the occipital bone, on the level of (DU)GV 16 Fengfu, in the depression between the upper ends of the sterno-cleidomastoid and trapezius muscles.

GB 21 Jianjing
On the shoulder, directly above the nipple, at the midpoint of the line connecting (DU)GV 14 Dazhui and the acromion.

GB 29 Juliao
On the hip, at the midpoint of the line connecting the anteriosuperior iliac spine and the prominence of the great trochanter.

GB 35 Yangjiao
On the lateral side of the leg, 7 Cun above the tip of the external malleolus, on the posterior border of the fibula.

(DU)GV 15 Yamen
On the nape, 0.5 Cun directly above the midpoint of the posterior hairline, below the 1st cervical vertebra.

(DU)GV 16 Fengfu
On the nape, 1 Cun directly above the midpoint of the posterior hairline, directly below the external occipital protuberance, in the depression between the trapezius muscle of both sides.

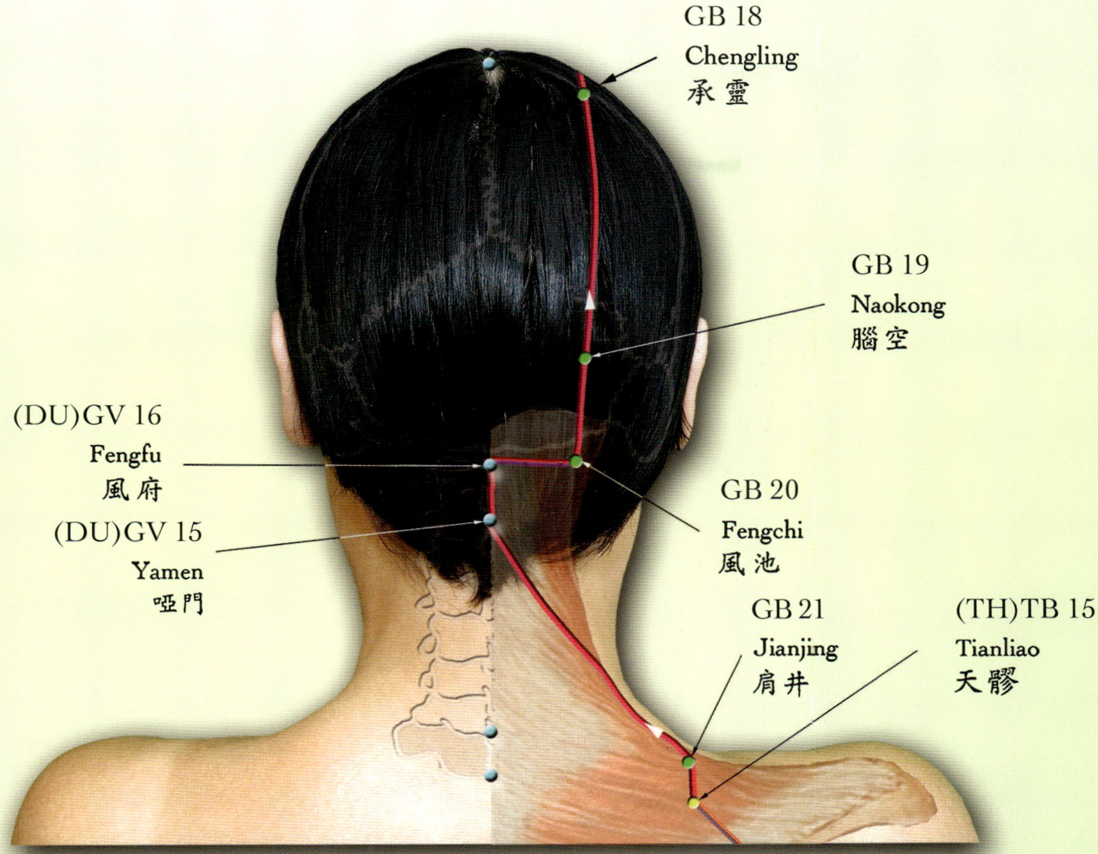

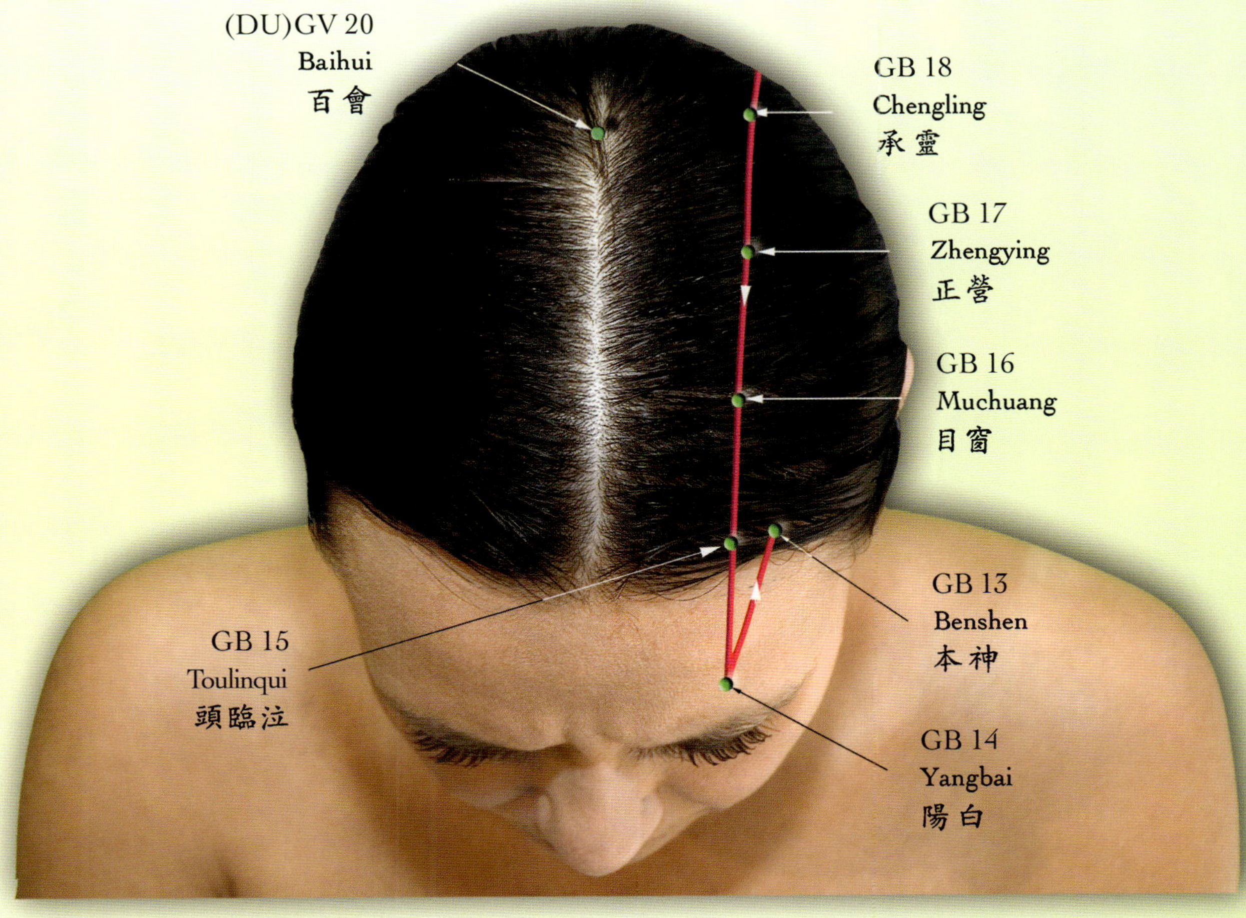

Variations in the cephalic course of Yang Wei Mai vessel

Dai Mai vessel (Belt vessel)

Master point: GB 41 Zulinqi

On the outer edge of the top the foot, behind the metatarsal-phalangeal joint of the fourth toe, in the depression that is formed at the side of the tendon of the extensor muscle of the fifth toe.

Course

It starts on the level of the 2nd lumbar vertebra.

It passes below the floating ribs.

It envelops the body on the lumbar and abdominal levels.

It passes by 3 points:

- GB 26 Daimai.
- GB 27 Wushu.
- GB 28 Weidao.

The number of 'belts' differs according to the authors:

- For Niboyet, the vessel possesses 3 belts, each taking one complete turn of the body and passing by the 3 following points: GB 26, GB 27, GB 28.

- For Nguyen Van Nghi, the vessel has only 1 belt passing, from top to bottom and on each side, by the 3 following points: GB 26, GB 27, GB 28. It then ascends to the level of the umbilicus on the midline.

This vessel is closely related on the level of the lumbar column with points (DU)GV 4 Mingmen and BL 23 Shenshu.

Other writers have it come into relation with (LR)LV 13 Zhangmen.

Points of the Dai Mai vessel

GB 26 Daimai
On the lateral side of the abdomen, 1.8 below (LR)LV 13 Zhangmen, at the crossing point of a vertical line through the free end of the 11th rib and a horizontal line through the umbilicus.

GB 27 Wushu
On the lateral side of the abdomen, anterior to the anterio-superior iliac spine, 3 Cun below the horizontal level of the umbilicus.

GB 28 Weidao
On the lateral side of the abdomen, anterior and inferior to GB 27 Wushu.

LR 13 Zhangmen
On the lateral side of the abdomen, below the free end of the 11th rib.

GB 41 Zulinqi
On the lateral side of the instep of the foot, posterior to the 4th metatarso-phalangeal joint, in the depression lateral to the tendon of the extensor muscle of the little toe.

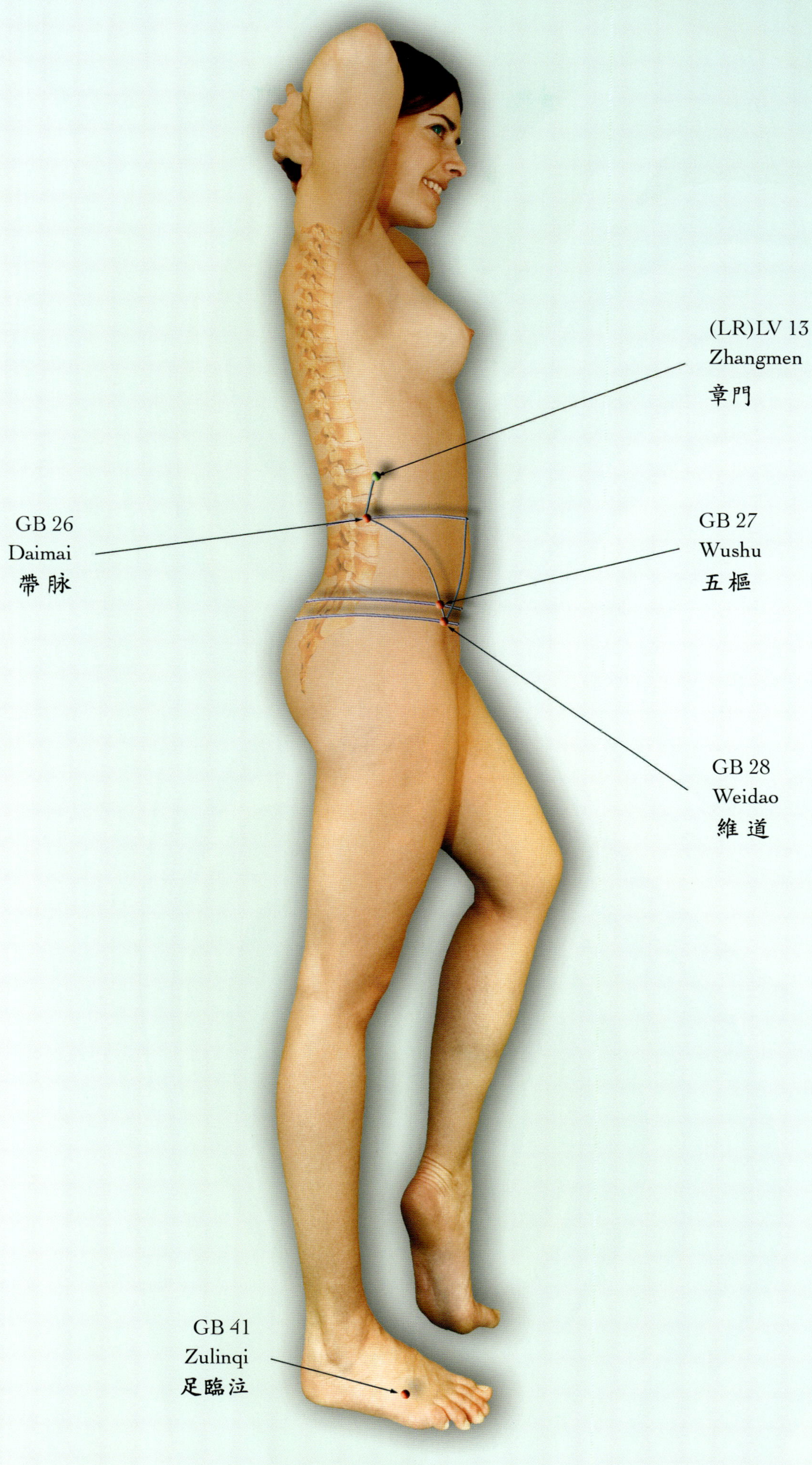

Course of Dai Mai vessel

GB 26 Daimai
On the lateral side of the abdomen, 1.8 below Zhangmen (LR)LV 13, at the crossing point of a vertical line through the free end of the 11th rib and a horizontal line through the umbilicus.

GB 27 Wushu
On the lateral side of the abdomen, anterior to the anteriosuperior iliac spine, 3 Cun below the level of the umbilicus.

GB 28 Weidao
On the lateral side of the abdomen, anterior and inferior to the anteriosuperior iliac spine, 0.5 Cun anterior and inferior to GB 27 Wushu.

(LR)LV 13 Zhangmen
On the lateral side of the abdomen, below the free end of the 11th rib.

GB 41 Zulinqi
On the lateral side of the instep of the foot, posterior to the 4th metatarsophalangeal joint, in the depression lateral to the tendon of the extensor muscle of the little toe.

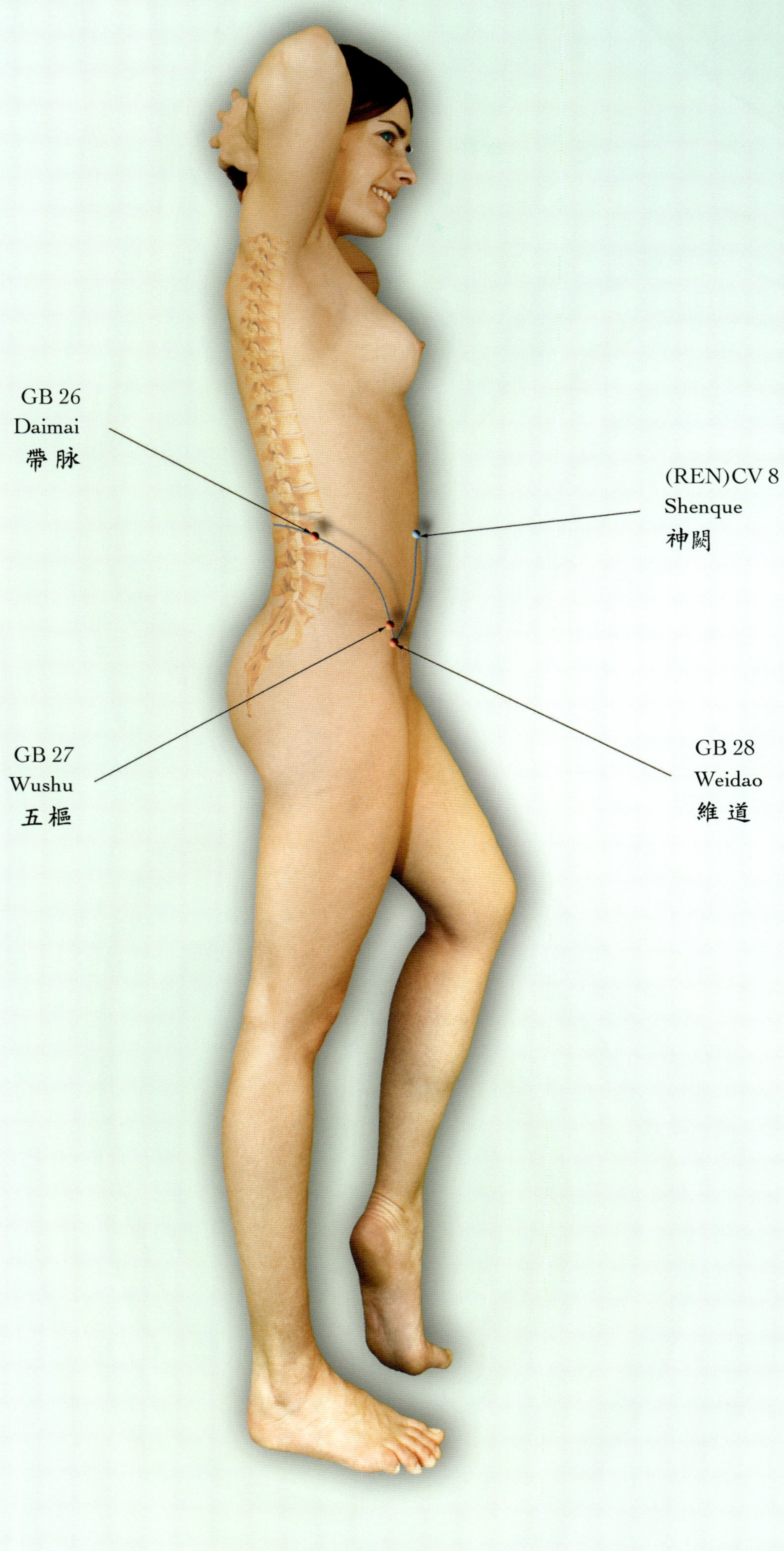

Course variations of Dai Mai vessel

Chapter II

The points of the different anatomical regions

Points of the hand

- Posterior side (dorsal)
  - Points of the Main channels
    - Shu Yang Ming (Large Intestine)
    - Shu Tai Yang (Small Intestine)
    - Shu Shao Yang (Triple Heater or Triple Burner)
  - Extra channel points
    - Curious points
    - New points
    - Points of Master Tong
- Anterior side (palmar)
  - Points of the Main channels
    - Shu Shao Yin (Heart)
    - Shu Jue Yin (Master of the Heart or Pericardium)
    - Shu Tai Yin (Lung)
  - Extra channel points
    - Curious points
    - New points
    - Points of Master Tong

Dorsal side

Points of the Main channels

Shu Tai Yin channel (Lung)

LU 11 Shaoshang
On the radial side of the distal segment of the thumb, 2 mm from the corner of the fingernail.

Shu Shao Yin channel (Heart)

(HT)HE 9 Shaochong
On the radial side of the distal segment of the little finger, 0.1 Cun from the corner of the nail.

Shu Jue Yin channel (Master of the Heart or Pericardium)

(MH)PC 9 Zhongchong
At the centre of the tip of the middle finger.

Shu Yang Ming channel (Large Intestine)

LI 1 Shangyang
On the radial side of the distal segment of the index finger, 0.1 Cun from the corner of the nail.

LI 2 Erjian
In the depression of the radial side, distal to the 2nd metacarpo-phalangeal joint when a loose fist is made.

LI 3 Sanjian
In the depression of the radial side, proximal to the 2nd metacarpo-phalangeal joint when a loose fist is made.

LI 4 Hegu
On the dorsum of the hand, between the 1st and 2nd metacarpal bones, and on the radial side of the midpoint of the 2nd metacarpal bone.

LI 5 Yangxi
At the radial end of the crease of the wrist, in the depression between the tendons of the short extensor and long extensor muscles of the thumb when the thumb is tilted upwards.

Shu Tai Yang channel (Small Intestine)

SI 1 Shaoze
On the ulnar side of the distal segment of the little finger, 0.1 Cun from the corner of the nail.

SI 2 Qiangu
At the 'junction of the red and white skin', along the ulnar border of the hand, at the ulnar end of the crease of the 5th metacarpo-phalangeal joint, when a loose fist is made.

SI 3 Houxi
At the 'junction of the red and white skin', along the ulnar border of the hand, at the ulnar end of the distal palmar crease, proximal to the 5th metacarpo-phalangeal joint, when a hollow fist is made.

SI 4 Wangu
On the ulnar border of the hand, in the depression between the proximal end of the 5th metacarpal bone and hamate bone, at the 'junction of the red and white skin'.

SI 5 Yanggu
On the ulnar border of the wrist, in the depression between the styloid process of the ulna and triangular bone.

Shu Shao Yang channel (Triple Heater or Triple Burner)

(TH)TB 1 Guanchong
On the ulnar side of the distal segment of the 4th finger, 0.1 Cun from the corner of the nail.

(TH)TB 2 Yemen
On the dorsum of the hand, between the 4th and 5th fingers, at the 'junction of the red and white skin', proximal to the margin of the web.

(TH)TB 3 Zhongzhu
On the dorsum of the hand, proximal to the 4th metacarpophalangeal joint, in the depression between the 4th and 5th metacarpal bones.

(TH)TB 4 Yangchi
At the midpoint of the dorsal crease of the wrist, in the depression of the ulnar side of the tendon of the extensor muscle of the fingers.

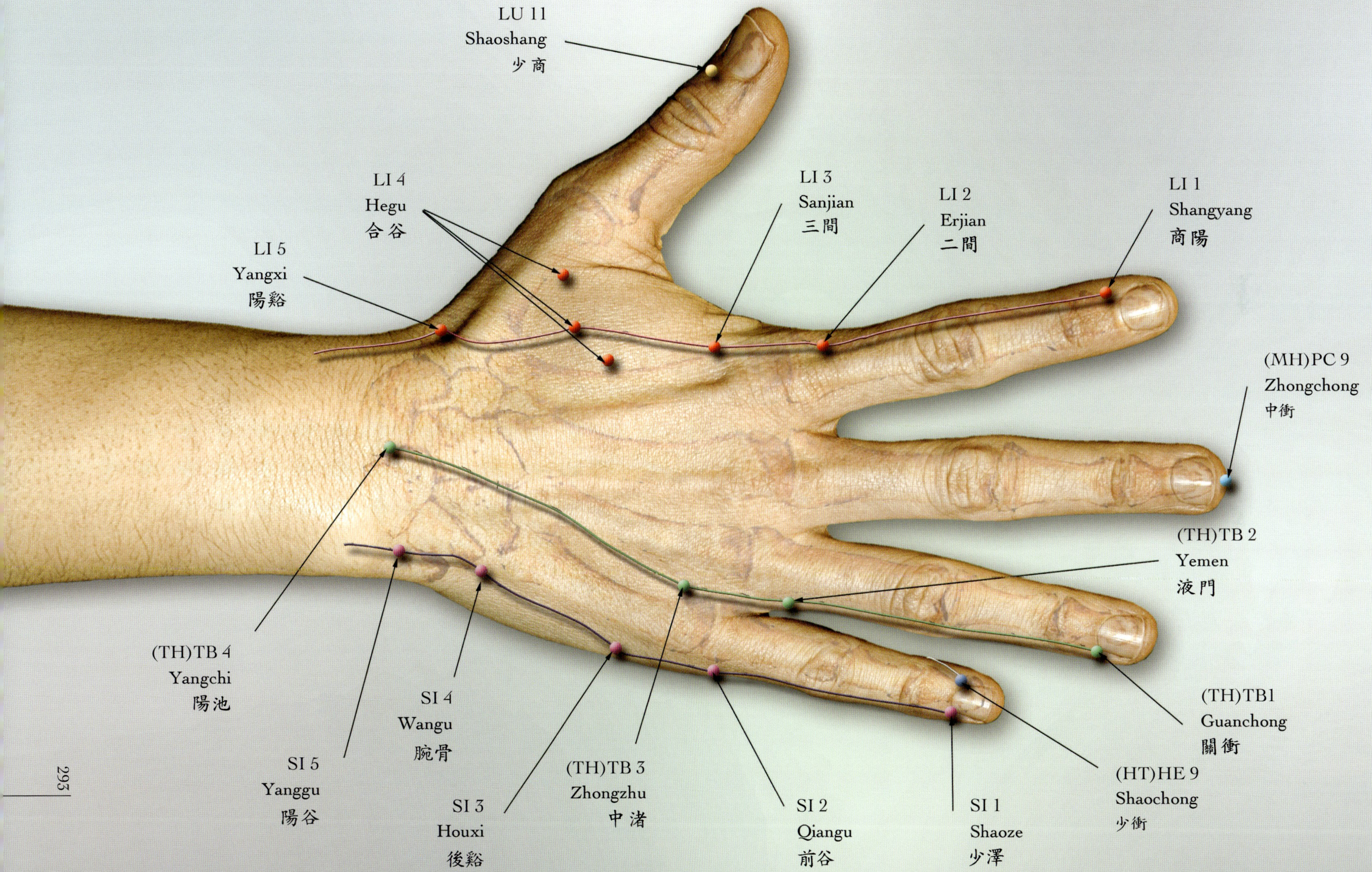
LU 11
Shaoshang
少商
LI 4
Hegu
合谷
LI 3
Sanjian
三間
LI 2
Erjian
二間
LI 1
Shangyang
商陽
LI 5
Yangxi
陽谿
(MH)PC 9
Zhongchong
中衝
(TH)TB 2
Yemen
液門
(TH)TB 4
Yangchi
陽池
SI 4
Wangu
腕骨
(TH)TB1
Guanchong
關衝
(HT)HE 9
Shaochong
少衝
SI 5
Yanggu
陽谷
(TH)TB 3
Zhongzhu
中渚
SI 3
Houxi
後谿
SI 2
Qiangu
前谷
SI 1
Shaoze
少澤

Extra channel points (MN – PC – PN)

1 MN
In the anatomical snuffbox of the (TH)TB, at the midpoint of the dorsal side of the wrist.

3 MN external Laogong
On the dorsum of the hand, in the 3rd intermetacarpian space, midpoint of the vertical line going down from the tip of the 3rd metacarpian (big bone). This point is opposite (MH)PC 8 Laogong.

4 MN Houtoudian
On the lateral side of the metacarpophalangeal joint of the little finger, on the line separating the red and white skins.

5 MN Sciatic point
On the internal surface of the metacarpo-phalangeal joint of the ring-finger on the dorsum of the hand.

6 MN Point of the pharynx
On the dorsum of the hand, on the latero-internal border of the metacarpo-phalangeal joint of the middle-finger.

7 MN Point of the neck and of the nape
On the dorsum of the hand, on the internal border of the metacarpo-phalangeal joint of the fore-finger.

8 MN Thoracic point
On the internal border of the metacarpo-phalangeal joint of the thumb on the dorsal side.

9 MN Malleolar point
On the external border of the thumb on the level of the metacarpo-phalangeal joint.

10 MN Shoulder point
On the external border of the metacarpo-phalangeal joint of the fore-finger, beside the Main channel of the Large Intestine.

12 MN Point of the Apex
On the internal and lateral side of the interphalangeal joint proximal to the ring-finger.

13 MN Ocular point
On the border of the interphalangeal joint of the thumb, on the side opposite to the Lung channel.

14 MN Qiantoudian
On the external border of the interphalangeal joint proximal to the thumb, beside the Large Intestine channel.

15 MN Point of the Vertex
On the latero-external side of the interphalangeal joint proximal to the middle finger.

16 MN Meeting point of the Yin
On the latero-external side of the interphalangeal joint proximal to the little finger.

18 MN Antipyretic point
In the external ungueal angle of the middle finger.

99PC Sanshang
Composed of three points.

- Old merchant: 0.1 Cun from the external ungueal angle of the thumb.
- Middle-aged merchant: 0.1 Cun from the nail, on the dorsal half of the thumb.
- Young merchant: corresponds to LU 11 Shaoshang.

105 PC Xiaoshixie
On the dorsal face, at the highest point situated in the métacarpo-phalangienne joint of the little finger.

106 PC Quanjian
On the dorsal face, in the most protuding point of the superior extremity of the third metacarpal.

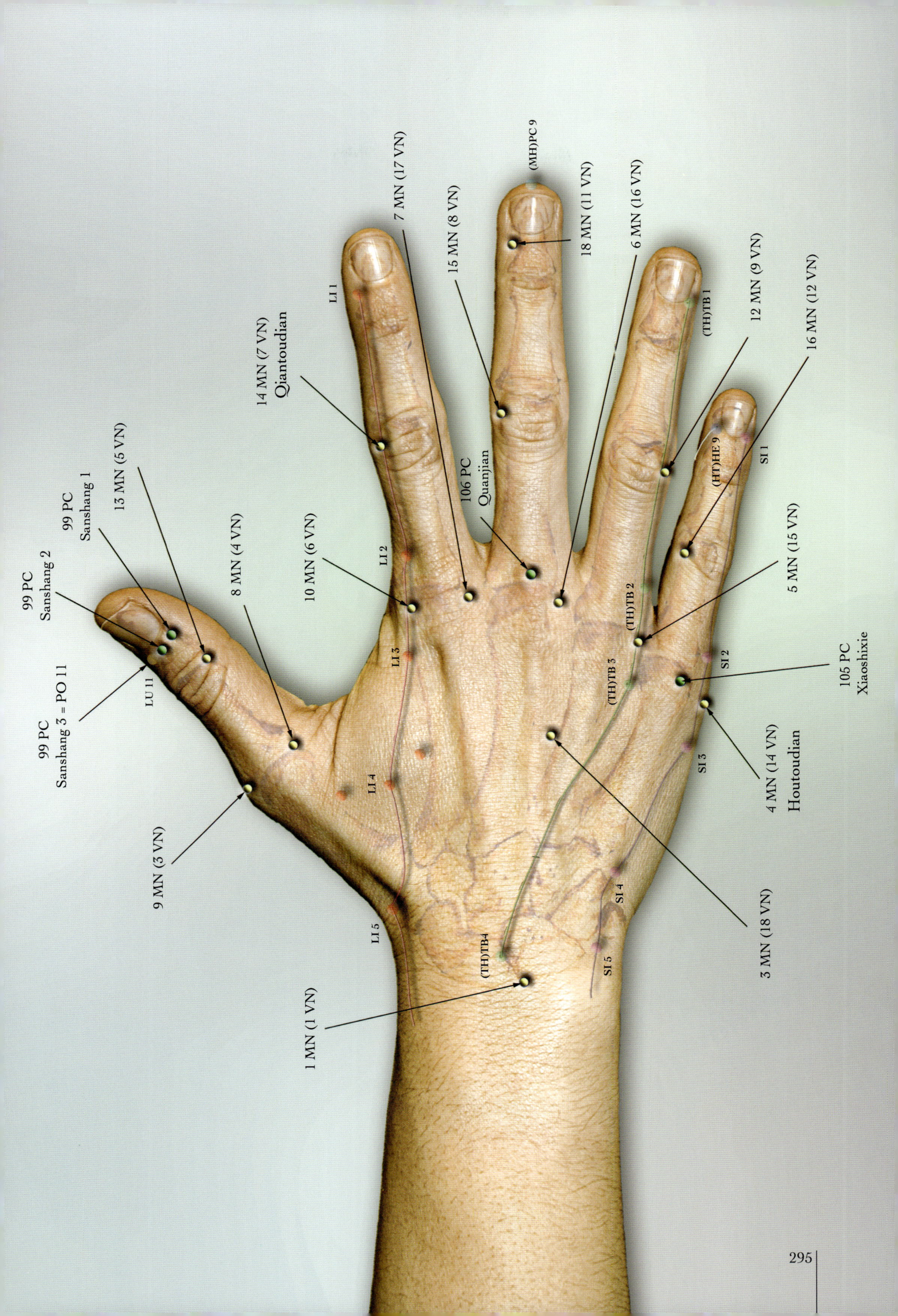

99 PC
Sanshang 2
99 PC
Sanshang 1
13 MN (5 VN)
99 PC
Sanshang 3 = PO 11
LU 11
8 MN (4 VN)
9 MN (3 VN)
14 MN (7 VN)
Qiantoudian
LI 1
LI 2
LI 3
LI 4
LI 5
10 MN (6 VN)
7 MN (17 VN)
15 MN (8 VN)
106 PC
Quanjian
(MH)PC 9
18 MN (11 VN)
6 MN (16 VN)
(TH)TB 1
(TH)TB 2
(TH)TB 3
(TH)TB4
12 MN (9 VN)
16 MN (12 VN)
(HT)HE 9
SI 1
SI 2
SI 3
SI 4
SI 5
5 MN (15 VN)
105 PC
Xiaoshixie
4 MN (14 VN)
Houtoudian
3 MN (18 VN)
1 MN (1 VN)

2 MN Yaotong 2
1.5 Cun from the crease of the wrist, beside the extensor tendons of the fore-finger and of the ring-finger, between the 5th and 4th metacarpians and between the 2nd and the 3rd metacarpians of the dorsal side. This point is double (2 MN for each hand).

86 PC Shixuan Sap Sun
At the tip of each finger (on the pad), 0.1 Cun below the nail.

87 PC Muzhijian
At the tip of the thumb, below the nail.

88 PC Xiaozhijian
At the tip of the little finger, below the nail.

100 PC Dagukong
On the dorsal side and midpoint of the interphalangeal line of the thumb.

101 PC Zhongkui Erjiaoming
On the dorsal side and midpoint of the interphalangeal line of the middle finger.

102 PC Xiaogukong
On the dorsal side, midpoint of the interphalangeal line distal to the little finger.

103 PC Yiachiamen
On the dorsal side, midpoint of the space separating the bases of the 2nd and 3rd fingers, in the interdigital space, on the junction between the red and white skin.

104 PC Erchianmen
On the dorsal side, midpoint of the space separating the bases of the 3rd and 4th fingers, (in the interdigital space between the 3rd and 4th fingers), on the junction between the red and white skin.

107 PC Baxie
On the dorsum, between the metacarpo-phalangeal joints.

108 PC Luozhen
Between the 2nd and 3rd metacarpians, 0.5 Cun from the metacarpo-phalangeal joint.

109 PC Wailaogong
On the dorsum of the hand, between the 2nd and 3rd intermetacarpian space, midpoint of the vertical line running from the wrist to the upper tip of the third metacarpian, opposite (MH)PC 8 Laogong.

110 PC Weiling
On the dorsum of the hand, between the channels of Large Intestine and Triple Heater, midpoint of the proximal angle formed by the 2nd and 3rd metacarpian.

111 PC Jingling
On the dorsum of the hand, on the line of 110 PC Weiling, in the proximal angle formed by the 3rd and 4th metacarpian.

118 PC Zhongquan
On the posterior side of the forearm, between points LI 5 Yangxi and (TH)TB 4 Yangchi.

61 PN Nuemen
On the dorsum of the hand, between the 3rd and 4th metacarpian, in the interphalangeal space, in front of point 107 PC Baxie.

62 PN Luolingwu
On the dorsum of the hand, 0.5 Cun above 108 PC Luozhen.

63 PN Shanghouxi
In the middle, between SI 3 Houxi and SI 4 Wangu of the hand.

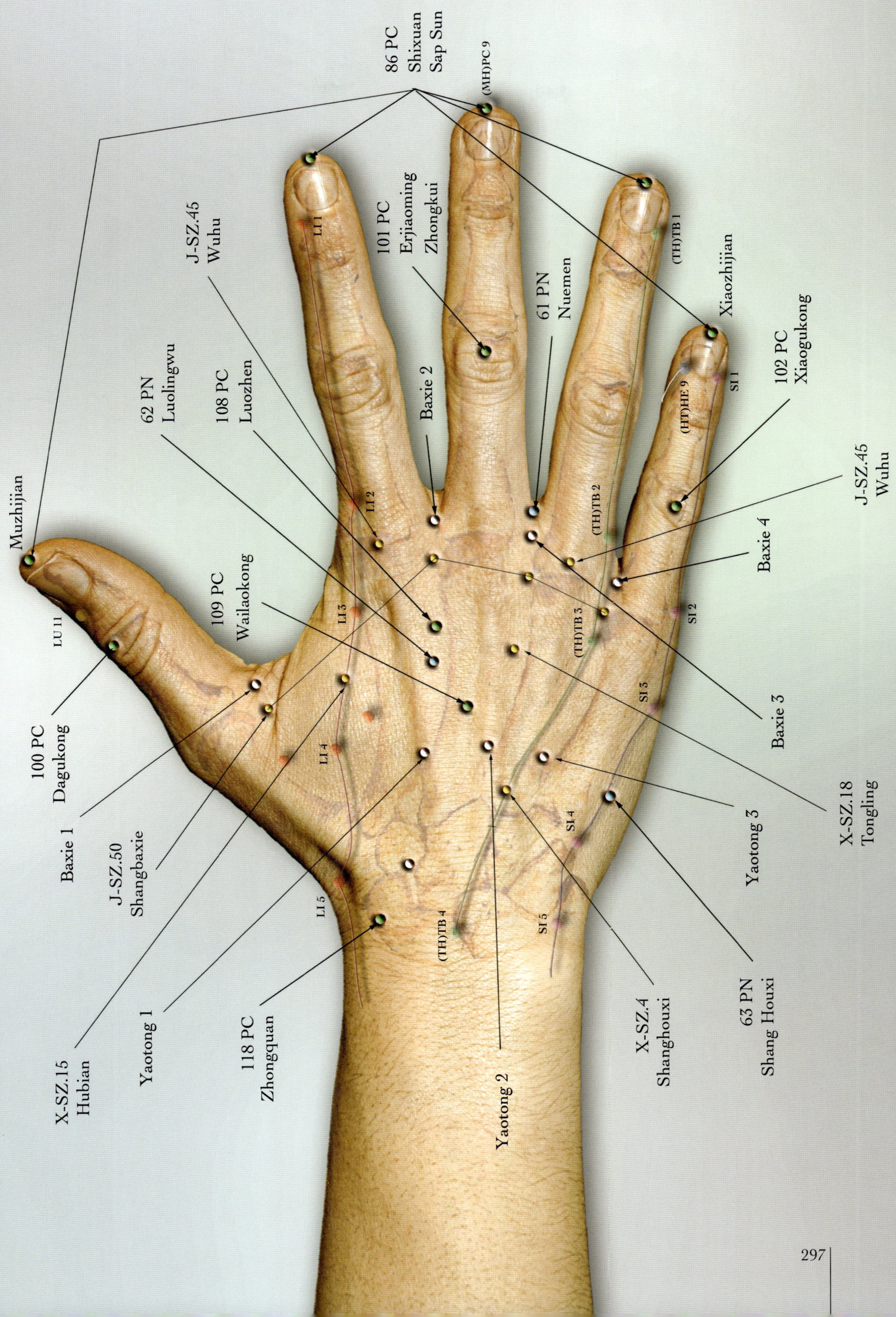

86 PC
Shixuan
Sap Sun
(MH)PC 9
LI 1
(TH)TB 1
101 PC
Erjiaoming
Zhongkui
J-SZ.45
Wuhu
Xiaozhijian
61 PN
Nuemen
102 PC
Xiaogukong
SI 1
(HT)HE 9
62 PN
Luolingwu
108 PC
Luozhen
Baxie 2
J-SZ.45
Wuhu
Muzhijian
LI 2
(TH)TB 2
Baxie 4
109 PC
Wailaokong
LI 3
(TH)TB 3
SI 2
LU 11
100 PC
Dagukong
SI 3
Baxie 3
LI 4
Baxie 1
X-SZ.18
Tongling
J-SZ.50
Shangbaxie
SI 4
Yaotong 3
LI 5
(TH)TB 4
SI 5
X-SZ.15
Hubian
Yaotong 1
118 PC
Zhongquan
Yaotong 2
X-SZ.4
Shanghouxi
63 PN
Shang Houxi

T 11.07 Zhi Si Ma
The hand being prone, draw a line passing by the midpoint of the middle joint of the fore-finger. From the centre of that line, the central point is situated 2 fen away in the direction of the internal (cubital) part. The next two points are 2 fen above and below the central point.

T 11.08 Zhi Wu Jin
The hand being prone, draw a line passing midpoint of the joint of the fore-finger.
Draw a second line 5 fen on the cubital side and divide it in three identical parts, the point being at the crossing of the most distal lines of the metacarpo-phalangeal joint.

T 11.09 Xinxi
The hand being prone, draw a line passing midpoint of the middle joint of the middle finger. From the centre of that line, the point is situated 4 fen on each side (radius and cubitus).

T 11.10 Mu Huo
The hand being prone, the point is midpoint of the middle joint of the middle finger.

T 11.11 Fei Xin
The hand being prone, draw a line passing midpoint of the middle joint of the middle finger.
The first point is 3 fen proximal to the middle joint-top joint articulation.
The second is 3 fen distal to the joint- middle joint articulation.

T 11.12 Er Jiao Ming
The hand being prone, draw a line passing midpoint of the middle-finger joint.
The first point is 3 fen proximal to the joint-middle joint articulation.
The second is 1 Cun distal to the metacarpo-phalangeal joint.

T 11.13 Dan
The hand being prone, draw a line passing midpoint of the joint of the middle-finger.
From the centre of that line, the two points are 3 fen on the right and on the left of the horizontal line.

T 11.14 Zhi San Zhong
The hand being prone, draw a line passing midpoint of the middle-joint of the ring-finger.
The points are on a line 2 fen from the midline, on the cubital side.
The first point is at the centre of that second line,
The other two points are 3 fen above and below.

T 11.15 Zhi Shen
The hand being prone, draw a line passing midpoint of the joint of the ring-finger. The lowest point is 1 Cun from the metacarpo-phalangeal joint. The other two points are 3 and 6 fen from the first point.

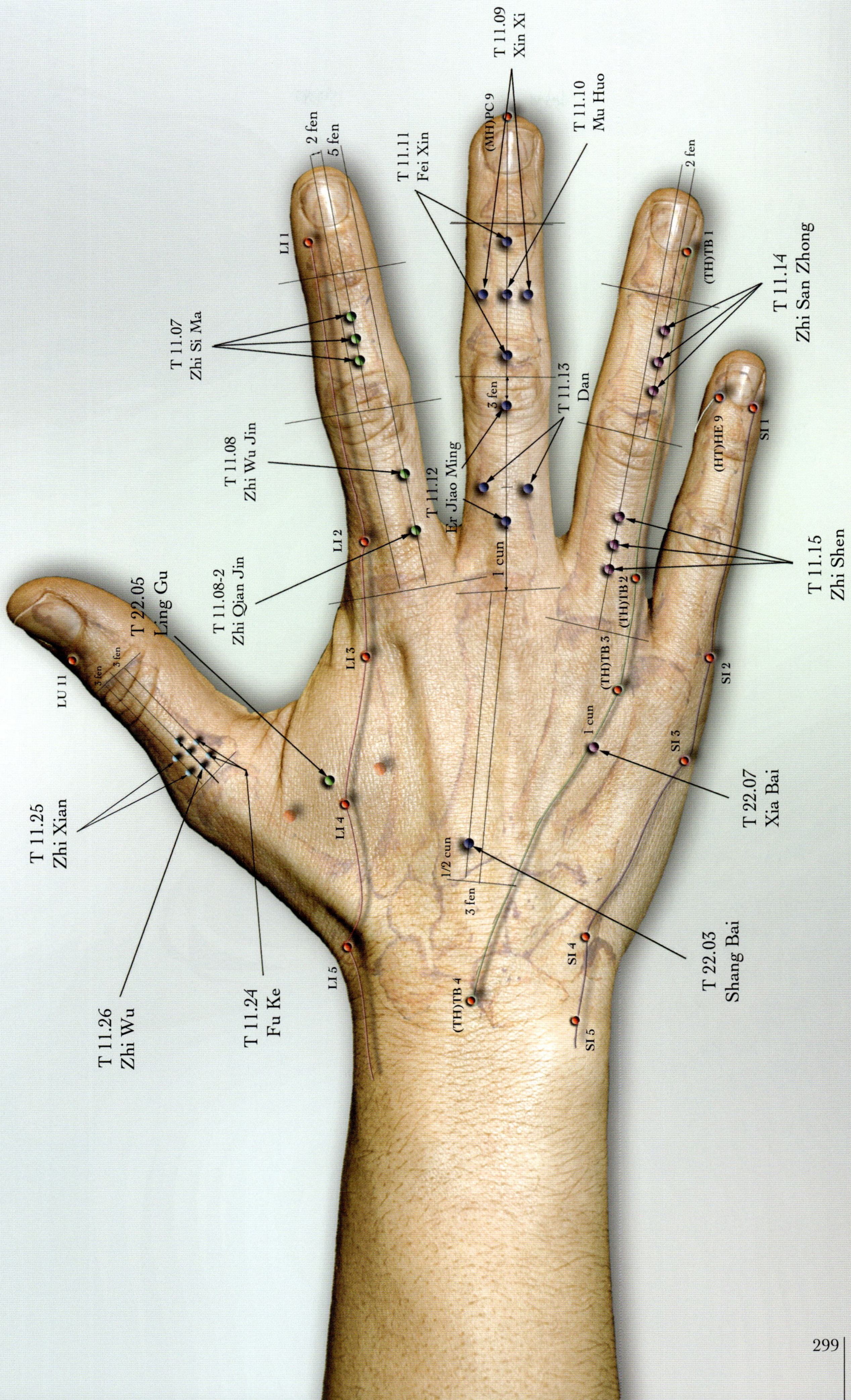

T 11.09
Xin Xi
T 11.10
Mu Huo
T 11.11
Fei Xin
(MH)PC 9
2 fen
5 fen
2 fen
LI 1
(TH)TB 1
T 11.14
Zhi San Zhong
T 11.07
Zhi Si Ma
3 fen
T 11.13
Dan
(HT)HE 9
SI 1
T 11.08
Zhi Wu Jin
T 11.12
Er Jiao Ming
LI 2
1 cun
T 11.15
Zhi Shen
T 11.08-2
Zhi Qian Jin
(TH)TB 2
T 22.05
Ling Gu
(TH)TB 3
LI 3
SI 2
LU 11
3 fen
3 fen
1 cun
SI 3
T 22.07
Xia Bai
LI 4
T 11.25
Zhi Xian
1/2 cun
3 fen
T 22.03
Shang Bai
SI 4
LI 5
(TH)TB 4
SI 5
T 11.26
Zhi Wu
T 11.24
Fu Ke

T 11.16 Huo Xi
With the hand pronated, the point is 2 fen from the internal angle of the root of the nail of the little finger.

T 11.24 Fu Ke
The hand being supine, draw a line, on the cubital side, 3 fen from the midline of the joint of the thumb.
The points are on that line, 3 and 6 fen distal from the metacarpophalangeal joint.

T 11.25 Zhi Xian
The hand being supine, draw a line, on the radial side, 3 fen from the midline of the joint of the thumb.
The points are on that line 2.5, 5, 7.5 fen distal from the metacarpo-phalangeal joint.

T 11.26 Zhi Wu
With the hand supine, draw the center line from the phalanx of the thumb. The points lie on this line at 2.5 fen, 5 fen, 7.5 fen distal to the metacarpophalangeal joint.

T 22.03 Shang Bai
The hand being prone, the point is 5 fen proximal to the metacarpo-phalangeal joint of the middle finger, 3 fen on the radial side.

T 22.04 Da Bai
On the dorsum of the hand, when a fist is made, the point corresponds to LI 3 Sanjian.

T 22.05 Ling Gu
When a fist is made, the point is in the angle between the first and the second metacarpian, 0.5 Cun above point LI 4 Hegu.

T 22.06 Zhong Bai
When a fist is made, the point is 1.5 Cun from Shang Bai (22.03) and 5 fen from the metacarpo-phalangeal joint. It corresponds to (TH)TB 3 Zhong Zhu.

T 22.07 Xia Bai
1 Cun above (TH)TB 3 Zhong Zhu.

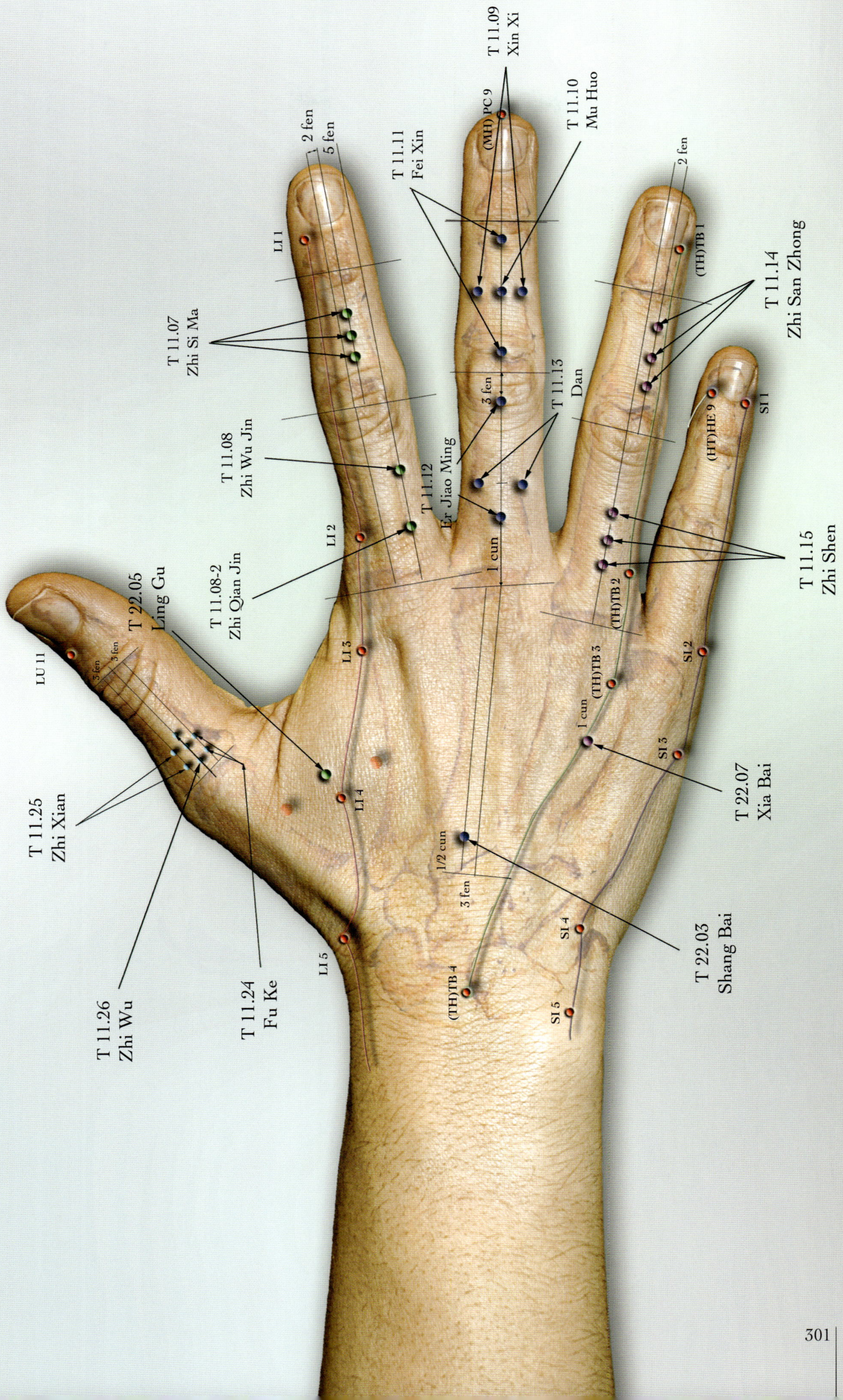
T 11.09
Xin Xi
T 11.10
Mu Huo
T 11.11
Fei Xin
(MH) PC 9
2 fen
5 fen
LI 1
T 11.07
Zhi Si Ma
T 11.14
Zhi San Zhong
(TH)TB 1
2 fen
T 11.13
Dan
3 fen
T 11.12
Er Jiao Ming
T 11.08
Zhi Wu Jin
SI 1
(HT)HE 9
T 11.08-2
Zhi Qian Jin
LI 2
1 cun
T 11.15
Zhi Shen
(TH)TB 2
T 22.05
Ling Gu
LU 11
3 fen
3 fen
LI 3
SI 2
(TH)TB 3
1 cun
SI 3
T 22.07
Xia Bai
T 11.25
Zhi Xian
LI 4
1/2 cun
3 fen
SI 4
T 22.03
Shang Bai
LI 5
(TH)TB 4
SI 5
T 11.26
Zhi Wu
T 11.24
Fu Ke

Palmar side

Points of the channels

Shu Tai Yin channel (Lung)

LU 8 Jingqu
On the radial side of the palmar surface of the forearm, 1 Cun above the crease of the wrist, in the depression between the styloid process of the radius and the radial artery.

LU 9 Taiyuan
On the radial end of the crease of the wrist, where the pulsation of the radial artery is palpable. (Do NOT needle directly into the artery!)

LU 10 Yuji
On the palm, on the midpoint of the 1st metacarpal bone, at the junction of 'the red and white skin'.

LU 11 Shaoshang
On the radial side of the distal segment of the thumb, at 2 mm back from the corner of the fingernail.

Shu Jue Yin channel (Master of the Heart or Pericardium)

(MH)PC 7 Daling
At the midpoint of the crease of the wrist, between the tendons of the little palmar muscles and radial flexor muscle of the wrist.

(MH)PC 8 Laogong
At the centre of the palm, between the 2nd and 3rd metacarpal bones, but close to the latter, and in the part touching the tip of the middle finger when a fist is made.

(MH)PC 9 Zhongchong
At the centre of the tip of the middle finger.

Shu Shao Yin channel (Heart)

(HT)HE 6 Yinxi
On the palmar side of the forearm and on the radial side of the tendon of the ulnar flexor muscle of the wrist, 0.5 Cun proximal to the crease of the wrist.

(HT)HE 7 Shenmen
On the wrist, at the ulnar end of the crease of the wrist, in the depression of the radial side of the tendon of the ulnar flexor muscle of the wrist.

(HT)HE 8 Shaofu
In the palm, between the 4th and 5th metacarpal bones, at the part of the palm touching the tip of the little finger, when a fist is made.

(HT)HE 9 Shaochong
On the radial side of the distal segment of the little finger, 0.1 Cun from the corner of the nail.

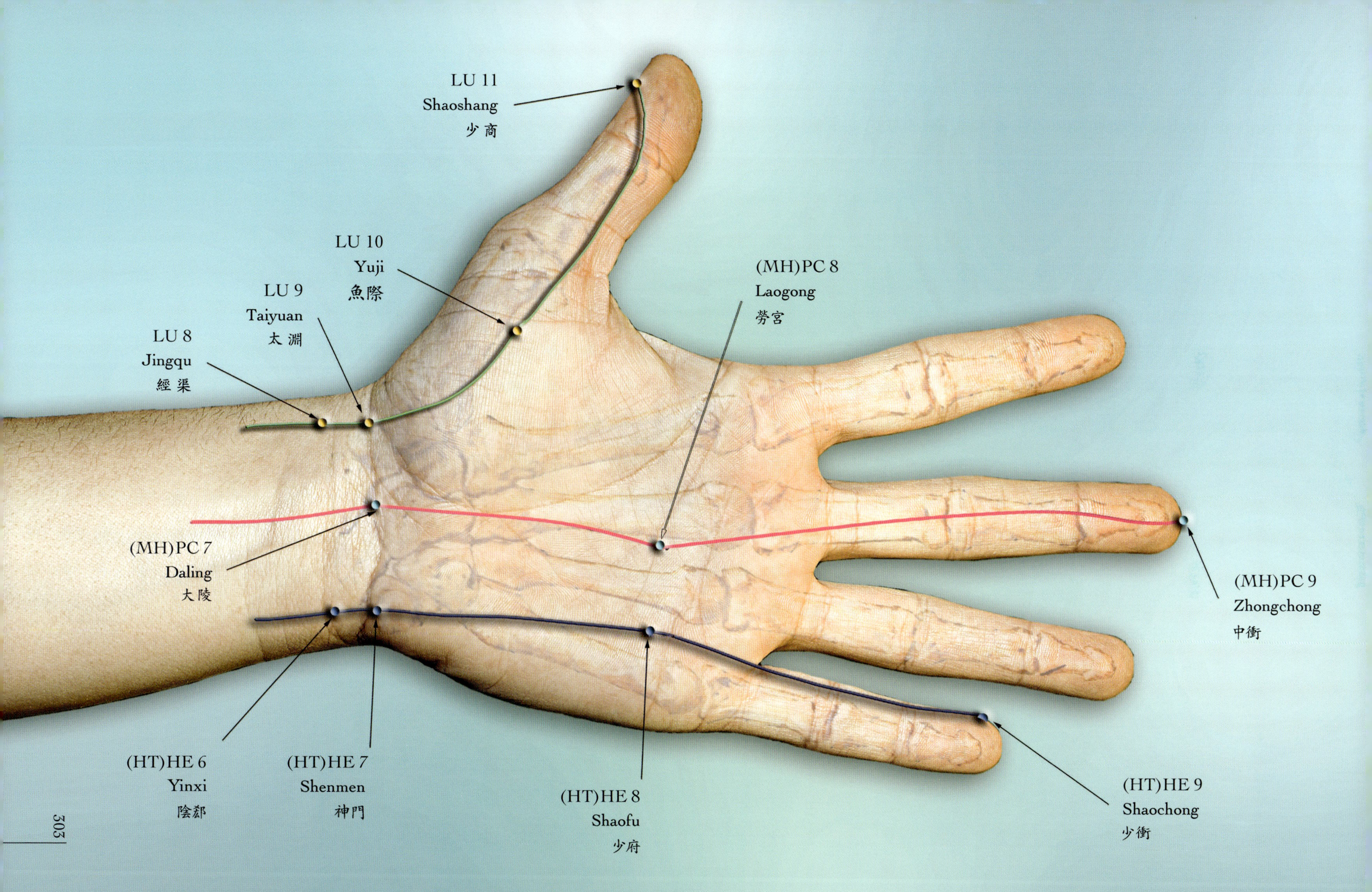
LU 11
Shaoshang
少商
LU 10
Yuji
魚際
LU 9
Taiyuan
太淵
LU 8
Jingqu
經渠
(MH)PC 8
Laogong
勞宮
(MH)PC 7
Daling
大陵
(MH)PC 9
Zhongchong
中衝
(HT)HE 6
Yinxi
陰郄
(HT)HE 7
Shenmen
神門
(HT)HE 8
Shaofu
少府
(HT)HE 9
Shaochong
少衝

Extra channel points (MN – PC – PN)

20 MN Tonsil point
Midpoint of the thenar, midpoint of the latero-internal side of the 1st metacarpal.

22 MN Xiaotianxin = J-SZ.47
In the palm of the hand, right where thenar and heel meet, 2/3 of the distance between PC 7 Daling and 21 MN (96 PC Panglaogong) vertically from (MH)PC 8 Laogong.

23 MN Heel point
1/3 of the distance between PC 7 Daling, MN 21 and PC 96 Panglaogong.

24 MN Toothache point
On the palmar side, on the external side of the 1st joint of the 4th finger.

25 MN Cough and asthma point
On the palmar side, on the internal border of the 1st joint of the 2nd finger.

26 MN Pollakiury point
On the palmar side of the hand, midpoint of the articular interline of the interphalangeal joint proximal to the little finger.

89 PC Fengquan
On the palmar side, midpoint of the joint of the 2nd finger.

90 PC Jiudianfeng
On the palmar side, midpoint of the 2nd joint of the 3rd finger.

92 PC Fangyan
On the external border,where the palmar and dorsal interphalangeal lines of the thumb meet.

93 PC Mingyan
On the palmar side of the thumb, at the internal interphalangeal tip, opposite 92 PC Fangyan, in the direction of the 2nd finger.

94 PC Sifeng
Four points on each hand, on the palmar side, midpoint of the first interphalangeal lines of the 2nd, 3rd, 4th and 5th fingers.

95 PC Shouzhongping
On the palmar side, midpoint of the metacarpo-phalangeal line of the 3rd finger.

96 PC Panglaogong = 21 MN
On the palmar side, 1.5 Cun from point (MH)PC 8 Laogong, on the radial side.

97 PC Neiyangchi
On the palmar side, 1 Cun from the central point of the crease of the wrist, that is to say opposite and 1 Cun below point (TH)TB 4 Yangchi, on the Pericardium channel.

98 PC Banmen
On the palmar side, 1 Cun from point LU 10 Yuji, in the direction of the thenar prominence.

60 PN Yatong
On the palmar side, between the 3rd and 4th metacarpian, 1 Cun above the interdigital groove.

X-SZ.16 Zhizhang
On the palmar side, between the metacarpo-phalangeal joints of the 3rd and 4th fingers.

X-SZ.17 Neihegu
On the palmar side, proximal to the head of the 2nd metacarpian.

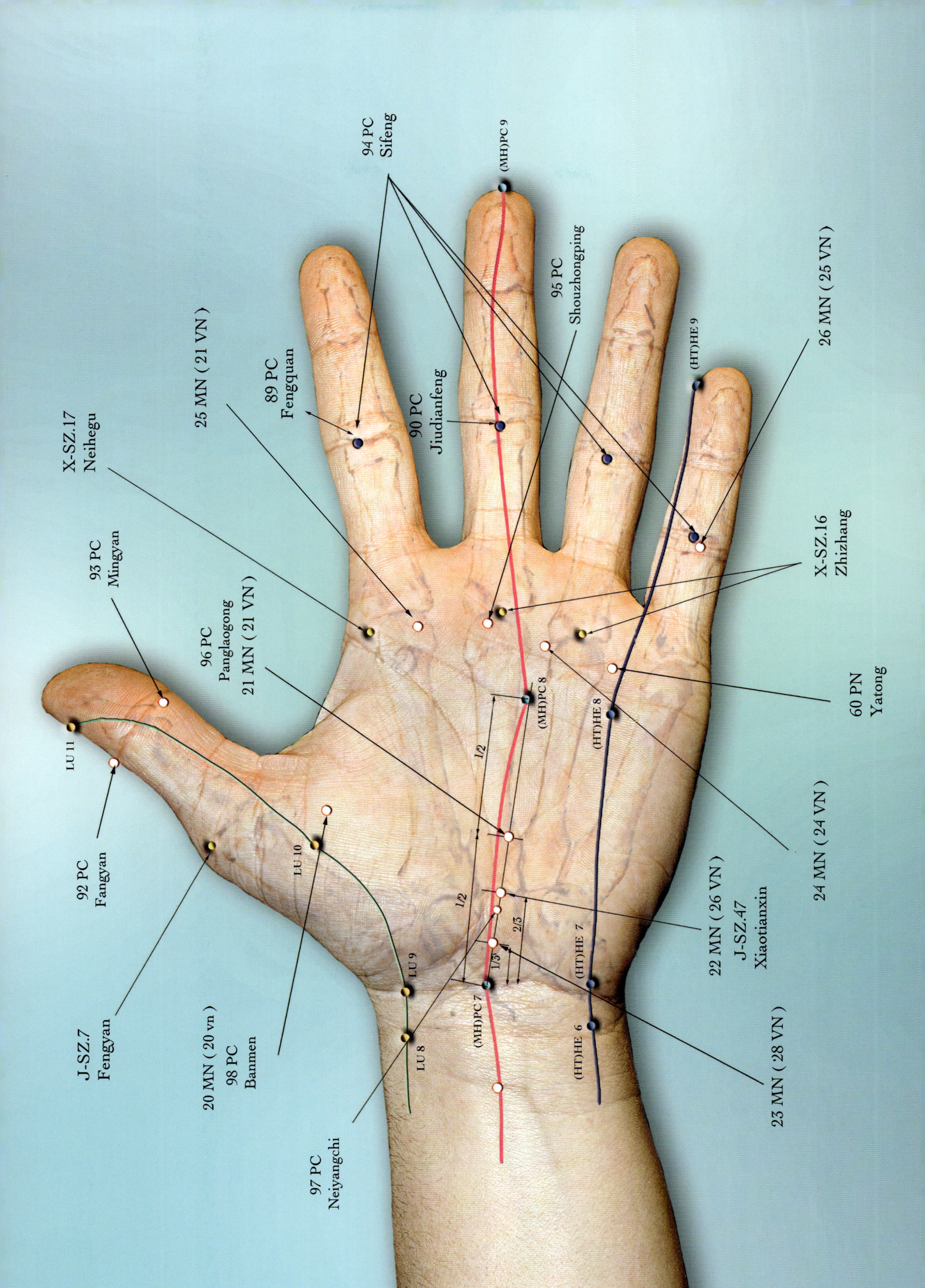
94 PC
Sifeng
(MH)PC 9
95 PC
Shouzhongping
26 MN (25 VN)
(HT)HE 9
90 PC
Jiudianfeng
89 PC
Fengquan
25 MN (21 VN)
X-SZ.17
Neihegu
93 PC
Mingyan
96 PC
Panglaogong
21 MN (21 VN)
X-SZ.16
Zhizhang
60 PN
Yatong
(MH)PC 8
(HT)HE 8
24 MN (24 VN)
LU 11
92 PC
Fangyan
LU 10
1/2
1/2
2/3
1/3
22 MN (26 VN)
J-SZ.47
Xiaotianxin
(HT)HE 7
LU 9
(MH)PC 7
J-SZ.7
Fengyan
20 MN (20 vn)
98 PC
Banmen
LU 8
(HT)HE 6
23 MN (28 VN)
97 PC
Neiyangchi

T 11.01 Da Jian
The hand being supine, draw a line passing midpoint of the phalanx of the forefinger. From the centre of that line, the point is 3 fen to the external part (radial-thumb).

T 11.02 Xiao Jian
The hand being supine, draw a line passing midpoint of the phalanx of the forefinger. From the centre of that line, find the point Da Jian which is 3 fen to the external part (radial) and 2 fen lower is point Xiao Jian (in the direction of the extremity).

T 11.03 Fu Jian
The hand being supine, draw a line passing midpoint of the middle phalanx of the forefinger. The point is 3 fen to the external part (radial) and 3 fen from the distal joint of the middle phalanx.

T 11.04 Wai Jian
The hand being supine, draw a line passing midpoint of the middle phalanx of the forefinger. The point is 3 fen to the external part (radial) and 6 fen from the distal joint of the middle phalanx. It is 3 fen above Fu Jian.

T 11.05 Zhong Jian
The hand being supine, draw a line passing midpoint of the phalanx of the forefinger. The point is at the centre of that line.

T 11.06 Huan Chao
The hand being supine, draw a line passing midpoint of the middle phalanx of the ring finger. From the centre of that line, the point is 5 fen in the internal direction (cubital).

T 11.17 Mu Xue
The hand being supine, draw a line passing 4 fen from the midline on the cubital side of the forefinger. Find the central point of that line. The 2 points are 3 fen above and below that point.

T 11.18 Pi Zhong
The hand being supine, find the midpoint of the middle phalanx of the middle finger. The two points are 2 fen above and below that central point.

T 11.19 Xin Chang
The hand being supine, draw a line 2 fen from the midline on the cubital side of the middle finger phalanx. Find the centre of that line. The 2 points are 2 fen above and below that central point.

T 11.20 Mu Yan
The hand being supine, draw a line 2 fen from the midline on the cubital side of the middle phalanx of the ring finger.
Find the centre of that line. The 2 points are 2 fen above and below that central point.

T 11.21 San Yan
The hand being supine, the point is 3 fen from the midline, on the radial side of the phalanx of the ring finger and 7 fen below the metacarpo-phalangeal crease.

T 11.22 Fu Yuan
The hand being supine, draw a line 2 fen from the midline, on the radial side of the phalanx of the ring finger. The points are respectively situated 2.5 fen, 5 fen, 7.5 fen below the metacarpo-phalangeal crease.

T 11.23 Yan Huang
The hand being supine, find the centre of the middle phalanx of the little finger where the point is situated.

T.11.26 Zhi Wu
With the hand in supination, draw the center line from the phalanx of the thumb. The points lie on this line at 2.5 fen and 7.5 fen from the metacarpophalangeal joint.

T 11.27 Wu Hu
Find the junction between the dorsal skin and the palmar skin of the thumb. Measure 2, 4, 8 fen and 21 Cun from the metacarpo-phalangeal joint.

T 22.08 Wan Shun Yi = SI 3 Houxi

T 22.09 Wan Shun Er
1 Cun above T 22.08 Wan Shun Yi, that is to say SI 3 Houxi.

T 22.10 Shu Jie = Shaofu = (HT)HE 8

T 22.11 Tu Shui
The hand being supine, the 1st point is 1 Cun above the metacarpo-phalangeal joint of the thumb, on the radial side of the phalanx.
The 2nd is 1.5 Cun above the joint.
The 3rd is 2 Cun above the joint.

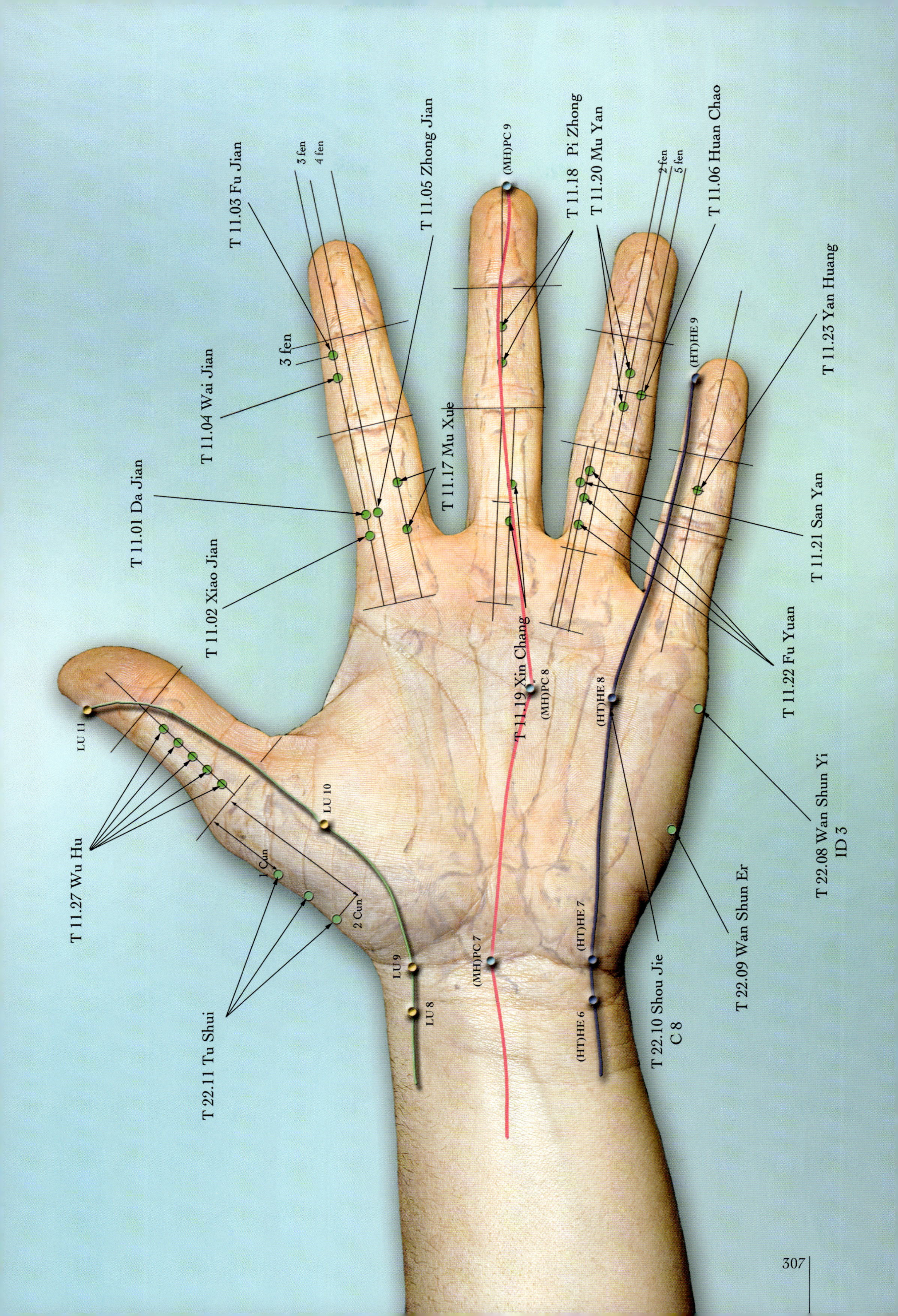

T 11.03 Fu Jian
3 fen
4 fen
T 11.05 Zhong Jian
(MH)PC 9
T 11.18 Pi Zhong
T 11.20 Mu Yan
2 fen
5 fen
T 11.06 Huan Chao
T 11.23 Yan Huang
(HT)HE 9
3 fen
T 11.04 Wai Jian
T 11.17 Mu Xue
T 11.01 Da Jian
T 11.02 Xiao Jian
T 11.21 San Yan
T 11.22 Fu Yuan
T 11.19 Xin Chang
(MH)PC 8
(HT)HE 8
T 22.08 Wan Shun Yi
ID 3
LU 11
LU 10
T 11.27 Wu Hu
1 Cun
2 Cun
T 22.09 Wan Shun Er
(HT)HE 7
T 22.10 Shou Jie
C 8
LU 9
(MH)PC 7
LU 8
(HT)HE 6
T 22.11 Tu Shui

POINTS OF THE FOREARM

- **Posterior side**
  - Points of the Main channels
    - Shu Yang Ming (Large Intestine)
    - Shu Tai Yang (Small Intestine)
    - Shu Shao Yang (Triple Heater or Triple Burner)
  - Extra channel points
    - Curious points
    - New points
    - Points of Master Tong
- **Anterior side**
  - Points of the Main channels
    - Shu Shao Yin (Heart)
    - Shu Jue Yin (Master of the Heart or Pericardium)
    - Shu Tai Yin (Lung)
  - Extra channel points
    - Curious points
    - New points
    - Points of Master Tong

Posterior side

Points of the Main channels

Shu Yang Ming channel (Large Intestine)

LI 5 Yangxi
At the radial end of the crease of the wrist, in the depression between the tendons of the short extensor and long extensor muscles of the thumb when the thumb is tilted upward.

LI 6 Pianli
With the elbow slightly flexed, on the radial side of the dorsal surface of the forearm and on the lines connecting LI 5 Yangxi and LI 11 Quchi, 3 Cun above the crease of the wrist.

LI 7 Wenliu
With the elbow flexed, on the radial side of the dorsal surface of the forearm and on the line connecting LI 5 Yangxi and LI 11 Quchi, 5 Cun above the crease of the wrist.

LI 8 Xialian
On the radial side of the dorsal surface of the forearm and on the line connecting LI 5 Yangxi and LI 11 Quchi, 4 Cun below the cubital crease.

LI 9 Shanglian
On the radial side of the dorsal surface of the forearm and on the line connecting LI 5 Yangxi and LI 11 Quchi, 3 Cun below the cubital crease.

LI 10 Shousanli
On the radial side of the dorsal surface of the forearm and on the line connecting LI 5 Yangxi and LI 11Quchi, 2 Cun below the cubital crease.

LI 11 Quchi
With the elbow flexed, at the lateral end of the cubital crease, at the midpoint of the line connecting LU 5 Chize and the external humeral epicondyle.

LI 12 Zhouliao
With the elbow flexed, on the lateral side of the upper arm, 1 Cun above LI 11 Quchi, on the border of the humerus.

Shu Shao Yang channel (Triple Heater or Triple Burner)

(TH)TB 4 Yangchi
At the midpoint of the dorsal crease of the wrist, in the depression on the ulnar side of the tendon of the extensor muscle of the fingers.

(TH)TB 5 Waiguan
On the dorsal side of the forearm and on the line connecting (TH)TB 4 Yangchi and the tip of the olecranon, 2 Cun proximal to the dorsal crease of the wrist, between the radius and ulna.

(TH)TB 6 Zhigou
On the dorsal side of the forearm and on the line connecting (TH)TB 4 Yangchi and the tip of the olecranon, 3 Cun proximal to the dorsal crease of the wrist, between the radius and ulna.

(TH)TB 7 Huizong
On the dorsal side of the forearm, 3 Cun proximal to the dorsal crease of the wrist, on the ulnar side of (TH)TB 6 Zhigou and on the radial border of the ulna.

(TH)TB 8 Sanyangluo
On the dorsal side of the forearm, 4 Cun proximal to the dorsal crease of the wrist, between the radius and ulna.

(TH)TB 9 Sidu
On the dorsal side of the forearm and on the line connecting (TH)TB 4 Yangchi and the tip of the olecranon, 5 Cun distal to the tip of the olecranon, between the radius and ulna.

(TH)TB 10 Tiangjing
On the lateral side of the upper arm, in the depression 1 Cun proximal to the tip of the olecranon when the elbow is flexed.

Shu Tai Yang channel (Small Intestine)

SI 4 Wangu
On the ulnar border of the hand, in the depression between the proximal end of the 5th metacarpal bone and hamate bone, and at the 'junction of the red and white skin'.

SI 5 Yanggu
On the ulnar border of the wrist, in the depression between the styloid process of the ulna and triangular bone.

SI 6 Yanglao
On the ulnar side of the posterior surface of the forearm, in the depression proximal to and on the radial side of the head of the ulna.

SI 7 Zhizheng
On the ulnar side of the posterior surface of the forearm and on the line connecting SI 5 Yanggu and SI 8 Xiaohai, 5 Cun proximal to the dorsal crease of the wrist.

SI 8 Xiaohai
On the medial side of the elbow, in the depression between the olecranon of the ulna and the medial epicondyle of the humerus.

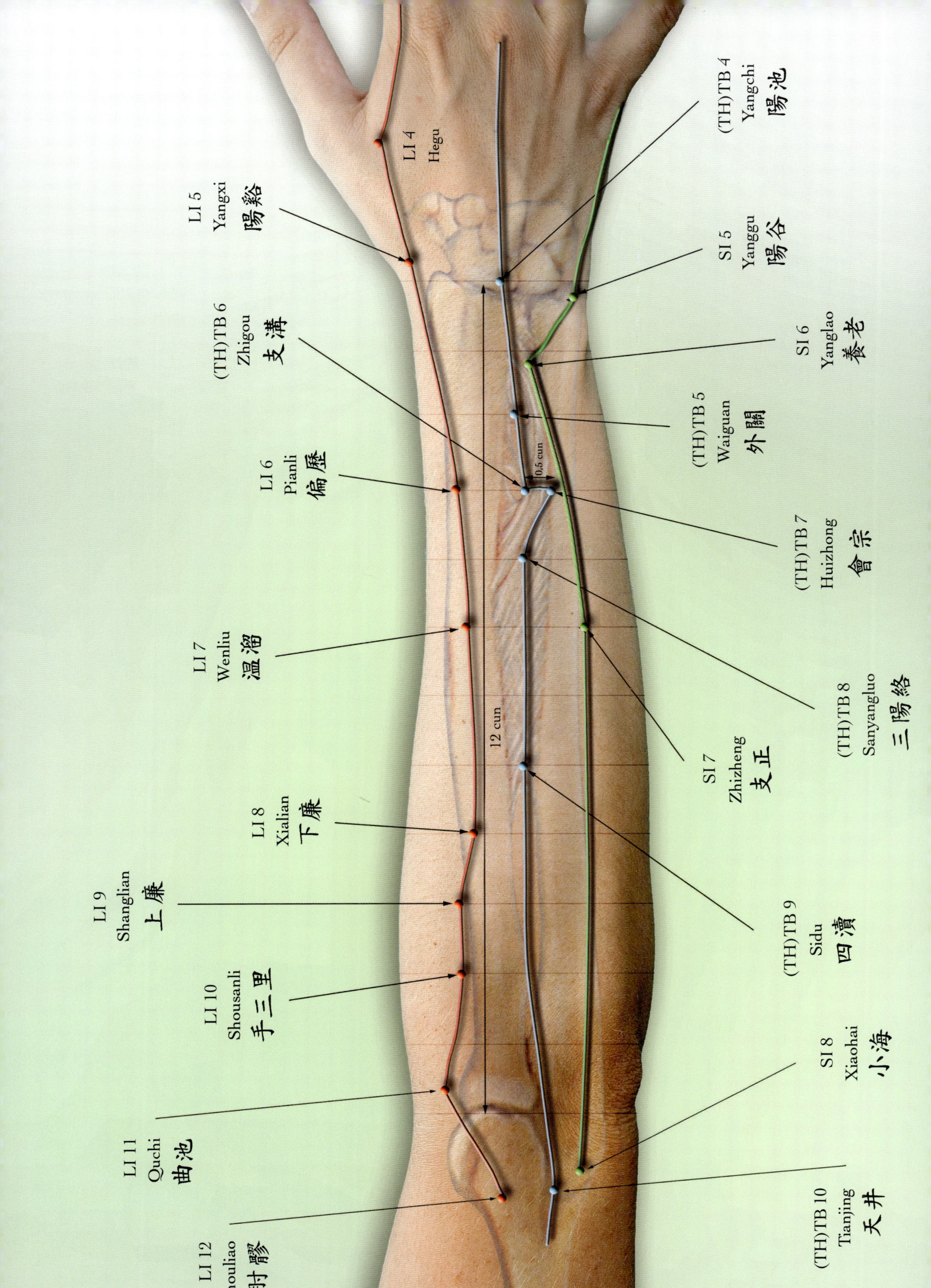

Points of 3 upper channels Yang

Extra channel points (PC – PN)

118 PC Zhongquan
On the posterior side of the forearm, between points LI 5 Yangxi and (TH)TB 4 Yangchi.

119 PC Cunping
On the posterior side of the forearm, 1 Cun above the previous point, 118 PC Zhongquan.

120 PC Xiawenliu
On the posterior side of the forearm, on the channel of the Large Intestine, 2 Cun from the radial head, on the level of (TH)TB 5 Waiguan.

121 PC Chinao
On the posterior side of the forearm, 6 Cun from the crease of the wrist and 1 Cun below (TH)TB 9 Sidu, between radius and ulna.

122 PC Quyangwei
In a depression, slightly outside the external tip of the elbow, when it is flexed, on the channel of the Large Intestine, 0.5 Cun below LI 11 Quchi.

123 PC Sanchi
Composed of 2 PC points and LI 11 Quchi.

- Chang Tcheu (Superior Sea), 1 Cun above LI 11 Quchi on the channel of the Large Intestine.

- Sha Tcheu (Inferior Sea), 1 Cun below LI 11 Quchi on the channel of the Large Intestine.

- Kou Tcheu (Central Sea) that corresponds to LI 11 Quchi.

124 PC Zhoushu
On the posterior part of the elbow, at the tip of the olecranon, between radius and humerus.

65 PN Luoshang
3 Cun above (TH)TB 5 Waiguan.

66 PN Yingxia
3 Cun below the tip of the olecranon, between radius and ulna.

67 PN Niushangxue
At ¼ above the line which unites the LI 11 Quchi and the (TH)TB 4 Yangchi.

J-SZ.39 Zhoushu
Between the olecranon and the humeral epicondyle.

X-SZ.20 Sanliwai
On the forearm, 1 Cun to the side and 2 Cun under LI 11 Quchi.

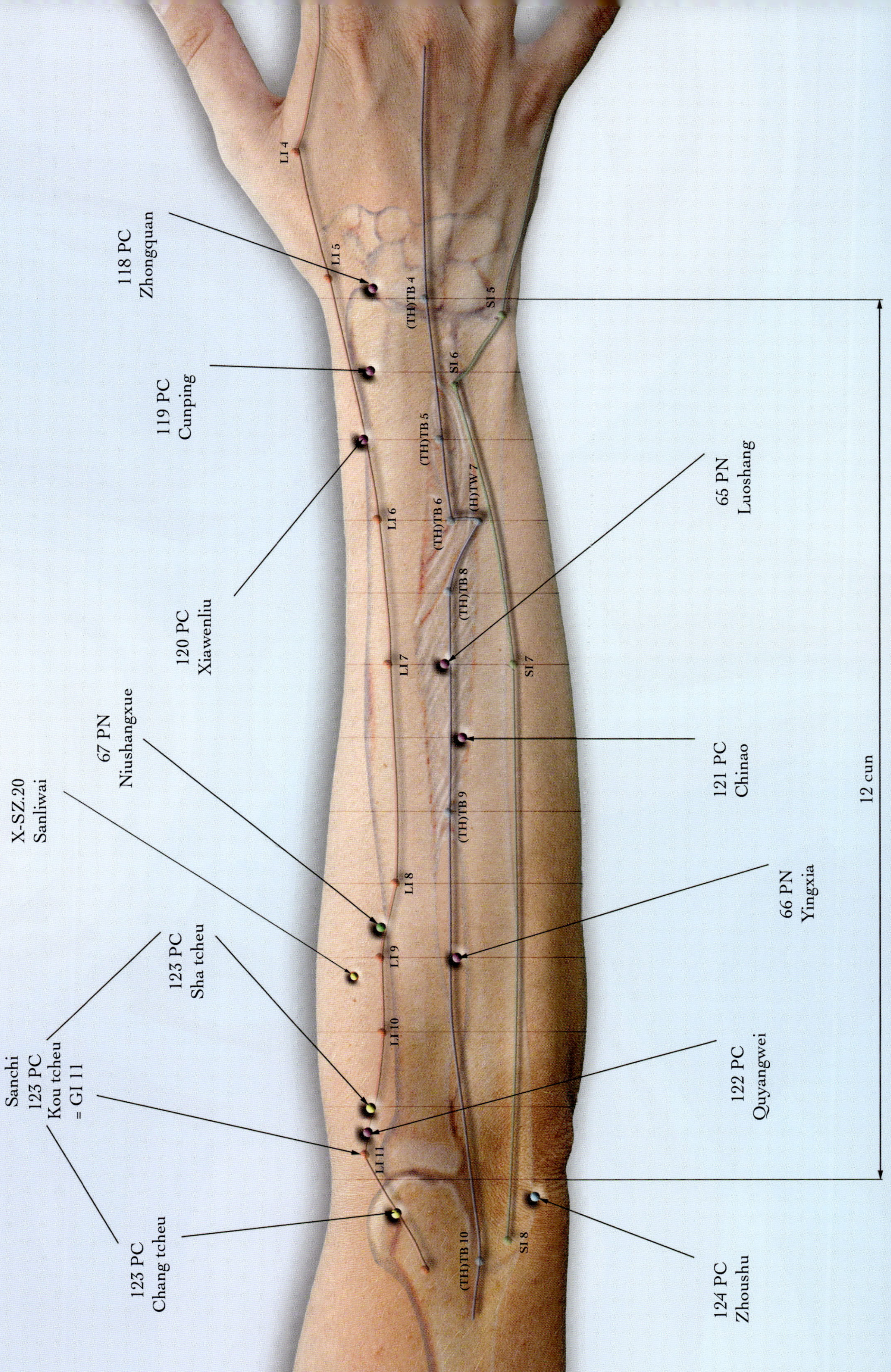

118 PC
Zhongquan
119 PC
Cunping
120 PC
Xiawenliu
67 PN
Niushangxue
X-SZ.20
Sanliwai
123 PC
Sha tcheu
Sanchi
123 PC
Kou tcheu
= GI 11
123 PC
Chang tcheu
65 PN
Luoshang
121 PC
Chinao
66 PN
Yingxia
122 PC
Quyangwei
124 PC
Zhoushu
12 cun
LI 4
LI 5
LI 6
LI 7
LI 8
LI 9
LI 10
LI 11
(TH)TB 4
(TH)TB 5
(TH)TB 6
(H)TW 7
(TH)TB 8
(TH)TB 9
(TH)TB 10
SI 5
SI 6
SI 7
SI 8

T 33.01 Qi Men
Hand and forearm supine, the point is 2 Cun above the wrist, on the lateral side of the radius.

T 33.02 Qi Jiao
Hand and forearm supine, the point is 4 Cun above the wrist, on the lateral side of the radius. It is also situated 2 Cun above T 33.01 Qi Men.

T 33.03 Qi Zheng
Hand and forearm supine, the point is 6 Cun above the wrist, on the lateral side of the radius. It is also situated 2 Cun above T 33.02 Qi Jiao.

T 33.04 Huo Chuan
Midpoint of the dorsal side of the forearm, 3 Cun above the flex crease of the wrist. It corresponds to (TH)TB 6 Zhigou.

T 33.05 Huo Ling
On the dorsal side of the forearm, with the hand on the chest, measure 5 Cun above the flex crease of the wrist (2 Cun above T 33.04 Huo Chuan).

T 33.06 Huo Shan
On the dorsal side of the forearm, with the hand on the chest, measure 6.5 Cun above the flex crease of the wrist.

T 33.07 Huo Fu Hai
2 Cun below LI 11 Quchi.
It corresponds to LI 10 Shousanli.

T 33.08 Shu Wu Jin
On the posterior side of the forearm, with the hand on the chest, measure 6.5 Cun above the pisifom on the lateral side of the ulna.

T 33.09 Shu Qian Jin
On the lateral side of the forearm, with the hand on the chest, measure 8 Cun above the pisiform on the lateral side of the ulna, 1.5 Cun from T 33.08 Shu Wu Jin.

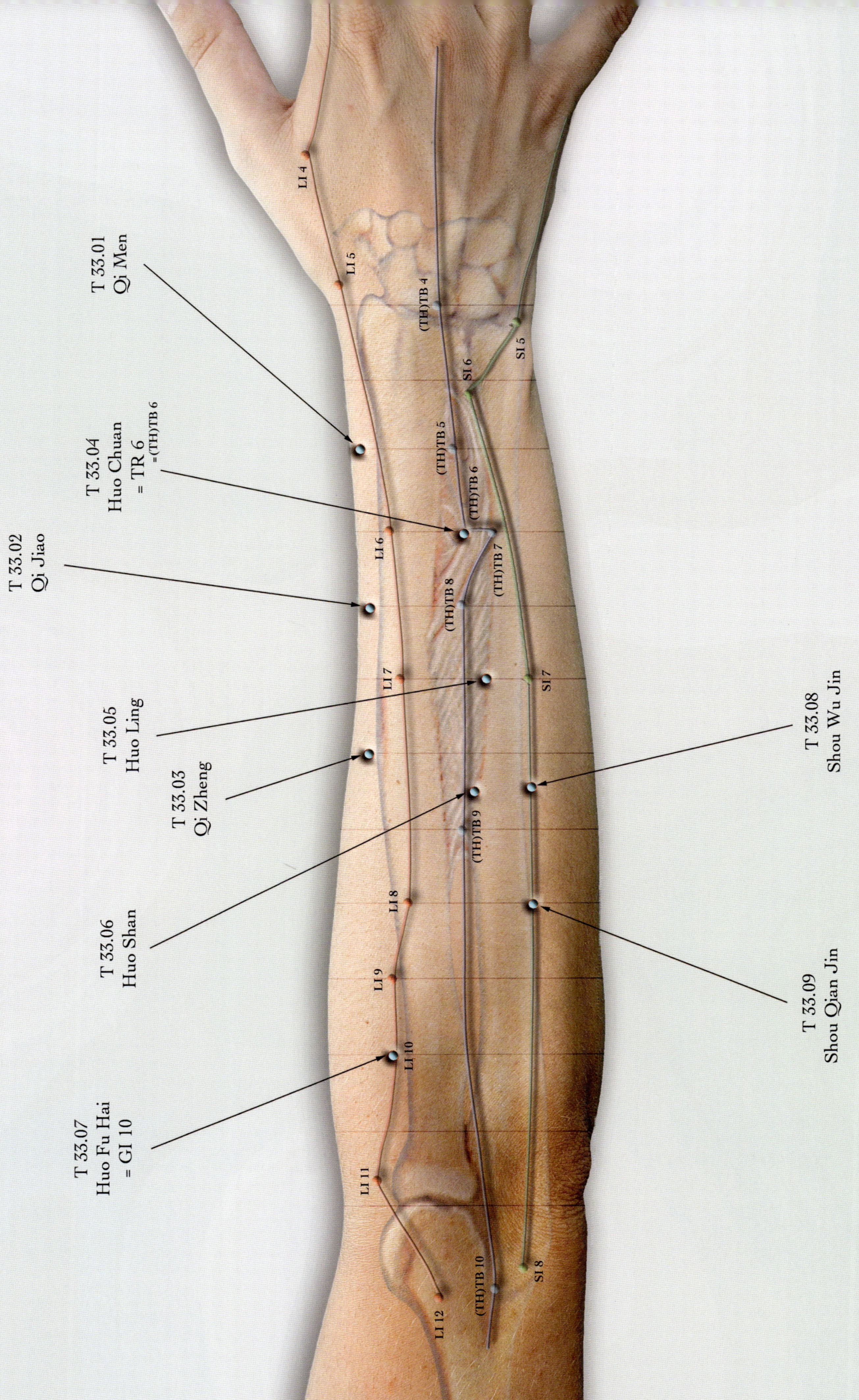
T 33.01
Qi Men
T 33.02
Qi Jiao
T 33.03
Qi Zheng
T 33.04
Huo Chuan
= TR 6
=(TH)TB 6
T 33.05
Huo Ling
T 33.06
Huo Shan
T 33.07
Huo Fu Hai
= GI 10
T 33.08
Shou Wu Jin
T 33.09
Shou Qian Jin
LI 4
LI 5
LI 6
LI 7
LI 8
LI 9
LI 10
LI 11
LI 12
(TH)TB 4
(TH)TB 5
(TH)TB 6
(TH)TB 7
(TH)TB 8
(TH)TB 9
(TH)TB 10
SI 5
SI 6
SI 7
SI 8

Anterior Side points of the Main channels

Shu Shao Yin channel (Heart)

(HT)HE 3 Shaohai
With the elbow flexed, at the midpoint of the line connecting the medial end of the cubital crease and the medial epicondyle of the humerus.

(HT)HE 4 Lingdao
On the palmar side of the forearm and on the radial side of the tendon of the ulnar flexor muscle of the wrist, 1.5 Cun proximal to the crease of the wrist.

(HT)HE 5 Tongli
On the palmar side of the forearm and on the radial side of the tendon of the ulnar flexor muscle of the wrist, 1 Cun proximal to the crease of the wrist.

(HT)HE 6 Yinxi
On the palmar side of the forearm and on the radial side of the tendon of the ulnar flexor muscle of the wrist, 0.5 Cun proximal to the crease of the wrist.

(HT)HE 7 Shenmen
On the wrist, at the ulnar end of the wrist, in the depression of the radial side of the tendon of the ulnar flexor muscle of the wrist.

Shu Jue Yin channel (Master of the Heart or Pericardium)

(MH)PC 3 Quze
At the midpoint of the cubital crease, on the ulnar side of the tendon of the biceps muscle of the arm.

(MH)PC 4 Ximen
On the palmar side of the forearm and on the line connecting (MH)PC 3 Quze and (MH)PC 7 Daling, 5 Cun above the crease of the wrist.

(MH)PC 5 Jianshi
On the palmar side of the forearm and on the line connecting (MH)PC 3 Quze and (MH)PC 7 Daling, 3 Cun above the crease of the wrist, between the tendons of the long palmar muscle and radial flexor muscle of the wrist.

(MH)PC 6 Neiguan
On the palmar side of the forearm and on the line connecting (MH)PC 3 Quze and (MH)PC 7 Daling, 2 Cun above the crease of the wrist, between the tendons of the long palmar muscle and radial flexor muscle of the wrist.

(MH)PC 7 Daling
At the midpoint of the crease of the wrist, between the tendons of the long palmar muscle and radial flexor muscle of the wrist.

Shu Tai Yin channel (Lung)

LU 3 Tianfu
On the medial side of the upper arm and on the radial border of the biceps muscle of the arm, 3 Cun below the anterior end of the axillary fold.

LU 4 Xiabai
On the medial side of the upper arm and on the radial border of the biceps muscle of the arm, 4 Cun below the anterior end of the axillary fold, or 5 Cun above the cubital crease.

LU 5 Chize
In the cubital crease, in the depression of the radial side of the tendon of the biceps muscle of the arm.

LU 6 Kongzu
On the radial side of the palmar surface of the forearm, and on the line connecting LU 5 Chize and LU 9 Taiyuan, 7 Cun above the cubital crease.

LU 7 Lieque
On the radial side of the forearm, proximal to the styloid process of the radius, 1.5 Cun above the crease of the wrist, between the brachioradial muscle and the tendon of the long abductor muscle of the thumb.

LU 8 Jingqu
On the radial side of the palmar surface of the forearm, 1 Cun above the crease of the wrist, in the depression between the styloid process of the radius and radial artery.

LU 9 Taiyuan
On the radial end of the crease of the wrist, where the pulsation of the radial artery is palpable. (Do NOT needle directly into the artery!)

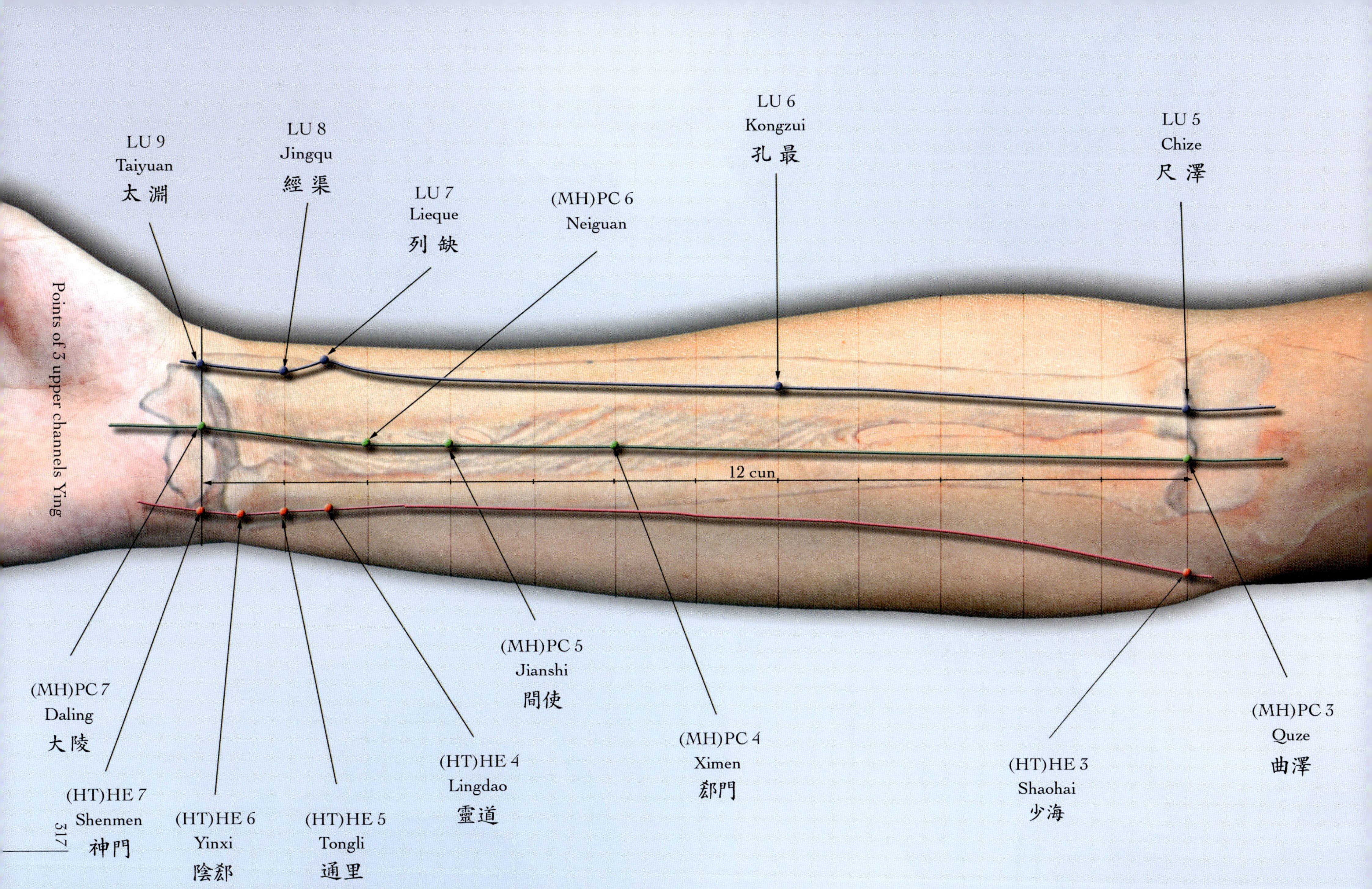

Points of 3 upper channels Ying

EXTRA CHANNEL POINTS (PC – PN)

112 PC Shoujinmen
On the anterior side of the forearm, 3.5 Cun from the crease of the wrist, that is to say 1.5 Cun from (MH)PC 6 Neiguan between the tendons of the long palmar muscle and short palmar muscle.

113 PC Dingshu
On the anterior face of the forearm, 4 Cun from point (HT)HE 7 Shenmen.

114 PC Erbai
It is a set of two points, on each forearm, that form only one point because it is imperative to use them together. It is situated on the anterior face of the forearm, 4 Cun from the midpoint of the crease of the wrist, 2 Cun from (MH)PC 6 Neiguan. One of the two points is outside the tendon, on the side of the radius.

115 PC Bizhong
On the anterior side of the forearm, midpoint of the line connecting the crease of the wrist with the tip of the elbow, between the ulna and the radius.

116 PC Zexia
On the anterior side of the forearm, 2 Cun below point LU 5 Chize, on the Lung channel.

117 PC Zequian
On the anterior side of the forearm, 1 Cun outside LU 5 Chize.

64 PN Xishang
3 Cun above MH 4 Ximen

J-SX.36 Cirao
6 Cun above midpoint of the crease of the wrist.

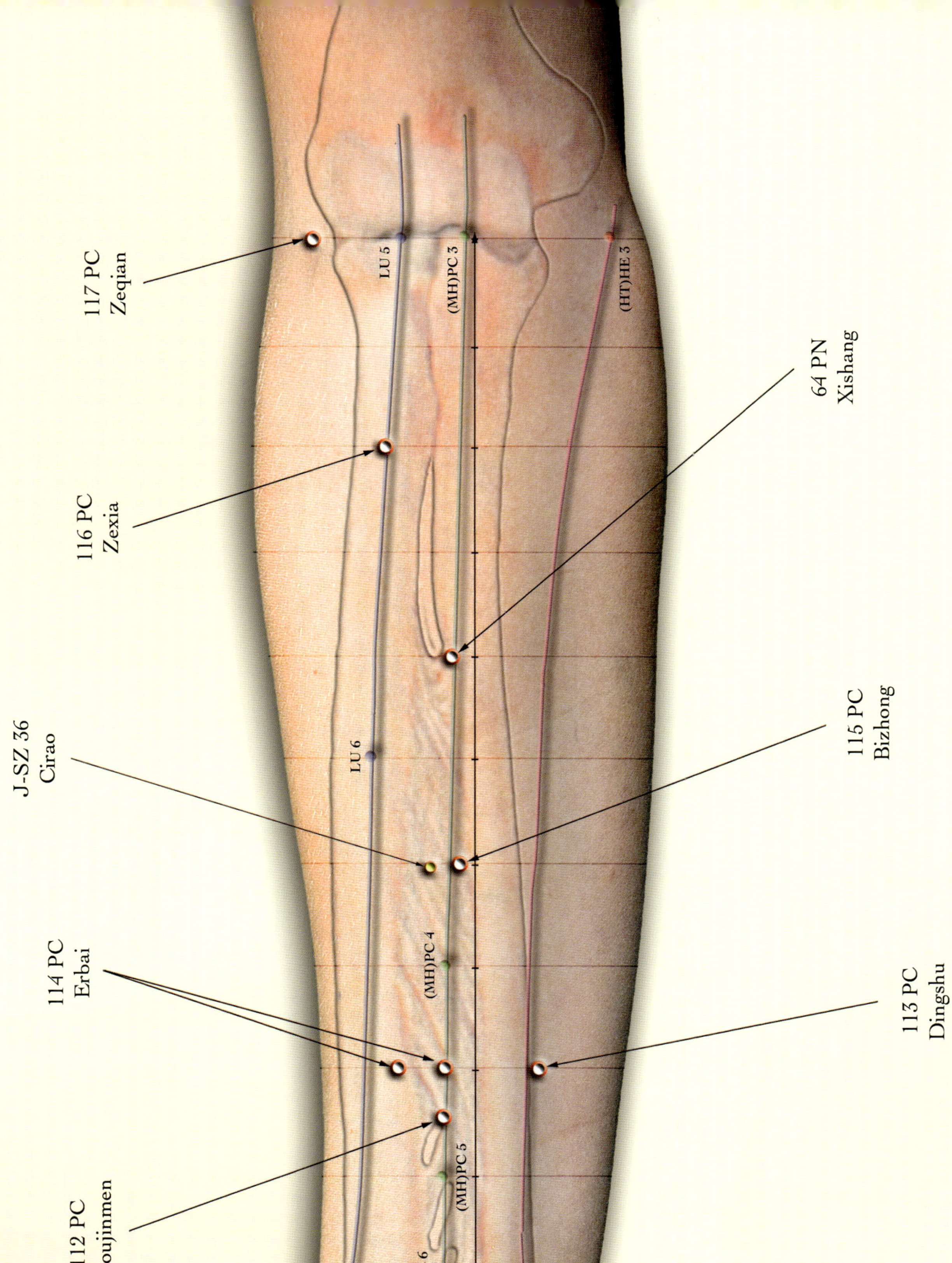
117 PC
Zeqian
116 PC
Zexia
64 PN
Xishang
J-SZ 36
Cirao
115 PC
Bizhong
114 PC
Erbai
113 PC
Dingshu
112 PC
Shoujinmen
LU 5
(MH)PC 3
(HT)HE 3
LU 6
(MH)PC 4
(MH)PC 5
(MH)PC 6
LU 7
LU 8
LU 9
(HT)HE 4
(HT)HE 5
(HT)HE 6
(HT)HE 7
(MH)PC 7

T 33.10 Chang Men
On the anterior part of the forearm, measure 3 Cun above the pisiform midpoint of the ulna, 0.5 Cun above (HT)HE 4 Lingdao.

T 33.11 Gan Men
On the anterior part of the forearm, measure 6 Cun above the pisiform midpoint of the ulna.

T 33.12 Xin Men
On the anterior part of the forearm, measure 1.5 Cun below the olecranon, midpoint of the ulna.

T 33.13 Ren Shi
On the anterior part of the forearm, 4 Cun above the crease of the wrist, on the lateral side of the radius.

T 33.14 Di Shi
On the anterior part of the forearm, 7 Cun above the crease of the wrist on the anteriolateral side of the radius.

T 33.15 Tian Shi
On the anterior part of the forearm, 10 Cun above the crease of the wrist on the anterio-lateral side of the radius.
It is also found 3 Cun above T 33.14 Di Shi.

T 33.16 Qu Ling
On the crease of the wrist, on the radial side of the biceps. Slightly flex the elbow to situate the point.
It corresponds to LU 5 Chize.

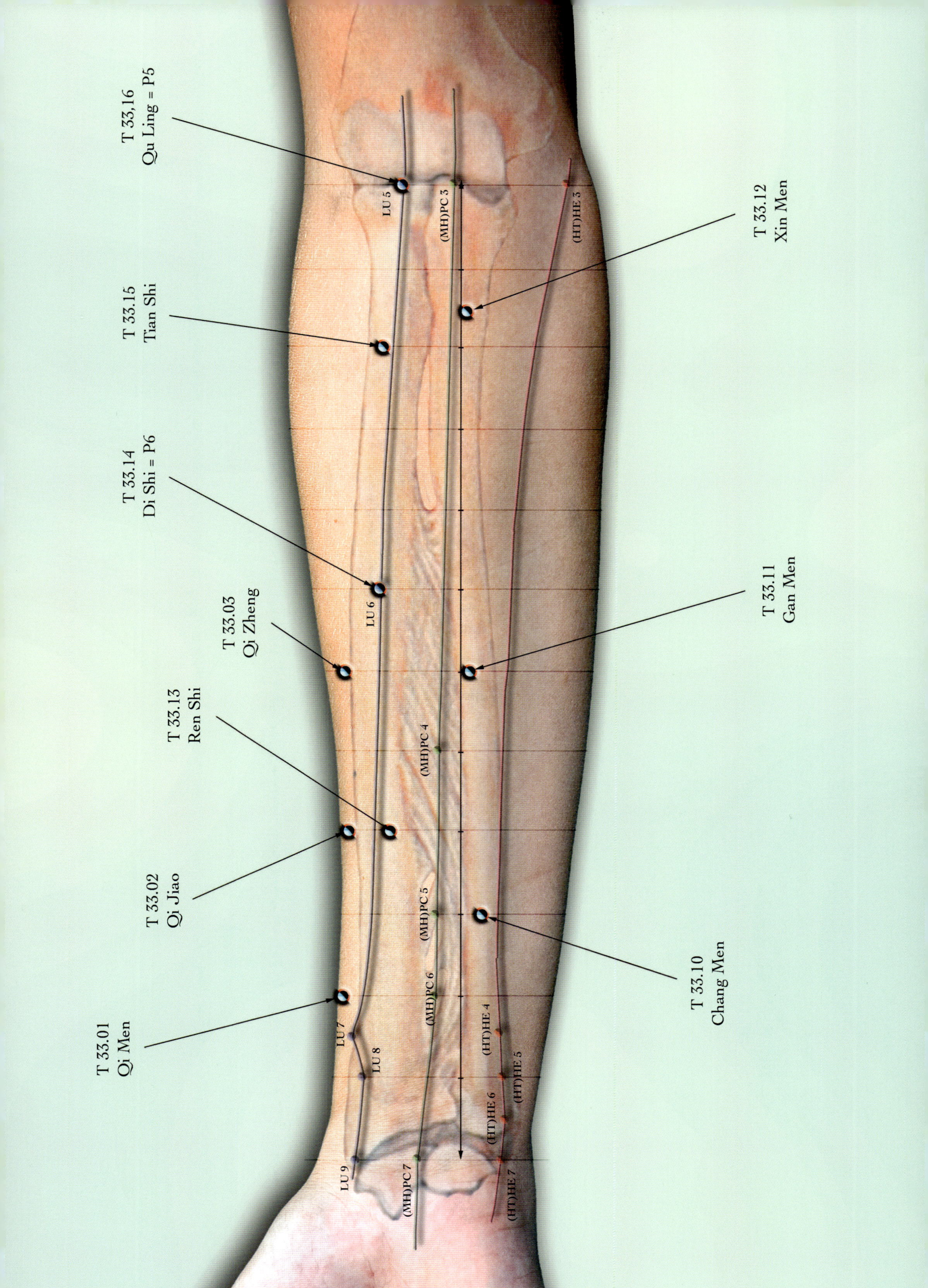
T 33,16
Qu Ling = P5
T 33.15
Tian Shi
T 33.14
Di Shi = P6
T 33.03
Qi Zheng
T 33.13
Ren Shi
T 33.02
Qi Jiao
T 33.01
Qi Men
T 33.12
Xin Men
T 33.11
Gan Men
T 33.10
Chang Men
LU 5
(MH)PC 3
(HT)HE 3
LU 6
(MH)PC 4
(MH)PC 5
(MH)PC 6
LU 7
LU 8
LU 9
(MH)PC 7
(HT)HE 4
(HT)HE 5
(HT)HE 6
(HT)HE 7

Points of the arm

- **Anterior side**
  - –Points of the Main channels
    - Shu Yang Ming (Large Intestine)
    - Shu Tai Yang (Small Intestine)
    - Shu Shao Yang (Triple Heater or Triple Burner)
  - –Extra channel points
    - Curious points
    - New points
    - Points of Master Tong
- **Posterior and lateral side**
  - –Points of the Main channels
    - Shu Shao Yin (Heart)
    - Shu Jue Yin (Master of the Heater or Pericardium)
    - Shu Tai Yin (Lung)
  - –Extra channel points
    - Curious points
    - New points
    - Points of Master Tong

Anterior side

Points of the Main channels

Shu Tai Yin channel (Lung)

LU 1 Zhongfu
In the superior lateral part of the anterior thoracic wall, 1 Cun below LU 2 Yunmen, on the level of the 1st intercostal space, 6 Cun lateral to the anterior midline.

LU 2 Yunmen
In the superior lateral part of the anterior thoracic wall, superior to the coracoid process of the scapula, in the depression odf the infra-clavicular fossa, 6 Cun lateral to the anterior midline.

LU 3 Tianfu
On the internal side of the upper arm and on the radial border of the biceps muscle of the arm, 3 Cun below the anterior end of the axillary fold.

LU 4 Xiabai
On the medial side of the upper arm and on the radial border of the biceps muscle of the arm, 4 Cun below the anterior end of the axillary fold, or 5 Cun above the cubital crease.

LU 5 Chize
In the cubital crease, in the depression of the radial side of the tendon of the biceps muscle of the arm.

Shu Jue Yin channel (Master of the Heart or Pericardium)

(MH)PC 1 Tianchi
On the chest, in the 4th intercostal space, 1 Cun lateral to the nipple and 5 Cun lateral to the anterior midline.

(MH)PC 2 Tianquan
On the internal side of the arm, 2 Cun below the anterior end of the axillary fold, between the long and short heads of the biceps muscle of the arm.

(MH)PC 3 Quze
At the midpoint of the cubital crease, on the ulnar side of the tendon of the biceps muscle of the arm.

Shu Shao Yin channel (Heart)

(HT)HE 1 Jiquan
At the apex of the axillary fossa, where the pulsation of the axillary artery is palpable. (Do NOT needle directly into the artery!)

(HT)HE 2 Qingling
On the medial side of the arm and on the line connecting (HT)HE 1 Jiquan and (HT)HE 3 Shaohai, 3 Cun above the cubital crease, in the groove medial to the biceps muscle of the arm.

(HT)HE 3 Shaohai
With the elbow flexed, at the midpoint of the line connecting the medial end of the cubital crease and the medial epicondyle of the humerus.

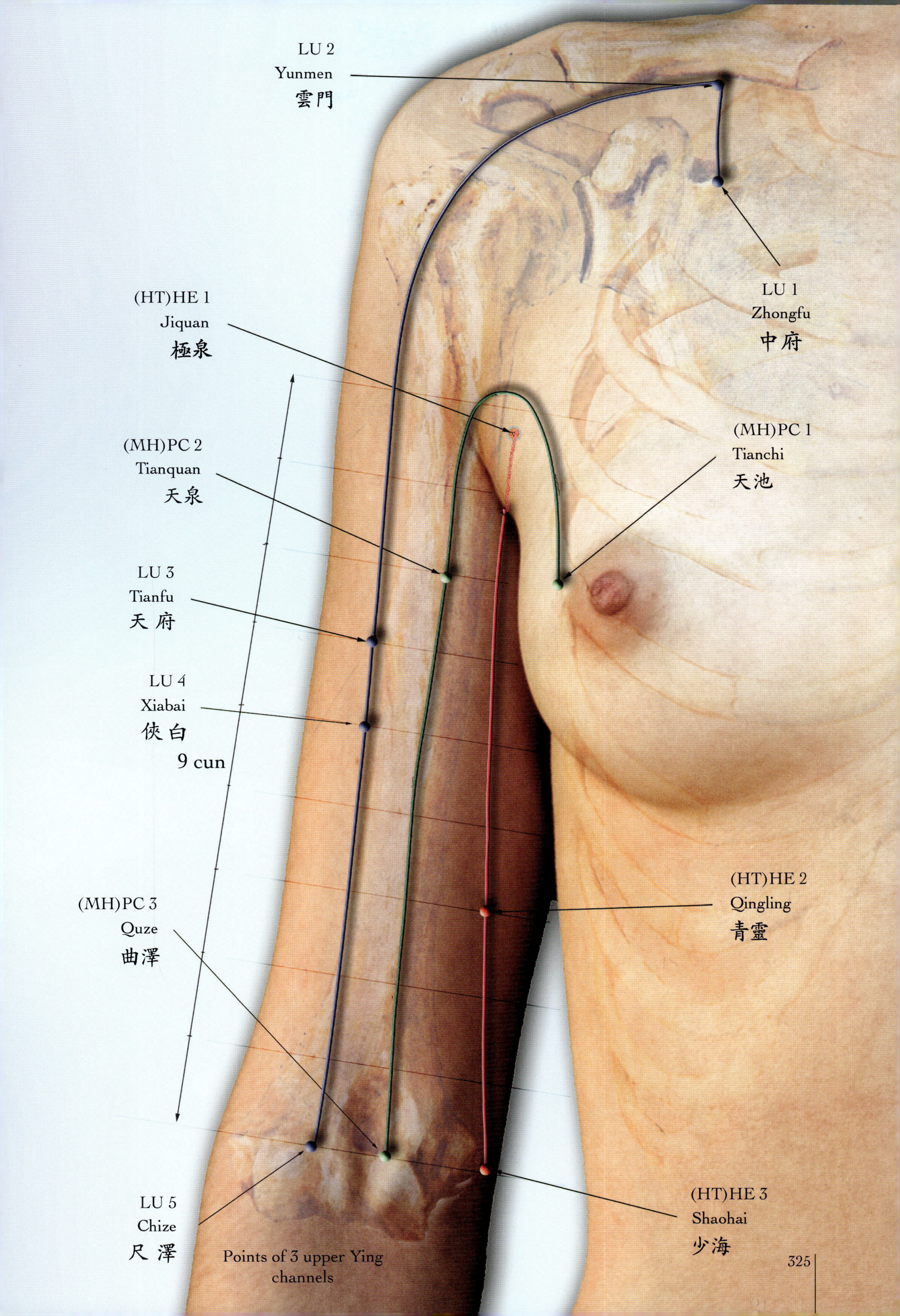

Points of 3 upper Ying channels

Extra channel points (PC – PN)

125 PC Yeling
In the angle formed by the thorax and the arm, in front of the axillary fold 0.5 Cun above the anterior extremity of the axillary fold.

126 PC Tianling
0.5 Cun to the thorax, outside the line formed by the extremity of the axillary fold and 1 Cun above that line.

127 PC Jianyu = X-SZ.27 Zhitan
Midpoint of the line connecting points LI 15 Jianyu and LU 2 Yunmen.

128 PC Yuqian
Situated between 127 PC Jianyu and LI 15 Jianyu, in a hollow.

68 PN Guangzhong
2.5 Cun below (MH)PC 2 Tianquan.

X-SZ.9 Gongzhong
On the arm, 2.5 Cun below LU 2 Yunmen.

X-SZ.25 Xiaxiabai
On the arm, 3 Cun below LU 4 Xiabai.

X-SZ.26 Xiaokuai
At the apical extremity of the anterior axillary fold.

J-SZ.48 Jianneiling
On the shoulder, equidistant between LI 15 Jianyu and the axillary fold.

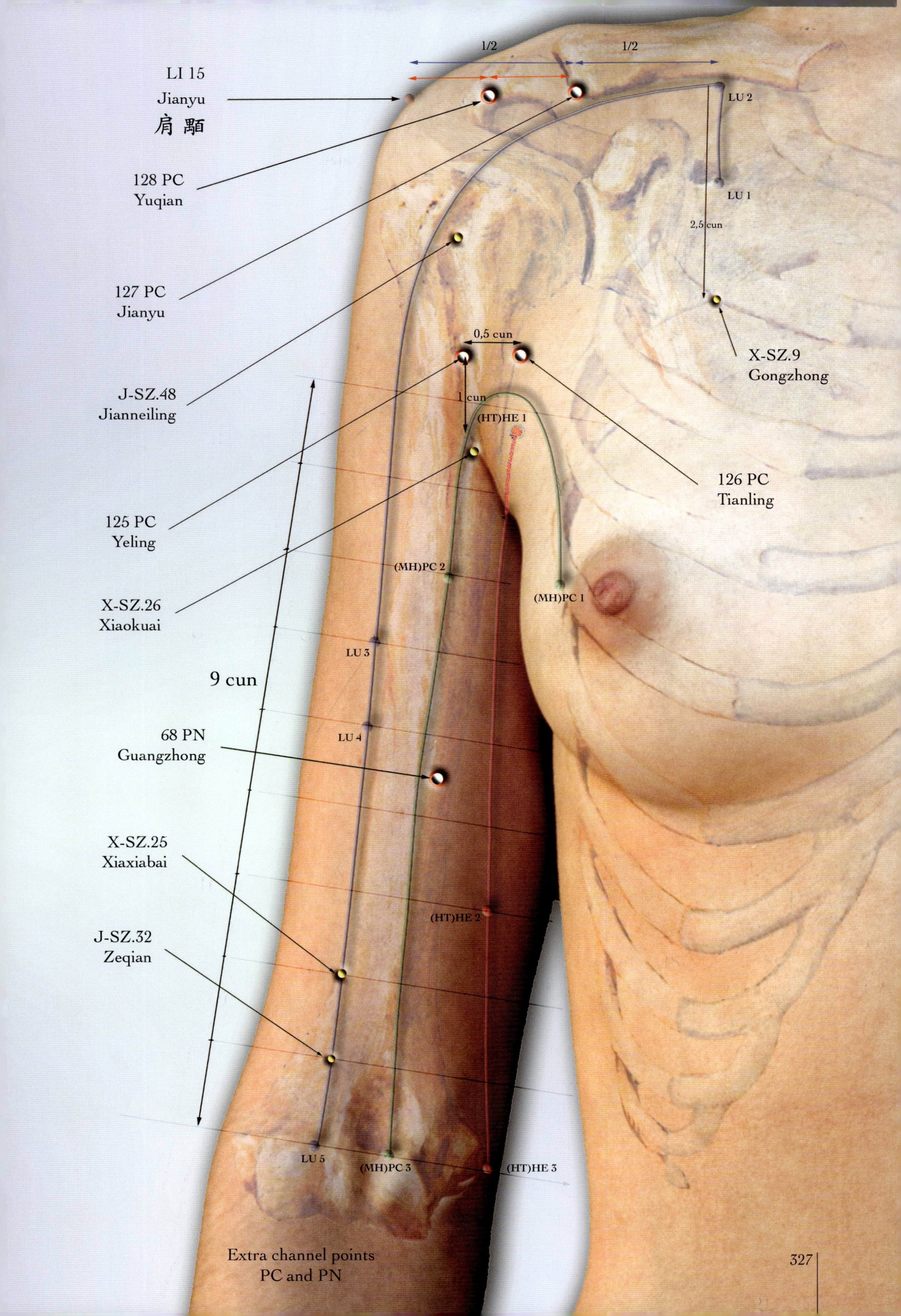

Extra channel points
PC and PN

T 44.01 Fen Jin
On the channel of the Lung, the point is situated 1.5 Cun above the cubital crease.
It is situated 3.5 Cun below LU 4 Xiabai, and 1.5 Cun above LU 5 Chize.

T 44.08 Ren Zong
On the anterior part of the arm, measure 3 Cun above the cubital crease, between the 2 parts of the biceps muscle.

T 44.09 Di Zong
On the anterior part of the arm, 6 Cun above the cubital crease, 3 Cun above T 44.08 Ren Zong.

T 44.10 Tian Zong
On the anterior side of the arm, measure 3 Cun above T 44.09 Di Zong or 9 Cun above the cubital crease between the 2 parts of the biceps muscle.

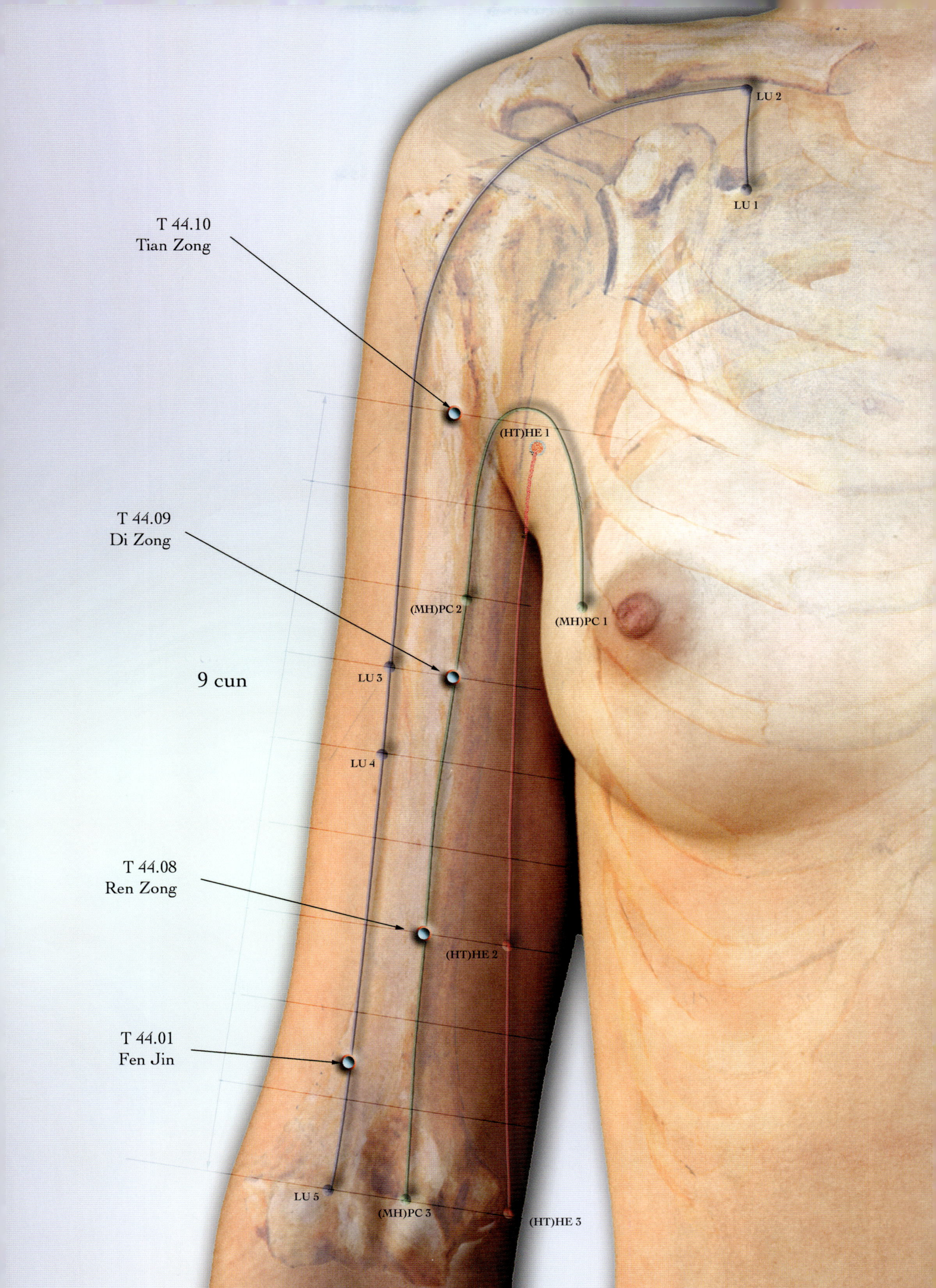
LU 2
LU 1
T 44.10
Tian Zong
(HT)HE 1
T 44.09
Di Zong
(MH)PC 2
(MH)PC 1
9 cun
LU 3
LU 4
T 44.08
Ren Zong
(HT)HE 2
T 44.01
Fen Jin
LU 5
(MH)PC 3
(HT)HE 3

Posterior side

Points of the Main channels

Shu Yang Ming channel (Large Intestine)

LI 11 Quchi
With the elbow flexed, at the external end of the cubital crease, at the midpoint of the line connecting LU 5 Chize and the external humeral epicondyle.

LI 12 Zhouliao
With the elbow flexed, on the lateral side of the upper arm, 1 Cun above LI 11 Quchi, on the border of the humerus.

LI 13 Shouwouli
On the external side of the upper arm and on the line connecting LI 11 Quchi and LI 15 Jianyu, 3 Cun above LI 11 Quchi.

LI 15 Jianyu
On the shoulder, superior to the deltoid muscle, in the depression anterior and inferior to the acromion, when the arm is abducted or raised on the level of the shoulder.

LI 16 Jugu
On the shoulder, in the depression between the acromial extremity of the clavicle and scapular spine.

Shu Shao Yang channel (Triple Heater or Triple Burner)

(TH)TB 10 Tianjing
On the external side of the upper arm, in the depression, 1 Cun proximal to the tip of the olecranon, when the elbow is flexed.

(TH)TB 11 Qinglengyuan
With the elbow flexed, on the external side of the upper arm, 2 Cun above the tip of the olecranon and 1 Cun above (TH)TB 10 Tianjing.

(TH)TB 12 Xiaoluo
On the external side of the upper arm, at the midpoint of the line connecting (TH)TB 11 Qinglenyuan and (TH)TB 13 Naohui.

(TH)TB 13 Naohui
On the lateral side of the upper arm and on the line connecting the tip of the olecranon and (TH)TB 14 Jianliao, 3 Cun below (TH)TB 14 Jianliao, and on the posterio-inferior border of the deltoid muscle.

(TH)TB 14 Jianliao
On the shoulder, posterior to LI 15 Jianyu, in the depression inferior and posterior to the acromion when the arm is abducted.

(TH)TB 15 Tianliao
On the scapula, at the midpoint between GB 21 Jianjing and SI 13 Quyuan, at the superior angle of the scapula.

Shu Tai Yang channel (Small Intestine)

SI 8 Xiaohai
On the medial side of the elbow, in the depression between the olecranon of the ulna and the medial epicondyle of the humerus.

SI 9 Jianzhen
Posterior and inferior to the shoulder joint, 1 Cun above the posterior end of the axillary fold with the arm abducted.

SI 10 Naoshu
On the shoulder, above the posterior end of the axillary fold, in the depression below the lower border of the scapular spine.

SI 11 Tianzong
On the scapula, in the depression of the centre of the subscapular fossa, and on the level of the 4th thoracic vertebra.

SI 12 Bingfeng
On the scapula, at the centre of the suprascapular fossa, directly above SI 11 Tianzong, in the depression found when the arm is raised.

SI 13 Quyuan
On the scapula, at the internal end of the supra-scapular fossa, at the midpoint of the line connecting SI 10 Naoshu and the spinous process of the 2nd thoracic vertebra.

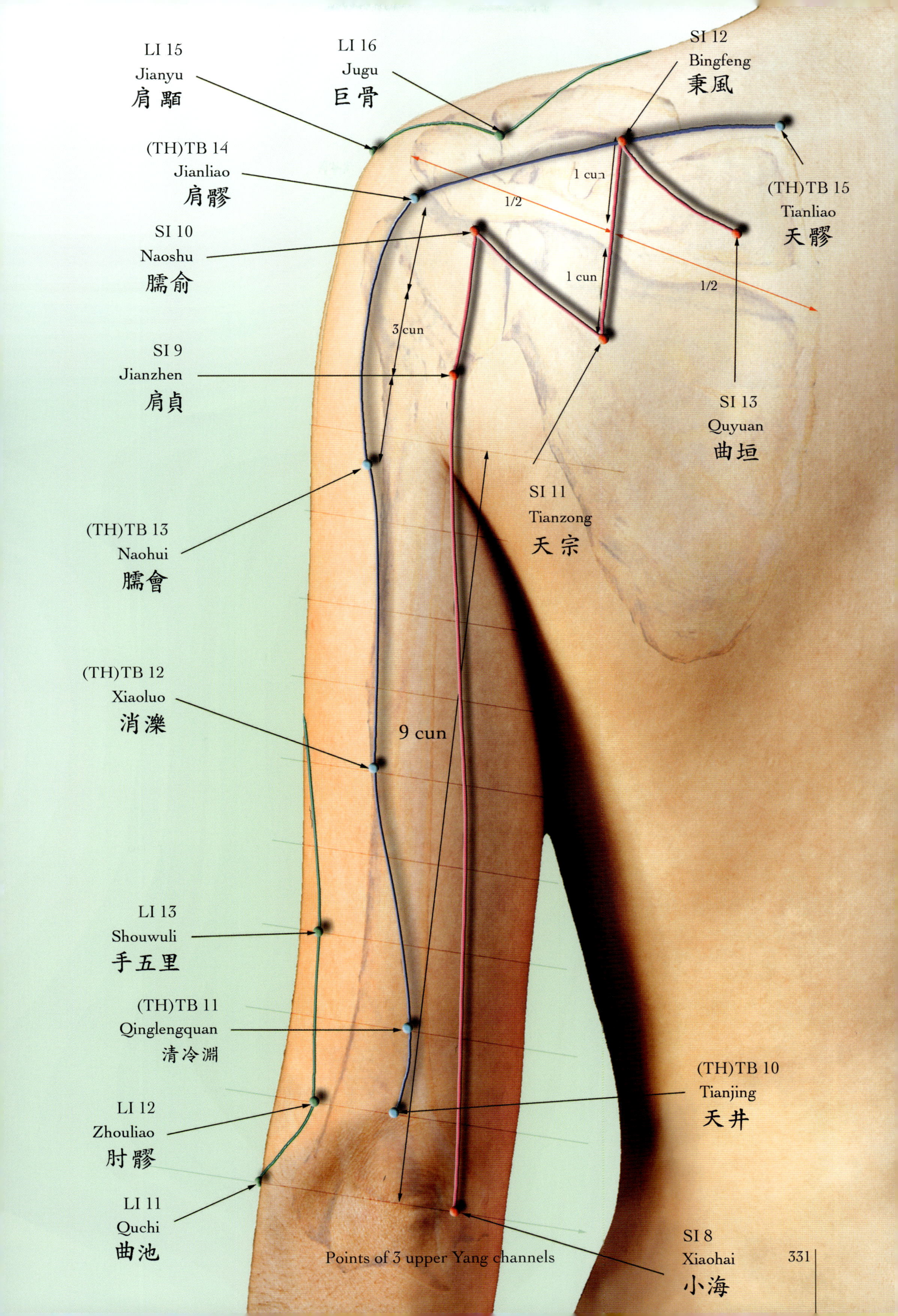

Points of 3 upper Yang channels

Extra channel points (PC – PN)

129 PC Houye
Right at the posterior extremity of the axillary fold.

69 PN
2 Cun below point 70 PN Taijian.

70 PN Taijian
1.5 Cun below the acromion.

71 PN Yingshang
4 Cun above the apex of the olecranon.

X-SZ.24 Shenzhou
On the cubital part of the arm, 1.5 Cun above the olecranon.

X-SZ.21 Xinquchi
0.5 Cun above LI 11 Quchi.

X-SZ.22 Shangquchi
1.5 Cun directly above LI 11 Quchi.

X- BY.19 Juguxia
2 Cun below LI 16 Jugu.

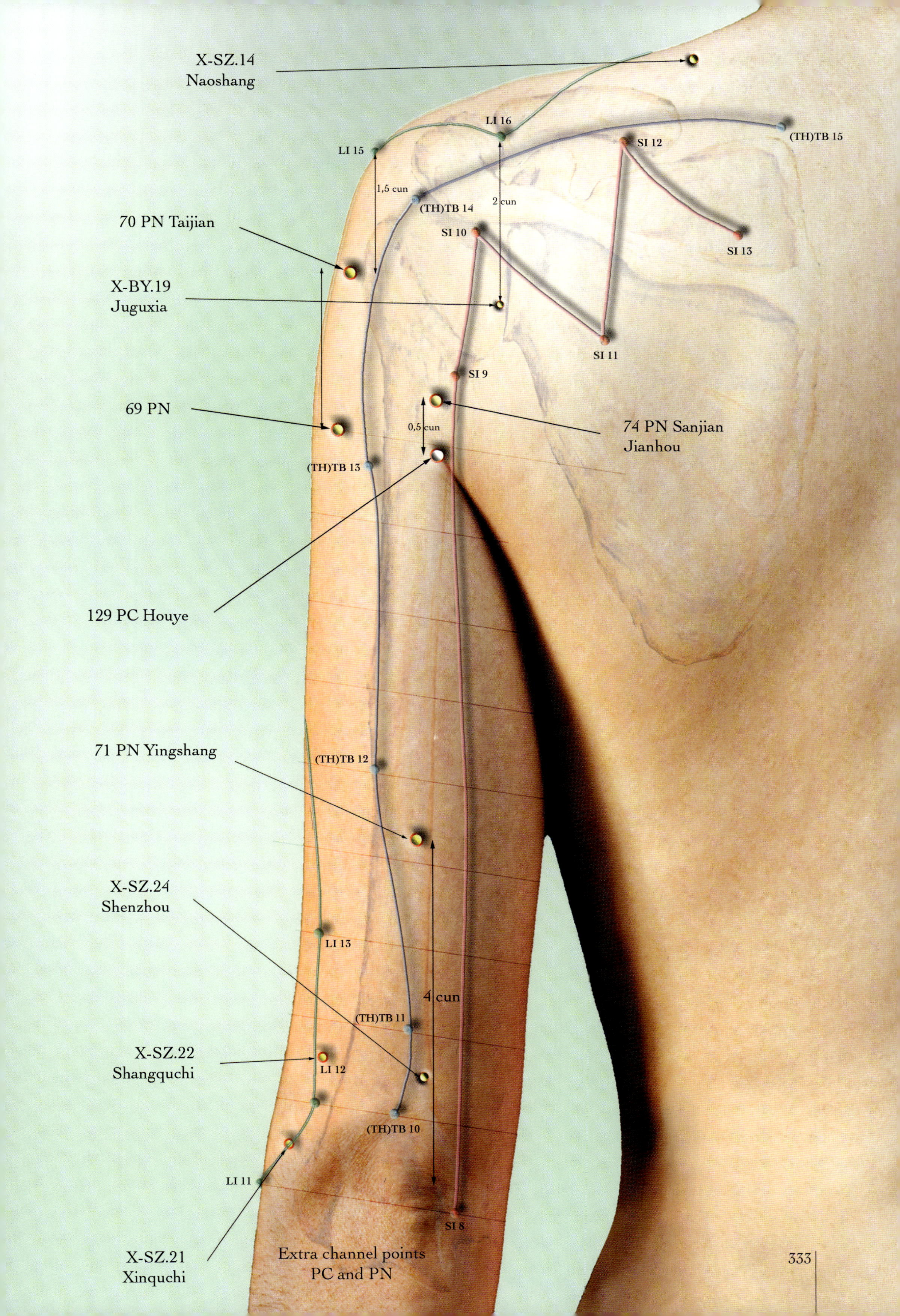

Extra channel points
PC and PN

T 44.02 Hou Zhui
2.5 Cun above the cubital crease, on the posterior side of the arm. 5 fen above (TH)TB 11 Qinglengyuan on the Triple Heater or Triple Burner channel.

T 44.03 Shou Ying
4.5 Cun above the cubital crease, on the posterior side of the arm.
4.5 Cun above the posterior cubital crease on the Triple Heater or Triple Burner channel or 2 Cun above T 44.02 Hou Zhui.
It is also near (TH)TB 12 Xiaoluo.

T 44.04 Fu Ding
2.5 Cun above T 44.03 Shou Ying or on the Triple Heater or Triple Burner channel, 2 Cun above (TH)TB 12 Xiaoluo.

T 44.05 Hou Zhi
8 Cun above the cubital joint.
Also 3 Cun above (TH)TB 12 Xiaoluo and 0.5 Cun below (TH)TB 13 Naohui.

T 44.07 Bei Mian
With the arm horizontal and totally abducted, the point is in a depression below the acromion.

T 44.13 Zhi Tong
The point is 4.5 Cun above the cubital crease, on the posterior part of the arm, 1 Cun internally from T 44.03 Shou Ying.

T 44.14 Luo Tong
The point is 7 Cun above the cubital crease, on the posterior part of the arm, 1 Cun internally from T 44.04 Fu Ding.

T 44.15 Xia Qu
8 Cun above the cubital crease, on the posterior part of the arm, 1 Cun internally from T 44.05 Hou Zhi.

T 44.16 Shang Qu
1 Cun posterior to T 44.06 Jian Zhong and 3 Cun below the acromion process.

T 44.17 Shui Yu 2.5
Cun above the acromion process and 2 Cun posterior to T 44.11 Yun Bai.
Has the same situation as SI 10 Naoshu.

DT.01 Fen Zhi Shang
1 Cun below the acromion, that is to say 1 Cun below SI 10 Naoshu.

DT.02 Fen Zhi Xia
Located 1,5 Cun below and 5 fen inside from DT.01 Fen Zhi Shang.

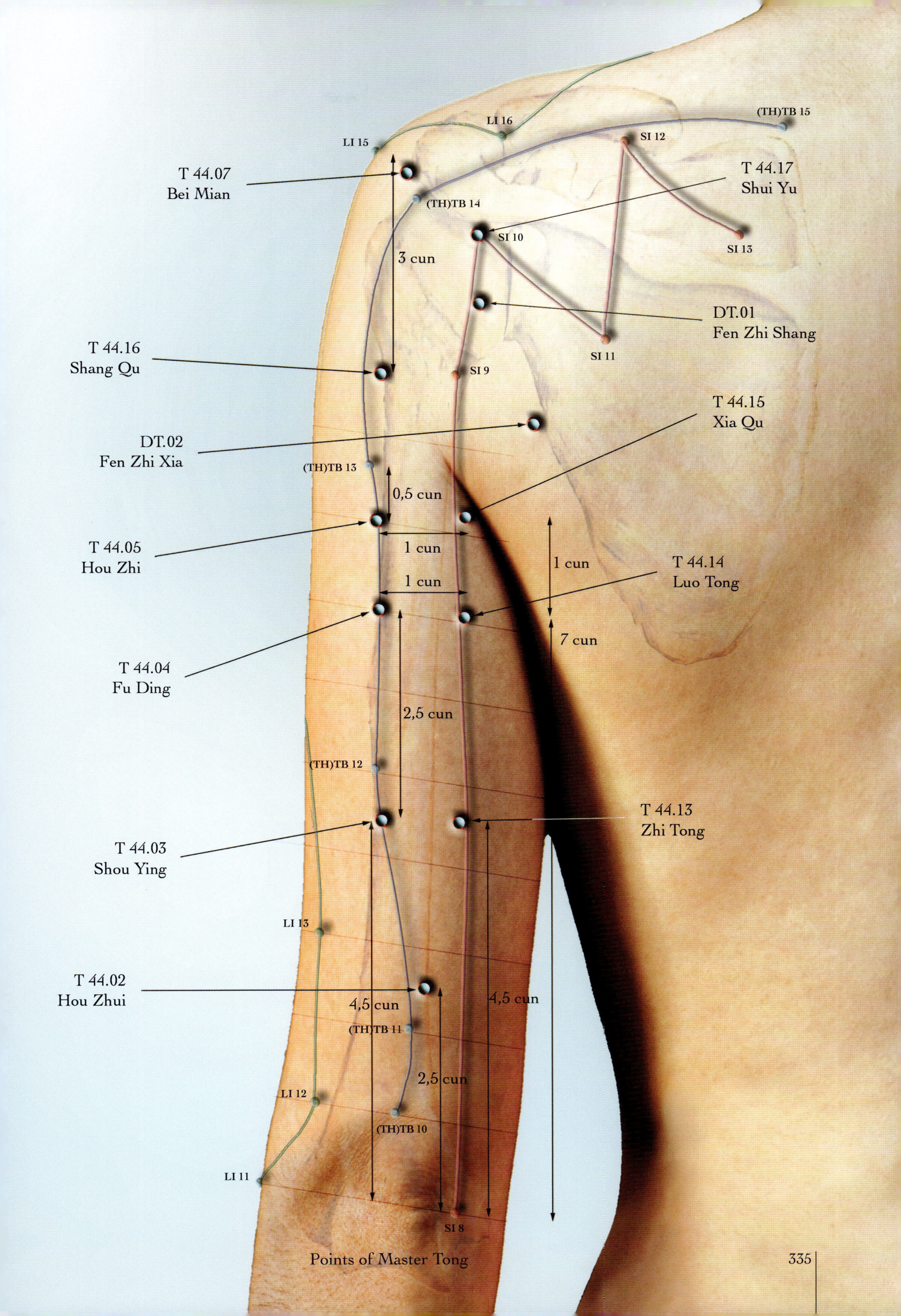
LI 15
LI 16
(TH)TB 15
SI 12
T 44.07
Bei Mian
T 44.17
Shui Yu
(TH)TB 14
SI 10
SI 13
3 cun
DT.01
Fen Zhi Shang
T 44.16
Shang Qu
SI 11
SI 9
T 44.15
Xia Qu
DT.02
Fen Zhi Xia
(TH)TB 13
0,5 cun
T 44.05
Hou Zhi
1 cun
1 cun
T 44.14
Luo Tong
1 cun
7 cun
T 44.04
Fu Ding
2,5 cun
(TH)TB 12
T 44.13
Zhi Tong
T 44.03
Shou Ying
LI 13
T 44.02
Hou Zhui
4,5 cun
4,5 cun
(TH)TB 11
2,5 cun
LI 12
(TH)TB 10
LI 11
SI 8

Lateral side

Points of the Main channels

Shu Yang Ming channel
(Large Intestine)

LI 11 Quchi
With the elbow flexed, at the lateral end of the cubital crease, at the midpoint of the line connecting LU 5 Chize and the external humeral epicondyle.

LI 12 Zhouliao
With the elbow flexed, on the lateral side of the upper arm, 1 Cun above LI 11 Quchi, on the border of the humerus.

LI 13 Shouwouli
On the lateral side of the upper arm and on the line connecting LI 11 Quchi and LI 15 Jianyu, 3 Cun above LI 11 Quchi.

LI 14 Binao
On the lateral side of the arm, at the insertion of the deltoid muscle and on the line connecting LI 11 Quchi and LI 15 Jianyu, 7 Cun above LI 11 Quchi.

LI 15 Jianyu
On the shoulder, superior to the deltoid muscle, in the depression anterior and inferior to the acromion when the arm is abducted or raised on the level of the shoulder.

Shu Shao Yang channel
(Triple Heater or Triple Burner)

(TH)TB 13 Naohui
On the lateral side of the upper arm and on the line connecting the tip of the olecranon and SI 14 Jianliao, 3 Cun below SI 14 Jianliao, and on the posterio-inferior border of the deltoid muscle.

Extra channel points

Curious points and new points

70 PN Taijian
1.5 Cun below the acromion.

72PN Jianming
On the posterio-external part of the arm, at the apex of the triangle formed by the muscles of the arm.

73 PN Rushang
On the posterio-external part of the arm, in the centre of the triangle formed by the muscles of the arm.

74 PN Sanjian
3 points:

- Jianyu = LI 15
- Jianqian or external Jian 1 Cun above the anterior extremity of the axillary fold.
- Jianhou or posterior Jian 0.5 Cun above the posterior extremity of the axillary fold.

X-SZ.21 Xinquchi
0.5 Cun above LI 11 Quchi.

X-SZ.22 Shangquchi
1.5 Cun above LI 11 Quchi

X-SZ.23 Zhiyang
1 Cun directly above LI 12 Zhouliao

X-SZ.10 Jubi
In the scapular area, 2 Cun below 70 PN Taijian.

Points of Master Tong

T44.06 Jian Zhong
In the centre of the deltoid muscle, 3 Cun below the acromion process.

T 44.11 Yun Bai
When the arm is hanging, the point is 5 fen above and 1.5 Cun in front of T 44.06 Jian Zhong.

T 44.12 Li Bai
First find T 44.11 Yun Bai, the patient being in the anatomical position, the point is 2 Cun below T 44.11 Yun Bai.

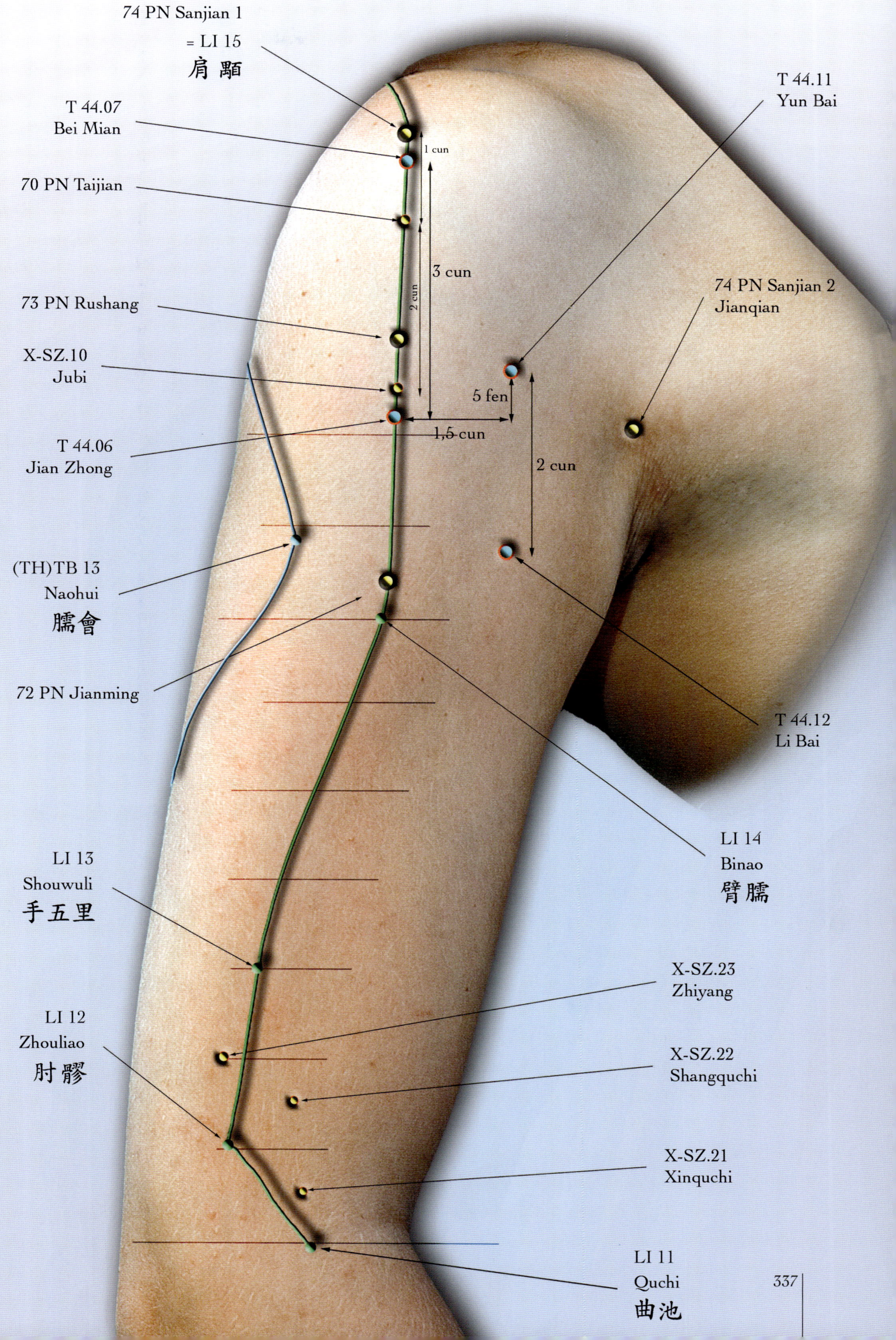

74 PN Sanjian 1
= LI 15
肩髃
T 44.07
Bei Mian
1 cun
70 PN Taijian
3 cun
2 cun
73 PN Rushang
X-SZ.10
Jubi
T 44.06
Jian Zhong
5 fen
1,5 cun
2 cun
T 44.11
Yun Bai
74 PN Sanjian 2
Jianqian
(TH)TB 13
Naohui
臑會
72 PN Jianming
T 44.12
Li Bai
LI 14
Binao
臂臑
LI 13
Shouwuli
手五里
X-SZ.23
Zhiyang
LI 12
Zhouliao
肘髎
X-SZ.22
Shangquchi
X-SZ.21
Xinquchi
LI 11
Quchi
曲池

Points of the head

- Anterior and posterior side
  - Points of the 3 upper Yang

    Shu Yang Ming (Large Intestine)

    Shu Tai Yang (Small Intestine)

    Shu Shao Yang (Triple Heater or Triple Burner)
  - Points of the 3 lower Yang

    Zu Yang Ming (Stomach)

    Zu Tai Yang (Bladder)

    Zu Shao Yang (Gall Bladder)
  - Extra channel points

    Curious points, New points, Points of Master Tong
- Posterior and lateral side
  - Extra channel points

    Curious points, New points, Points of Master Tong
- Points of the tongue
- A summary of the points of the neck
  - Anterior side
  - Lateral side

Antero-lateral side

Shu Yang Ming channel (Large Intestine)

LI 17 Tianding
On the lateral side of the neck, at the posterior border of the sterno-cleido-mastoid muscle beside the laryngeal protuberance, at the midpoint of the line connecting LI 18 Futu and ST 12 Quepen.

LI 18 Futu
On the lateral side of the neck, beside the laryngeal protuberance, between the anterior and posterior borders of the sternocleidomastoid muscle.

LI 19 Zouliao (Kouheliao)
On the upper lip, directly below the lateral border of the nostril, on the level of (DU)GV 26 Shuigou (Renzhong).

LI 20 Yingxiang
In the naso-labial groove, beside the midpoint of the lateral border of the nasal ala.

Shu Tai Yang channel (Small Intestine)

SI 16 Tianchuang
On the lateral side of the neck, posterior to the sterno-cleido-mastoid muscle and LI 18 Futu, on the level of the laryngeal protuberance.

SI 17 Tianrong
On the lateral side of the neck, posterior to the mandibular angle, in the depression of the anterior border of the sterno-cleido-mastoid muscle.

SI 18 Quanliao
On the face, directly below the outer canthus, in the depression below the zygomatic bone.

SI 19 Tinggong
On the face, anterior to the tragus and posterior to the mandibular condyloid process, in the depression found when the mouth is open.

Shu Shao Yang channel (Triple Heater or Triple Burner)

(TH)TB 16 Tianyou
On the lateral side of the neck, directly below the posterior border of the mastoid process, on the level of the mandibular angle, and on the posterior border of the sterno-cleido-mastoid muscle.

(TH)TB 17 Yifeng
Posterior to the ear lobe, in the depression between the mastoid process and mandibular angle.

(TH)TB 18 Chimai
On the head, at the centre of the mastoid process, and at the junction of the middle third and lower third of the line connecting (TH)TB 20 Jiaosun and (TH)TB 17 Yifeng along the curve of the ear helix.

(TH)TB 19 Luxi
On the head, at the junction of the upper third and middle third of the line connecting (TH)TB 20 Jiaosun and (TH)TB 17 Yifeng along the curve of the ear helix.

(TH)TB 20 Jiaosun
On the head, in the scalp, when the the ear is folded forward, directly above the top of the fold thus formed.

(TH)TB 21 Ermen
On the face, anterior to the supratragic notch of the tragus, in the depression behind the posterior border of the condyloid process of the mandible, when the mouth is open.

(TH)TB 22 Erheliao
On the lateral side of the head, on the posterior margin of the temples, anterior to the anterior border of the root of the ear auricle and posterior to the superficial temporal artery.

(TH)TB 23 Sizhukong
On the face, in the depression of the lateral end of the eyebrow.

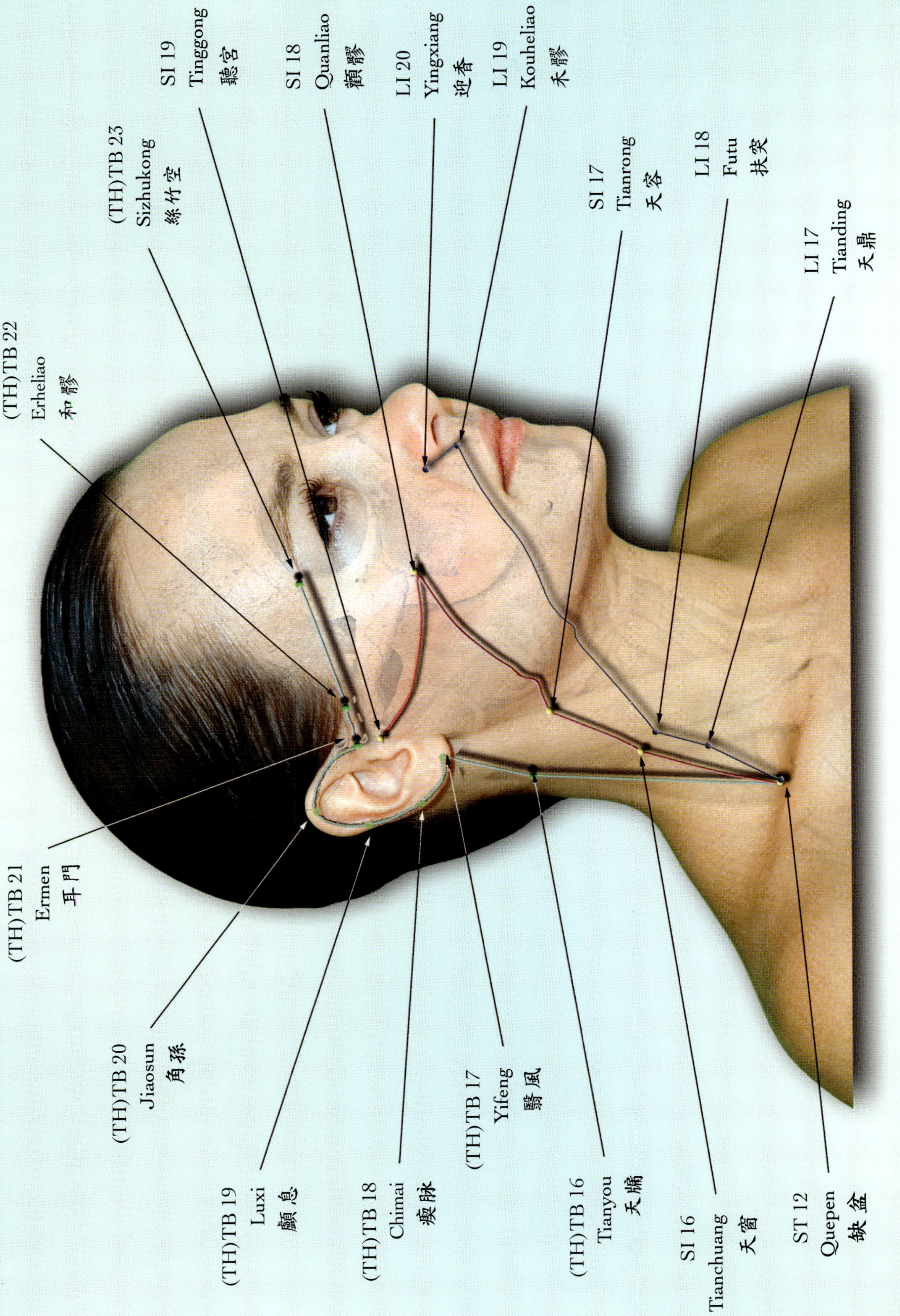

(TH)TB 22
Erheliao
和髎
(TH)TB 23
Sizhukong
絲竹空
SI 19
Tinggong
聽宮
SI 18
Quanliao
顴髎
LI 20
Yingxiang
迎香
LI 19
Kouheliao
禾髎
SI 17
Tianrong
天容
LI 18
Futu
扶突
LI 17
Tianding
天鼎
(TH)TB 21
Ermen
耳門
(TH)TB 20
Jiaosun
角孫
(TH)TB 19
Luxi
顱息
(TH)TB 18
Chimai
瘈脉
(TH)TB 17
Yifeng
翳風
(TH)TB 16
Tianyou
天牖
SI 16
Tianchuang
天窗
ST 12
Quepen
缺盆

Zu Yang Ming channel (Stomach)

ST 1 Chengqi
On the face, directly below the pupil, between the eyeball and the infra-orbital ridge.

ST 2 Sibai
On the face, directly below the pupil, in the depression of the infraorbital foramen, below ST 1 Chengqi point.

ST 3 Juliao
On the face, directly below the pupil, on the level of the lower border of the nasal ala, beside the nasolabial groove.

ST 4 Dicang
On the face, directly below the pupil, beside the mouth angle.

ST 5 Daying
Anterior to the mandibular angle, on the anterior border of the masséter muscle, where the pulsation of the facial artery is palpable. (Do NOT needle directly into the artery!)

ST 6 Jiache
On the cheek, one finger breadth (middle finger) anterior and superior to the mandibular angle, in the depression where the masseter muscle is prominent.

ST 7 Xiaguan
On the face, anterior to the ear, in the depression between the zygomatic arch and inferior mandibular notch.

ST 8 Touwei
On the lateral side of the head, 0.5 Cun above the anterior hairline at the corner of the forehead, and 4.5 Cun lateral to the midline of the head.

ST 9 Renying
On the neck, beside the laryngeal protuberance, and on the anterior border of the sterno-cleido-mastoid muscle, where the pulsation of the common carotid artery is palpable. (Do NOT needle directly into the artery!)

ST 10 Shuitu
On the neck and on the anterior border of the sternocleidomastoid muscle, at the midpoint of the line connecting ST 9 Renying and ST 11 Qishe.

ST 11 Qishe
On the neck and on the upper border of the medial end of the clavicle, between the sternal and clavicular heads of the sterno-cleido-mastoid muscle.

ST 12 Quepen
At the centre of the supra-clavicular fossa, 4 Cun lateral to the anterior midline.

Zu Tai Yang channel (Bladder)

BL1 Jingming
On the face, in the depression slightly above the inner canthus.

BL 2 Cuanzhu
On the face, in the depression of the medial end of the eyebrow, at the supra-orbital notch.

BL 3 Meichong
On the head, directly above BL 2 Cuanzhu, 0.5 Cun above the anterior hairline, on the line connecting (DU)GV 24 Shenting and BL 4 Qucha.

BL 4 Qucha
On the head, 0.5 Cun directly above the midpoint of the anterior hairline and 1.5 Cun lateral to the midline, at the junction of the medial third and middle third of the line connecting (DU)GV 24 Shenting and ST 8 Touwei.

BL 5 Wuchu
On the head, 1 Cun directly above the midoint of the anterior hairline and 1.5 Cun lateral to the midline.

BL 6 Chengguang
On the head, 2.5 Cun directly above the midpoint of the anterior hairline and 1.5 Cun lateral to the midline.

BL 7 Tongtian
On the head, 4 Cun directly above the midpoint of the anterior hairline and 1.5 Cun lateral to the midline.

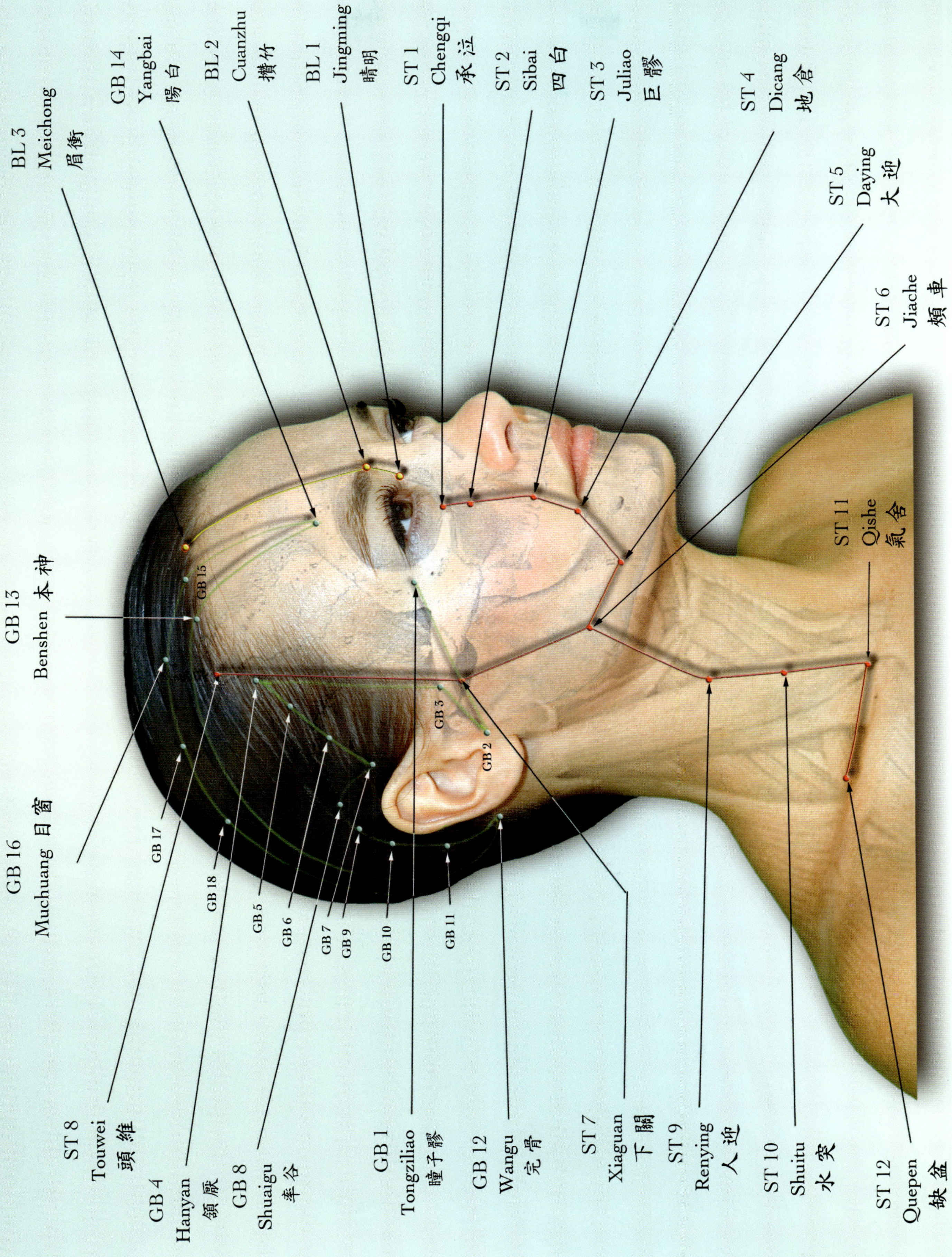
BL 3 Meichong 眉衝
GB 14 Yangbai 陽白
BL 2 Cuanzhu 攢竹
BL 1 Jingming 睛明
ST 1 Chengqi 承泣
ST 2 Sibai 四白
ST 3 Juliao 巨髎
ST 4 Dicang 地倉
ST 5 Daying 大迎
ST 6 Jiache 頰車
GB 13 Benshen 本神
GB 15
GB 3
GB 2
ST 11 Qishe 氣舍
GB 16 Muchuang 目窗
GB 17
GB 18
GB 5
GB 6
GB 7
GB 9
GB 10
GB 11
ST 8 Touwei 頭維
GB 4 Hanyan 頷厭
GB 8 Shuaigu 率谷
GB 1 Tongziliao 瞳子髎
GB 12 Wangu 完骨
ST 7 Xiaguan 下關
ST 9 Renying 人迎
ST 10 Shuitu 水突
ST 12 Quepen 缺盆

Zu Shao Yang channel (Gall Bladder)

GB 1 Tongziliao
On the face, lateral to the outer canthus, on the lateral border of the orbit.

GB 2 Tinghui
On the face, anterior to the intertragic notch, in the depression posterior to the condyloid process of the mandible when the mouth is open.

GB 3 Shangguan
Anterior to the ear, directly above ST 7 Xiaguan, in the depression above the upper border of the zygomatic arch.

GB 4 Hanyan
On the head, in the hair above the temples, at the junction of the upper fourth and lower three fourths of the curved line connecting ST 8 Touwei and GB 7 Qubin.

GB 5 Xuanlu
On the head, in the hair above the temples, at the midpoint of the curved line connecting ST 8 Touwei and GB 7 Qubin.

GB 6 Xuanli
On the head, in the hair above the temples, at the junction of the upper three fourths and lower fourth of the curved line connecting ST 8 Touwei and GB 7 Qubin.

GB 7 Qubin
On the head, at the crossing point of the vertical posterior border of the temples and the horizontal line through the ear apex.

GB 8 Shuaigu
On the head, directly above the ear apex, 1.5 Cun above the hairline, directly above (TH)TB 20 Jiaosun.

GB 9 Tianchong
On the head, directly above the posterior border of the ear root, 2 Cun above the hairline and 0.5 Cun posterior to GB 8 Shuaigu.

GB 10 Fubai
On the head, posterior and superior to the mastoid process, at the junction of the middle third and upper third of the curved line connecting GB 9 Tianchong and GB 12 Wangu.

GB 11 Touqiaoyin
On the head, posterior and superior to the mastoid process, at the junction of the middle third and lower third of the curved line connecting GB 9 Tianchong and GB 12 Wangu.

GB 12 Wangu
On the head, in the depression posterior and inferior to the mastoid process.

GB 13 Benshen
On the head, 0.5 Cun above the anterior hairline, 3 Cun lateral to (DU)GV 24 Shenting, at the junction of the medial two thirds and lateral third of the line connecting (DU)GV 24 Shenting and ST 8 Touwei.

GB 14 Yangbai
On the forehead, directly above the pupil, 1 Cun above the eyebrow.

GB 15 Toulinqi
On the head, directly above the pupil and 0.5 Cun above the anterior hairline, at the midpoint of the line connecting (DU)GV 24 Shenting and ST 8 Touwei.

GB 16 Muchuang
On the head,1.5 Cun above the anterior hairline and 2.25 Cun lateral to the midline of the head.

GB 17 Zhengying
On the head, 2.5 Cun above the anterior hairline and 2.25 Cun lateral to the midline of the head.

GB 18 Chengling
On the head, 4 Cun above the anterior hairline and 2.25 Cun lateral to the midline of the head.

Du Mai (Governor vessel)

(DU)GV 20 Baihui
On the head, 5 Cun directly above the midpoint of the anterior hairline, at the midpoint of the line connecting the apexe of both ears.

(DU)GV 21 Qianding
On the head, 3.5 Cun directly above the midpoint of the anterior hairline and 1.5 Cun anterior to GV 20 Baihui.

(DU)GV 22 Xinhui
On the head, 2 Cun directly above the midpoint of the anterior hairline and 3 Cun anterior to (DU)GV 20 Baihui.

(DU)GV 23 Shangxing
On the head, 1 Cun directly above the midpoint of the anterior hairline.

(DU)GV 24 Shenting
On the head, 0.5 Cun directly above the midpoint of the anterior hairline.

(DU)GV 25 Suliao
On the face, at the centre of the nose apex.

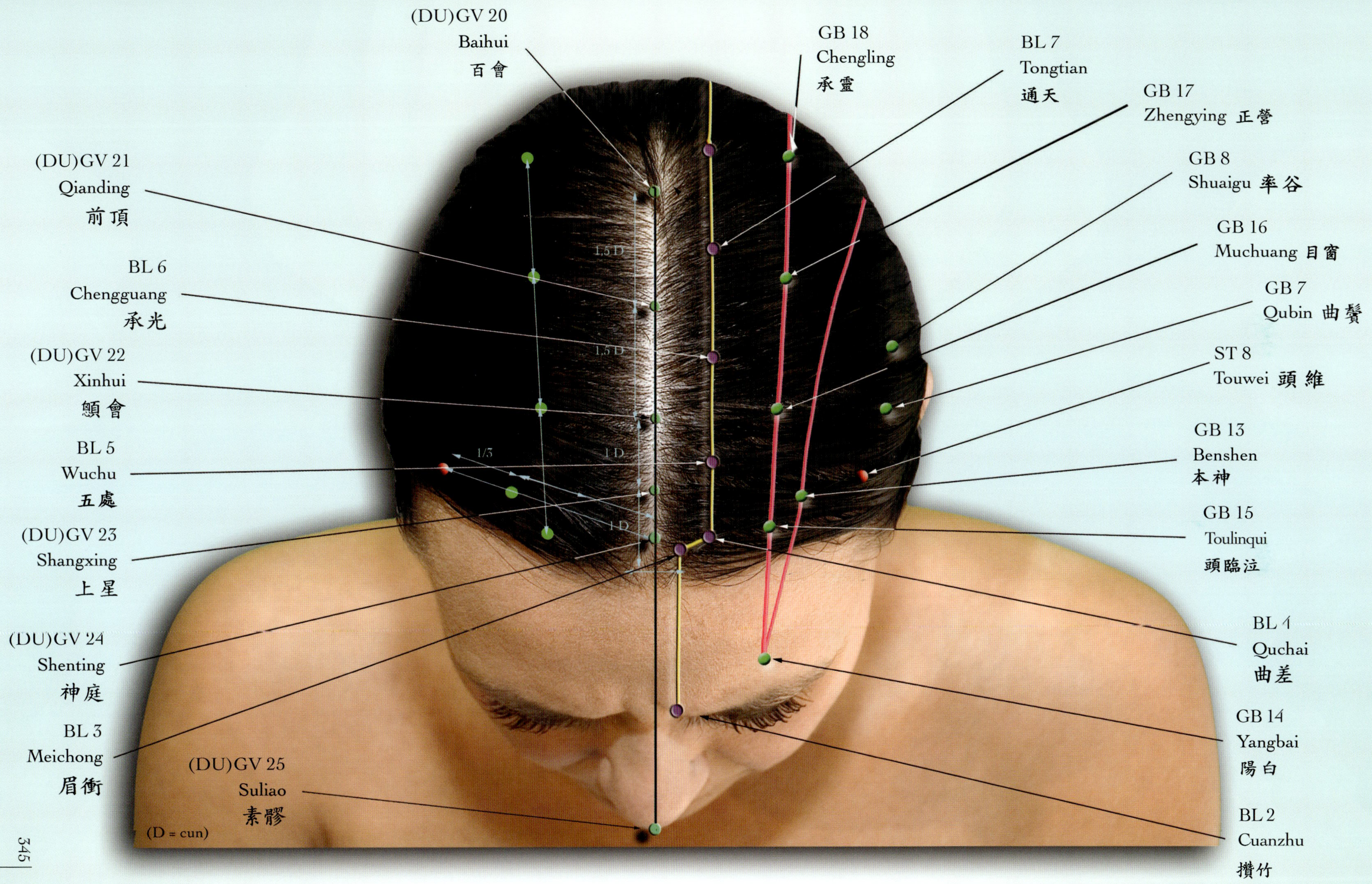
(DU)GV 20
Baihui
百會
GB 18
Chengling
承靈
BL 7
Tongtian
通天
GB 17
Zhengying 正營
GB 8
Shuaigu 率谷
GB 16
Muchuang 目窗
GB 7
Qubin 曲鬢
ST 8
Touwei 頭維
GB 13
Benshen
本神
GB 15
Toulinqui
頭臨泣
BL 4
Quchai
曲差
GB 14
Yangbai
陽白
BL 2
Cuanzhu
攢竹
(DU)GV 21
Qianding
前頂
BL 6
Chengguang
承光
(DU)GV 22
Xinhui
顖會
BL 5
Wuchu
五處
(DU)GV 23
Shangxing
上星
(DU)GV 24
Shenting
神庭
BL 3
Meichong
眉衝
(DU)GV 25
Suliao
素髎
1.5 D
1.5 D
1 D
1 D
1/3
(D = cun)

Extra channel points (PC – PN)

6 PC Yuyao
At the midpoint of the eyebrow, vertical to the iris when the eyes are looking straight forwards when one is right in front of the face.

7 PC Yuwei
At the outer canthus, between that canthus and GB 1 Tongziliao.

8 PC Qiuhou
Horizontal to ST 1 Chengqi and 0.5 Cun outside.

9 PC Taiyang
In the temporal area, midpoint of the temple, on the temporal artery between GB 1 Tongziliao and the limit of the whiskers.

10 PC Erjian
On the apex of the auricle; fold the ear forward to find the point.

11 PC Lungxue
At the midpoint of the line connecting the points SI 19 Tinggong and (TH)TB 21 Ermen.

14 PC Shangyingxiang
5 Cun below the inner canthus, on the level of the centre of the bridge of the nose.

15 PC Jiabi
Below 14 PC Shangyingxiang, where the cartilage and the bone of the nose meet.

16 PC Biliu
At the entrance of the nostrils, midpoint between the septum and the nasal ala.

17 PC Sanjiao
At the midpoint of the nasogenial groove, outside and below LI 20 Yingxian.

18 PC Heliao
Where the vertical line passing by ST 4 Dicang and the horizontal line passing by (REN)CV 24 Chengjiang cross.

19 PC Dihe
It is the highest point of the chin.
It is just on the tip and does not coincide with (REN)CV 24 Chengjiang.

21 PC Shanglianquan
1 Cun above point (REN)CV 23 Lianquan, in a depression above the thyroid cartilage.

22 PC Waijinjin – Waiyuyue
0.3 Cun outside and above (REN)CV 23 Lianquan.

23 PC Hongyin
On the horizontal line of (REN)CV 23 Lianquan, 0.5 Cun from that point.

24 PC Panglianquan
Midpoint of the line connecting (REN)CV 23 Lianquan and the anterior border of the sternocleidomastoid muscle

25 PC Yaxue
4 points: 2 on the anterior side of the neck, 2 on the posterior side.
The anterior points are between ST 9 Renyin and LI 18 Futu, where the pulsation of the artery is palpable.
The posterior points are 2.5 Cun above point GB 20 Fengchi and on the same horizontal line, internally, 1 Cun from 13 PC Yiming.

26 PC Biantao
Below the lower border of the angle of the inferior maxilla, just before the jugular vein, near SI 17 Tianrong.

27 PC Luojing
On the posterior border of the sterno-cleido-mastoid muscle, on its upper third, on the horizontal line passing by 26 PC Biantao.

J-TJ.32 Muming
Above the pupil, at the limit of the hairline, or 0.5 Cun below GB 15 Toulinqui.

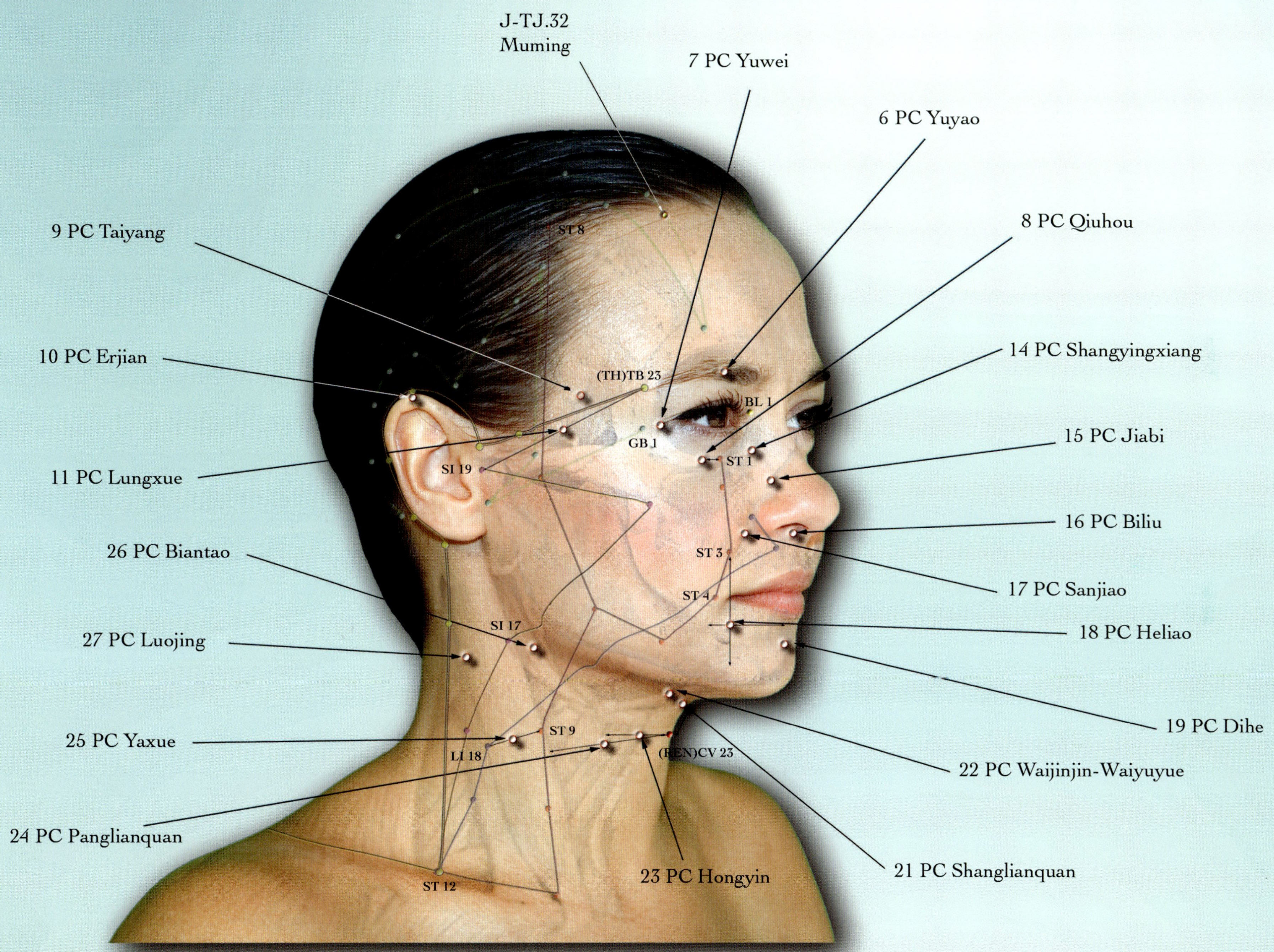
J-TJ.32
Muming
7 PC Yuwei
6 PC Yuyao
8 PC Qiuhou
9 PC Taiyang
ST 8
10 PC Erjian
(TH)TB 23
14 PC Shangyingxiang
BL 1
GB 1
ST 1
15 PC Jiabi
11 PC Lungxue
SI 19
16 PC Biliu
26 PC Biantao
ST 3
17 PC Sanjiao
ST 4
18 PC Heliao
SI 17
27 PC Luojing
19 PC Dihe
25 PC Yaxue
ST 9
LI 18
(REN)CV 23
22 PC Waijinjin-Waiyuyue
24 PC Panglianquan
ST 12
23 PC Hongyin
21 PC Shanglianquan

A SUMMARY OF THE POINTS OF THE FACE (PC)

2 PC Erzhong
At the limit of the forehead, on the anterior midline, 2 Cun from 3 PC Yintang.

3 PC Yintang
At the midpoint of the line connecting the inner extremities of the eyebrows (a link between the BL 2 Cuanzhu points).

4 PC Shangen
At the midpoint of the line connecting the inner canthuses on the bridge.

5 PC Touquangming
Slightly above point 6 PC Yuyao.

6 PC Yuyao
At the midpoint of the eyebrow, vertical to the iris when one is right in front of the face.

7 PC Yuwei
At the outer canthus, between that canthus and GB 1 Tongziliao.

8 PC Qiuhou
Horizontal to ST 1 Chengqi and 0.5 Cun outside.

14 PC Shangyingxiang
5 Cun below the inner canthus, on the level of the centre of the bridge of the nose.

15 PC Jiabi
Below 14 PC Shangyingxiang, where the cartilage and the bone of the nose meet.

16 PC Biliu
At the entrance of the nostrils, at the midpoint between the septum and the nasal ala.

17 PC Sanjiao
At the midpoint of the nasogenial groove, outside and below LI 20.

18 PC Heliao
Where the vertical line passing by ST 4 Dicang and the horizontal line passing by (REN)CV 24 Chengjiang cross.

19 PC Dihe
It is the highest point on the chin. It is just on the tip and does not coincide with (REN)CV 24 Chengjiang.

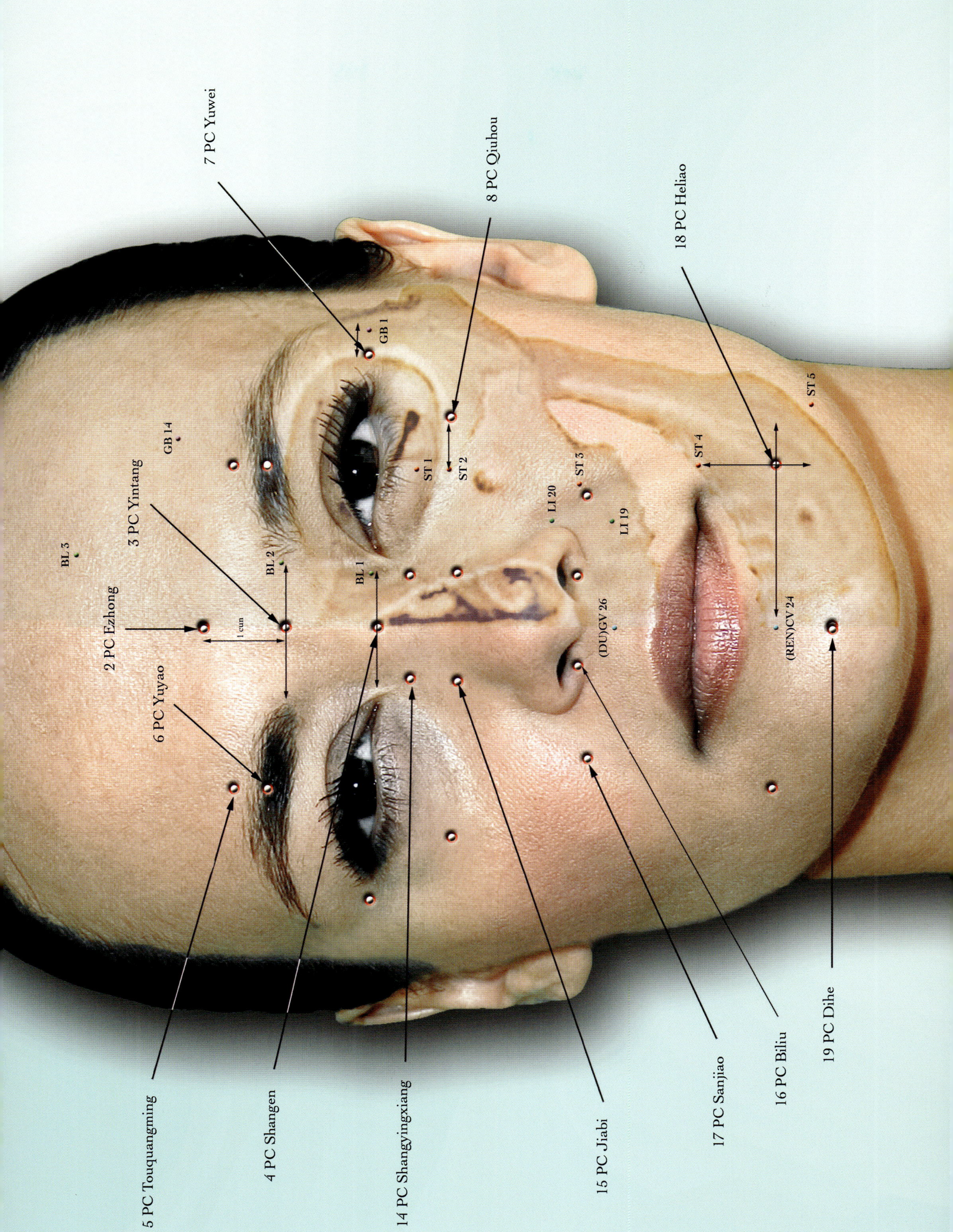

Summary of PC points on the face

Extra channel points (PC – PN)

13 PN Tingxue
At the midpoint between SI 19 Tinggong and GB 2 Tinghui.

14 PN Tingling
At the midpoint between 13 PC Tingxue and GB 2 Tinghui, or where the lower fourth of the distance between GB 2 Tinghui and SI 18 Quanliao meet the upper three fourths.

15 PN Tingcong
0.2 Cun below GB 2 Tinghui.

16 PN Tingmin
On the lower part of the groove formed by the ear lobe and the upper branch of the lower maxilla.

25 PN Qianzhengxue
0.5 Cun before the root of the ear lobe.

J-TJ.18 Jiachengjiang
In the mandibular foramen, directly below ST 4 Dicang.

J-TJ.34 Yankou
In the corner of the lips.

J-TJ.35 Neiyingxiang
On the upper border of the nasal mucus membrane, near the nostril.

X-TJ.3 Jianming
0.4 Cun below BL 1 Jingming, tangential to the inferior border of the orbital cavity.

X-TJ.24 Ronghou
Posterior to SI 17 Tianrong and 1.5 Cun below (TH)TB 17 Yifeng.

X-TJ.31 Tounie
1 Cun posterior to 9 PC Taiyang.

X-TJ.32 Dingshen
Between the lower third and the middle third of the philtrum, about 0.4 Cun, below the nose.

X-TJ.33 Guancai
0.2 Cun before (TH)TB 20 Jiaosun.

X-TJ.35 Tongming
0.5 Cun below GB 1 Tongziliao.

X-TJ.36 Tingxiang
0.1 Cun above (TH)TB 21 Ermen.

X-TJ.37 Shanglong
Between (TH)TB 21 Ermen and SI 19 Tinggong.

X-38 Tinglongqian
Between X-TJ.8 Tingxue and SI 19 Tinggong.

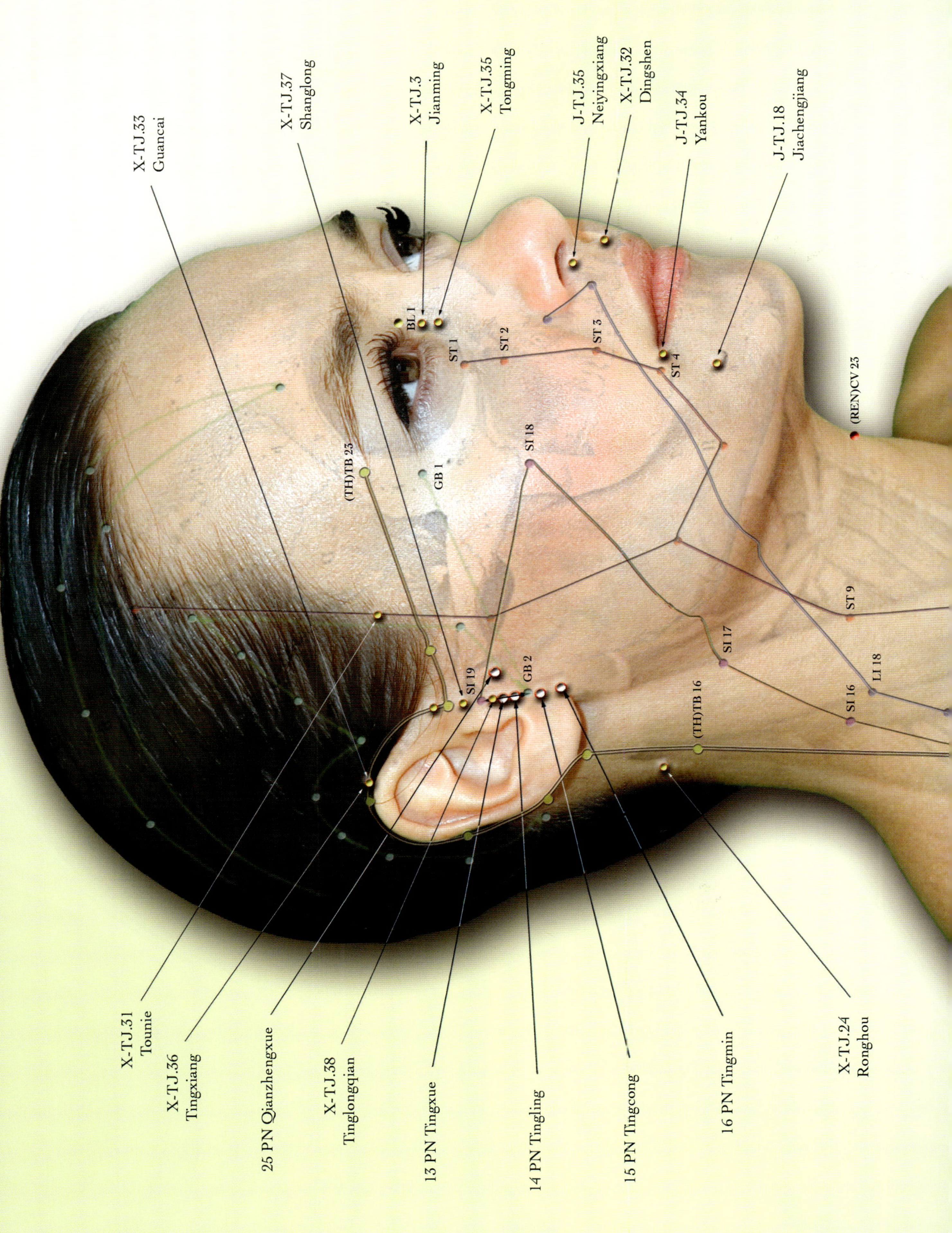

X-TJ.33 Guancai
X-TJ.37 Shanglong
X-TJ.3 Jianming
X-TJ.35 Tongming
J-TJ.35 Neiyingxiang
X-TJ.32 Dingshen
J-TJ.34 Yankou
J-TJ.18 Jiachengjiang
BL 1
ST 1
ST 2
ST 3
ST 4
(REN)CV 23
SI 18
GB 1
(TH)TB 23
ST 9
SI 17
LI 18
SI 16
(TH)TB 16
SI 19
GB 2
X-TJ.31 Tounie
X-TJ.36 Tingxiang
25 PN Qianzhengxue
X-TJ.38 Tinglongqian
13 PN Tingxue
14 PN Tingling
15 PN Tingcong
16 PN Tingmin
X-TJ.24 Ronghou

Extra channel points

Lateral side of the head

13 PC Yiming
On the inferior border of the mastoid, on the tip, 5 fen behind the attachment of the ear lobe, between points (TH)TB 17 Yifeng at the front and GB 12 Wangu behind.

16 PN Tingmin
On the inferior side of the groove formed by the ear lobe and the upper branch of the lower maxilla.

17 PN Shangergen
In the groove formed by the helix and the temporal bone.

18 PN Houtinggong
When the ear is folded forward, a tendinous mass appears on the ring formed by the base of the ear. The point is situated in the depression under that mass.
It is on the same level as SI 19 Tinggong situated in front of the ear.

19 PN Houtingxue
Midpoint between 18 PN Houtinggong and 20 PN Houtinghui, on the same level as 13 PN Tingxue situated in front of the ear.

20 PN Houtinghui
Behind the ear, 0.5 Cun below (TH)TB 17 Yifeng.

21 PN Hougong
Midpoint between 17 PN Shangergen and 18 PN Houtianggong. The point is situated below the tendinous mass, in a depression.

23 PN Yimingxia
0.5 Cun below 13 PC Yiming.

25 PN Qianzhengxue
0.5 Cun in front of the root of the ear lobe.

27 PN Anmian 1
Midpoint between (TH)TB 17 Yifeng and 13 PC Yiming.

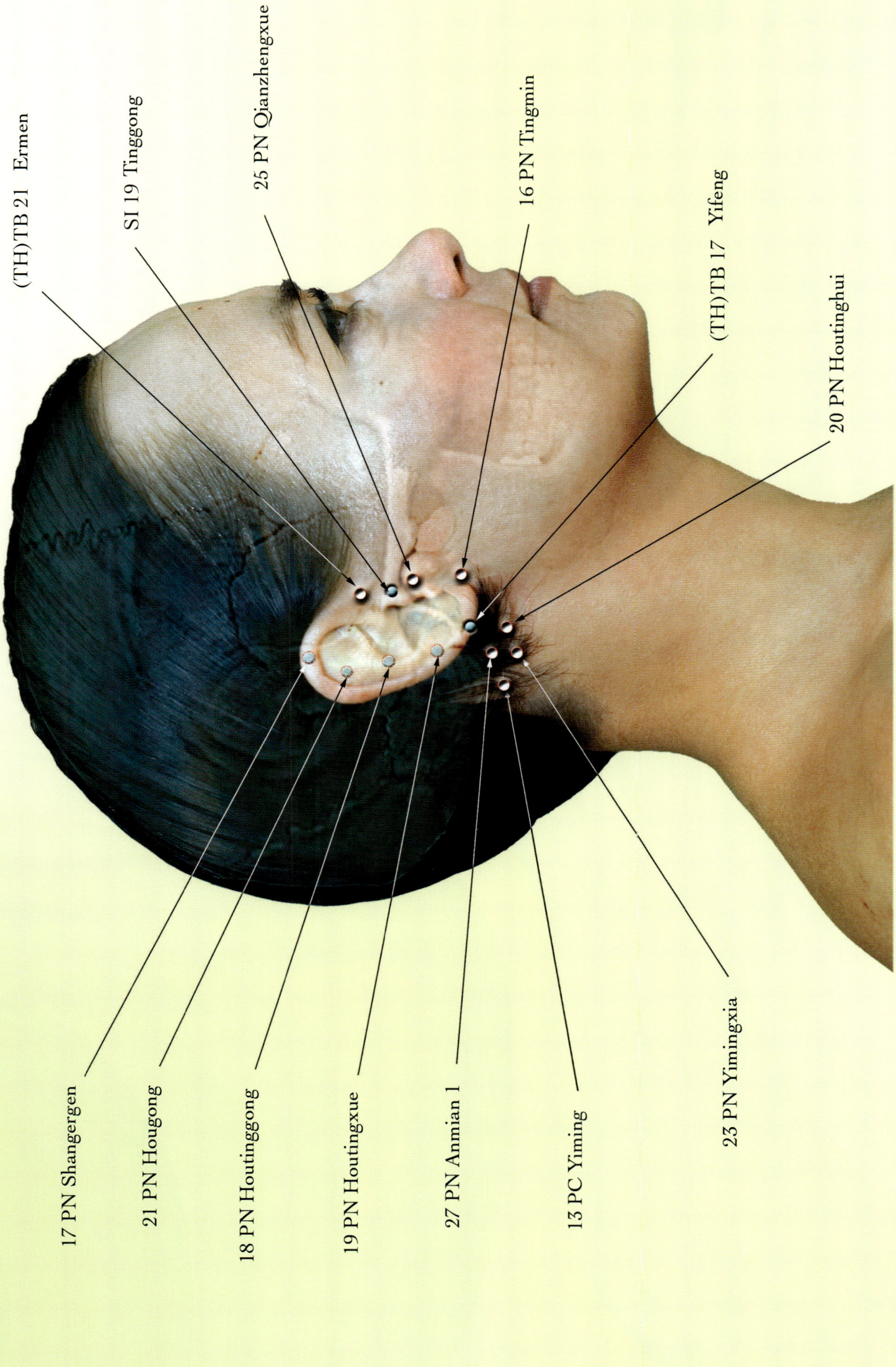

Points around the ear

A summary of the points of the face (PN)

1 PN Shangjingming
0.2 Cun above BL 1 Jingming.

2 PN Xiajingming
0.2 Cun below BL 1 Jingming.

3 PN Jiangming
0.2 Cun below 2 PN Xiajingming.

4 PN Jiangming n° 1
Between 3 PN Jiangming and ST 1 Chengqi.

5 PN Jiangming n° 2
Between 8 PC Qiuhou and ST 1 Chengqi.

6 PN Jiangming n° 3
0.3 Cun below the inner canthus, slightly outside.

7 PN Jiangming n° 4
0.3 Cun below the outer canthus, slightly outside.

8 PN Shangming
0.2 Cun below 6 PC Yuyao or below the rim of the orbital cavity.

9 PN Zengming n° 1
0.2 Cun inside 8 PN Shangming.

10 PN Zengming n° 2
0.2 Cun outside 8 PN Shangming.

11 PN Waiming
0.3 Cun above the outer canthus.

12 PN Bitong
In a depression, on the lower side of the nose, on the upper part of the philtrum, above and inside LI 20 Yingxiang.

35 PN Jianei
On the inside of the cheek, opposite the upper first bicuspid.

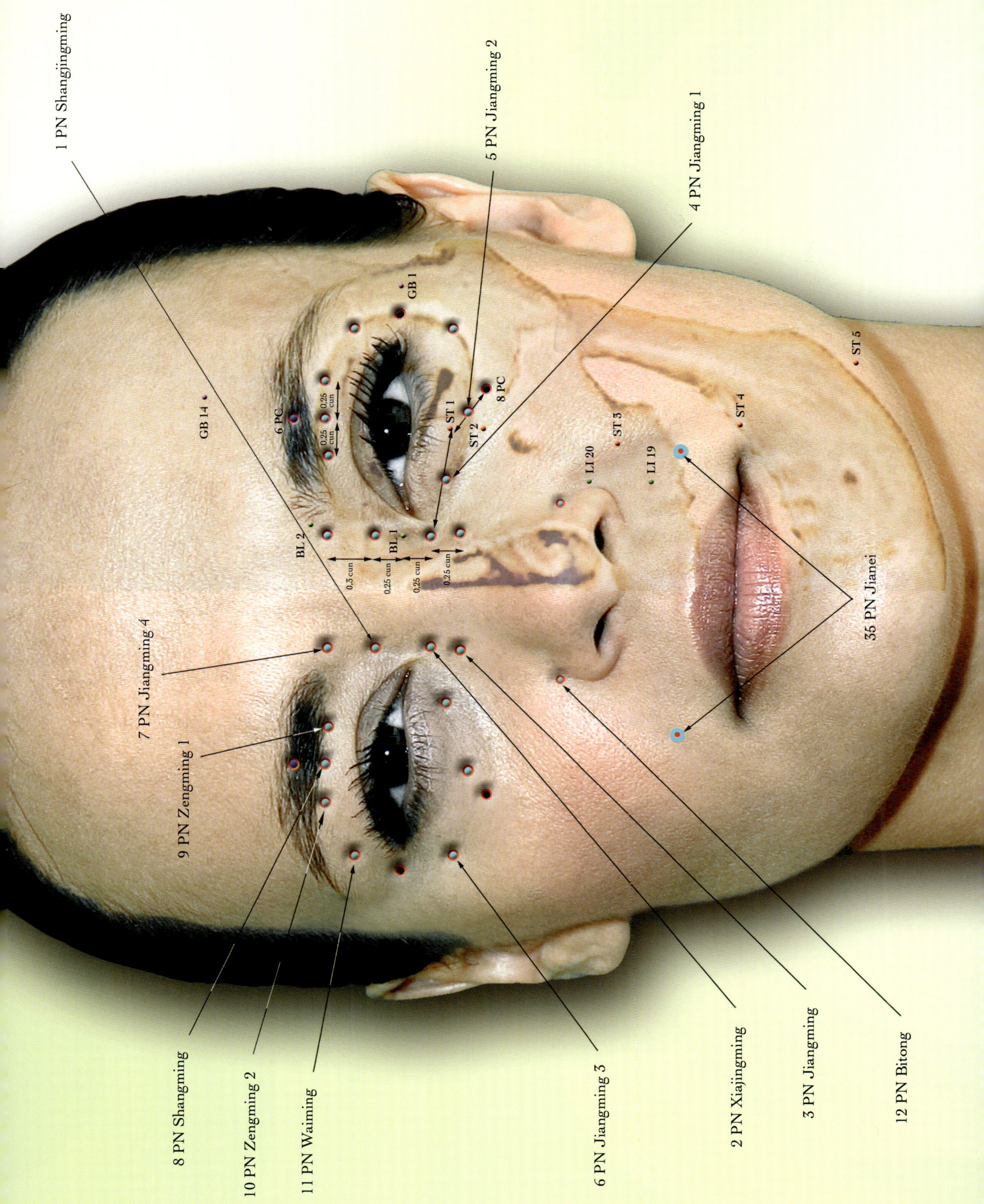

Summary of the points of the face

T 1010.10 Si Fu Er
2 fen above the centre of the eyebrow.
Corresponds to 6 PC Yuyao.

T 1010.11 Si Fu Yi
In the depression at the external extremity of the eyebrow.
Corresponds to (TH)TB 23 Sizhukong.

T 1010.12 Zheng Ben
At the tip of the nose.
Corresponds to (DU)GV 25 Suliao.

T 1010.13 Ma Jin Shui
On the inferior side of the maxillomalar joint.
Corresponds to SI 18 Quanliao.

T 1010.14 Ma Kuai Shui
4 fen below T 1010.13 Ma Jin Shui.

T 1010.15 Fu Kuai
5 fen lateral to the alae of the nose.
Corresponds to LI 20 Yingxiang.

T 1010.16 Liu Kuai
On the horizontal line passing by (DU)GV 26 Shuigou, 1.4 Cun on each side.

T 1010.17 Qi Kuai
Horizontally, the point is 5 fen outside the angle of the mouth.
Corresponds to ST 4 Dicang.

T 1010.18 Mu Zhi
In the depression situated at the lower border of the zygomatic arch, before the maxillary condyle.
Corresponds to ST 7 Xiaguan.

T 1010.19 Shui Tong
4 fen below the angle of the mouth.

T 1010.20 Shui Jin
4 fen below the angle of the mouth and 5 fen toward the midline. 5 fen inside T 1010.19 Shui Tong.

T 1010.21 Yu Huo
0.5 Cun below ST 1 Chengqi, in the depression where the infraorbital hole is.
Corresponds to ST 2 Sibai.

T 1010.22 Bi Yi
In the upper depression of the nasal ala.

T 1010.23 Zhou Huo
1.5 Cun above the auricle, vertically (below GB 8 Shuaigu).

T 1010.24 Zhou Jin
1.5 Cun above and 1 Cun behind the posterior border of the auricle, vertically.

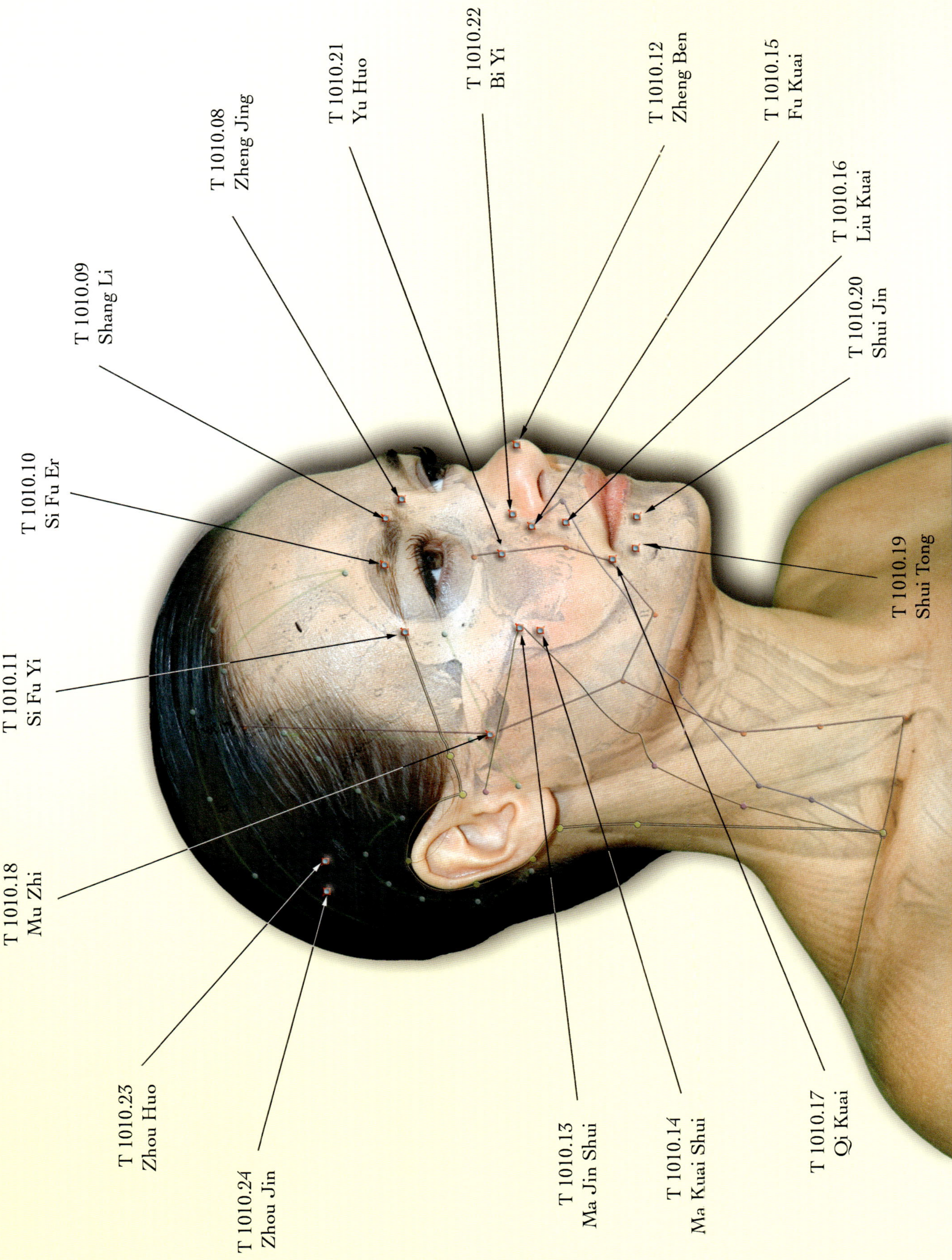
T 1010.10
Si Fu Er
T 1010.09
Shang Li
T 1010.08
Zheng Jing
T 1010.21
Yu Huo
T 1010.22
Bi Yi
T 1010.12
Zheng Ben
T 1010.15
Fu Kuai
T 1010.16
Liu Kuai
T 1010.20
Shui Jin
T 1010.19
Shui Tong
T 1010.11
Si Fu Yi
T 1010.18
Mu Zhi
T 1010.23
Zhou Huo
T 1010.24
Zhou Jin
T 1010.13
Ma Jin Shui
T 1010.14
Ma Kuai Shui
T 1010.17
Qi Kuai

Points of Master Tong
(Cranium)

T 1010.01 Zheng Hui
On the midline of the cranium, 7 Cun above the posterior hairline, midpoint of a line connecting the apexes of the ears.
Corresponds to (DU)GV 20 Baihui.

T 1010.02 Zhou Yuan
3 Cun from the coronal suture, 1.5 Cun outside Du Mai vessel.
Corresponds to BL 7 Tongtian.

T 1010.03 Zhou Kun
In the lambda suture, 1.5 Cun behind point T 1010.02 Zhou Yuan.
Corresponds to BL 8 Luoque.

T 1010.04 Zhou Lun
1.5 Cun from BL 5 Wuchu, 1.5 Cun outside Du Mai vessel. 1.5 Cun before T 1010.02 Zhou Yuan. Corresponds to BL 6 Chengguang.

T 1010.05 Qian Hui
On the midline of the cranium, 1.5 Cun in front of point (DU)GV 20 Baihui. 1.5 Cun in front of T 1010.01 Zhen Hui.
Corresponds to (DU)GV 21 Qianding.

T 1010.08 Zheng Jing
3 fen above the point that is situated midpoint of the eyes.
Corresponds to 3 PC Yin Tang.

T 1010.09 Shang Li
At the root of the eyebrows, in the supraorbital notch. Corresponds to BL 2 Cuanzhu.

T 1010.10 Si Fu Er
2 fen above the centre of the eyebrow.
Corresponds to BL 2 Cuanzhu.

T 1010.11 Si Fu Yi
In the depression at the external extremity of the eyebrow.
Corresponds to (TH)TB 23 Sizhukong.

T 1010.12 Zheng Ben
At the tip of the nose.
Corresponds to (DU)GV 25 Suliao.

1 PC Sishencong
A set of 4 points, forming a cross, 1 Cun from (DU)GV 20 Baihui.

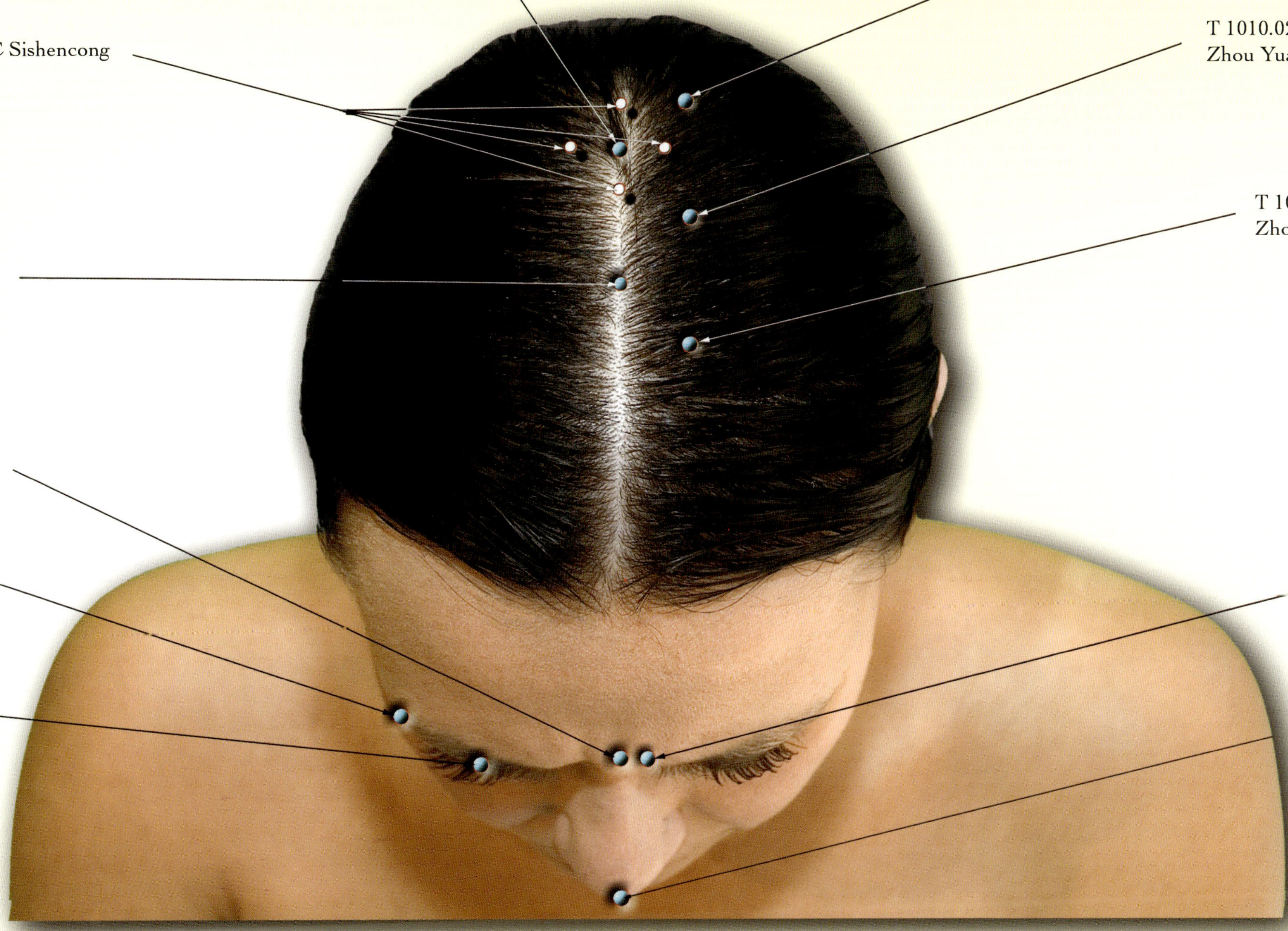
T 1010.01
Zheng Hui
T 1010.03
Zhou Kun
1 PC Sishencong
T 1010.02
Zhou Yuan
T 1010.04
Zhou Lun
T 1010.05
Qian Hui
T 1010.08
Zheng Jing
T 1010.11
Si Fu Yi
T 1010.09
Shang Li
T 1010.10
Si Fu Er
T 1010.12
Zheng Ben

Posterior side

Extra channel points

22 PN Qiqian
5 Cun before GB 20 Fengchi.

24 PN Tianting
0.5 Cun below 28 PN Anmian 2.

28 PN Anmian 2
Between GB 20 Fengchi and 13 PC Yiming.

29 PN Xingfen
0.5 Cun below 28 PN Anmian 2.

42 PN Liu Jingzhui Pang
0.5 Cun outside the spinous process of the 6th cervical vertebra.

43 PN Xueya Dian
2 Cun outside the spinous process of the 6th cervical vertebra.

44 PN Qi Jingzhui Pang
0.5 Cun outside the spinous process of the 7th cervical vertebra.

47 PN Jiehe Xue
3.5 Cun outside point (DU)GV 14 Dazhui.

13 PC Yiming
On the inferior border of the mastoid, on the tip, 5 fen behind the attachment of the ear lobe, between points (TH)TB 17 Yifeng before and GB 12 Wangu behind.

25 PC Yaxue
Four points: 2 on the anterior side of the neck and 2 on the posterior side.
The anterior points are between ST 9 Renyin and LI 18 Futu, where the pulsation of the artery is palpable.
The posterior points are 2.5 Cun above point GB 20 Fengchi and on the same horizontal line internally 1 Cun from 13 PC Yiming.

28 PC Fengyan
¼ Cun behind and below 13 PC Yiming.

29 PC Xinxi
1.5 Cun outside and below the 3rd cervical vertebra, on the level of its spinous process.

30 PC Bailiao
2 Cun above and 1 Cun outside point (DU)GV 14 Dazhui, in the space between the 6th and 7th cervical vertebrae.

31 PC Chonggu
2 Cun above point (DU)GV 14 Dazhui, in the space between the 6th and 7th cervical vertebrae.

X-TJ.40 Chixia
0.5 Cun below GB 20 Fengchi.

X-TJ.46 Fuyamen
1 Cun below and 0.5 Cun lateral to (DU)GV 15 Yamen.

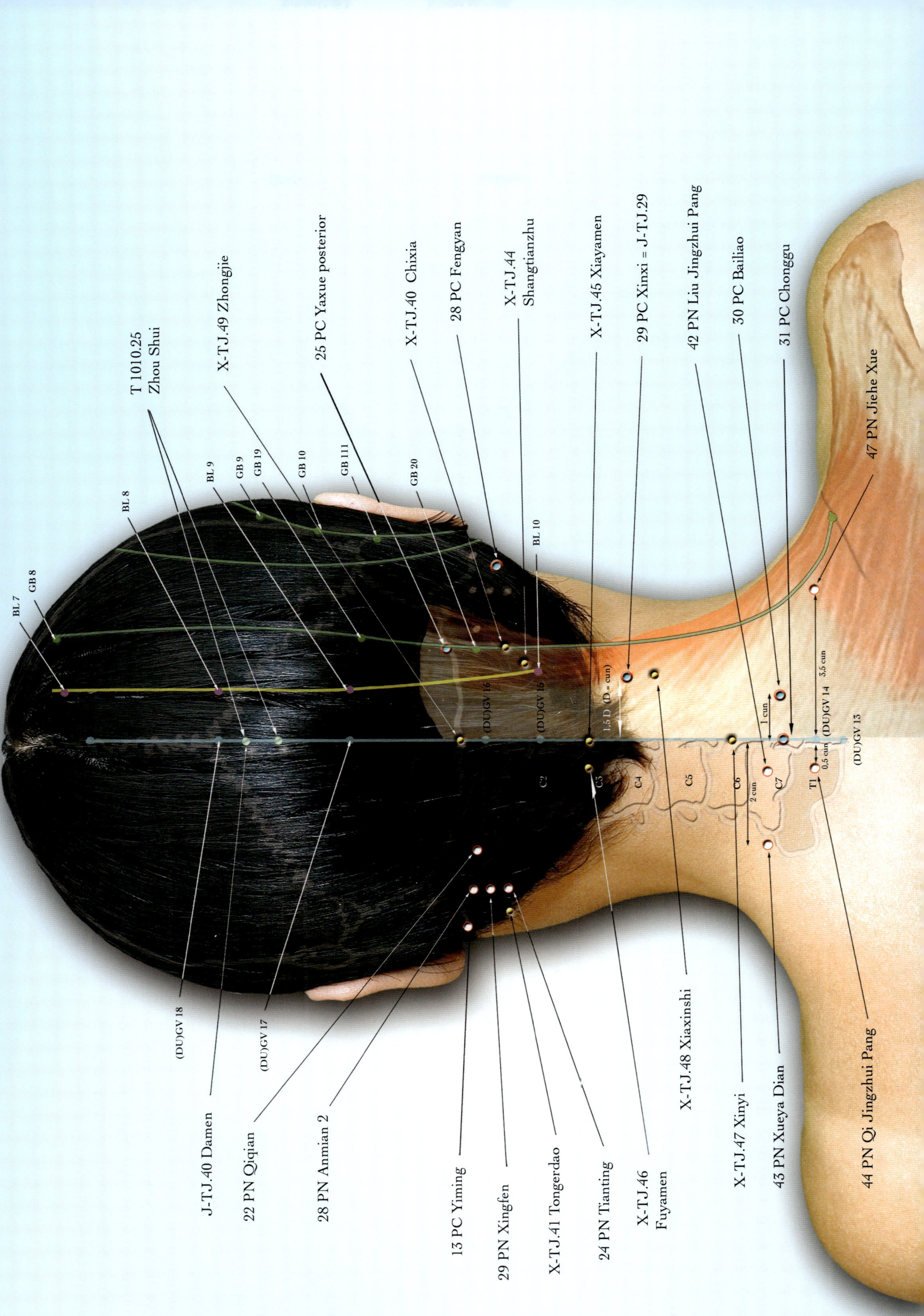
T 1010.25 Zhou Shui
X-TJ.49 Zhongjie
25 PC Yaxue posterior
X-TJ.40 Chixia
28 PC Fengyan
X-TJ.44 Shangtianzhu
X-TJ.45 Xiayamen
29 PC Xinxi = J-TJ.29
42 PN Liu Jingzhui Pang
30 PC Bailiao
31 PC Chonggu
47 PN Jiehe Xue
BL 7
GB 8
BL 8
BL 9
GB 9
GB 19
GB 10
GB 111
GB 20
BL 10
(DU)GV 16
(DU)GV 15
1,5 D (D = cun)
1 cun
(DU)GV 14
3,5 cun
(DU)GV 13
0,5 cun
2 cun
C2
C3
C4
C5
C6
C7
T1
(DU)GV 18
(DU)GV 17
J-TJ.40 Damen
22 PN Qiqian
28 PN Anmian 2
13 PC Yiming
29 PN Xingfen
X-TJ.41 Tongerdao
24 PN Tianting
X-TJ.46 Fuyamen
X-TJ.48 Xiaxinshi
X-TJ.47 Xinyi
43 PN Xueya Dian
44 PN Qi Jingzhui Pang

X –TJ.41 Tongerdao
1 Cun below Yiming.

X-TJ.44 Shangtianzhu
0.5 Cun above BL 10 Tianzhu.

X-TJ.45 Xiayamen
1 Cun below GV 15 Yamen.

X-TJ.47 Xinyi
Between the spinous processes of the 5th and 6th cervical vertebrae.

X-TJ.48 Xiaxinshi
0.5 Cun below 29 PC Xinxi.

X-TJ.49 Zhongjie
0.5 Cun above (DU)GV 16 Fengu.

J-TJ.40 Damen
On the vertex, 0.5 Cun behind (DU)GV 18 Qiangjian.

T 1010.25 Zhou Shui
On the midline of the posterior side of the head, the 1st point is on the occipital protuberance, the 2nd point is 8 fen above.

Points of the tongue

20 PC Jinjin – Yuye
On the inferior side of the tongue, on the sublingual veins. The point on the left is called Jinjin, the one on the right Yuye.

J-TJ.36 Juquan
At the midpoint of the back of the tongue.

J-TJ.37 Haiquan
At the midpoint of the frenum of the tongue.

J-TJ.38 Sheshu
Where the frenum and the fold of the tongue meet.

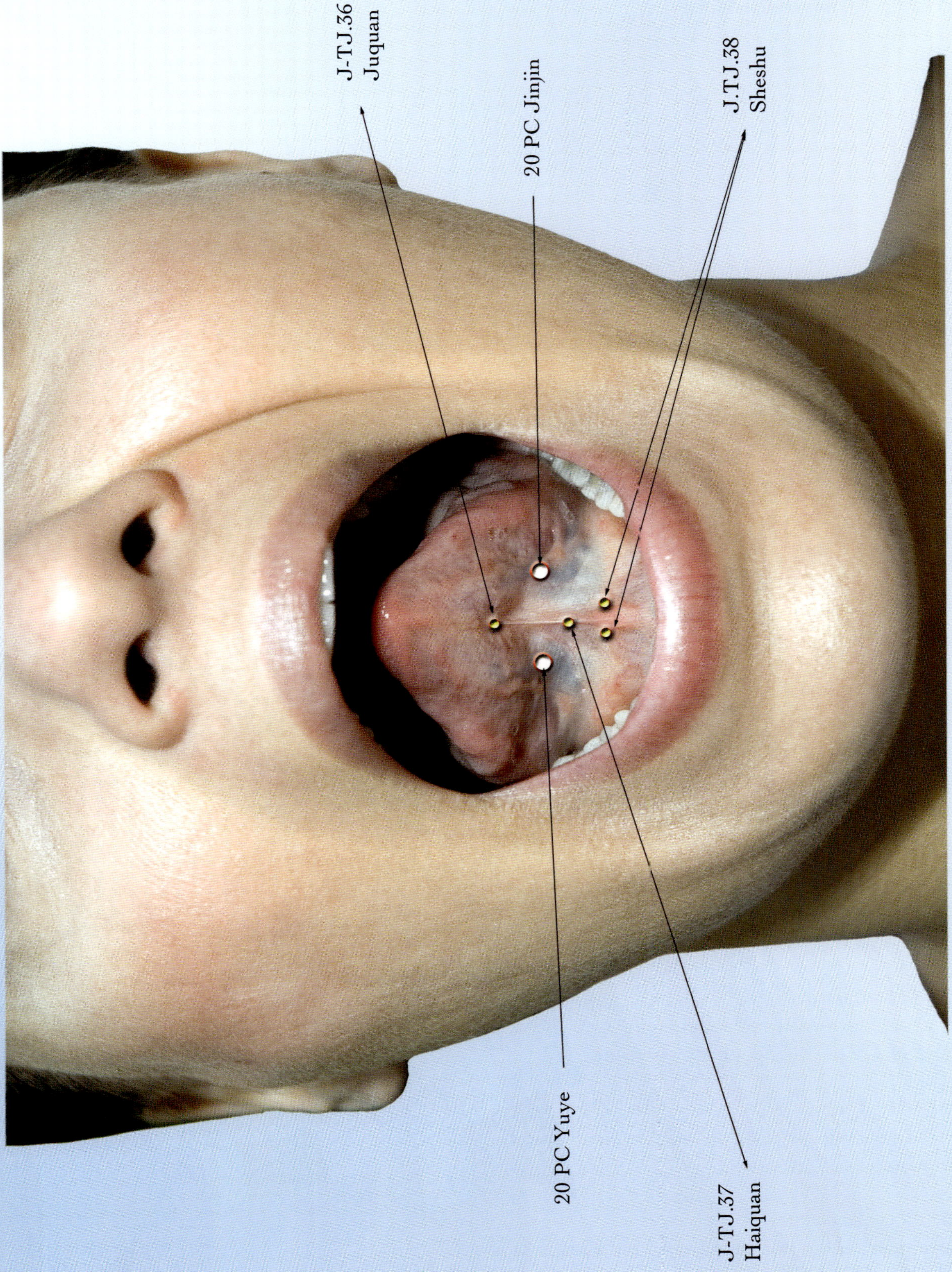
J-TJ.36
Juquan
20 PC Jinjin
J.TJ.38
Sheshu
20 PC Yuye
J-TJ.37
Haiquan

A summary of the points of the neck
(Anterior side)

21 PC Shanglianquan
1 Cun above point (REN)CV 23 Lianquan, in a depression above the thyroid cartilage.

22 PC Waijinjin – Waiyuyue
0.3 Cun outside and above (REN)CV 23 Lianquan.

23 PC Hongyin
On the horizontal line of (REN)CV 23 Lianquan, 0.5 Cun laterally.

24 PC Panglianquan
Midpoint of the line connecting (REN)CV 23 Lianquan and the anterior border of the sterno-cleido-mastoid.

25 PC Yaxue
4 points: 2 on the anterior, 2 on the posterior side of the neck. The anterior points are between ST 9 Renyin and LI 18 Futu, where the pulsation of the artery is palpable.
The posterior points are 2.5 Cun above GB 20 Fengchi and on the same horizontal line, internally, 1 Cun from 13 PC Yiming

26 PC Biantao
On the inferior border of the angle of the inferior maxilla, just before the jugular vein, near SI 17 Tianrong.

27 PC Luojing
On the posterior border of the sterno-cleido-mastoid muscle on the upper 1/3, on the horizontal line passing by 26 PC Biantao.

32 PC Chixue
1 Cun horizontally and outside point (REN)CV 21 Xuanji, on the sternum.

31 PN Qiangyin
2 Cun outside (REN)CV 23 Lianquan, behind and above ST 9 Renyin.

32 PN Zhenyin
Midpoint of the line contiguous to the angle of the inferior maxilla and Adam's apple.

X-TJ.43 Qiying
0.2 Cun before and above ST 10 Shuitu.

X-TJ.53 Zhiou
Midpoint between (REN)CV 23 Lianquan and (REN)CV 22 Tiantu.

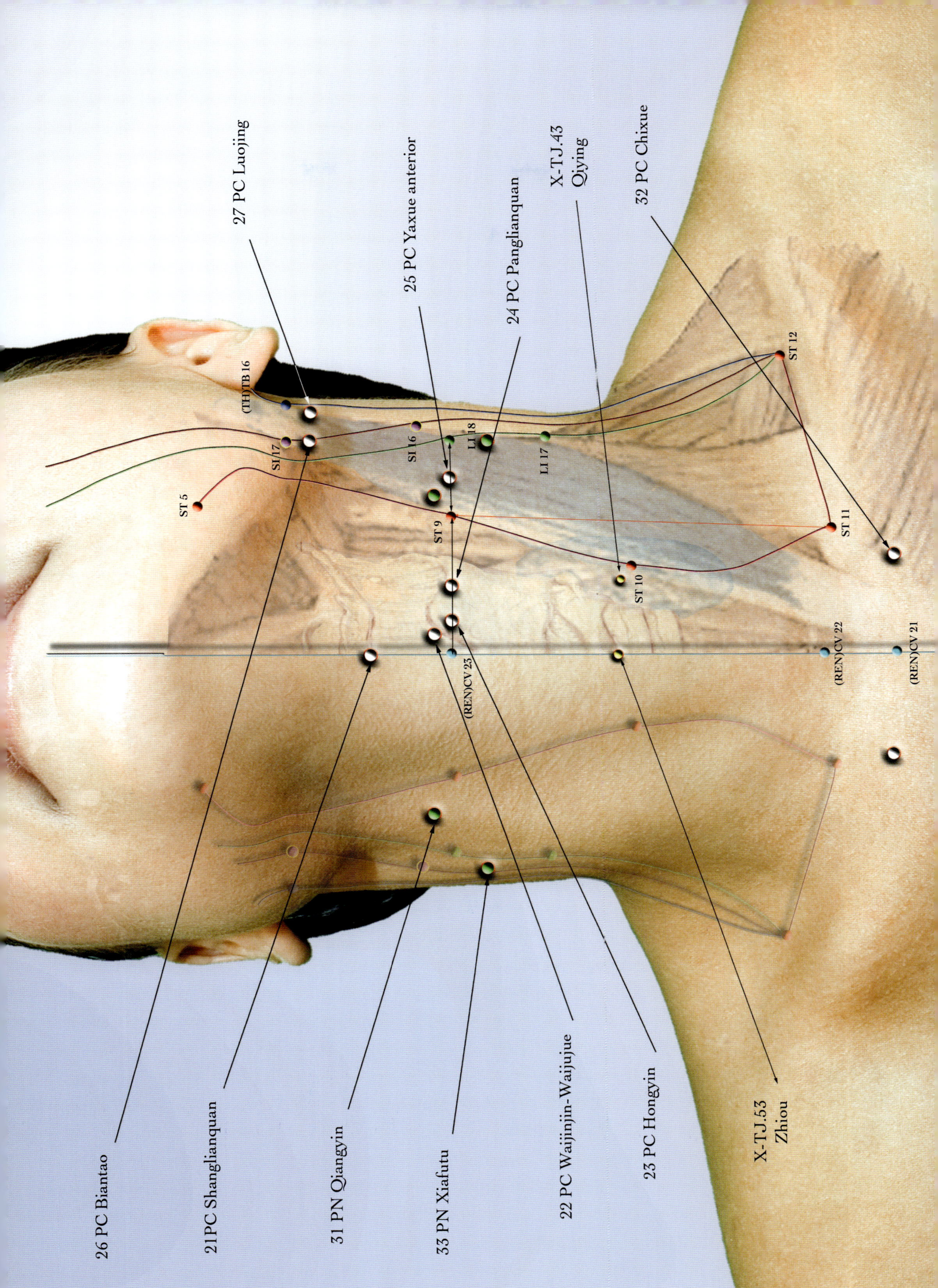

Summary of neck points, anterior side

A summary of the points of the neck
(Lateral side)

13 PC Yiming
On the inferior border of the mastoid, on its tip, 5 fen behind the attachment of the ear lobe, between (TH)TB 17 Yifeng before and GB 12 Wangu behind.

21 PC Shanglianquan
1 Cun above point (REN)CV 23 Lianquan, in a depression above the thyroid cartilage.

22 PC Waijinjin-Waiyuyue
0.3 Cun outside and above (REN)CV 23 Lianquan.

23 PC Hongyin
On the horizontal line of (REN)CV 23 Lianquan, 0.5 Cun laterally.

24 PC Panglianquan
At the midpoint of the line connecting (REN)CV 23 Lianquan and the anterior border of the sternocleidomastoid muscle.

25 PC Yaxue
4 Points: 2 on the anterior, 2 on the posterior side of the neck.
The anterior points are between ST 9 Renyin and LI 18 Futu, where the pulsation of the artery is palpable.
The posterior points are 2.5 Cun above GB 20 Fengchi, on the same horizontal line, internally and 1 Cun from 13 PC Yiming.

26 PC Biantao
Below the inferior border of the angle of the lower maxilla, just before the jugular vein, near SI 17 Tianrong.

27 PC Luojing
On the upper border of the sternocleidomastoid muscle, on the upper 1/3 of the muscle, on the horizontal line passing by 26 PC Biantao.

28 PC Fengyan
0.25 Cun behind and below 13 PC Yiming.

32 PC Chixue
1 Cun horizontally and outside point (REN)CV 21 Xuanji, on the base of the sternum.

31 PN Qiangyin
2 Cun outside (REN)CV 23 Lianquan, behind and above ST 9 Renying.

32 PN Zhenyin
Midpoint of the line contiguous to the angle of the inferior maxilla and of Adam's apple, before the sterno-cleido-mastoid muscle.

33 PN Xiafutu
0.5 Cun below LI 18 Futu.

34 PN Jingzhong
2 Cun below point 28 PN Anmian 2 on the posterior border of the sterno-cleido-mastoid muscle.

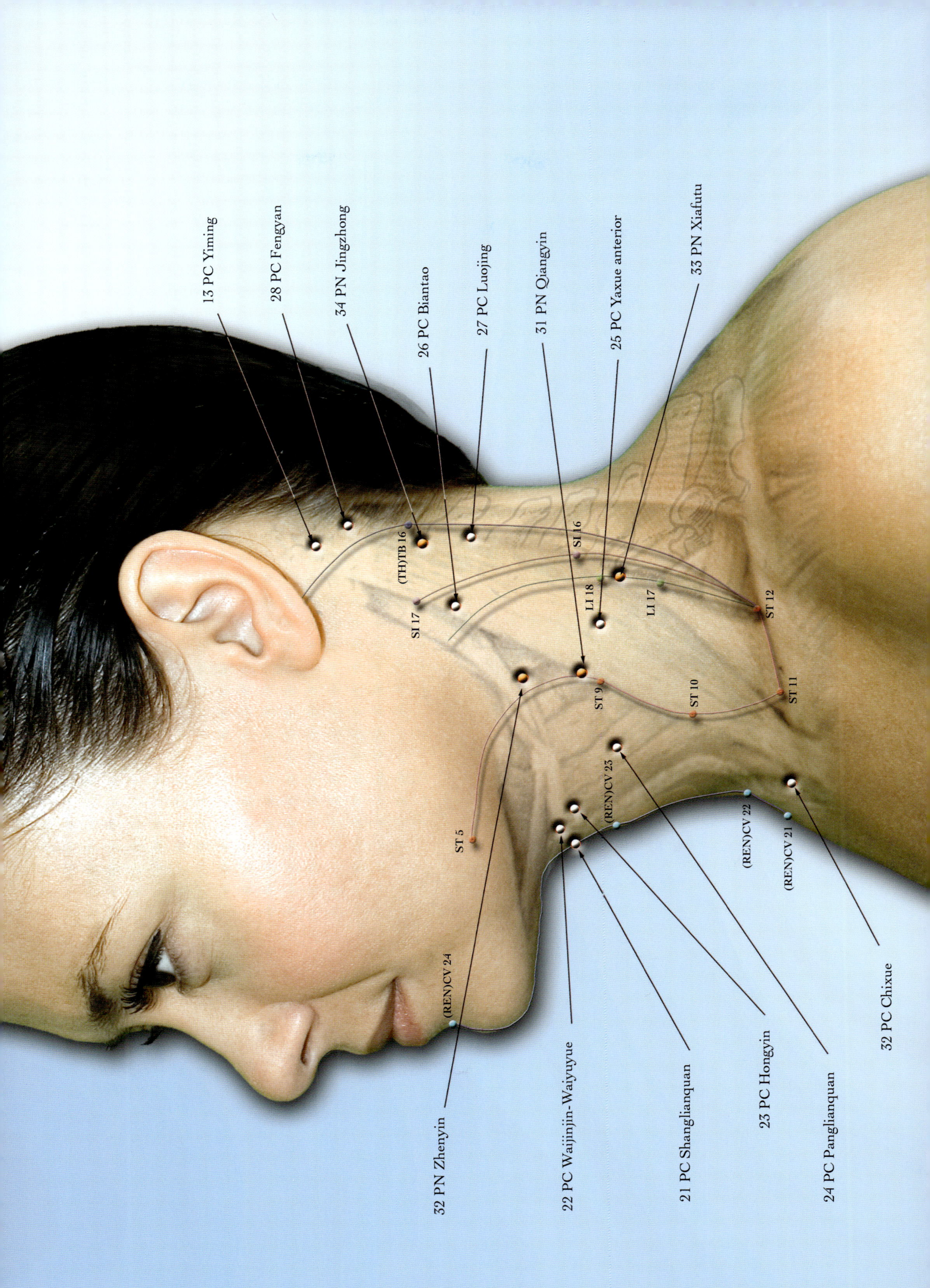

13 PC Yiming
28 PC Fengyan
34 PN Jingzhong
26 PC Biantao
27 PC Luojing
31 PN Qiangyin
25 PC Yaxue anterior
33 PN Xiafutu
(TH)TB 16
SI 16
SI 17
LI 18
LI 17
ST 12
ST 11
ST 10
ST 9
ST 5
(REN)CV 23
(REN)CV 22
(REN)CV 21
(REN)CV 24
32 PN Zhenyin
22 PC Waijinjin-Waiyuyue
21 PC Shanglianquan
23 PC Hongyin
24 PC Panglianquan
32 PC Chixue

Chest

- Points of the Main channels
  - –Ren Mai (Conception vessel)
  - –Zu Shao Yin (Kidney)
  - –Zu Yang Ming (Stomach)
  - –Zu Tai Yin (Spleen)
  - –Zu Jue Yin (Liver)
  - –Zu Shao Yang (Gall Bladder)
  - –Shu Jue Yin (Master or the Heart or Pericardium)
- Extra channel points
  - –Curious points
  - –New points
  - –Points of Master Tong
- Lateral side of the chest
  - –Points of the channels

Ren Mai (Conception vessel)

(REN)CV 14 Juque
On the upper abdomen and on the anterior hairline, 6 Cun above the umbilicus.

(REN)CV 15 Jiuwei
On the upper abdomen and on the anterior midline, 1 Cun below the xiphisternal synchondrosis.

(REN)CV 16 Zhongting
On the chest and on the anterior midline, on the level of the 5th intercostal space, on the xiphisternal synchondrosis.

(REN)CV 17 Danzhong
On the chest and on the anterior midline, on the level of the 4th intercostal space, at the midpoint of the line connecting both nipples.

(REN)CV 18 Yutang
On the chest and on the anterior midline, on the level of the 3rd intercostal space.

(REN)CV 19 Zigong
On the chest and on the anterior midline, on the level of the 2nd intercostal space.

(REN)CV 20 Huagai
On the chest and on the anterior midline, on the level of the 1st intercostal space.

(REN)CV 21 Xuanji
On the chest and on the anterior midline, 1 Cun below CV 22 Tiantu.

(REN)CV 22 Tieantu
On the neck and on the anterior midline, at the centre of the suprasternal fossa.

Zu Yang Ming channel (Stomach)

ST 12 Quepen
At the centre of the supra-clavicular fossa, 4 Cun lateral to the anterior midline.

ST 13 Qihu
On the chest, below the midpoint of the lower border of the clavicle, 4 Cun lateral to the anterior midline.

ST 14 Kufang
On the chest, in the 1st intercostal space, 4 Cun lateral to the anterior midline.

ST 15 Wuyi
On the chest, in the 2nd intercostal space, 4 Cun lateral to the anterior midline.

ST 16 Yingchuang
On the chest, in the 3rd intercostal space, 4 Cun lateral to the anterior midline.

ST 17 Ruzhong
On the chest, in the 4th intercostal space, at the centre of the nipple, 4 Cun lateral to the anterior midline.

ST 18 Rugen
On the chest, directly below the nipple, on the lower border of the breast, in the 5th intercostal space, 4 Cun lateral to the anterior midline.

ST 19 Burong
On the upper abdomen, 6 Cun above the umbilicus and 2 Cun lateral to the anterior midline.

Shu Jue Yin channel (Master of the Heart or Pericardium)

(MH)PC 1 Tianchi
On the chest, in the 4th intercostal space, 1 Cun lateral to the nipple and 5 Cun lateral to the anterior midline.

Zu Jue Yin channel (Liver)

(LR)LV 13 Zhangmen
On the lateral side of the abdomen, below the free end of the 11th rib.

(LR)LV 14 Qimen
On the chest, directly below the nipple, in the 6th intercostal space, 4 Cun lateral to the anterior midline.

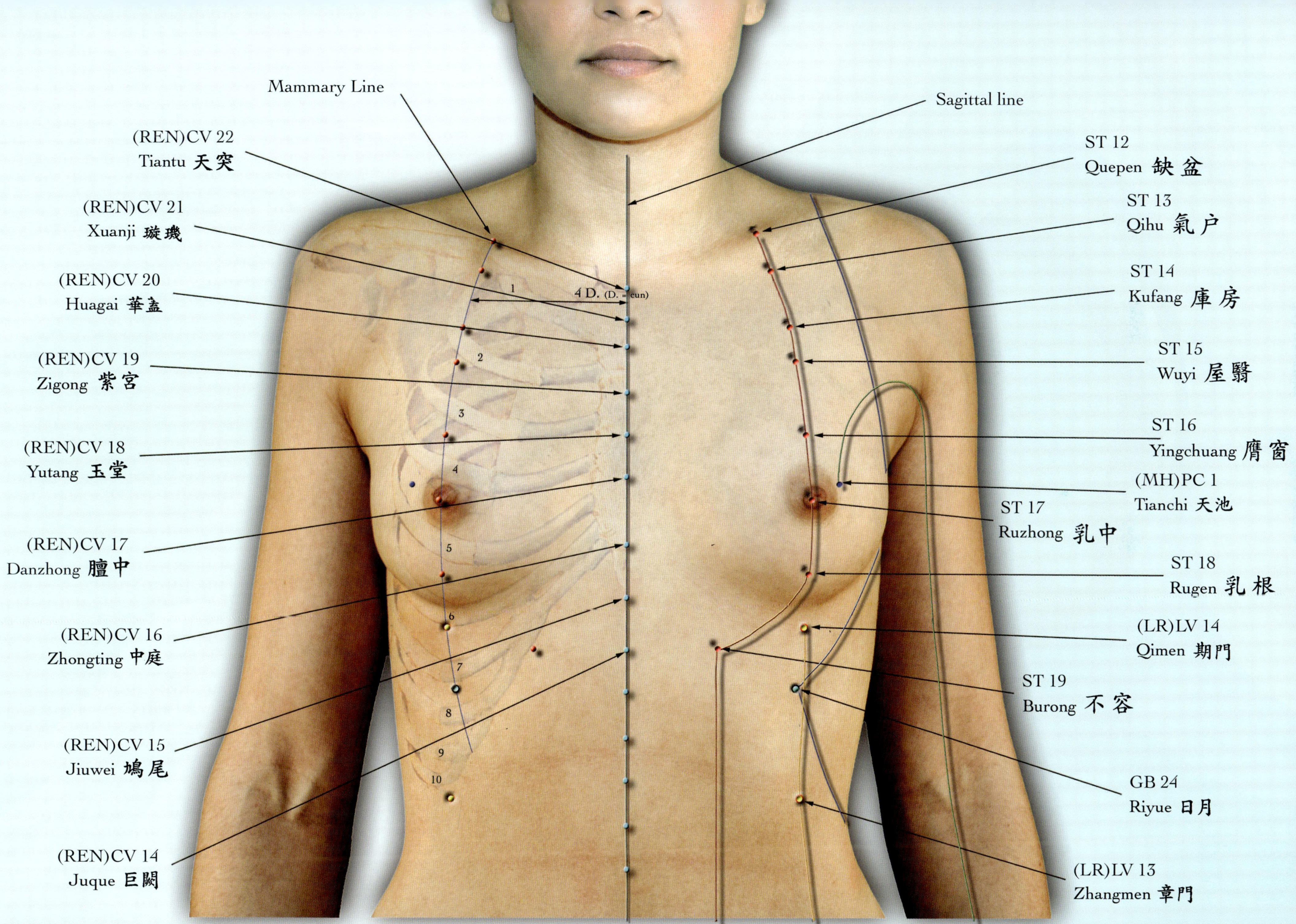
Mammary Line
Sagittal line
(REN)CV 22
Tiantu 天突
(REN)CV 21
Xuanji 璇璣
(REN)CV 20
Huagai 華蓋
(REN)CV 19
Zigong 紫宮
(REN)CV 18
Yutang 玉堂
(REN)CV 17
Danzhong 膻中
(REN)CV 16
Zhongting 中庭
(REN)CV 15
Jiuwei 鳩尾
(REN)CV 14
Juque 巨闕
4 D. (D. = cun)
1
2
3
4
5
6
7
8
9
10
ST 12
Quepen 缺盆
ST 13
Qihu 氣户
ST 14
Kufang 庫房
ST 15
Wuyi 屋翳
ST 16
Yingchuang 膺窗
(MH)PC 1
Tianchi 天池
ST 17
Ruzhong 乳中
ST 18
Rugen 乳根
(LR)LV 14
Qimen 期門
ST 19
Burong 不容
GB 24
Riyue 日月
(LR)LV 13
Zhangmen 章門

Zu Shao Yang channel (Gall Bladder)

GB 24 Riyue
On the upper abdomen, directly below the nipple, in the 7th intercostal space, 4 Cun lateral to the anterior midline.

Ren Mai (Conception vessel)

(REN)CV 14 Juque
On the upper abdomen and on the anterior midline, 6 Cun above the centre of the umbilicus.

(REN)CV 15 Jiuwei
On the upper abdomen and on the anterior midline, 1 Cun below the xiphisternal synchondrosis.

(REN)CV 16 Zhongting
On the chest and on the anterior midline, on the level of the 5th intercostal space, on the xiphisternal synchondrosis.

(REN)CV 17 Danzhong
On the chest and on the anterior midline, on the level of the 4th intercostal space, at the midpoint of the line connecting both nipples.

(REN)CV 18 Yutang
On the chest and on the anterior midline, on the level of the 3rd intercostal space.

(REN)CV 19 Zigong
On the chest and on the anterior midline, on the level of the 2nd intercostal space.

(REN)CV 20 Huagai
On the chest and on the anterior midline, on the level of the 1st intercostal space.

(REN)CV 21 Xuanji
On the chest and on the anterior midline, 1 Cun below (REN)CV 22 Tiantu.

(REN)CV 22 Tiantu
On the neck and on the anterior midline, at the centre of the supra-sternal fossa.

Zu Shao Yin channel (Kidney)

KI 21 Youmen
On the upper abdomen, 6 Cun above the umbilicus and 0.5 Cun lateral to the anterior midline.

KI 22 Bulang
On the chest, in the 5th intercostal space, 2 Cun lateral to the anterior midline.

KI 23 Shenfeng
On the chest, in the 4th intercostal space, 2 Cun lateral to the anterior midline.

KI 24 Lingxu
On the chest, in the 3rd intercostal space, 2 Cun lateral to the anterior midline.

KI 25 Shencang
On the chest, in the 2nd intercostal space, 2 Cun lateral to the anterior midline.

KI 26 Yuzhong
On the chest, in the 1st intercostal space, 2 Cun lateral to the anterior midline.

KI 27 Shufu
On the chest, below the border of the clavicle, 2 Cun lateral to the midline.

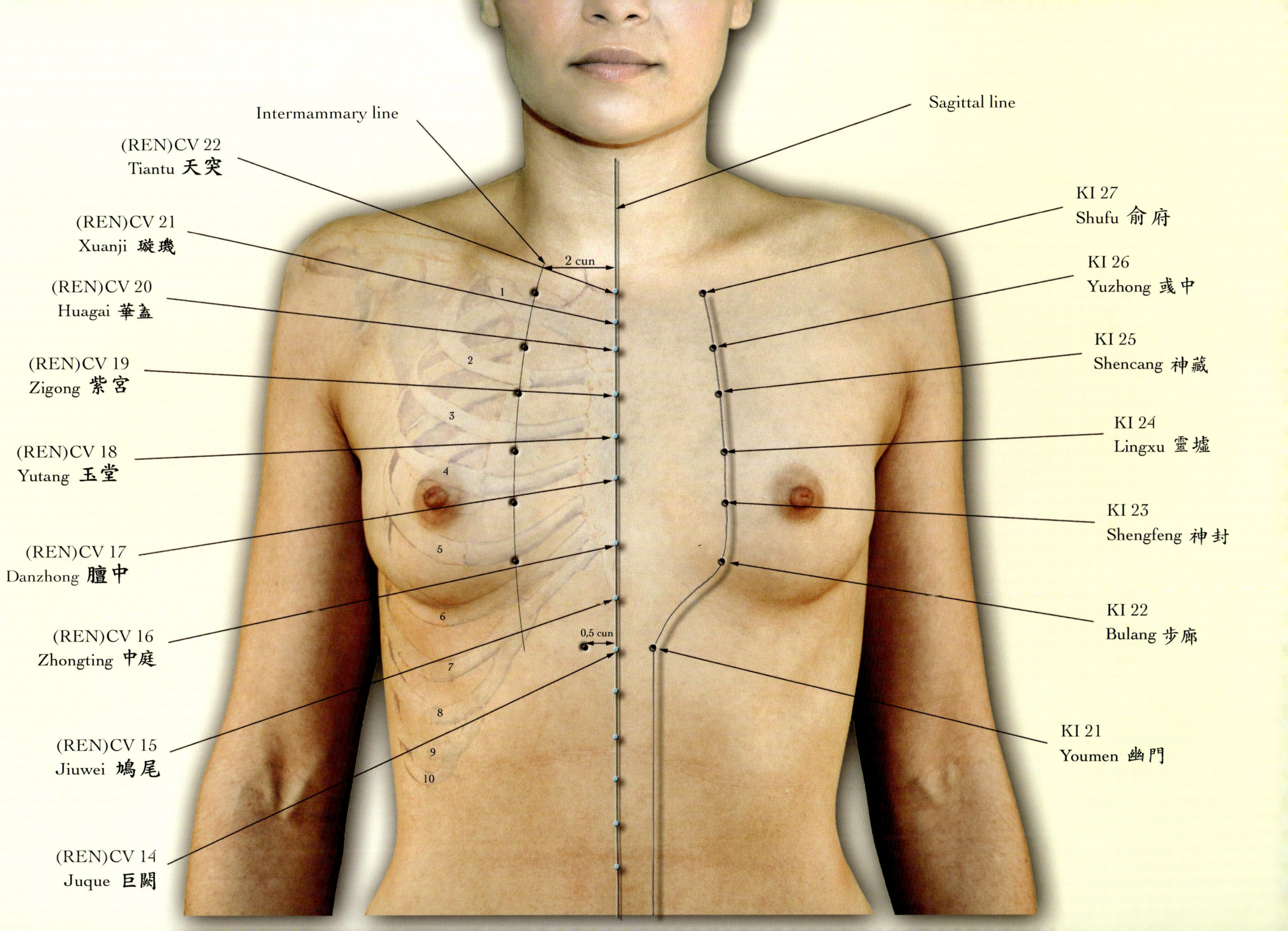
Intermammary line
Sagittal line
(REN)CV 22
Tiantu 天突
(REN)CV 21
Xuanji 璇璣
(REN)CV 20
Huagai 華蓋
(REN)CV 19
Zigong 紫宮
(REN)CV 18
Yutang 玉堂
(REN)CV 17
Danzhong 膻中
(REN)CV 16
Zhongting 中庭
(REN)CV 15
Jiuwei 鳩尾
(REN)CV 14
Juque 巨闕
KI 27
Shufu 俞府
KI 26
Yuzhong 彧中
KI 25
Shencang 神藏
KI 24
Lingxu 靈墟
KI 23
Shengfeng 神封
KI 22
Bulang 步廊
KI 21
Youmen 幽門
2 cun
0,5 cun
1
2
3
4
5
6
7
8
9
10

Ren Mai (Conception vessel)

(REN)CV 14 Juque
On the upper abdomen and on the anterior midline, 6 Cun above the centre of the umbilicus.

(REN)CV 15 Jiuwei
On the upper abdomen and on the anterior midline, 1 Cun below the xiphisternal synchondrosis.

(REN)CV 16 Zhongting
On the chest and on the anterior midline, on the level of the 5th intercostal space, on the xiphisternal synchondrosis.

(REN)CV 17 Danzhong
On the chest and on the anterior midline, on the level of the 4th intercostal space, at the midpoint of the line connecting both nipples.

(REN)CV 18 Yutang
On the chest and on the anterior midline, on the level of the 3rd intercostal space.

(REN)CV 19 Zitong
On the chest and on the anterior midline, on the level of the 2nd intercostal space.

(REN)CV 20 Huagai
On the chest and on the anterior midline, on the level of the 1st intercostal space.

(REN)CV 21 Xuanji
On the chest and on the anterior midline, 1 Cun below (REN)CV 22 Tiantu.

(REN)CV 22 Tiantu
On the neck and on the anterior midline, at the centre of the supra-sternal fossa.

Zu Tai Yin channel (Spleen)

SP 16 Fu'ai
On the upper abdomen, 3 Cun above the umbilicus, and 4 Cun lateral to the anterior midline.

SP 17 Shidu
On the lateral side of the chest and in the 5th intercostal space, 6 Cun lateral to the anterior midline.

SP 18 Tianxi
On the lateral side of the chest and in the 4th intercostal space, 6 Cun lateral to the anterior midline.

SP 19 Xiongxiang
On the lateral side of the chest and in the 3rd intercostal space, 6 Cun lateral to the anterior midline.

SP 20 Zhourong
On the lateral side of the chest and in the 2nd intercostal space, 6 Cun lateral to the anterior midline.

SP 21 Dabao
On the lateral side of the chest and on the middle axillary line, in the 6th intercostal space.

Shu Tai Yin channel (Lung)

LU 1 Zhongfu
In the superior external part of the anterior thoracic wall, 1 Cun below LU 2 Yunmen, on the level of the 1st intercostal space, 6 Cun lateral to the anterior midline.

LU 2 Yunmen
In the superior external part of the anterior thoracic wall, superior to the coracoid process of the scapula, in the depression of the infraclavicular fossa, 6 Cun to the anterior midline.

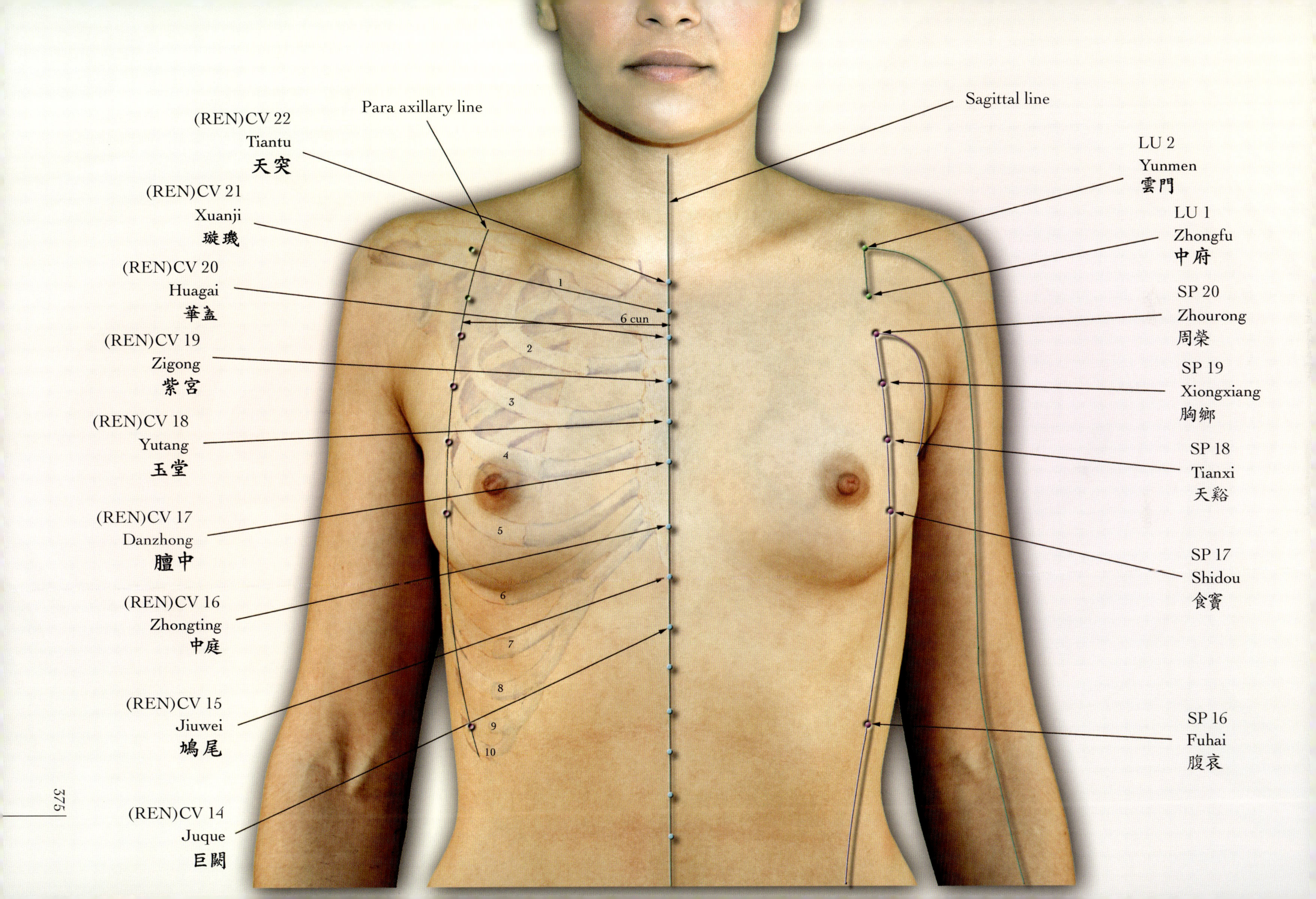
(REN)CV 22
Tiantu
天突
(REN)CV 21
Xuanji
璇璣
(REN)CV 20
Huagai
華蓋
(REN)CV 19
Zigong
紫宮
(REN)CV 18
Yutang
玉堂
(REN)CV 17
Danzhong
膻中
(REN)CV 16
Zhongting
中庭
(REN)CV 15
Jiuwei
鳩尾
(REN)CV 14
Juque
巨闕
Para axillary line
Sagittal line
6 cun
1
2
3
4
5
6
7
8
9
10
LU 2
Yunmen
雲門
LU 1
Zhongfu
中府
SP 20
Zhourong
周榮
SP 19
Xiongxiang
胸鄉
SP 18
Tianxi
天谿
SP 17
Shidou
食竇
SP 16
Fuhai
腹哀

Extra channel points

32 PC Chixue
1 Cun horizontally and outside point (REN) CV 21 Xuanji, at the base of the sternum.

33 PC Tanchuan
2 Cun horizontally and outside ST 16 Yingchuang. It is found on the same horizontal line as (REN)CV 18 Yutang, KI 24 Lingxu and ST 16 Yingchuang.

34 PC Longhan
1.5 Cun vertically above point (REN)CV 15 Jiuwei.

35 PC Zuoyi – Youyi
Situated on the same vertical line as 33 PC Tanchuan, on the same horizontal line as ST 18 Rugen, 1 Cun outside this point.

J-XF.21 Eni
On the 7th intercostal space, just below the nipple.

J-XF.22 Xiaoerjixiong
Three pairs of symmetrical points on the 2nd, 3rd and 4th intercostal space, 2.5 Cun laterally to the anterior midline.

X-XF.8 Ganshi
On the 6th intercostal space, below the nipple.

X-XF.9 Chuangxinmen
4 Cun lateral and 3 Cun above the umbilicus, on the inferior border of the ribs.

Points of Master Tong

VT.02 Shi Er Hou
2 lines of 3 points below the clavicle.
The 1st line is situated 1.3 Cun below the clavicle and the 2nd line 2.8 Cun.
3 points to each line, 1 in the centre and 2 on each side, at 1.5 Cun.

VT.03 Jin Wu
This line of 5 points is situated vertically on the anterior midline.
The 1st, Jin Gan, is in the depression between the manubrium and the sternum.
The 4 other points, Jin Yin, Jin Yang, Jin Chuan, Jin Jiao, are situated 1 Cun below each other.

VT.04 Wei Mao Qi
The inferior point is below the tip of the Xyphoid. The 2 next points are found 1 and 2 Cun above the 1st.
The other points are on an horizontal line, 1 Cun on each side of the 1st and 2nd points.

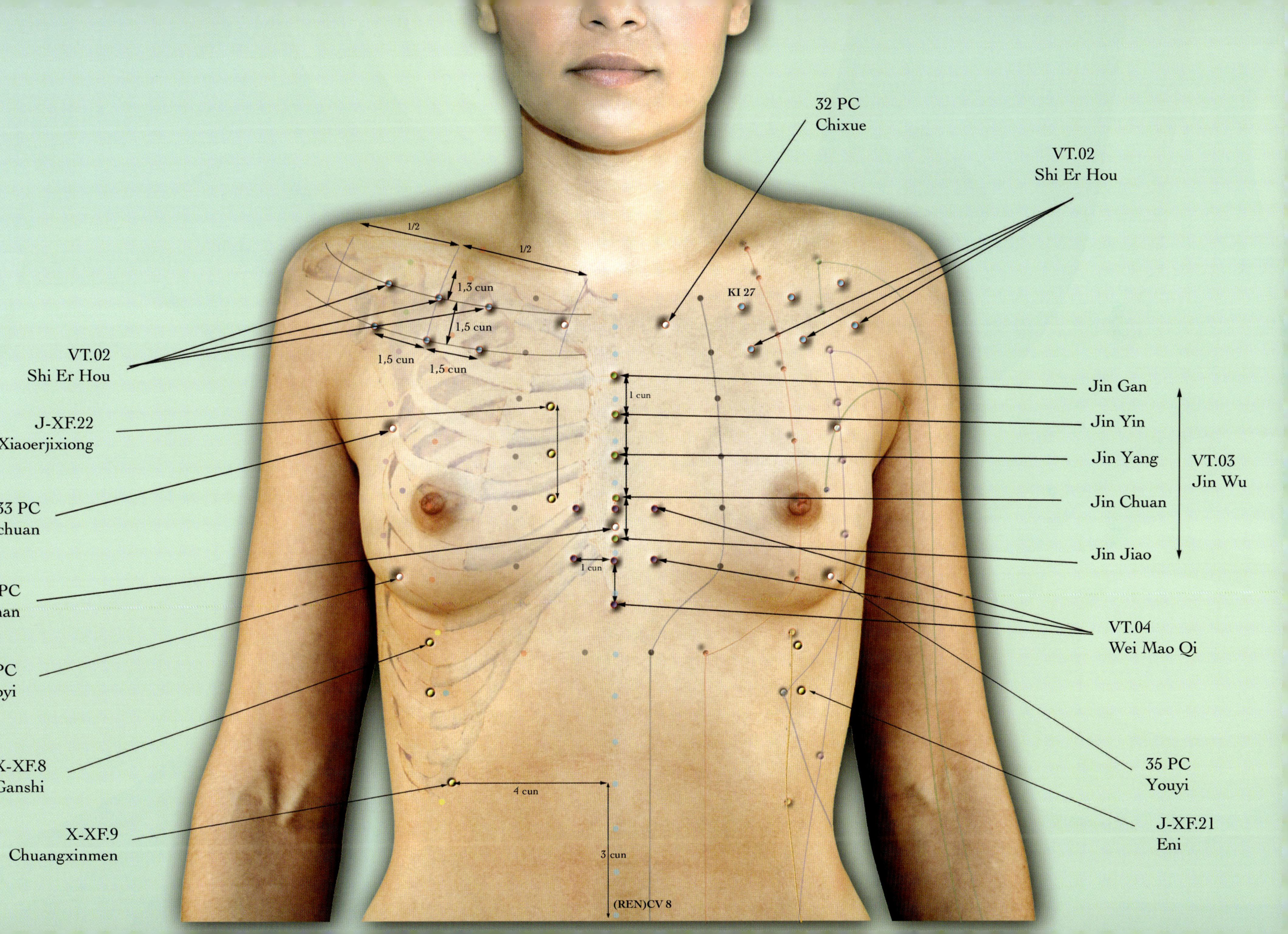

32 PC
Chixue
VT.02
Shi Er Hou
1/2
1/2
1,3 cun
1,5 cun
1,5 cun
1,5 cun
KI 27
VT.02
Shi Er Hou
1 cun
Jin Gan
Jin Yin
Jin Yang
Jin Chuan
Jin Jiao
VT.03
Jin Wu
J-XF.22
Xiaoerjixiong
33 PC
Tanchuan
34 PC
Longhan
1 cun
VT.04
Wei Mao Qi
35 PC
Zuoyi
X-XF.8
Ganshi
4 cun
35 PC
Youyi
X-XF.9
Chuangxinmen
3 cun
(REN)CV 8
J-XF.21
Eni

Zu Shao Yang channel (Gall Bladder)

GB 22 Yuanye
On the lateral side of the chest, on the midaxillary line when the arm is raised, 3 Cun below the axilla, in the 4th intercostal space.

GB 23 Zhejin
On the lateral side of the chest, 1 Cun anterior to GB 22 Yuanye, on the level of the nipple, and in the 4th intercostal space.

GB 24 Riyue
On the upper abdomen, directly below the nipple, in the 7th intercostal space 4 Cun lateral to the anterior midline.

GB 25 Jingmen
On the lateral side of the waist, 1.8 Cun posterior to (LR)LV 13 Zhangmen, below the free end of the 12th rib.

GB 26 Daimai
On the lateral side of the abdomen, 1.8 Cun below (LR)LV 13 Zhangmen, at the crossing point of a vertical line through the free end of the 11th rib and of a horizontal line through the umbilicus.

GB 27 Wushu
On the lateral side of the abdomen, anterior to the anteriosuperior iliac spine 3 Cun below the level of the umbilicus.

GB 28 Weidao
On the lateral side of the abdomen, anterior and inferior to the anteriosuperior iliac spine, 0.5 Cun anterior and inferior to GB 27 Wushu.

GB 29 Juliao
On the hip, at the midpoint of the line connecting the anterio-superior iliac spine and the prominence of the great trochanter.

GB 30 Huantiao
On the lateral side of the thigh, at the junction of the middle third and lateral third of the line connecting the prominence of the great trochanter and the sacral hiatus when the patient is in a lateral recumbent position with the thigh flexed.

Zu Tai Tin channel (Spleen)

SP 12 Chongmen
At the lateral end of the inguinal groove, 3.5 Cun lateral to the midpoint of the upper border of the symphysis pubis, lateral to the pulsating external iliac artery.

SP 13 Fushe
On the lower abdomen, 4 Cun below the centre of the umbilicus, 0.7 Cun above SP 12 Chongmen, and 4 Cun lateral to the anterior midline.

SP 14 Fujie
On the lower abdomen, 1.3 Cun below SP 15 Daheng, and 4 Cun lateral to the anterior midline.

SP 15 Daheng
On the middle abdomen, 4 Cun lateral to the umbilicus.

SP 16 Fu'ai
On the upper abdomen, 3 Cun above the centre of the umbilicus, and 4 Cun lateral to the anterior midline.

SP 17 Shidu
On the lateral side of the chest and in the 5th intercostal space, 6 Cun lateral to the anterior midline.

SP 18 Tianxi
On the lateral side of the chest and in the 4th intercostal space, 6 Cun lateral to the anterior midline.

SP 19 Xiongxiang
On the lateral side of the chest and in the 3rd intercostal space, 6 Cun lateral to the anterior midline.

SP 20 Zhourong
On the lateral side of the chest and in the 2nd intercostal space, 6 Cun lateral to the anterior midline.

SP 21 Dabao
On the lateral side of the chest and on the middle axillary line, in the 6th intercostal space.

Zu Jue Yin channel (Liver)

(LR)LV 12 Jimai
Lateral to the pubic tubercle, lateral and inferior to ST 30 Qichong, in the inguinal groove where the pulsation of the femoral artery is palpable, 2.5 Cun lateral to the anterior midline. (Do NOT needle directly into the artery!)

(LR)LV 13 Zhangmen
On the lateral side of the abdomen, below the free end of the 11th rib.

(LR)LV 14 Qimen
On the chest, directly below the nipple, in the 6th intercostal space, 4 Cun lateral to the anterior midline.

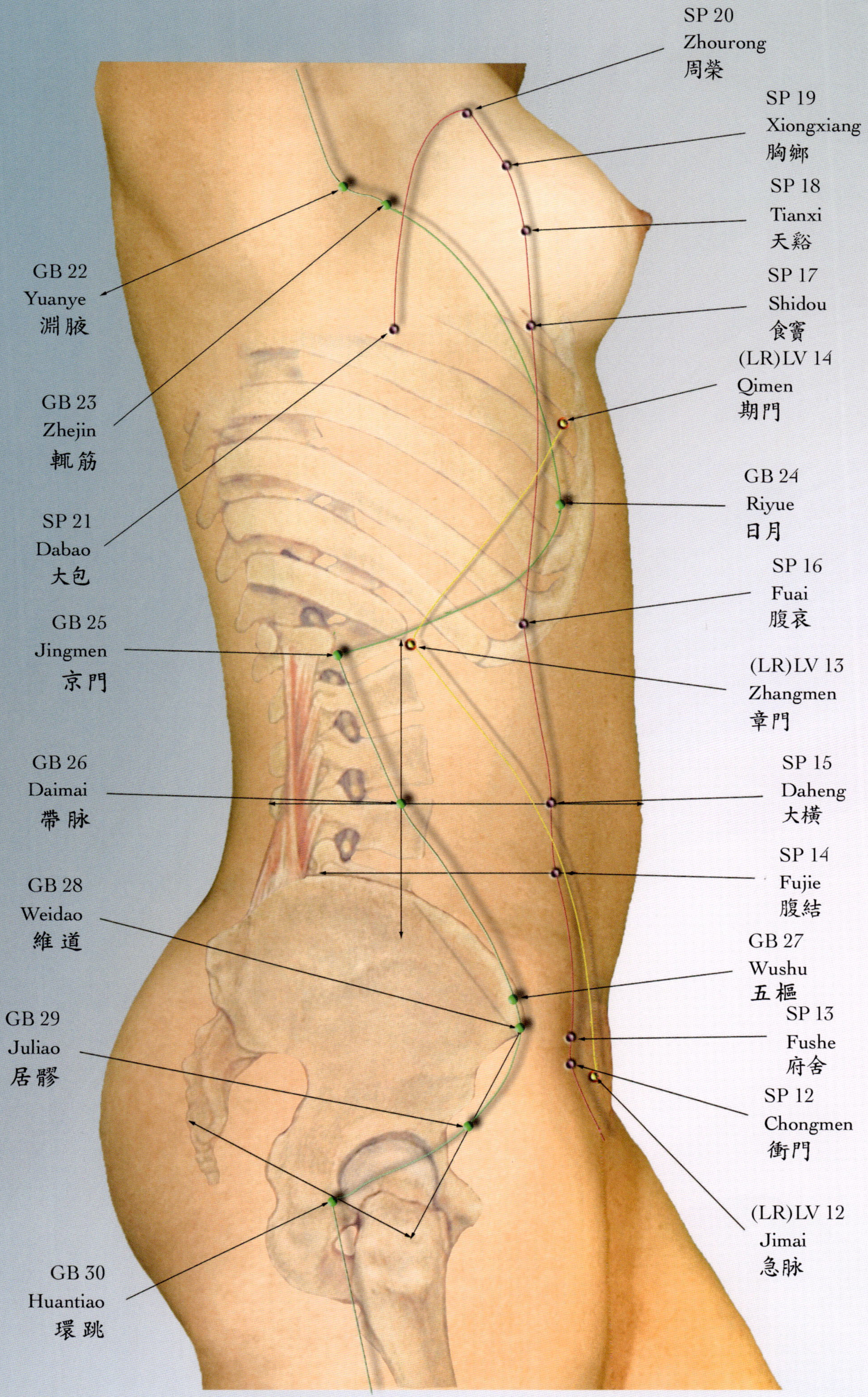

SP 20
Zhourong
周榮
SP 19
Xiongxiang
胸鄉
SP 18
Tianxi
天谿
SP 17
Shidou
食竇
(LR)LV 14
Qimen
期門
GB 24
Riyue
日月
SP 16
Fuai
腹哀
(LR)LV 13
Zhangmen
章門
SP 15
Daheng
大橫
SP 14
Fujie
腹結
GB 27
Wushu
五樞
SP 13
Fushe
府舍
SP 12
Chongmen
衝門
(LR)LV 12
Jimai
急脈
GB 22
Yuanye
淵腋
GB 23
Zhejin
輒筋
SP 21
Dabao
大包
GB 25
Jingmen
京門
GB 26
Daimai
帶脈
GB 28
Weidao
維道
GB 29
Juliao
居髎
GB 30
Huantiao
環跳

Back

- Points of the Main channels
  - Du Mai (Governor vessel)
  - Zu Tai Yang (Bladder)
  - Shu Tai Yang (Small Intestine)
  - Shu Shao Yang (Triple Heater or Triple Burner)
  - Shu Yang Ming (Large Intestine)
- Extra channel points
  - Curious points
  - New points
  - Points of Master Tong

Du Mai (Governor vessel)

(DU)GV 6 Jizhong
On the back and on the posterior midline, in the depression below the spinous process of the 11th thoracic vertebra.

(DU)GV 7 Zhongshu
On the back and on the posterior midline, in the depression below the spinous process of the 10th thoracic vertebra.

(DU)GV 8 Jinsuo
On the back and on the posterior midline, in the depression below the spinous process of the 9th thoracic vertebra.

(DU)GV 9 Zhiyang
On the back and on the posterior midline, in the depression below the spinous process of the 7 the thoracic vertebra.

(DU)GV 10 Lingtai
On the back and on the posterior midline, in the depression below the spinous process of the 6th thoracic vertebra.

(DU)GV 11 Shendao
On the back and on the posterior midline, in the depression below the spinous process of the 5th thoracic vertebra.

(DU)GV 12 Shenzhu
On the back and on the posterior midline, in the depression below the spinous process of the 3rd thoracic vertebra.

(DU)GV 13 Taodao
On the back and on the posterior midline, in the depression below the spinous process of the 1st thoracic vertebra.

(DU)GV 14 Dazhui
On the posterior midline, in the depression below the 7th thoracic vertebra.

Shu Tai Yang channel (Small Intestine)

SI 9 Jianzhen
Posterior and inferior to the shoulder joint, 1 Cun above the posterior end of the axillary fold with the arm adducted.

SI 10 Naoshu
On the shoulder, above the posterior end of the axillary fold, in the depression below the lower border of the scapular spine.

SI 12 Bingfeng
On the scapula, at the centre of the supra-scapular fossa, directly above SI 11 Tianzong, in the depression found when the arm is raised.

SI 13 Quyuan
On the scapula, at the medial end of the supra-scapular fossa, at the midpoint of the line connecting SI 10 Naoshu and the spinous process of the 2nd thoracic vertebra.

SI 14 Jianwaishu
On the back, below the spinous process of the 1st thoracic vertebra, 3 Cun lateral to the posterior midline.

SI 15 Jianzhongshu
On the back, below the spinous process of the 7th cervical vertebra, 2 Cun lateral to the posterior midline.

Shu Shao Yang channel (Triple Heater or Triple Burner)

(TH)TB 13 Naoshui
On the lateral side of the upper arm and on the line connecting the tip of the olecranon and (TH)TB 14 Jianliao, 3 Cun below (TH)TB 14 Jianliao, and on the posterio-inferior border of the deltoid muscle.

(TH)TB 14 Jianliao
On the shoulder, posterior to LI 15 Jianyu, in the depression inferior and posterior to the acromion when the arm is abducted.

(TH)TB 15 Tianliao
On the scapula, at the midpoint between GB 21 Jianjing and SI 13 Quyuan, at the superior angle of the scapula.

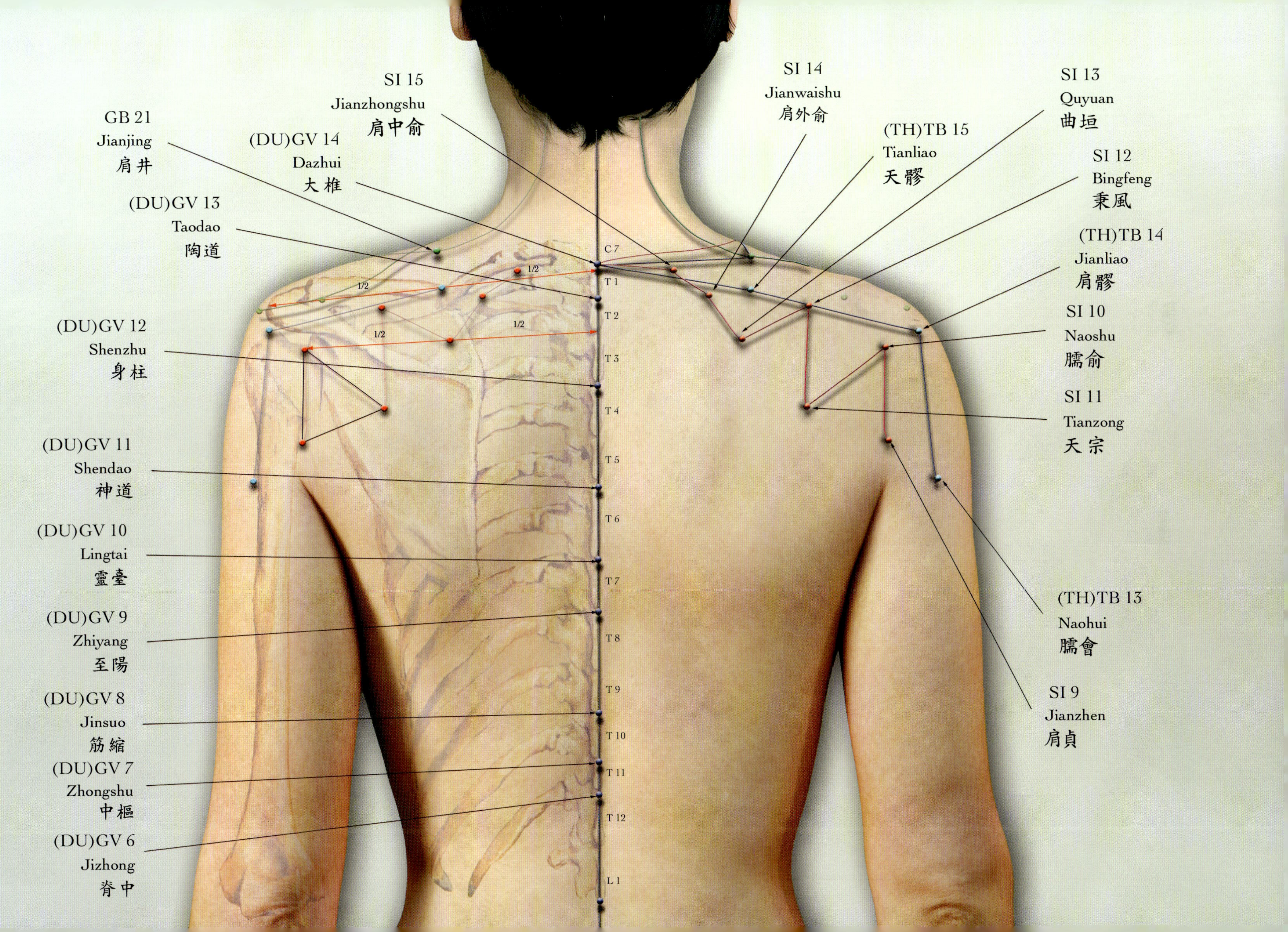
GB 21
Jianjing
肩井
(DU)GV 14
Dazhui
大椎
SI 15
Jianzhongshu
肩中俞
(DU)GV 13
Taodao
陶道
(DU)GV 12
Shenzhu
身柱
(DU)GV 11
Shendao
神道
(DU)GV 10
Lingtai
靈臺
(DU)GV 9
Zhiyang
至陽
(DU)GV 8
Jinsuo
筋縮
(DU)GV 7
Zhongshu
中樞
(DU)GV 6
Jizhong
脊中
SI 14
Jianwaishu
肩外俞
(TH)TB 15
Tianliao
天髎
SI 13
Quyuan
曲垣
SI 12
Bingfeng
秉風
(TH)TB 14
Jianliao
肩髎
SI 10
Naoshu
臑俞
SI 11
Tianzong
天宗
(TH)TB 13
Naohui
臑會
SI 9
Jianzhen
肩貞
C7
T1
T2
T3
T4
T5
T6
T7
T8
T9
T10
T11
T12
L1
1/2
1/2
1/2
1/2

Zu Tai Yang channel (Bladder)

BL 11 Dazhu
On the back, below the spinous process of the 1st thoracic vertebra, 1.5 Cun lateral to the posterior midline.

BL 12 Fengmen
On the back, below the spinous process of the 2nd thoracic vertebra, 1.5 Cun lateral to the posterior midline.

BL 13 Feishu
On the back, below the spinous process of the 3rd thoracic vertebra, 1.5 Cun lateral to the posterior midline.

BL 14 Jueyinshu
On the back, below the spinous process of the 4th thoracic vertebra, 1.5 Cun lateral to the posterior midline.

BL 15 Xinshu
On the back, below the spinous process of the 5th thoracic vertebra, 1.5 Cun lateral to the posterior midline.

BL 16 Dushu
On the back, below the spinous process of the 6th thoracic vertebra, 1.5 Cun lateral to the posterior midline.

BL 17 Geshu
On the back, below the spinous process of the 7th thoracic vertebra, 1.5 Cun lateral to the posterior midline.

BL 18 Ganshu
On the back, below the spinous process of the 9th thoracic vertebra, 1.5 Cun lateral to the posterior midline.

BL 19 Danshu
On the back, below the spinous process of the 10th thoracic vertebra, 1.5 Cun lateral to the posterior midline.

BL 20 Pishu
On the back, below the spinous process of the 11th thoracic vertebra, 1.5 Cun lateral to the posterior midline.

BL 41 Fufen
On the back, below the spinous process of the 2nd thoracic vertebra, 3 Cun lateral to the posterior midline.

BL 42 Pohu
On the back, below the spinous process of the 3rd thoracic vertebra, 3 Cun lateral to the posterior midline.

BL 43 Gaohuang
On the back, below the spinous process of the 4th thoracic vertebra, 3 Cun lateral to the posterior midline.

BL 44 Shentang
On the back, below the spinous process of the 5th thoracic vertebra, 3 Cun lateral to the posterior midline.

BL 45 Yixi
On the back, below the spinous process of the 6th thoracic vertebra, 3 Cun lateral to the posterior midline.

BL 46 Geguan
On the back, below the spinous process of the 7th thoracic vertebra, 3 Cun lateral to the posterior midline.

BL 47 Hunmen
On the back, below the spinous process of the 9th thoracic vertebra, 3 Cun lateral to the posterior midline.

BL 48 Yanggang
On the back, below the spinous process of the 10th thoracic vertebra, 3 Cun lateral to the posterior midline.

BL 49 Yishe
On the back, below the spinous process of the 11th thoracic vertebra, 3 Cun lateral to the posterior midline.

Shu Yang Ming channel (Large Intestine)

LI 15 Jianyu
On the shoulder, superior to the deltoid muscle, in the depression anterior and inferior to the acromion when the arm is abducted or raised on the level of the shoulder.

LI 16 Jugu
On the shoulder, in the depression between the acromial extremity of the clavicle and scapular spine.

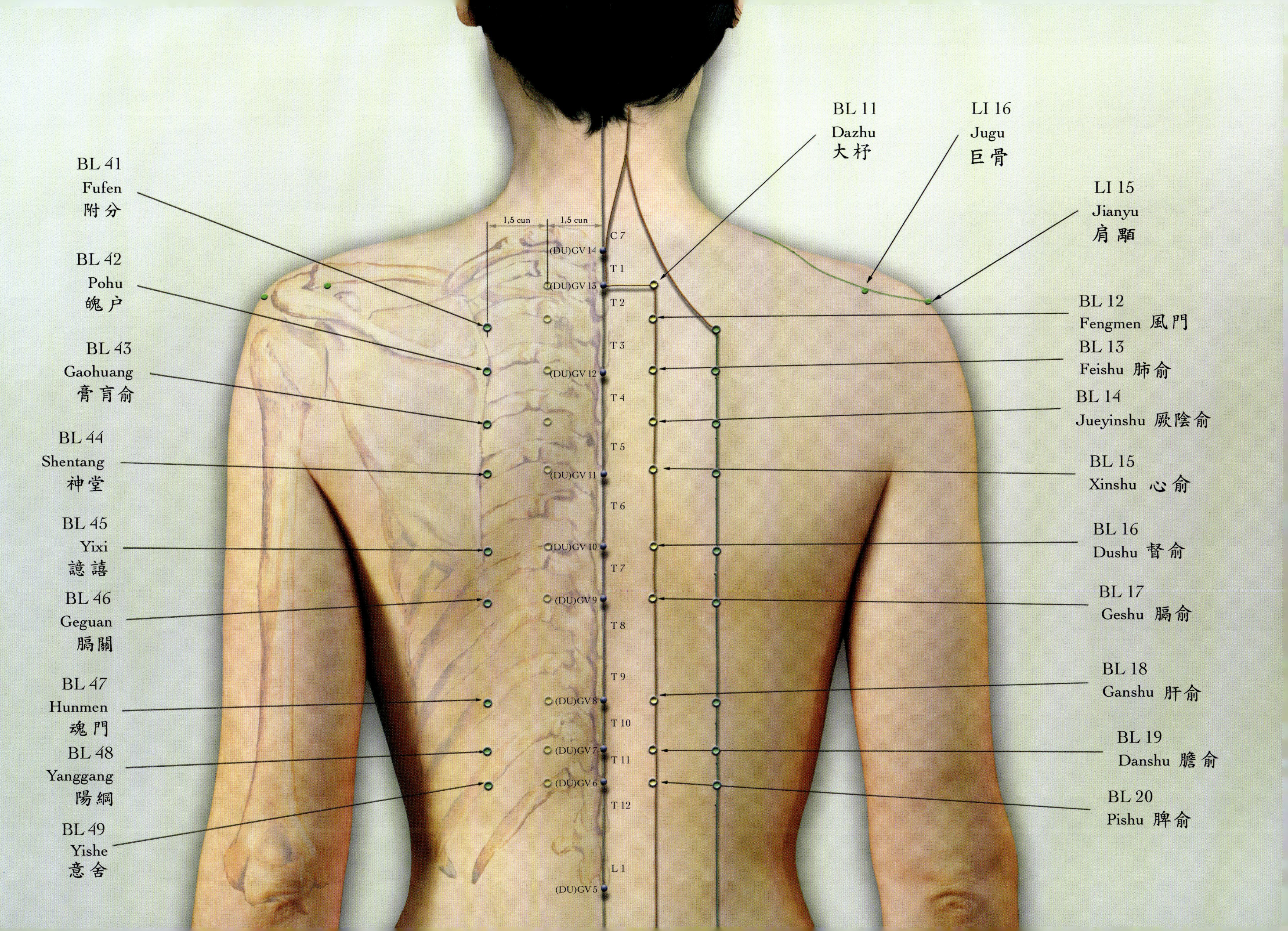

BL 11
Dazhu
大杼
LI 16
Jugu
巨骨
LI 15
Jianyu
肩顒
BL 41
Fufen
附分
BL 42
Pohu
魄户
BL 43
Gaohuang
膏肓俞
BL 44
Shentang
神堂
BL 45
Yixi
譩譆
BL 46
Geguan
膈關
BL 47
Hunmen
魂門
BL 48
Yanggang
陽綱
BL 49
Yishe
意舍
BL 12
Fengmen 風門
BL 13
Feishu 肺俞
BL 14
Jueyinshu 厥陰俞
BL 15
Xinshu 心俞
BL 16
Dushu 督俞
BL 17
Geshu 膈俞
BL 18
Ganshu 肝俞
BL 19
Danshu 膽俞
BL 20
Pishu 脾俞
1,5 cun
1,5 cun
C 7
T 1
T 2
T 3
T 4
T 5
T 6
T 7
T 8
T 9
T 10
T 11
T 12
L 1
(DU)GV 14
(DU)GV 13
(DU)GV 12
(DU)GV 11
(DU)GV 10
(DU)GV 9
(DU)GV 8
(DU)GV 7
(DU)GV 6
(DU)GV 5

Extra channel points (PC)

51 PC Chuanxi
Between the 7th cervical and 1st thoracic vertebrae, 1 Cun from the midline.

52 PC Bai Zhong Feng
Between the channel of the Small Intestine and the one of the Triple Heater or Triple Burner, 2.5 Cun horizontally from point GV 14 Dazhui.

53 PC Bahua
Take a piece of wire, its length being the distance between the nipples (8 Cun). Take away ¼ of its length. With the ¾ left, form an equilateral triangle. The apex of that triangle is laid on (DU)GV 14 Dazhui. The other 2 angles determine 2 new points.
In the middle of the horizontal line joining the two lower points, apply again the apex of the triangle to determine 2 new points.
Repeat the operation to obtain a total of 8 points, 'the 8 flowers'.

54 PC Jiujia
On the apex of the scapula, internally, between the channels of Small Intestine and Triple Heater or Triple Burner.

55 PC Liaosen
Just below the spinous process of T 2.

56 PC Zhouzhu
Between Du Mai vessel and the first chain of Bladder channel below 54 PC Jiujia 1.5 Cun horizontally from the spinous process of T 3.

57 PC Juqureshu
Below the spinous process of T 4.

58 PC Liangyan
1.5 Cun from the spinous process of T 6, just midpoint of the line joining BL 15 Xinshu and BL 16 Dushu.

59 PC Qichuan
2 Cun horizontally from T 7.

60 PC Yinkou
When a horizontal line is drawn from the 7th thoracic vertebra to the scapula, the point is in the inferior angle of the scapula.

61 PC Bazhuixia
Just below the spinous process of T 8.

62 PC Yishu
1.5 Cun from the spinous process of T 8.

63 PC Shubian
On the level of the 10th thoracic, 1 Cun horizontally from its spinous process.

64 PC Zhuogu
Between the 1st and 2nd chain of Bladder, on an horizontal line with 63 PC Shubian and 2.5 Cun horizontally from the spinous process of T 10.

65 PC Jiagu
Below the spinous process of T 12.

66 PC Pigen
3.5 Cun horizontally from the spinous process of L 1.

67 PC Xuechou
Above the spinous process of L 2.

68 PC Changfeng
1 Cun horizontally from the spinous process of L 2.

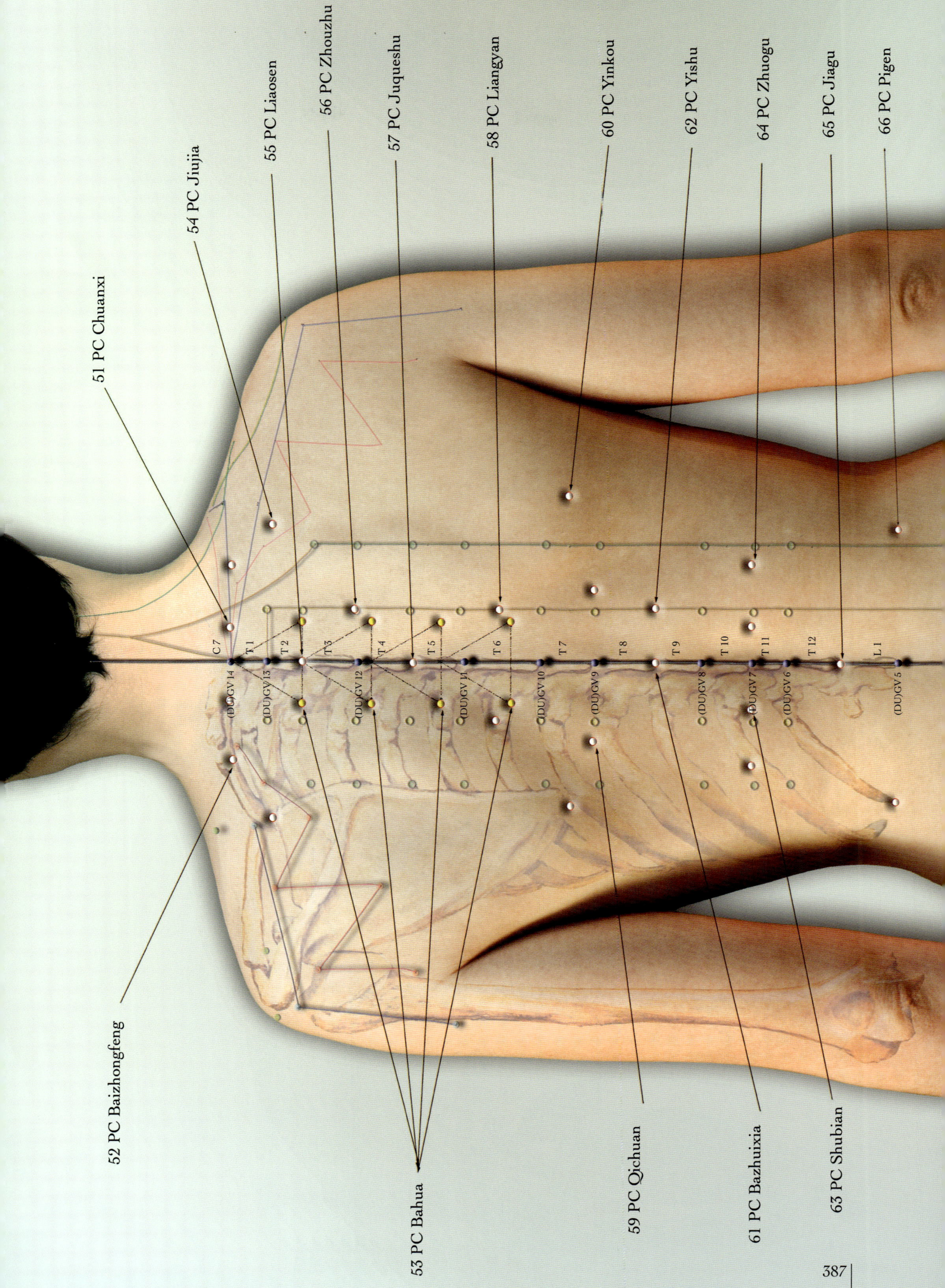
51 PC Chuanxi
52 PC Baizhongfeng
53 PC Bahua
54 PC Jiujia
55 PC Liaosen
56 PC Zhouzhu
57 PC Juqueshu
58 PC Liangyan
59 PC Qichuan
60 PC Yinkou
61 PC Bazhuixia
62 PC Yishu
63 PC Shubian
64 PC Zhuogu
65 PC Jiagu
66 PC Pigen
C 7
T 1
T 2
T 3
T 4
T 5
T 6
T 7
T 8
T 9
T 10
T 11
T 12
L 1
(DU)GV 14
(DU)GV 13
(DU)GV 12
(DU)GV 11
(DU)GV 10
(DU)GV 9
(DU)GV 8
(DU)GV 7
(DU)GV 6
(DU)GV 5

Extra channel points (PN)

44 PN Qi Jingzhui Pang
0.5 Cun outside the spinous process of the 7th cervical vertebra.

45 PN Dingchuan
0.5 Cun outside point (DU)GV 14 Dazhui.

46 PN Wai Dingchuan
1.5 Cun outside (DU)GV 14 Dazhui.

47 PN Jiehe Xue
3.5 Cun outside (DU)GV 14 Dazhui.

48 PN Weire Xue
0.5 Cun outside and below the 4th thoracic vertebra.

49 PN Jiantong Dian
Midpoint of the external border of the scapula.

50 PN Zhongchuan
On the external border of the spinous process of the 5th thoracic vertebra.

51 PN Pirexue
0.5 Cun outside the spinous process of the 6th thoracic vertebra. Point classified in the Huatuojiaji.

52 PN Shenrexue
0.5 Cun outside the spinous process of the 7th thoracic vertebra. Point classified in the Huatuojiaji.

53 PN Jianming
1.5 Cun outside the spinous process of the 9th thoracic vertebra and 0.5 Cun below BL 18 Ganshu.

54 PN Kuishang Xue
2 Cun ouside point BL 50 Weicang, that is to say, 5 Cun outside the spinous process of the 12th thoracic vertebra.

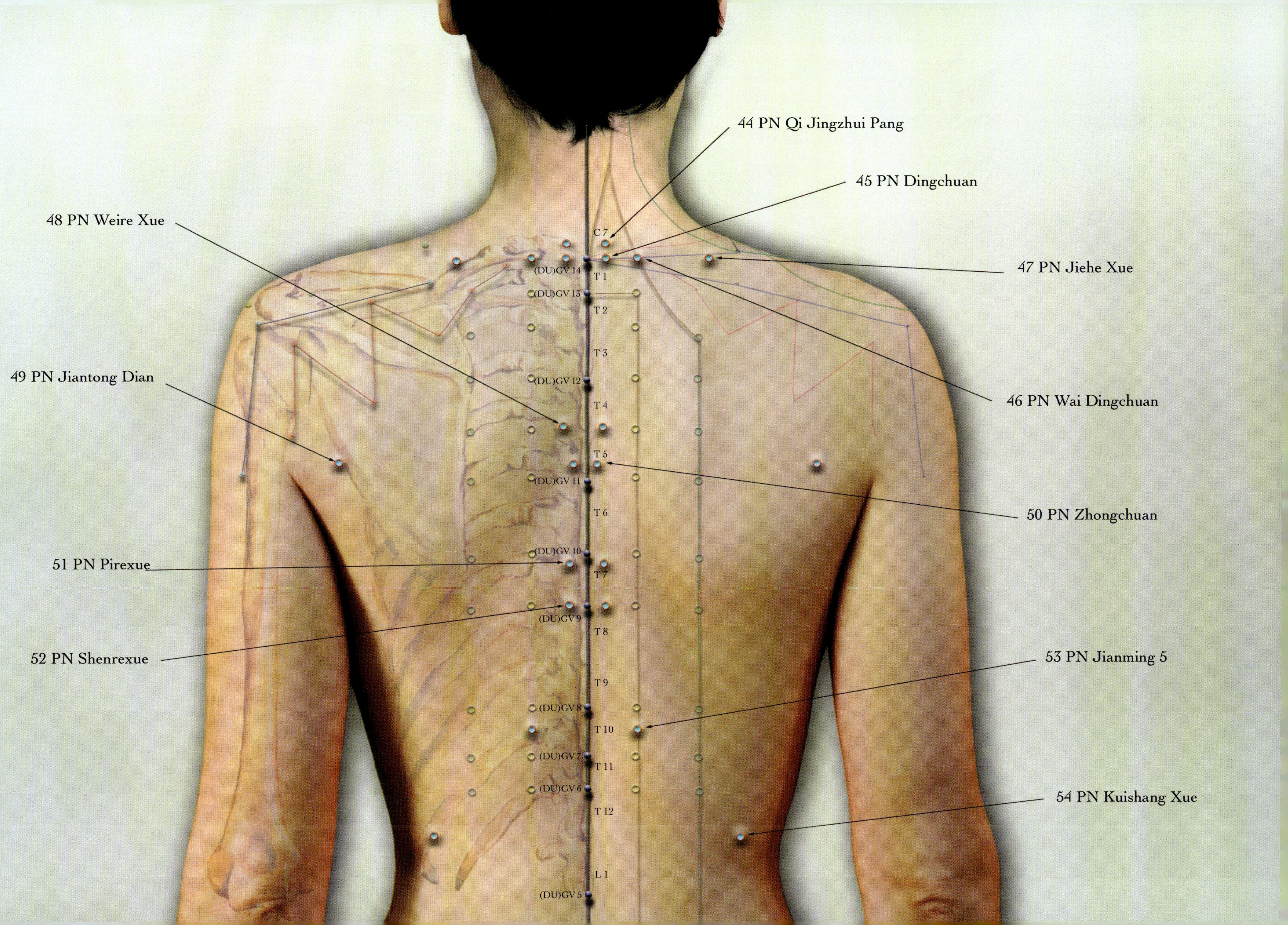
44 PN Qi Jingzhui Pang
45 PN Dingchuan
46 PN Wai Dingchuan
47 PN Jiehe Xue
48 PN Weire Xue
49 PN Jiantong Dian
50 PN Zhongchuan
51 PN Pirexue
52 PN Shenrexue
53 PN Jianming 5
54 PN Kuishang Xue
C 7
T 1
T 2
T 3
T 4
T 5
T 6
T 7
T 8
T 9
T 10
T 11
T 12
L 1
(DU)GV 14
(DU)GV 13
(DU)GV 12
(DU)GV 11
(DU)GV 10
(DU)GV 9
(DU)GV 8
(DU)GV 7
(DU)GV 6
(DU)GV 5

J-BY.3 Liuhua
It is a set of three pairs of points, between the second and the fourth thoracic, 1 Cun lateral from Governor vessel.

J-BY.5 Erzhuixia
Below the spinous process of the 2nd thoracic.

J-BY.15 Jiegu
Below the spinous process of the 12th thoracic.

J-BY.35 = 85 PC Huatuojiaji
A set of 23 points disposed symetrically along the vertebral column between the 2nd cervical and the 5th lumbar, 0.5 Cun from the spinous processes.

J-BY.37 Jisanxue
A set of 3 paravertebral symmetrical points.

- The first is below (DU)GV 15 Yamen, 0.5 Cun laterally.
- The second is 0.5 Cun laterally from the transversal process of the 2nd thoracic.
- The third is 0.5 Cun laterally from the transversal process of the 2nd lumbar.

X-BY.8 Ganrexue
0.5 Cun from the spinous process of the 5th thoracic. It is considered as a Huatuojiaji.

X-BY.11 Shenrexue
0.5 Cun lateral from the spinous process of the 7th thoracic. It is considered as a Huatuojiaji.

X-BY.12 Erjianming
1.5 Cun lateral from the spinous process of the 9th thoracic, 0.5 Cun above BL 18 Ganshu.

X-BY.20 Fei Rexue
0.5 Cun lateral from the spinous process of the 3rd thoracic. It is considered as a Huatuojiaji.

X-BY.22 Zhongjiaoshu
2 Cun lateral from the spinous process of the 12th thoracic.

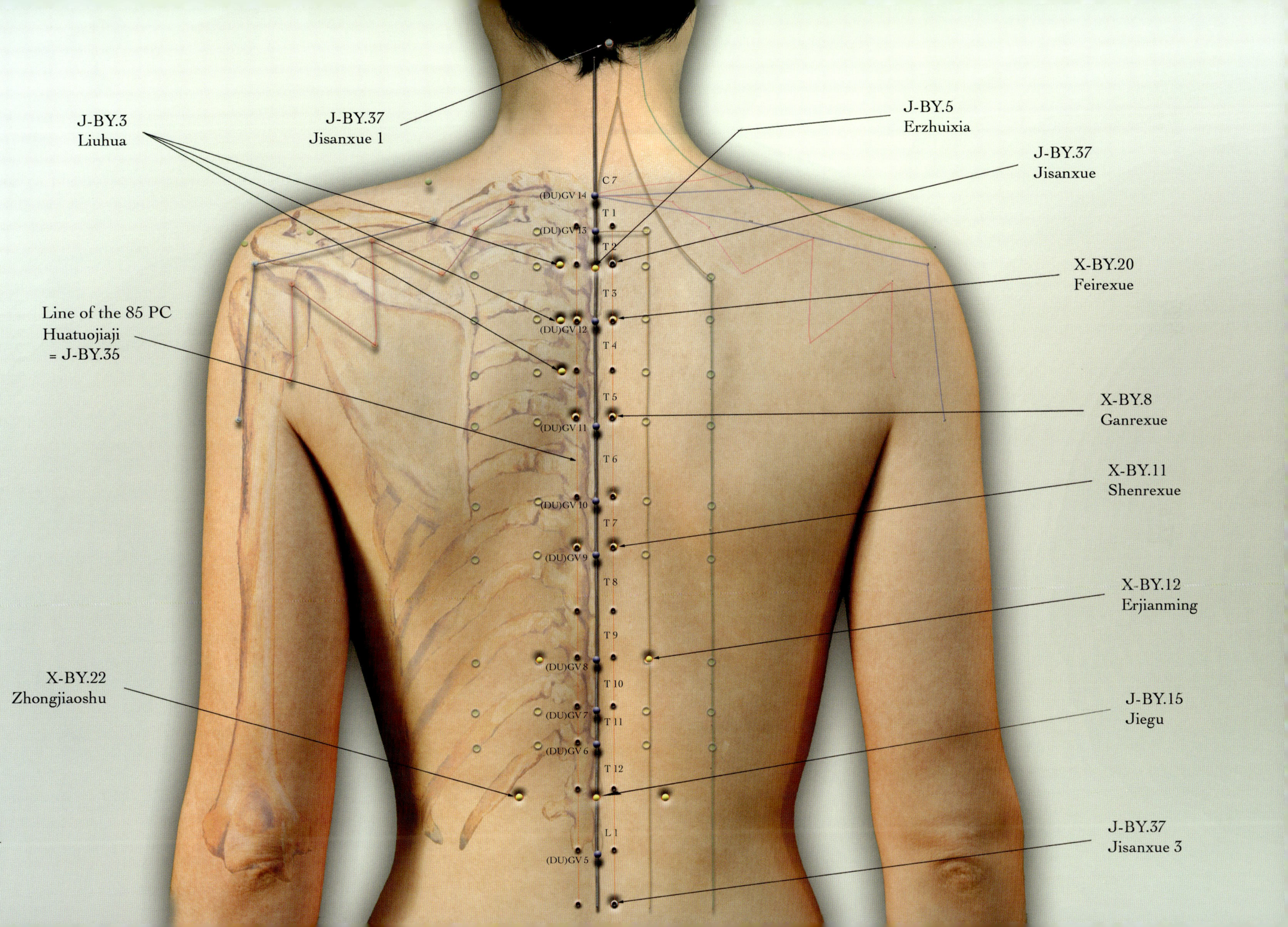
J-BY.3
Liuhua
J-BY.37
Jisanxue 1
J-BY.5
Erzhuixia
J-BY.37
Jisanxue
X-BY.20
Feirexue
Line of the 85 PC
Huatuojiaji
= J-BY.35
X-BY.8
Ganrexue
X-BY.11
Shenrexue
X-BY.12
Erjianming
X-BY.22
Zhongjiaoshu
J-BY.15
Jiegu
J-BY.37
Jisanxue 3
C 7
T 1
T 2
T 3
T 4
T 5
T 6
T 7
T 8
T 9
T 10
T 11
T 12
L 1
(DU)GV 14
(DU)GV 13
(DU)GV 12
(DU)GV 11
(DU)GV 10
(DU)GV 9
(DU)GV 8
(DU)GV 7
(DU)GV 6
(DU)GV 5

DT.01 Fen Zhi Shang
1 Cun below the acromion, that is to say 1 Cun below SI 10 Naoshu.

DT.02 Fen Zhi Xia
1.5 Cun below and 5 fen inside DT.01 Fen Zhi Shang.

DT.05 Shaung Feng
The first point is found on the level of T 2 and 1.5 Cun lateral and external (1st chain of Bladder channel). The others are 1 Cun below the ones from the others:

- Huo Feng
- Huo Fu
- Huo Miao
- Huo Chao
- Huo Cheng
- Huo Hua
- Huo Mi

DT.06 Jiu Hou
Same location as Huo Feng

- Huo Fu
- Huo Miao
- Jin Tang
- Jin Bei
- Jin Tou
- Jin Ji
- Jin Zhi
- Jin Jing

A total of 9 points.

DT.07 San Jin
Corresponds to

- Jin Tou
- Jin Chi
- Jin Ling

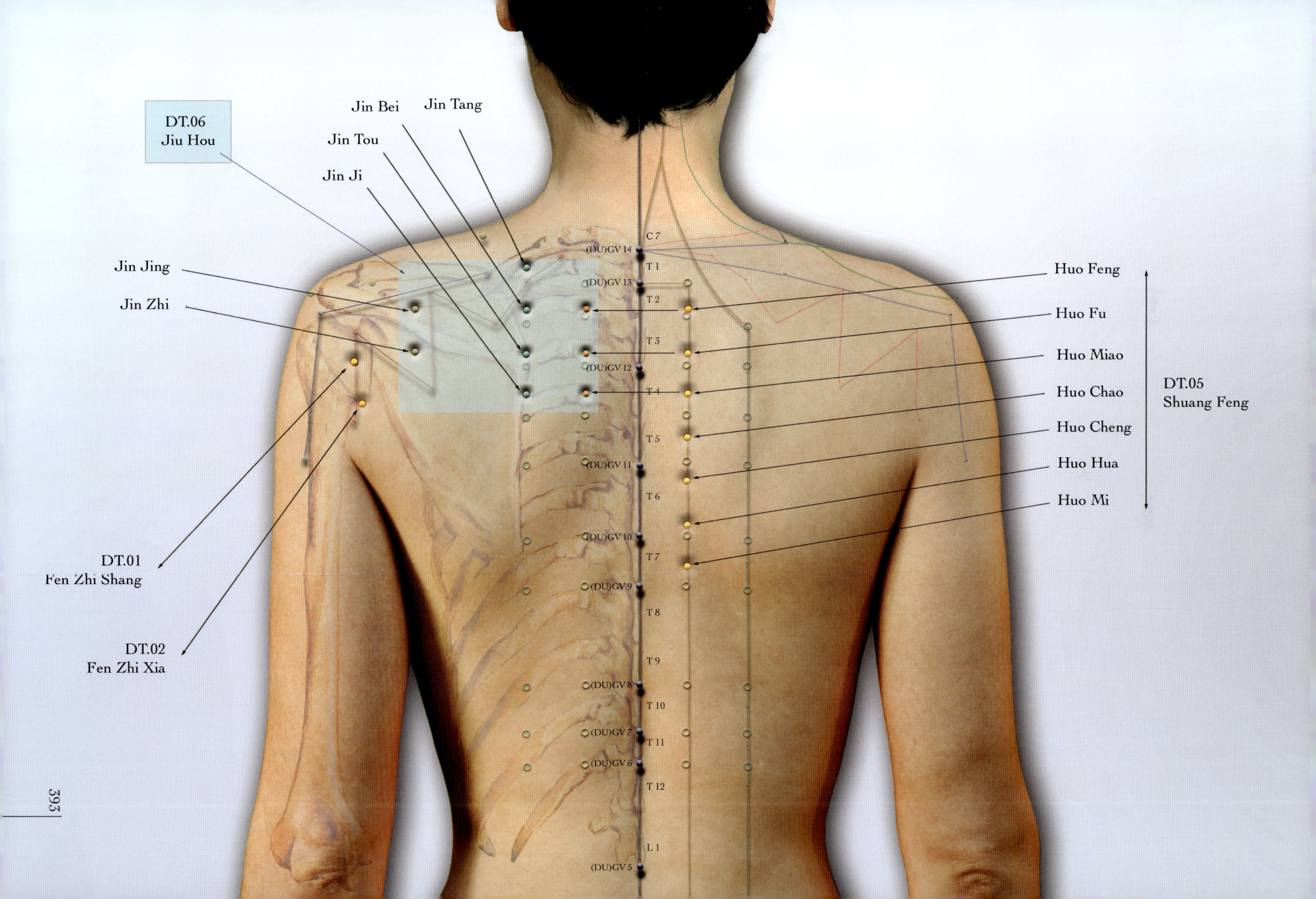
DT.06
Jiu Hou
Jin Bei
Jin Tang
Jin Tou
Jin Ji
Jin Jing
Jin Zhi
DT.01
Fen Zhi Shang
DT.02
Fen Zhi Xia
Huo Feng
Huo Fu
Huo Miao
Huo Chao
Huo Cheng
Huo Hua
Huo Mi
DT.05
Shuang Feng
C 7
T 1
T 2
T 3
T 4
T 5
T 6
T 7
T 8
T 9
T 10
T 11
T 12
L 1
(DU)GV 14
(DU)GV 13
(DU)GV 12
(DU)GV 11
(DU)GV 10
(DU)GV 9
(DU)GV 8
(DU)GV 7
(DU)GV 6
(DU)GV 5

DT.08 Jing Zhi
Corresponds to the association of:

- Jin Zhi
- Jin Jing

DT.09 Jin Lin
Corresponds to:

- Jin Tou
- Jin Chi
- Jin Ling

DT.10 Ding Zhu
Corresponds to:

- Jin Chi
- Jin Ling
- Huo Jin
- Mu Teng
- Mu Tu
- Jin Shen
- Mu Mei
- Mu Yuan
- Mu Tai
- Mu Ju
- Mu Song

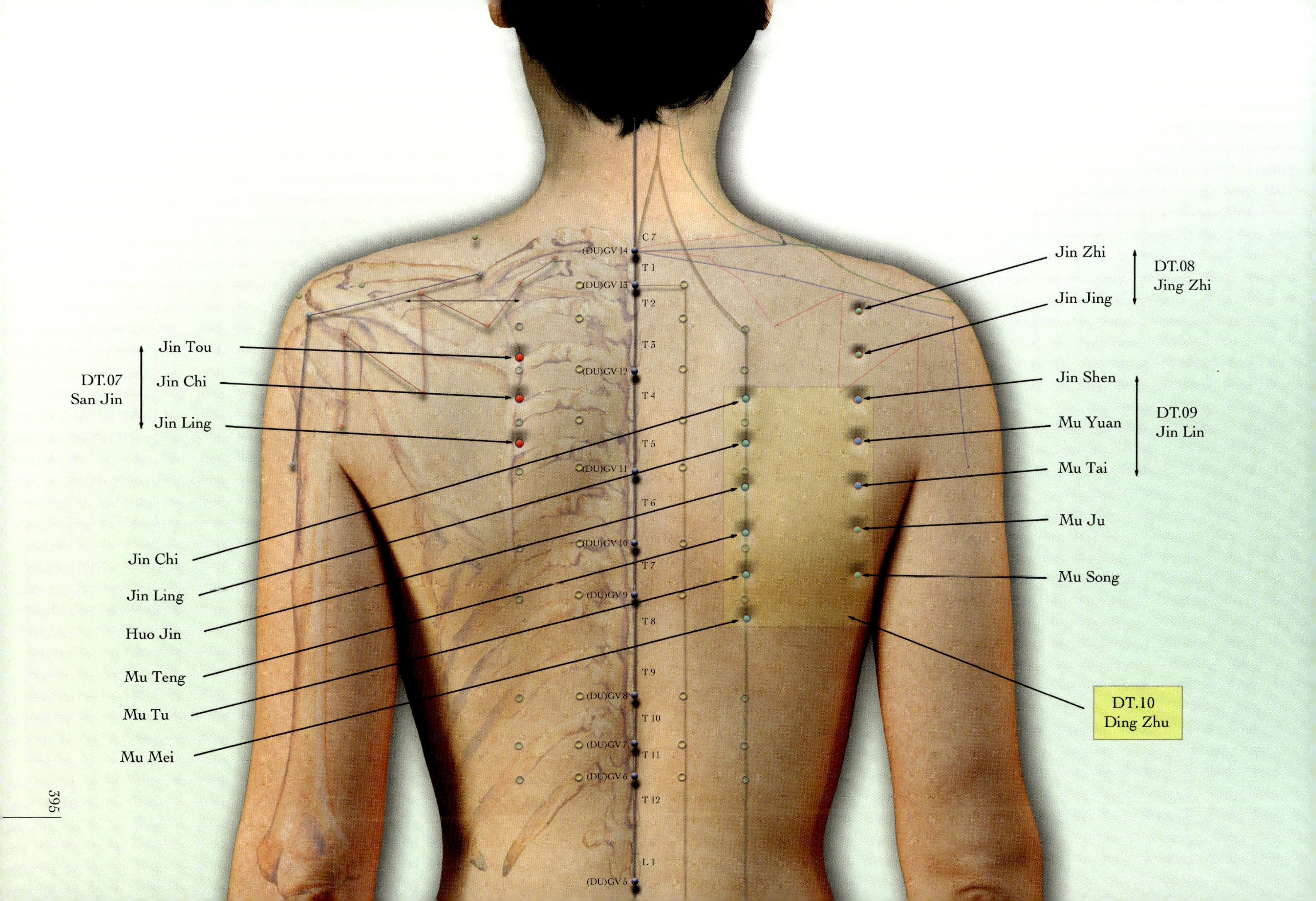

DT.07
San Jin
Jin Tou
Jin Chi
Jin Ling
Jin Chi
Jin Ling
Huo Jin
Mu Teng
Mu Tu
Mu Mei
C 7
(DU)GV 14
T 1
(DU)GV 13
T 2
T 3
(DU)GV 12
T 4
T 5
(DU)GV 11
T 6
(DU)GV 10
T 7
(DU)GV 9
T 8
T 9
(DU)GV 8
T 10
(DU)GV 7
T 11
(DU)GV 6
T 12
L 1
(DU)GV 5
Jin Zhi
Jin Jing
DT.08
Jing Zhi
Jin Shen
Mu Yuan
Mu Tai
DT.09
Jin Lin
Mu Ju
Mu Song
DT.10
Ding Zhu

DT 11 Hou Xin
Corresponds to:

- Huo Yun
- Huo Chang
- Huo Ming
- Huo Xiao
- Huo Men
- Tu Yue
- Huo Miao
- Huo Chao
- Huo Cheng
- Huo Hua
- Jin Chi
- Jin Ling
- Huo Jin

DT 12 Gan Mao San
Corresponds to:

- An Quan
- Jin Tou

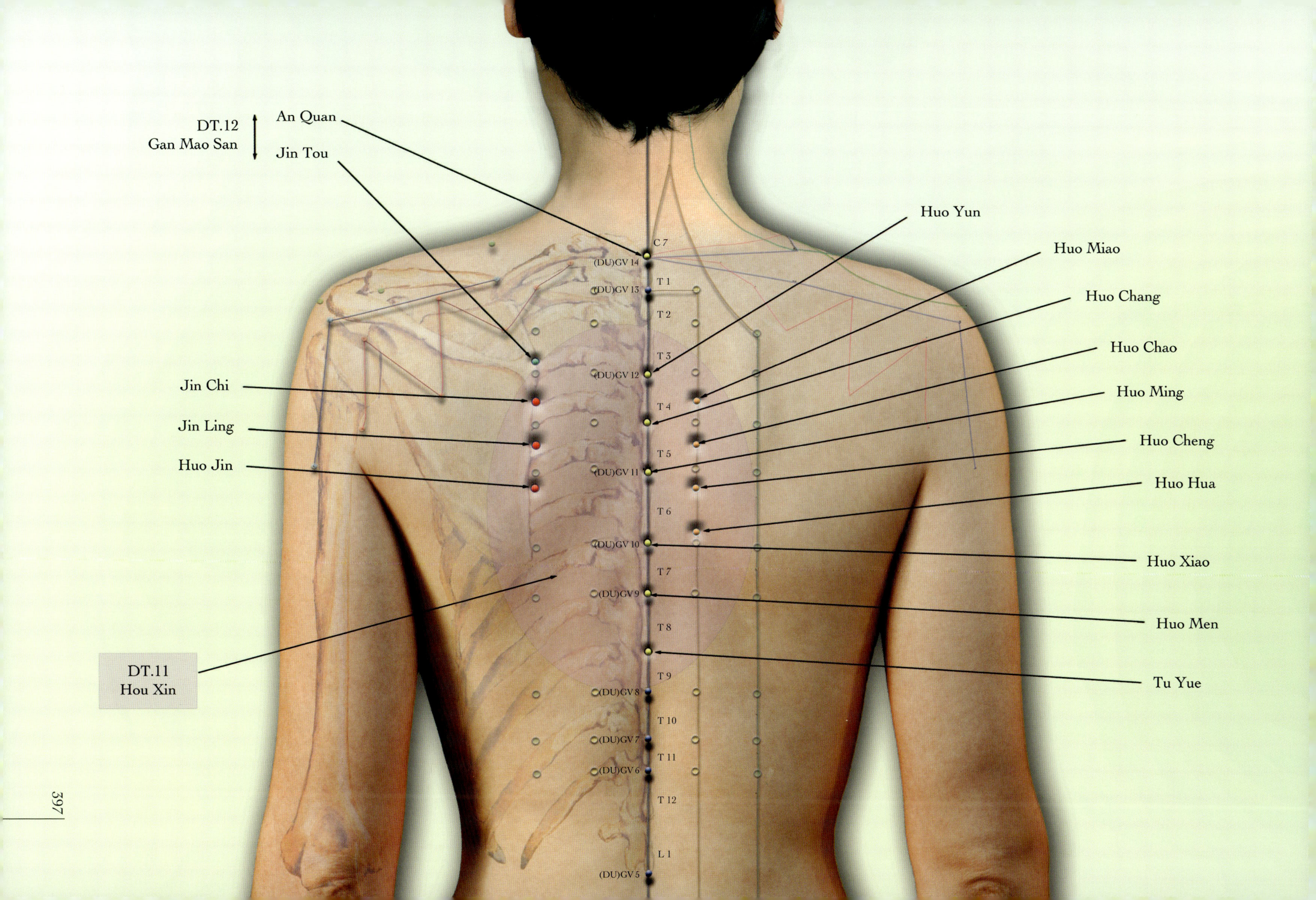
DT.12
Gan Mao San
An Quan
Jin Tou
Huo Yun
Huo Miao
Huo Chang
Huo Chao
Huo Ming
Huo Cheng
Huo Hua
Huo Xiao
Huo Men
Tu Yue
Jin Chi
Jin Ling
Huo Jin
DT.11
Hou Xin
C 7
T 1
T 2
T 3
T 4
T 5
T 6
T 7
T 8
T 9
T 10
T 11
T 12
L 1
(DU)GV 14
(DU)GV 13
(DU)GV 12
(DU)GV 11
(DU)GV 10
(DU)GV 9
(DU)GV 8
(DU)GV 7
(DU)GV 6
(DU)GV 5

DT.04 Wu Ling

A set of 25 points thus distributed:

– from Huo Chu to Huo Men: represent the Heart area (Huo = Fire).

– from Tu Yue to Tu Ke: represent the Spleen area (Tu = Earth).

– from Jin Tang to Huo Jin: represent the Lung and Heart area.

– from Huo Jin to Mu Mei: represent the Lung area on the left.

– from Huo Jin to Mu Mei: represent the Liver area on the right (Mu = Wood).

– from Jin Jing to Jin Shen: represent the Lung area (Jin = Metal).

– from Jin Shen to Mu Song: represent the Lung area and Spleen area on the left.

– from Jin Shen to Mu Song: represent the Lung area and Liver area on the right.

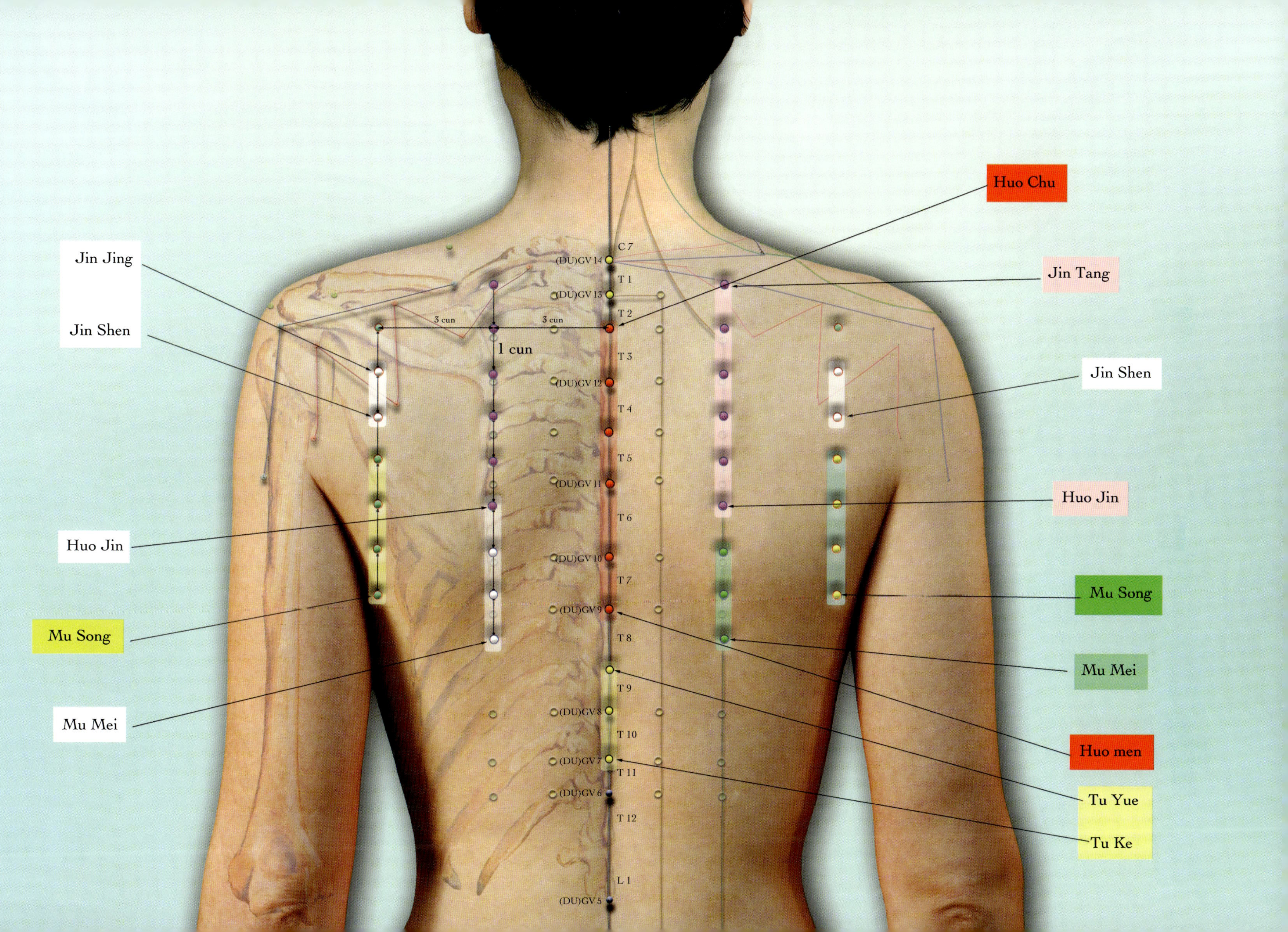

Huo Chu
Jin Tang
Jin Shen
Huo Jin
Mu Song
Mu Mei
Huo men
Tu Yue
Tu Ke
Jin Jing
Jin Shen
Huo Jin
Mu Song
Mu Mei
C 7
T 1
T 2
T 3
T 4
T 5
T 6
T 7
T 8
T 9
T 10
T 11
T 12
L 1
(DU)GV 14
(DU)GV 13
(DU)GV 12
(DU)GV 11
(DU)GV 10
(DU)GV 9
(DU)GV 8
(DU)GV 7
(DU)GV 6
(DU)GV 5
3 cun
3 cun
1 cun

Abdomen

- Points of the Main channels
  - Ren Mai (Conception vessel)
  - Zu Yang Ming (Stomach)
  - Zu Shao Yin (Kidney)
  - Zu Tai Yin (Spleen)
  - Zu Jue Yin (Liver)
  - Zu Shao Yang (Gall Bladder)
- Extra channel points
  - Curious points
  - New points
  - Points of Master Tong

Ren Mai (Conception vessel)

(REN)CV 2 Qugu
On the lower abdomen and on the anterior midline, at the midpoint of the upper border of the pubic symphysis.

(REN)CV 3 Zhongji
On the lower abdomen and on the anterior midline, 4 Cun below the umbilicus.

(REN)CV 4 Guanyuan
On the lower abdomen and on the anterior midline, 3 Cun below the umbilicus.

(REN)CV 5 Shimen
On te lower abdomen and on the anterior midline, 2 Cun below the umbilicus.

(REN)CV 6 Qihai
On the lower abdomen and on the anterior midline, 1.5 Cun below the umbilicus.

(REN)CV 7 Yinjiao
On the lower abdomen and on the anterior midline, 1 Cun below the umbilicus.

(REN)CV 8 Shenque
On the middle abdomen and at the centre of the umbilicus.

(REN)CV 9 Shuifen
On the upper abdomen and on the anterior midline, 1 Cun above the umbilicus.

(REN)CV 10 Xiawan
On the upper abdomen and on the anterior midline, 2 Cun above the umbilicus.

(REN)CV 11 Jianli
On the upper abdomen and on the anterior midline, 3 Cun above the umbilicus.

(REN)CV 12 Zhongwan
On the upper abdomen and on the anterior midline, 4 Cun above the umbilicus.

(REN)CV 13 Shangwan
On the upper abdomen and on the anterior midline, 5 Cun above the umbilicus.

(REN)CV 14 Juque
On the upper abdomen and on the anterior midline, 6 Cun above the umbilicus.

(REN)CV 15 Jiuwei
On the upper abdomen and on the anterior midline, 1 Cun below the xiphisternal synchondrosis.

Zu Yang Ming channel (Stomach)

ST 19 Burong
On the upper abdomen, 6 Cun above the umbilicus and 2 Cun lateral to the anterior midline.

ST 20 Chengman
On the upper abdomen, 5 Cun the umbilicus and 2 Cun lateral to the anterior midline.

ST 21 Liangmen
On the upper abdomen, 4 Cun above the centre of the umbilicus and 2 Cun lateral to the anterior midline.

ST 22 Guanmen
On the upper abdomen, 3 Cun above the umbilicus and 2 Cun lateral to the anterior midline.

ST 23 Taiyi
On the upper abdomen, 2 Cun above the umbilicus and 2 Cun lateral to the anterior midline.

ST 24 Huaroumen
On the upper abdomen, 1 Cun above the umbilicus and 2 Cun lateral to the anterior midline.

ST 25 Tianshu
On the middle abdomen, 2 Cun outside the umbilicus.

ST 26 Wailing
On the lower abdomen, 1 Cun below the umbilicus and 2 Cun lateral to the anterior midline.

ST 27 Daju
On the lower abdomen, 2 Cun below the umbilicus and 2 Cun lateral to the anterior midline.

ST 28 Shuidao
On the lower abdomen, 3 Cun below the umbilicus and 2 Cun lateral to the anterior midline.

ST 29 Guilai
On the lower abdomen, 4 Cun below the umbilicus and 2 Cun lateral to the anterior midline.

ST 30 Qichong
Slightly above the inguinal groove, 5 Cun below the umbilicus and 2 Cun lateral to the anterior midline.

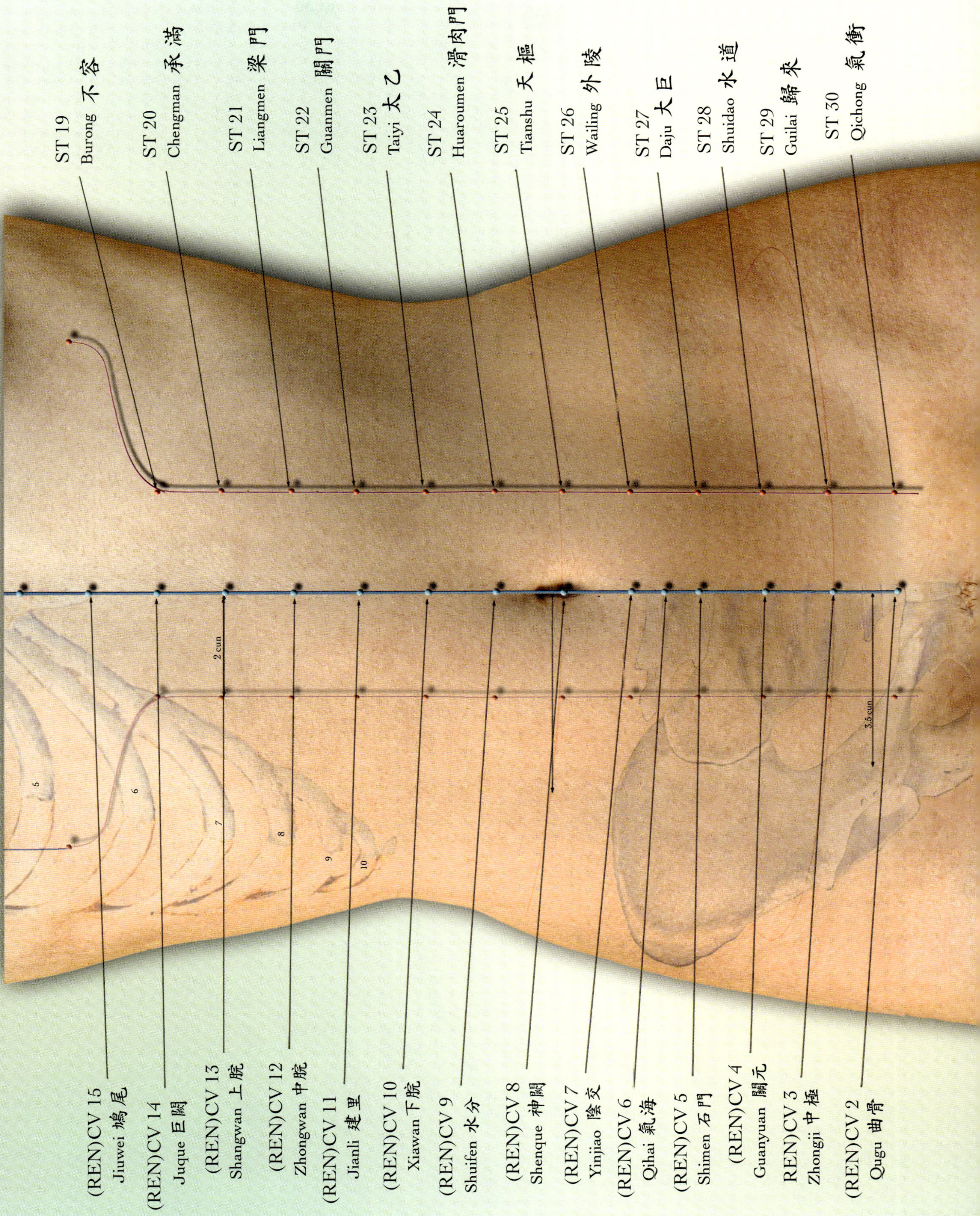

ST 19 Burong 不容
ST 20 Chengman 承滿
ST 21 Liangmen 梁門
ST 22 Guanmen 關門
ST 23 Taiyi 太乙
ST 24 Huaroumen 滑肉門
ST 25 Tianshu 天樞
ST 26 Wailing 外陵
ST 27 Daju 大巨
ST 28 Shuidao 水道
ST 29 Guilai 歸來
ST 30 Qichong 氣衝
2 cun
3,5 cun
5
6
7
8
9
10
(REN)CV 15 Jiuwei 鳩尾
(REN)CV 14 Juque 巨闕
(REN)CV 13 Shangwan 上脘
(REN)CV 12 Zhongwan 中脘
(REN)CV 11 Jianli 建里
(REN)CV 10 Xiawan 下脘
(REN)CV 9 Shuifen 水分
(REN)CV 8 Shenque 神闕
(REN)CV 7 Yinjiao 陰交
(REN)CV 6 Qihai 氣海
(REN)CV 5 Shimen 石門
(REN)CV 4 Guanyuan 關元
REN)CV 3 Zhongji 中極
(REN)CV 2 Qugu 曲骨

Zu Shao Yin channel (Kidney)

KI 11 Henggu
On the lower abdomen, 5 Cun below the umbilicus and 0.5 Cun lateral to the anterior midline.

KI 12 Dahe
On the lower abdomen, 4 Cun below the umbilicus and 0.5 Cun lateral to the anterior midline.

KI 13 Qixue
On the lower abdomen, 3 Cun below the umbilicus and 0.5 Cun lateral to the anterior midline.

KI 14 Siman
On the lower abdomen, 2 Cun below the umbilicus and 0.5 Cun lateral to the anterior midline.

KI 15 Zhongzhu
On the lower abdomen, 1 Cun below the umbilicus and 0.5 Cun lateral to the anterior midline.

KI 16 Huangshu
On the middle abdomen, 0.5 Cun lateral to the umbilicus.

KI 17 Shanqu
On the upper abdomen, 2 Cun above the umbilicus, and 0.5 Cun to the anterior midline.

KI 18 Shiguan
On the upper abdomen, 3 Cun above the umbilicus and 0.5 Cun to the anterior midline.

KI 19 Yindu
On the upper abdomen, 4 Cun above the umbilicus and 0.5 Cun to the anterior midline.

KI 20 Futonggu
On the upper abdomen, 5 Cun above the umbilicus and 0.5 Cun to the anterior midline.

KI 21 Youmen
On the upper abdomen, 6 Cun above the umbilicus and 0.5 Cun to the anterior midline.

Zu Tai Yin channel (Spleen)

SP 12 Chongmen
At the lateral end of the inguinal groove, 3.5 Cun lateral to the midpoint of the upper border of the symphysis pubis, lateral to the pulsating external iliac artery.

SP 13 Fushe
On the lower abdomen, 4 Cun below the umbilicus, 0.7 Cun above SP 12 Chongmen, and 4 Cun lateral to the anterior midline.

SP 14 Fujie
On the lower abdomen, 1.3 Cun below SP 15 Daheng and 4 Cun lateral to the anterior midline.

SP 15 Daheng
On the upper abdomen, 4 Cun lateral to the umbilicus.

SP 16 Fu'ai
On the upper abdomen, 3 Cun above the umbilicus, and 4 Cun lateral to the anterior midline.

Zu Jue Yin channel (Liver)

(LR)LV 12 Jimai
Lateral to the pubic tubercle, outside and inferior to ST 30 Qichong, in the inguinal groove where the pulsation of the femoral artery is palpable, 2.5 Cun to the anterior midline. (Do NOT needle directly into the artery!)

(LR)LV 13 Zhangmen
On the lateral side of the abdomen, below the free end of the 11th rib.

(LR)LV 14 Qimen
On the chest, directly below the nipple, in the 6th intercostal space, 4 Cun to the anterior midline.

Zu Shao Yang channel (Gall Bladder)

GB 24 Riyue
On the upper abdomen, directly below the nipple, in the 7th intercostal space, 4 Cun to the anterior midline.

GB 26 Daimai
On the lateral side of the abdomen, 1.8 Cun below (LR)LV 13 Zhangmen, at the crossing point of a vertical line through the free end of the 11th rib and a horizontal line through the umbilicus.

GB 27 Wushu
On the lateral side of the abdomen, anterior to the anteriosuperior iliac spine, 3 Cun below the level of the umbilicus.

GB 28 Weidao
On the lateral side of the abdomen, 0.5 Cun anterior and inferior to GB 27 Wushu.

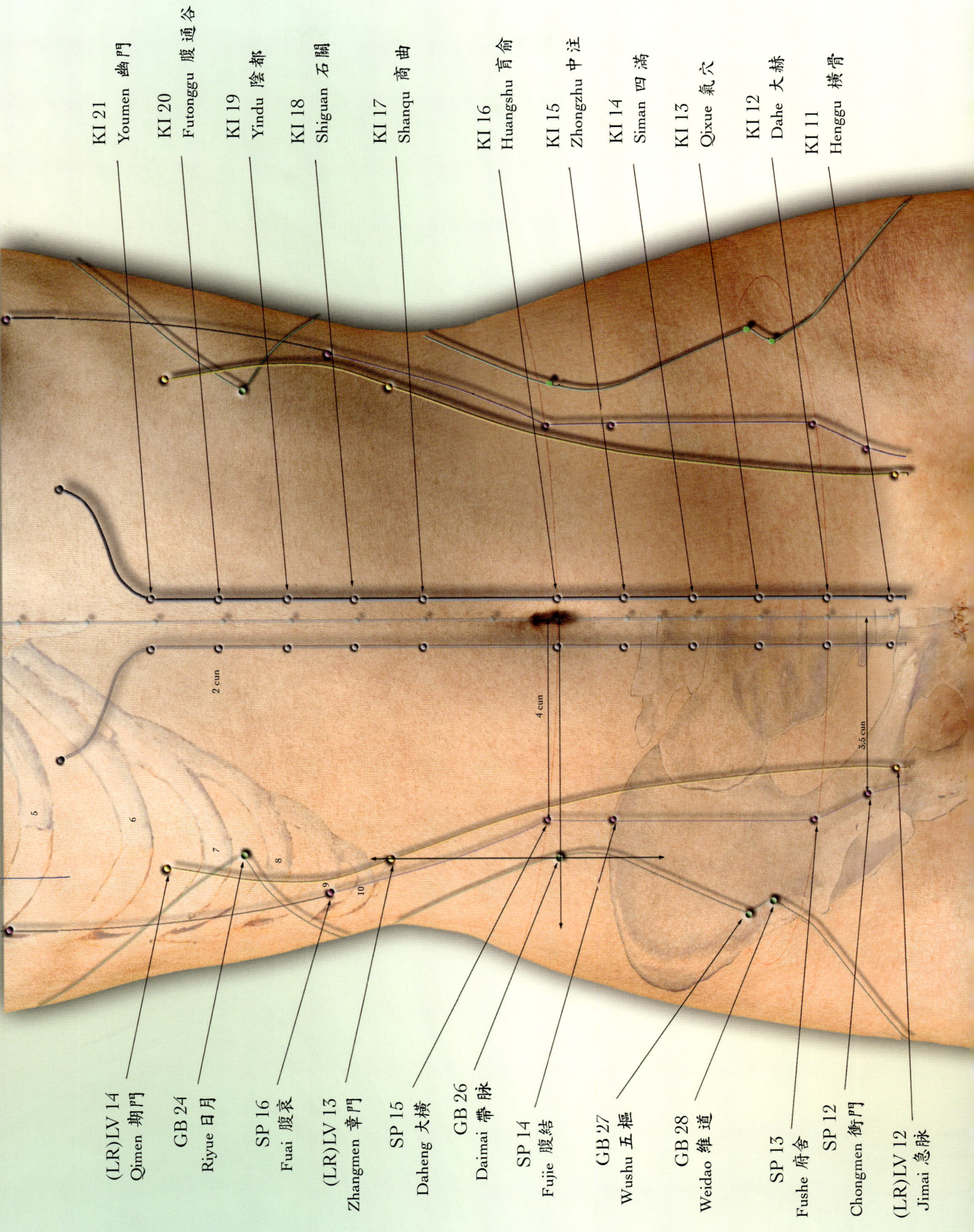
KI 21 Youmen 幽門
KI 20 Futonggu 腹通谷
KI 19 Yindu 陰都
KI 18 Shiguan 石關
KI 17 Shanqu 商曲
KI 16 Huangshu 肓俞
KI 15 Zhongzhu 中注
KI 14 Siman 四滿
KI 13 Qixue 氣穴
KI 12 Dahe 大赫
KI 11 Henggu 橫骨
2 cun
4 cun
3,5 cun
5
6
7
8
9
10
(LR)LV 14 Qimen 期門
GB 24 Riyue 日月
SP 16 Fuai 腹哀
(LR)LV 13 Zhangmen 章門
SP 15 Daheng 大橫
GB 26 Daimai 帶脈
SP 14 Fujie 腹結
GB 27 Wushu 五樞
GB 28 Weidao 維道
SP 13 Fushe 府舍
SP 12 Chongmen 衝門
(LR)LV 12 Jimai 急脈

Extra channel points (PC)

36 PC Maihua
Round point (REN)CV 12 Zhongwan and 0.5 Cun outside, above and below that point. These 4 points, with (REN)CV 12 Zhongwan, are called 'plum-tree flower'.

37 PC Shicang
Horizontally with (REN)CV 12 Zhongwan, 3 Cun from that point.

38 PC Shiguan
Between the Kidney and Stomach channels, 1 Cun from (REN)CV 11 Jianli.

39 PC Hunshe
Between the Kidney and Stomach channels, 1 Cun from (REN)CV 8 Shenque.

40 PC Hangwen
Between the Stomach and Spleen channels, 1.5 Cun horizontally and towards the inside of SP 15 Daheng.

41 PC Qizhong
Between the Stomach and Spleen channels, 1.5 Cun horizontally from (REN)CV 6 Qihai.

42 PC Jingzhong
Between the stomach and Spleen channels, 3 Cun horizontally from (REN)CV 6 Qihai.

43 PC Waisiman
Between the channels of Stomach and Kidney, horizontally 1 Cun from point KI 14 Siman and 0.5 Cun vertically and below 41 PC Qizhong, 3 Cun from the border of the pubis.

44 PC Jueyun
On the Conception vessel, 2.3 Cun vertically and below (REN)CV 5 Shimen.

45 PC Yijing
1 Cun horizontally from point (REN)CV 4 Guanyuan.

46 PC Qimen
Between the Spleen and Stomach channels, 3 Cun horizontally from point (REN)CV 4 Guanyuan.

47 PC Weibao
Between the spleen and gall bladder channels at 6 Cun horizontally from the point (REN)CV 4 Guanyuan.

48 PC Changyi
2,5 Cun horizontally from the point (REN)CV 3 Zhongji.

49 PC Zigong
3 Cun horizontally from the point (REN)CV 3 Zhongji, on the liver channel, to 1 cun above the symphysis pubis.

50 PC Tingtou
0.5 Cun under the KI 12 Dahe in a vertical line.

X-XF.12 Xinqixue
0.5 Cun above and 1.5 Cun to the side of (REN)CV 6 Qihai.

J-XF.24 Longmen
The lower edge of the symphysis pubis.

X-XF.13 Xiazhongji
0.5 Cun under (REN)CV 3 Zhongji.

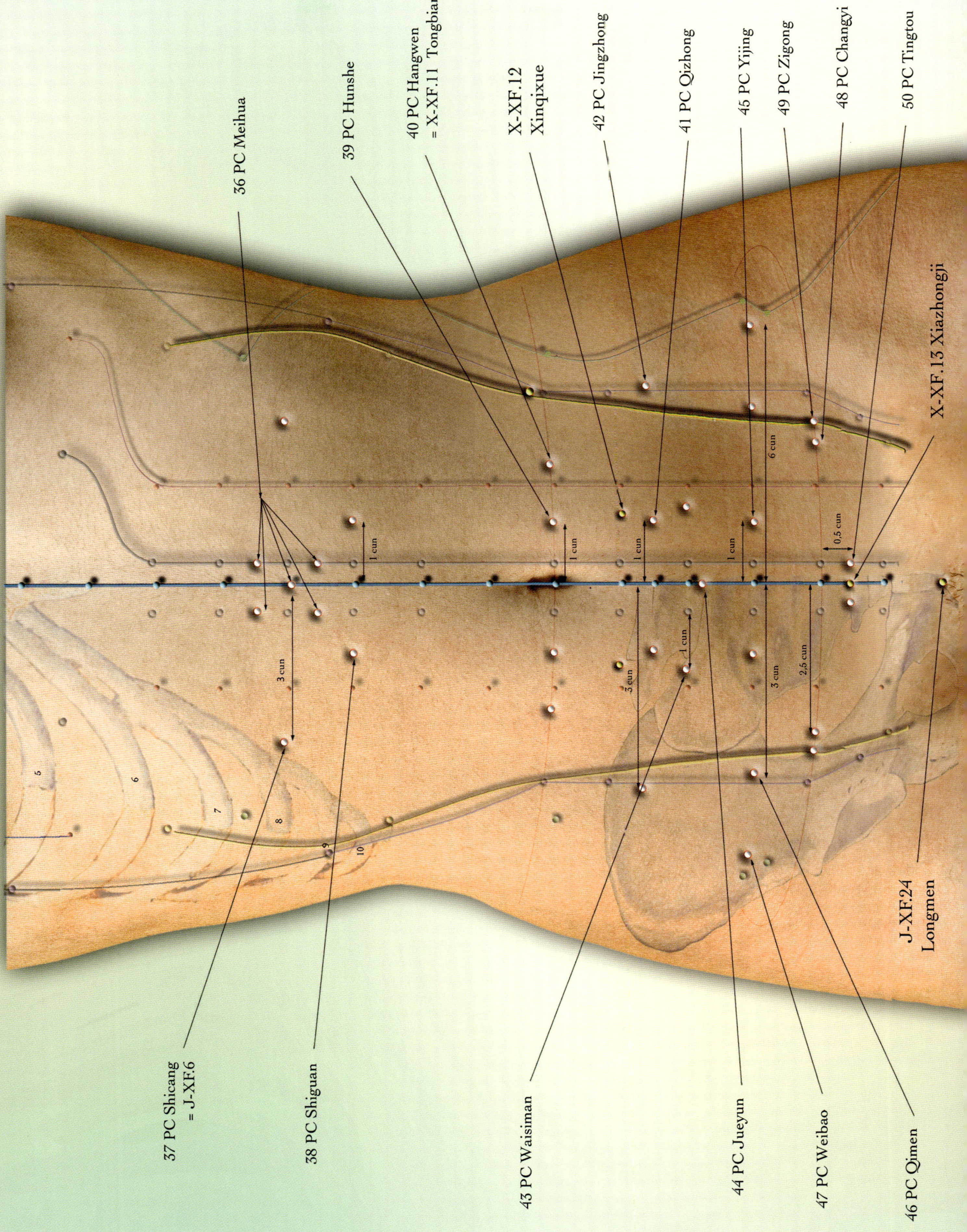
36 PC Meihua
39 PC Hunshe
40 PC Hangwen
= X-XF.11 Tongbian
X-XF.12
Xinqixue
42 PC Jingzhong
41 PC Qizhong
45 PC Yijing
49 PC Zigong
48 PC Changyi
50 PC Tingtou
X-XF.13 Xiazhongji
J-XF.24
Longmen
37 PC Shicang
= J-XF.6
38 PC Shiguan
43 PC Waisiman
44 PC Jueyun
47 PC Weibao
46 PC Qimen
1 cun
3 cun
6 cun
0,5 cun
2,5 cun

Extra channel points
(PN – Master Tong)

36 PN Shuishang
0.5 Cun above point (REN)CV 9 Shuifen.

37 PN Weile
4 Cun outside and 0.2 Cun above point (REN)CV 9 Shuifen.

38 PN Zhixie
2.5 Cun below the umbilicus.

39 PN Tituoxue
4 Cun outside point (REN)CV 4 Guanyuan.

40 PN Chongjian
3 Cun outside point (REN)CV 2 Qugu.

41 PN Shuxi
3 Cun below and 1 Cun lateral to Tituoxue.

X-XF.17 Yinbian
On the inferior border of the pubic symphisis, 0.5 Cun lateral to J-XF.24 Longmen.

X-XF.18 Weishangxue
3 Cun lateral to (REN)CV 10 Xiawan.

X-XF.14 Yeniao
1 Cun lateral to X-XF.13 Xiazhongji.

VT.05 Fu Chao Er Shi San
All these points are situated round the umbilicus, 1 Cun from each other, except for the umbilicus which hasn't a point.

The upper point is 2 Cun above the umbilicus, and the two inferior points 3 and 4 Cun below the umbilicus.

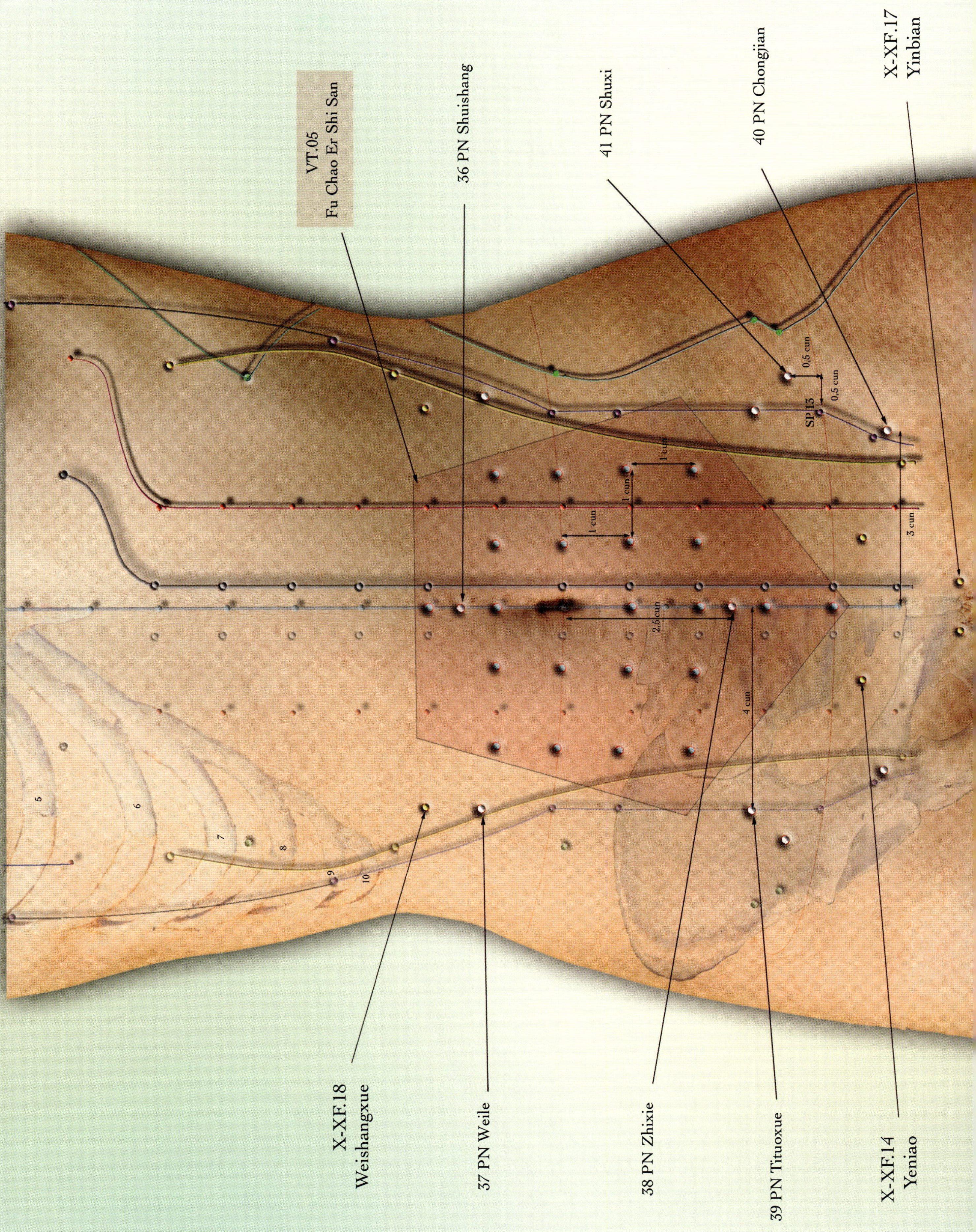

VT.05
Fu Chao Er Shi San
36 PN Shuishang
41 PN Shuxi
40 PN Chongjian
X-XF.17
Yinbian
X-XF.18
Weishangxue
37 PN Weile
38 PN Zhixie
39 PN Tituoxue
X-XF.14
Yeniao
SP 13
0,5 cun
1 cun
3 cun
2,5 cun
4 cun

Sacro-lumbar region

- Points of the Main channels
  - Du Mai (Governor vessel)
  - Zu Tai Yang (Bladder)
  - Zu Shao Yang (Gall Bladder)
- Extra channel points
  - Curious points
  - New points
  - Points of Master Tong

Du Mai (Governor vessel)

(DU)GV 1 Changqiang
Below the tip of the coccyx, at the midpoint of the line connecting the tip of the coccyx and anus.

(DU)GV 2 Yaoshu
On the sacrum and on the posterior midline, just at the sacral hiatus.

(DU)GV 3 Yaoyangguan
On the low back and on the posterior midline, in the depression below the spinous process of the 4th lumbar vertebra.

(DU)GV 4 Mingmen
On the low back and on the posterior midline, in the depression below the spinous process of the 2nd lumbar vertebra.

(DU)GV 5 Xuanshu
On the low back and on the posterior midline, in the depression below the spinous process of the 1st lumbar vertebra.

Zu Tai Yang channel (1st Bladder chain)

BL 21 Weishu
On the back, below the spinous process of the 12th thoracic vertebra, 1.5 Cun lateral to the posterior midline.

BL 22 Sanjiaoshu
On the low back, below the spinous process of the 1st lumbar vertebra, 1.5 Cun lateral to the posterior midline.

BL 23 Shenshu
On the low back, below the spinous process of the 2nd lumbar vertebra, 1.5 Cun lateral to the posterior midline.

BL 24 Qihaishu
On the low back, below the spinous process of the 3rd lumbar vertebra, 1.5 Cun lateral to the posterior midline.

BL 25 Dachangshu
On the low back, below the spinous process of the 4th lumbar vertebra, 1.5 Cun lateral to the posterior midline.

BL 26 Guanyuanshu
On the low back, below the spinous process of the 5th lumbar vertebra, 1.5 Cun lateral to the posterior midline.

BL 27 Xiaochangshu
On the sacrum and on the level of the 1st posterior sacral foramen, 1.5 Cun lateral to the median sacral crest.

BL 28 Pangguangshu
On the sacrum and on the level of the 2nd posterior sacral foramen, 1.5 Cun lateral to the median sacral crest.

BL 29 Zhongliishu
On the sacrum and on the level of the 3rd posterior sacral foramen, 1.5 Cun lateral to the median sacral crest.

BL 30 Baihuanshu
On the sacrum and on the level of the 4th posterior sacral foramen, 1.5 Cun lateral to the median sacral crest.

BL 31 Shangliao
On the sacrum, at the midpoint between the posterio-superior iliac spine and the posterior midline, just at the 1st posterior sacral foramen.

BL 32 Ciliao
On the sacrum, medial and inferior to the posterio-superior iliac spine, just at the 2nd posterior sacral foramen.

BL 33 Zhongliao
On the sacrum, medial and inferior to BL 32 Ciliao, just at the 3rd posterior sacral foramen.

BL 34 Xialiao
On the sacrum, medial and inferior to BL 33 Zhongliao, just at the 4th posterior sacral foramen.

BL 35 Huiyang
On the sacrum, 0.5 Cun lateral to the tip of the coccyx.

BL 36 Chengfu
On the posterior side of the thigh, at the midpoint of the inferior gluteal crease.

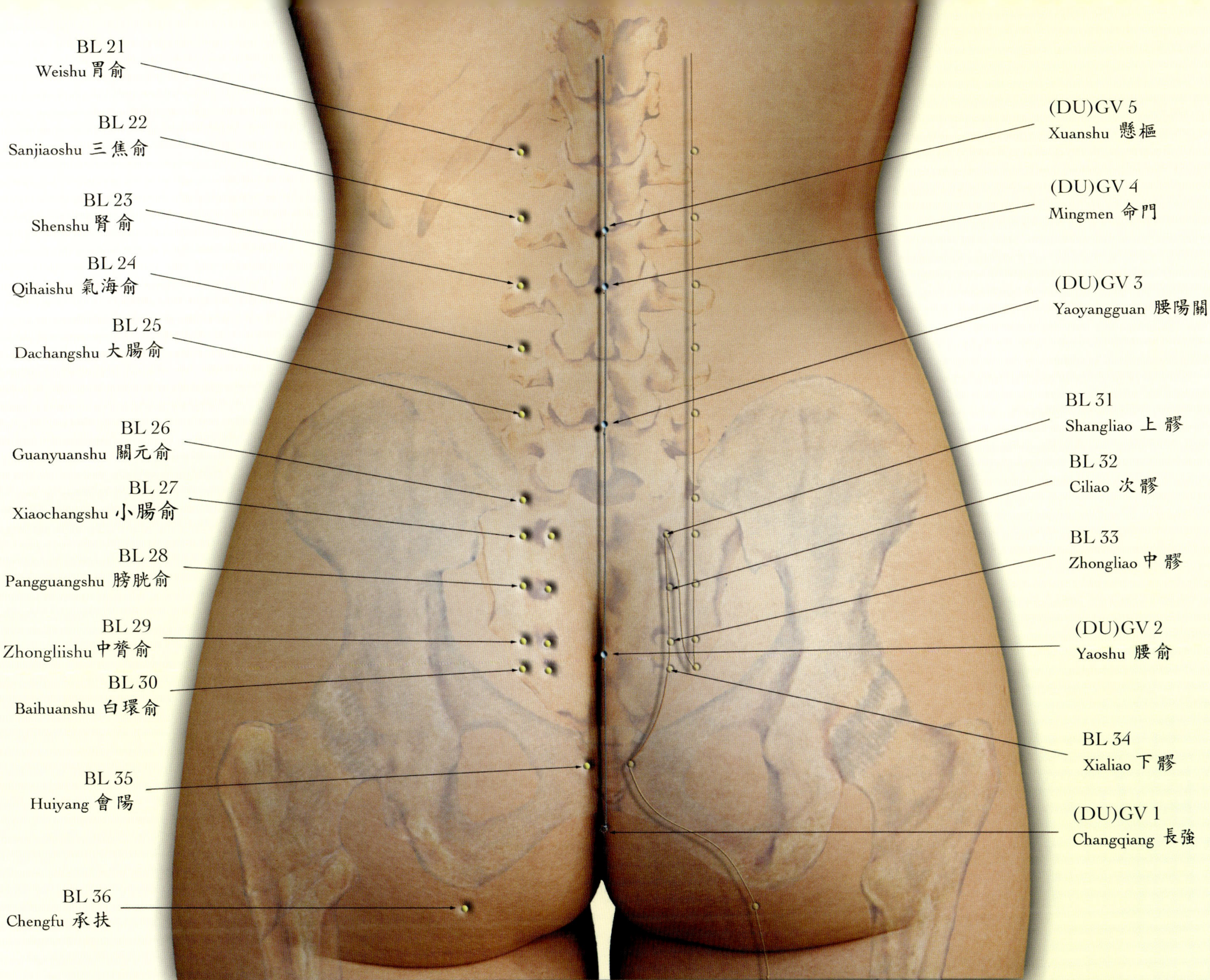
BL 21
Weishu 胃俞
BL 22
Sanjiaoshu 三焦俞
BL 23
Shenshu 腎俞
BL 24
Qihaishu 氣海俞
BL 25
Dachangshu 大腸俞
BL 26
Guanyuanshu 關元俞
BL 27
Xiaochangshu 小腸俞
BL 28
Pangguangshu 膀胱俞
BL 29
Zhongliishu 中膂俞
BL 30
Baihuanshu 白環俞
BL 35
Huiyang 會陽
BL 36
Chengfu 承扶
(DU)GV 5
Xuanshu 懸樞
(DU)GV 4
Mingmen 命門
(DU)GV 3
Yaoyangguan 腰陽關
BL 31
Shangliao 上髎
BL 32
Ciliao 次髎
BL 33
Zhongliao 中髎
(DU)GV 2
Yaoshu 腰俞
BL 34
Xialiao 下髎
(DU)GV 1
Changqiang 長強

Zu Tai Yang channel (2nd Bladder chain)

BL 50 Weicang
On the back, below the spinous process of the 12th thoracic vertebra, 3 Cun lateral to the posterior midline.

BL 51 Huangmen
On the low back, below the spinous process of the 1st lumbar vertebra, 3 Cun lateral to the posterior midline.

BL 52 Zhishi
On the low back, below the spinous process of the 2nd lumbar vertebra, 3 Cun lateral to the posterior midline.

BL 53 Baohuang
On the buttock and on the level of the 2nd posterior sacral foramen, 3 Cun lateral to the median sacral crest.

BL 54 Zhibian
On the buttock and on the level of the 4th posterior sacral foramen, 3 Cun lateral to the median sacral crest.

Zu Shao Yang channel (Gall Bladder)

GB 25 Jingmen
On the lateral side of the waist, 1.8 Cun posterior to (LR)LV 13 Zhangmen, below the free end of the 12th rib.

GB 30 Huantiao
On the lateral side of the thigh, at the junction of the middle third and lateral third of the line connecting the prominence of the great trochanter and the sacral hiatus, when the patient is in a lateral recumbent position with the thigh flexed.

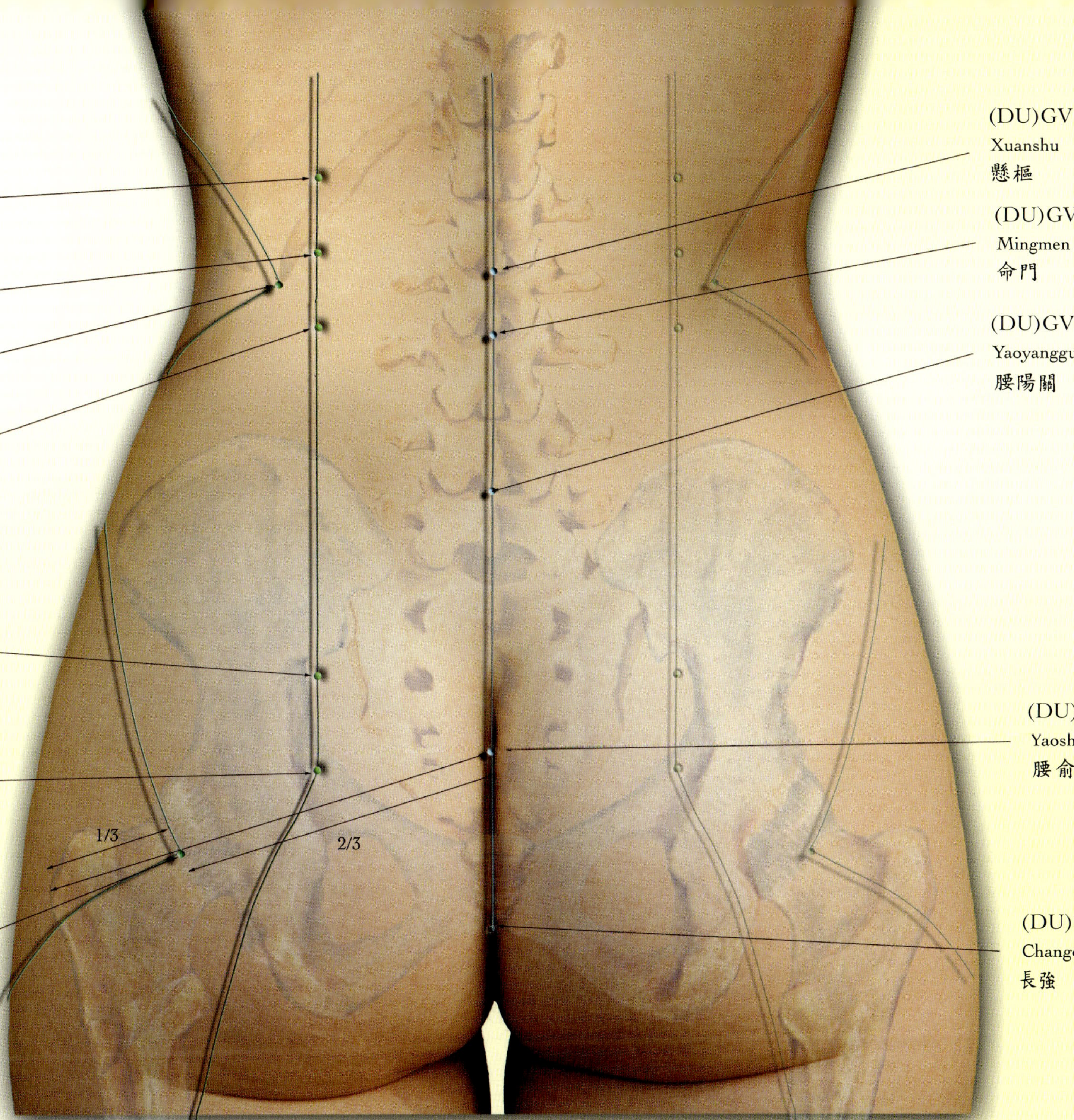
BL 50
Weigang
胃倉
BL 51
Huangmen
肓門
GB 25
Jingmen
京門
BL 52
Zhishi
志室
BL 53
Baohuang
胞肓
BL 54
Zhibian
秩邊
1/3
2/3
GB 30
Huantiao
環跳
(DU)GV 5
Xuanshu
懸樞
(DU)GV 4
Mingmen
命門
(DU)GV 3
Yaoyangguan
腰陽關
(DU)GV 2
Yaoshu
腰俞
(DU)GV 1
Changqiang
長強

Extra channel points (PC)

65 PC Jiagu
Below the spinous process of T 12.

66 PC Pigen
3.5 Cun horizontally from the spinous process of L 1.

67 PC Xuechou
Above the spinous process of L 2.

68 PC Changfeng
1 Cun horizontally from the spinous process of L 2.

69 PC Jijupikuai
4 Cun from L 2 outside the Bladder channel.

70 PC Zhuzhang
Above the spinous process of L 3.

71 PC Xiajishu
Just below the spinous process of L 3.

72 PC Huqimen
On the posterio-superior part of the iliac crest. Draw a line connecting the great trochanter on the level of GB 30 Huantiao and the tip of the coccyx. Midpoint of that line, a vertical line is drawn. The point is where that vertical line crosses the iliac crest.

73 PC Yaoyi
3 Cun horizontally from L 4.

74 PC Yaoyan
3.8 Cun horizontally and below L 4, between L 4 and the iliac crest.

75 PC Shiqizhuixia
Just below the spinous process of the 5th lumbar vertebra.

76 PC Zhonggong
3.5 Cun horizontally and below the spinous process of L 5.

77 PC Yaogen
On the external branch of the Bladder channel, 3 Cun horizontally from the spinous process of the 1st sacral vertebra.

78 PC Couyi Jiuqi
Below the spinous process of the 1st sacral vertebra.

79 PC Yaoqi
Below the spinous process of the 2nd sacral vertebra on the Du Mai vessel.

80 PC Xiazhui
Below the spinous process of the 3rd sacral vertebra on the Du Mai vessel.

81 PC Yutian
Below the 4th sacral vertebra.

82 PC Pinxueling
On the spinous process of the 5th sacral vertebra, below point 81 PC Yutian.

83 PC Tunzhong
At the apex of the equilateral triangle formed by the union of the great trochanter and of the isquion.

84 PC Huanzhong
Above the coxofemoral joint, midpoint of the line connecting points GB 30 Huantiao and (DU)GV 2 Yaoshu.

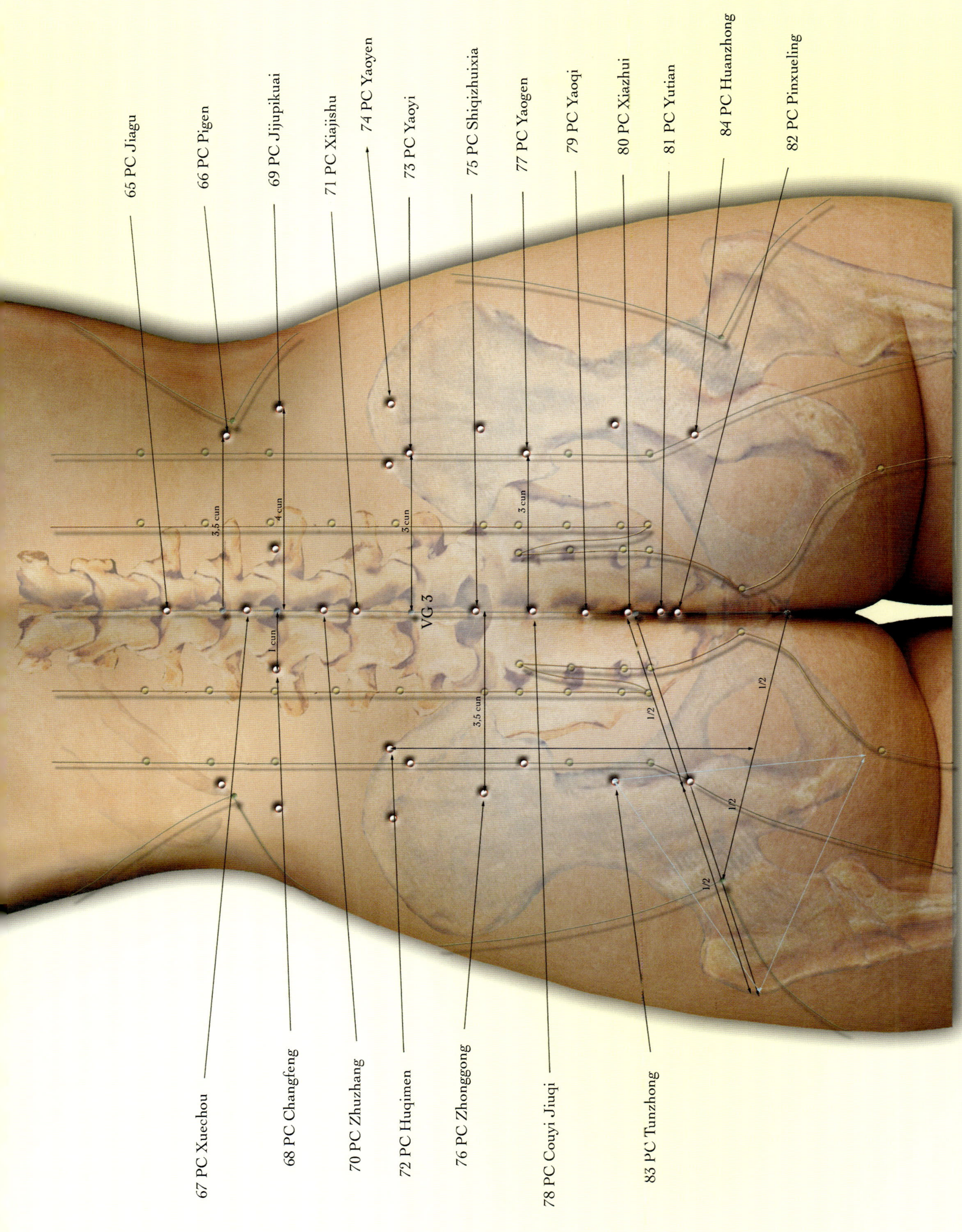
65 PC Jiagu
66 PC Pigen
69 PC Jijupikuai
71 PC Xiajishu
74 PC Yaoyen
73 PC Yaoyi
75 PC Shiqizhuixia
77 PC Yaogen
79 PC Yaoqi
80 PC Xiazhui
81 PC Yutian
84 PC Huanzhong
82 PC Pinxueling
3,5 cun
4 cun
3 cun
3 cun
VG 3
1 cun
3,5 cun
1/2
1/2
1/2
1/2
67 PC Xuechou
68 PC Changfeng
70 PC Zhuzhang
72 PC Huqimen
76 PC Zhonggong
78 PC Couyi Jiuqi
83 PC Tunzhong

Extra channel points (PN)

54 PN Kuishang Xue
2 Cun outside point BL 50 Weicang, that is to say 5 Cun outside the spinous process of the 12th thoracic vertebra.

55 PN Weishu
4.5 Cun outside and below the spinous process of the 2nd lumbar vertebra.

56 PN Shenji
0.5 Cun outside and below the spinous process of the 2nd lumbar process.

57 PN Tiao Yue
2 Cun below the iliac posterior spinous process.

58 PN Zuogu
Midpoint of the line connecting the great trochanter and the tip of the coccyx.

59 PN Pangqiang
1.5 Cun outside (DU)GV 1 Changqiang.

X-BY.23 Shenxin
0.5 Cun lateral to BL 23 Shenshu.

X-BY.24 Zhantan
2.5 Cun lateral to the inferior border of the spinous process of the 2nd lumbar vertebra.

X-BY.25 Xishang
0.5 Cun lateral to the inferior border of the spinous process of the 4th lumbar vertebra.

X-BY.31 Maigen
3 Cun lateral and 0.5 Cun below the 2nd sacral hole.

X-BY.32 Dayan
2.5 Cun lateral to and 0.5 Cun below (DU)GV 2 Yaoshu.

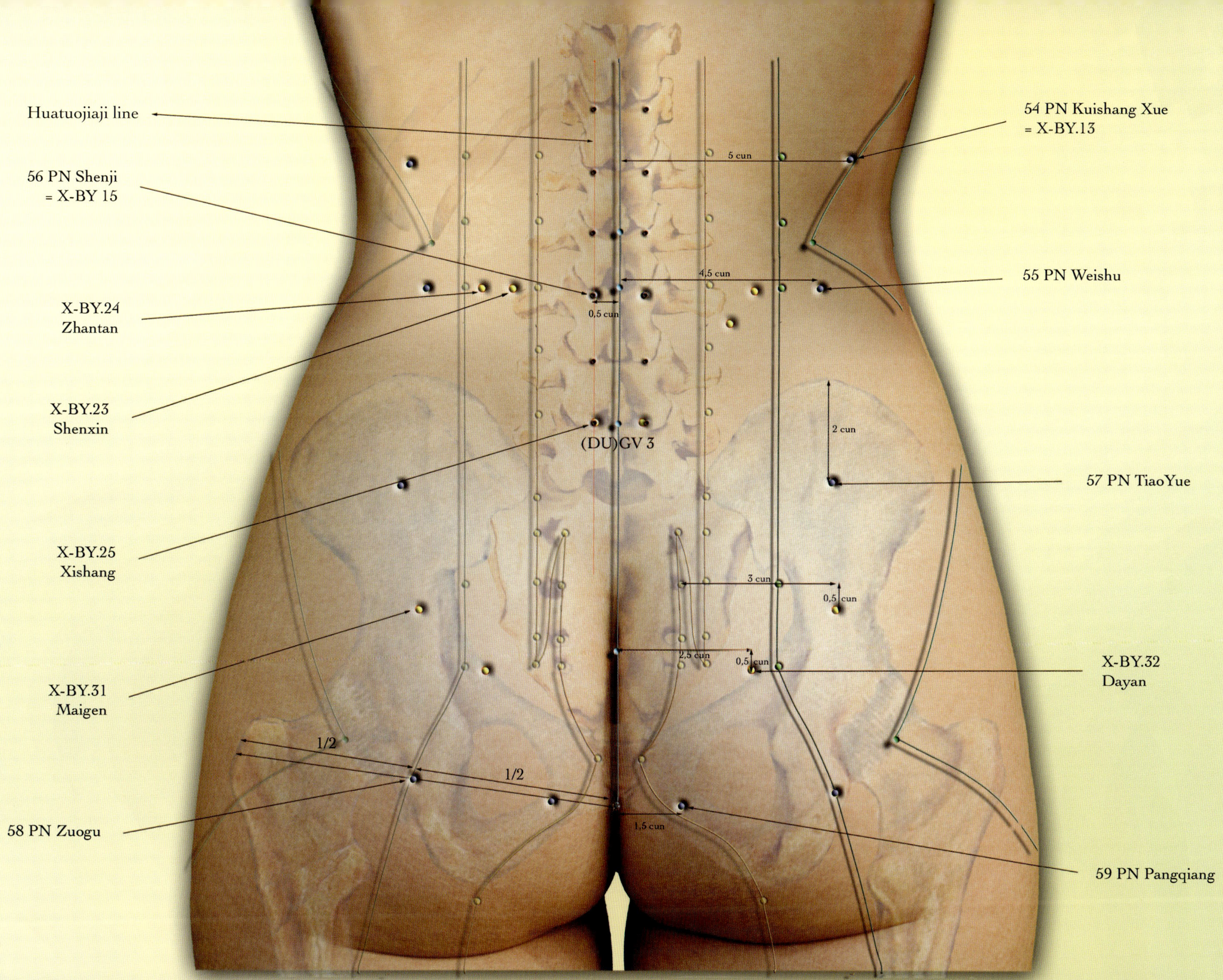
Huatuojiaji line
54 PN Kuishang Xue
= X-BY.13
5 cun
56 PN Shenji
= X-BY 15
4,5 cun
55 PN Weishu
X-BY.24
Zhantan
0,5 cun
X-BY.23
Shenxin
2 cun
(DU)GV 3
57 PN TiaoYue
X-BY.25
Xishang
3 cun
0,5 cun
X-BY.31
Maigen
2,5 cun
0,5 cun
X-BY.32
Dayan
1/2
1/2
58 PN Zuogu
1,5 cun
59 PN Pangqiang

DT.13 Shui Zhong
1.5 Cun on the lateral and external side, and below the 1st lumbar vertebra.

DT.14 Shui Fu
Between L 2 and L 3, 1.5 Cun from Du Mai vessel. Corresponds to BL 23 Shenshu.

DT.15 San Jiang
Composed of two lines of points.

In the first line, the points start below ths spinous process from L 1 to S 2:

- Fen Xian
- Shui Fen
- Shui Cheng
- Shui Guan
- Liu Zong
- Feng Chao
- Zhu Chao

The second line, bilateral, starts 3 Cun laterally to Du Mai vessel on the level of the 2nd point Shui Fen:

- Liu Wan
- Liu Man
- Liu Dao
- Hua Chao
- Huan Chao
- He Chao

DT.16 Shuang He
Corresponds to the second line of points, from Liu Wan to He Chao.

- Liu Wan
- Liu Man
- Liu Dao
- Hua Chao
- Huan Chao
- He Chao

DT.17 Chong Xiao
Found respectively below S 3, S 4, S 5:

- Miao Chao (S 3)
- Shang Tui (S 4)
- Shang Gao (S 5)

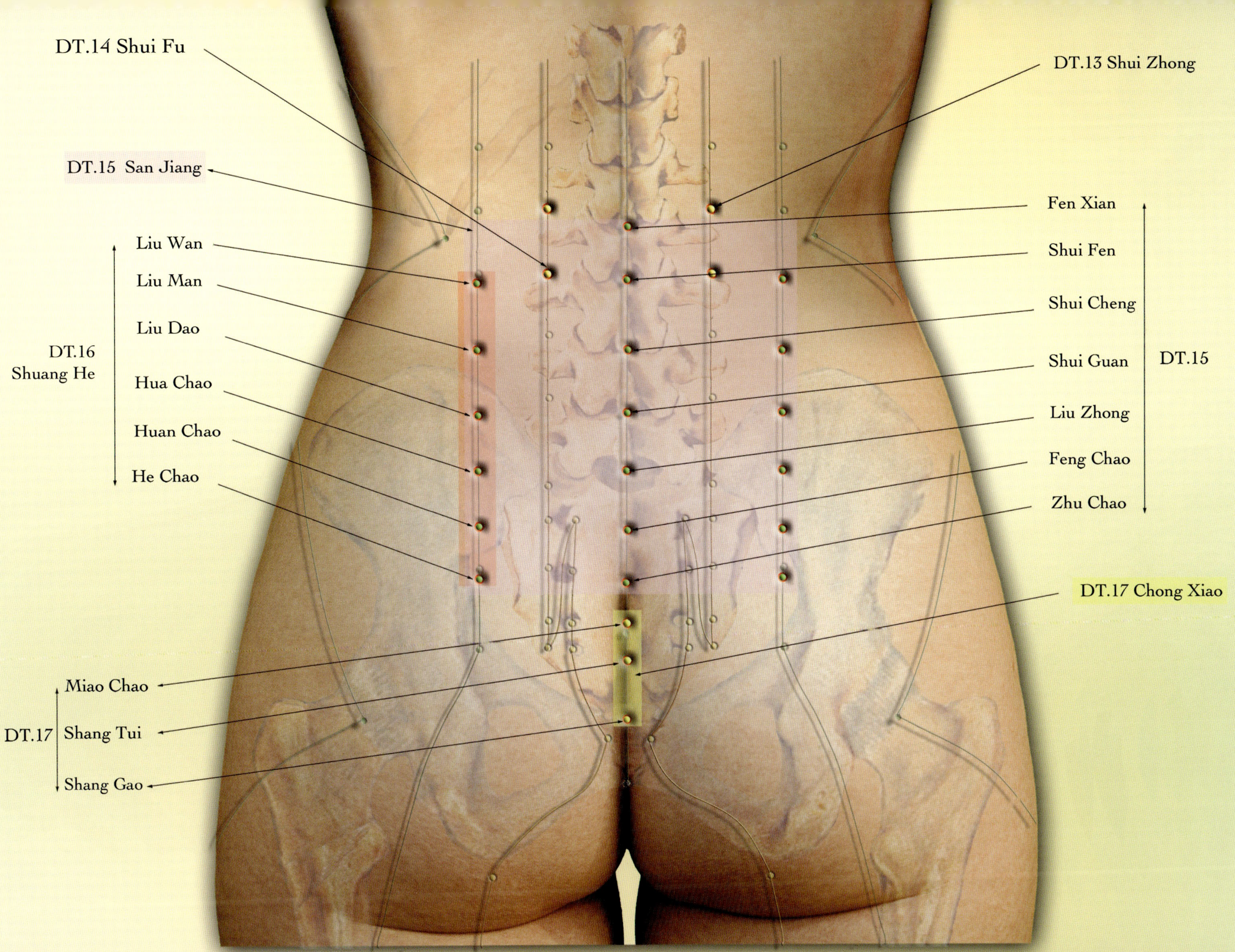
DT.14 Shui Fu
DT.15 San Jiang
Liu Wan
Liu Man
Liu Dao
Hua Chao
Huan Chao
He Chao
DT.16
Shuang He
Miao Chao
Shang Tui
Shang Gao
DT.17
DT.13 Shui Zhong
Fen Xian
Shui Fen
Shui Cheng
Shui Guan
Liu Zhong
Feng Chao
Zhu Chao
DT.15
DT.17 Chong Xiao

The thigh

- Anterior, internal, external, posterior side
  - Points of the Main channels

    Zu Yang Ming (Stomach)

    Zu Shao Yin (Kidney)

    Zu Jue Yin (Liver)

    Zu Tai Yin (Spleen)
  - Extra channel points

    Curious points

    New points

    Points of Master Tong.
- A synthesis of the points of the knee

    Anterior side

    Latero-external side

    Posterior side

Anterior side

Points of the Main channels

Zu Yang Ming channel (Stomach)

ST 30 Qichong
Slightly above the inguinal groove, 5 Cun below the centre of the umbilicus and 2 Cun lateral to the anterior midline.

ST 31 Biguan
On the anterior side of the thigh and on the line connecting the anteriosuperior iliac spine and the superiolateral corner of the patella, on the level of the perineum when the thigh is flexed, in the depression lateral to the sartorius muscle.

ST 32 Futu
On the anterior side of the thigh and on the line connecting the anterio-superior iliac spine and the superio-lateral corner of the patella, 6 Cun above this corner.

ST 33 Yinshi
On the anterior side of the thigh and on the line connecting the anteriosuperior iliac spine and the superio-lateral corner of the patella, 3 Cun above this corner.

ST 34 Liangqiu
With the knee flexed, on the anterior side of the thigh and on the line connecting the anteriosuperior iliac spine and the superio-lateral corner of the patella, 2 Cun above this corner.

Zu Tai Yin channel (Spleen)

SP 10 Xuehai
With the knee flexed, on the medial side of the thigh, 2 Cun above the superior medial corner of the patella, on the prominence of the medial head of the quadriceps muscle of the thigh.

SP 11 Jimen
On the medial side of the thigh and on the line connecting SP 10 Xuehai and SP 12 Chongmen, 6 Cun above SP 10 Xuehai.

SP 12 Chongmen
At the lateral end of the inguinal groove, 3.5 Cun lateral to the midpoint of the upper border of the symphysis pubis, lateral to the pulsating external iliac artery.

Zu Jue Yin channel (Liver)

(LR)LV 9 Yinbao
On the medial side of the thigh, 4 Cun above the medial epicondyle of the femur, between the medial vastus muscle and sartorius muscle.

(LR)LV 10 Zuwuli
On the medial side of the thigh, 3 Cun directly below ST 30 Qichong, at the proximal end of the thigh, below the pubic tubercle and on the lateral border of the long abductor muscle of the thigh.

(LR)LV 11 Yinlian
On the medial side of the thigh, 2 Cun directly below ST 30 Qichong, at the proximal end of the thigh, below the pubic tubercle and on the lateral border of the long abductor muscle of the thigh.

Zu Shao Yin channel (Kidney)

KI 11 Henngu
On the lower abdomen, 5 Cun below the umbilicus and 0.5 Cun lateral to the anterior midline.

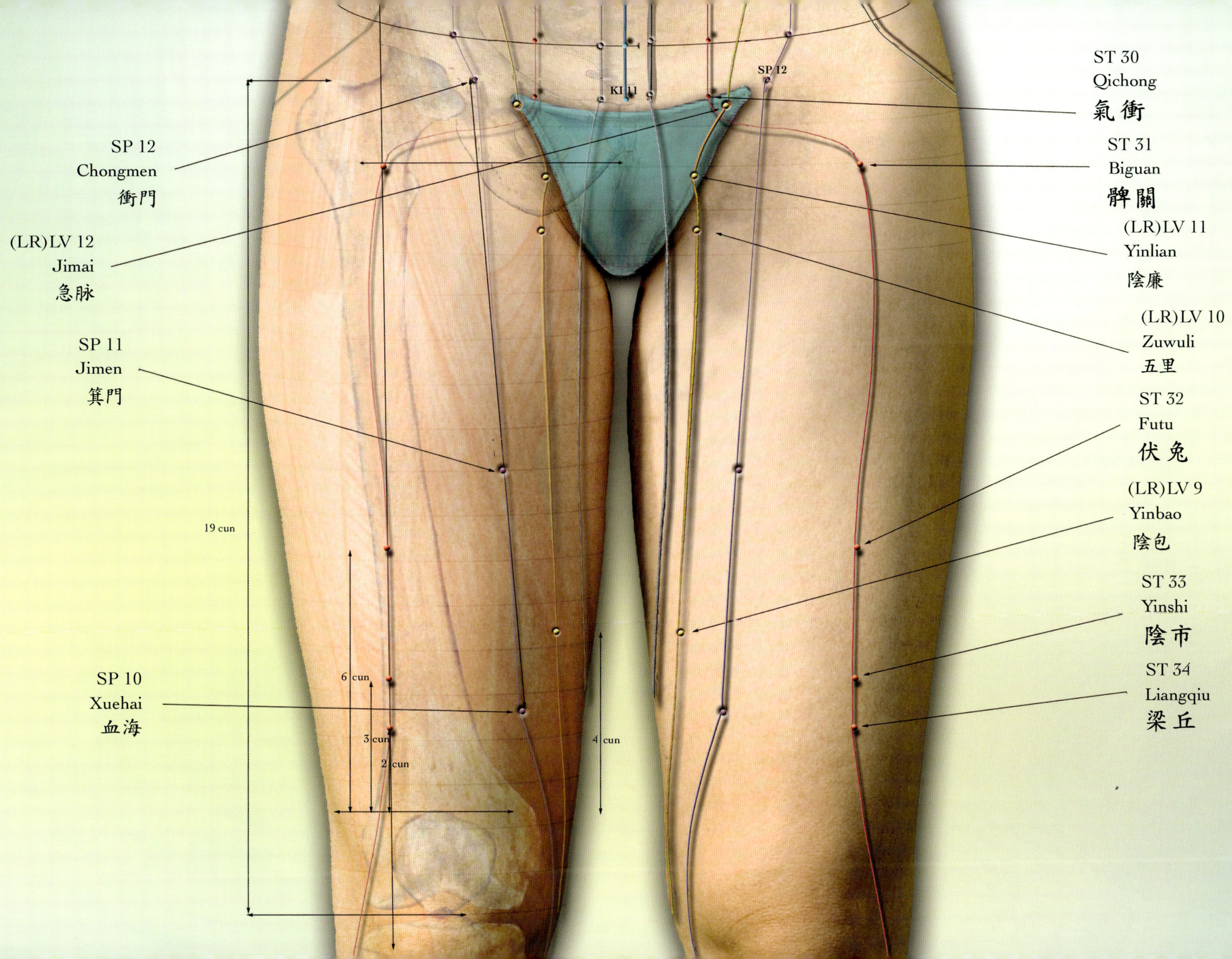
ST 30
Qichong
氣衝
ST 31
Biguan
髀關
(LR)LV 11
Yinlian
陰廉
(LR)LV 10
Zuwuli
五里
ST 32
Futu
伏兔
(LR)LV 9
Yinbao
陰包
ST 33
Yinshi
陰市
ST 34
Liangqiu
梁丘
SP 12
KI 11
SP 12
Chongmen
衝門
(LR)LV 12
Jimai
急脉
SP 11
Jimen
箕門
SP 10
Xuehai
血海
19 cun
6 cun
3 cun
2 cun
4 cun

156 PC Heding
At the midpoint of the superior border of the patella, in a depression. To find the point, have the knee flexed.

157 PC Kuangku
Right and left of point ST 34 Xialiao, 1.5 Cun from this one.

158 PC Shenxi
1 Cun below ST 32 Futu.

159 PC Quantu
On the anterior side of the knee, 0.5 Cun between ST 32 Futu and ST 31 Shangliao.

160 PC Liaoliao
On the internal side of the knee, on the most prominent part of the condyle of the femur.

161 PC Dalun
On the superiointernal border of the femur, 1 Cun from 160 PC Liaoliao.

162 PC Zuming
On the internal side of the thigh, 2 Cun above point 161 PC Dalun.

163 PC Baichongwo
On the internal side of the thigh, 1 Cun above point SP 10 Xuehai.

164 PC Zuluo
On the internal side of the thigh, 3 Cun from point 161 PC Dalun.

165 PC Changgu
On the external side of the knee, on the prominence of the condyle of the femur.

93 PN Jianxi
3 Cun above the upper border of the patella.

94 PN
4 Cun above the midpoint of the upper border of the patella.

95 PN Jixia
2 Cun below SP 11 Jimen.

96 PN Xin Futu
Futu 0.5 Cun outside ST 32 Futu.

99 PN Jiaoling
2 Cun below (LR)LV 10 Zuwuli.

105 PN Hou Xuehai
1.5 Cun below SP 10 Xuehai.

106 PN Jiejian
4 Cun above 105 PN Hou Xuehai.

X-XZ.42 Qiabinzhong
3 Cun above and 1 Cun lateral to ST 32 Futu.

X-XZ.41 Tankang
6 Cun above the superio-external angle of the knee.

X-XZ.19 Siqiang
4 Cun above the midpoint of the upper border of the patella.

X-XZ.40 Tanli
4 Cun above the superio-external angle of the patella.

X-XZ.22 Maibu
On the recto-femoral muscle, 3 Cun above ST 32 Futu.

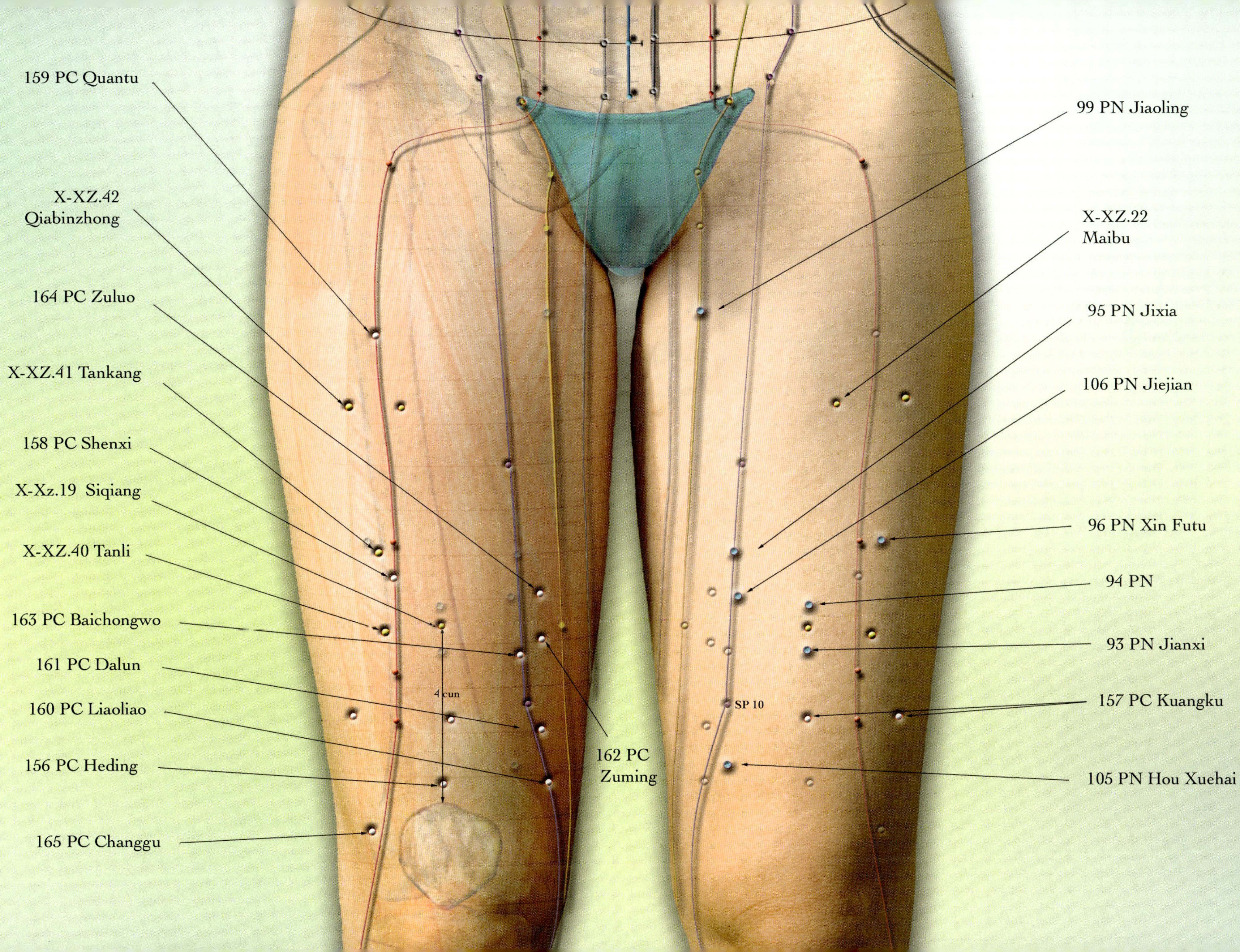

159 PC Quantu
X-XZ.42
Qiabinzhong
164 PC Zuluo
X-XZ.41 Tankang
158 PC Shenxi
X-Xz.19 Siqiang
X-XZ.40 Tanli
163 PC Baichongwo
161 PC Dalun
160 PC Liaoliao
156 PC Heding
165 PC Changgu
4 cun
162 PC
Zuming
SP 10
99 PN Jiaoling
X-XZ.22
Maibu
95 PN Jixia
106 PN Jiejian
96 PN Xin Futu
94 PN
93 PN Jianxi
157 PC Kuangku
105 PN Hou Xuehai

T 77.16 Xia Chun
1 Cun below T 77.15 Shang Chun.

T 88.01 Tong Guan
Find the midpoint of the upper border of the patella. From that point raise a line passing midpoint of the anterior side of the thigh. The point is 5 Cun above the patella.

T 88.02 Tong Shan
Find the midpoint of the upper border of the patella. From that point, raise a line passing midpoint of the anterior side of the thigh. The point is 7 Cun above the patella or 2 Cun above T 88.01 Tong Guan.

T 88.03 Tang Tian
Find the midpoint of the upper border of the patella. From that point, raise a line passing midpoint of the anterior side of the thigh. The point is 9 Cun above the patella or 2 Cun above T 88.02 Tong Shan.

T 88.04 Jie Mei Yi
First find T 88.03 Tong Tian. From that point measure 1 Cun on the latero-internal side and 1 Cun below to find point T 88.04 Jie Mei Yi.

T 88.05 Jie Mei Er
2.5 Cun above T 88.04 Jie Mei Yi.

T 88.06 Jie Mei San
2.5 Cun above T 88.05 Jie Mei Er.

T 88.07 Gan Mao Yi
First find T 88.05 Jie Mei Er. The wanted point is 0.5 Cun on the latero-internal side.

T 88.08 Gan Mao Er
First find T 88.06 Jie Mei San. The wanted point is 0.5 Cun on the latero-internal side.

T 88.09 Tong Shen
At the superio-internal angle of the patella.

T 88.10 Tong Wei
2 Cun above the superio-internal border of the patella. It is situated near SP 10 Xuehai.

T 88. 11 Tong Bei
4 Cun above the superio-internal border of the patella. 2 Cun above T 88.10 Tong Wei.

T 88.28 Jie
1 Cun above the upper border of the patella and 5 fen exterior.

T 88.29 Nei Tong Guan
Find T 88.01 Tong Guan. The wanted point is 5 fen inside on the same horizontal line.

T 88.30 Nei Tong Shan
2 Cun above T 88.29 Nei Tong Guan or 5 fen inside T 88.02 Tong Shan.

T 88.31 Neo Tong Tian
0,5 Cun Inside T 88.03 Tong Tian.

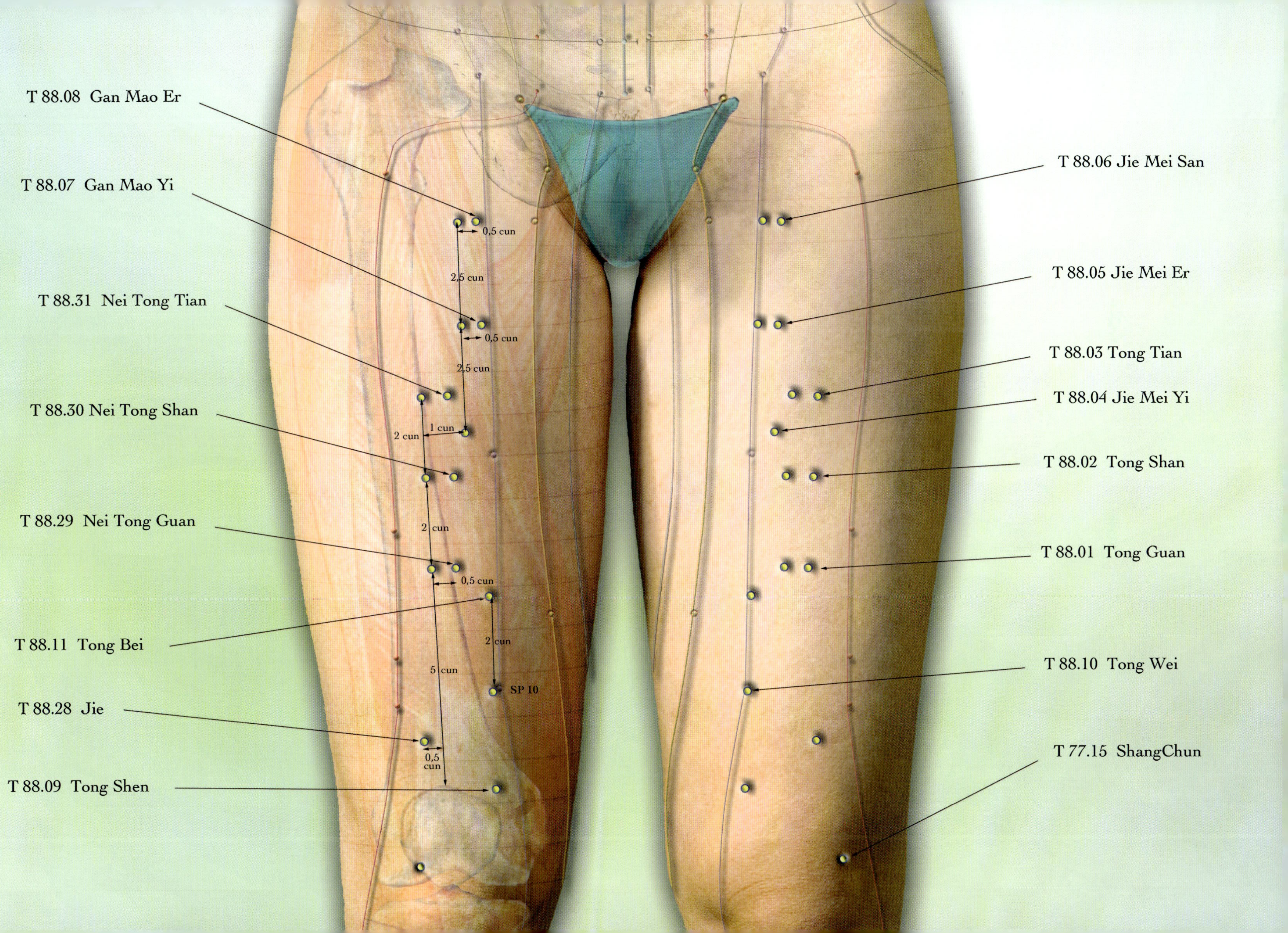

T 88.08 Gan Mao Er
T 88.07 Gan Mao Yi
T 88.31 Nei Tong Tian
T 88.30 Nei Tong Shan
T 88.29 Nei Tong Guan
T 88.11 Tong Bei
T 88.28 Jie
T 88.09 Tong Shen
0,5 cun
2,5 cun
0,5 cun
2,5 cun
1 cun
2 cun
2 cun
0,5 cun
2 cun
5 cun
SP 10
0,5 cun
T 88.06 Jie Mei San
T 88.05 Jie Mei Er
T 88.03 Tong Tian
T 88.04 Jie Mei Yi
T 88.02 Tong Shan
T 88.01 Tong Guan
T 88.10 Tong Wei
T 77.15 ShangChun

Medial side

Points of the Main channels

Zu Tai Yin channel (Spleen)

SP 10 Xuehai
With the knee flexed, on the medial side of the thigh, 2 Cun above the superior medial corner of the patella, on the prominence of the medial head of the quadriceps muscle of the thigh.

SP 11 Jimen
On the medial side of the thigh and on the line connecting SP 10 Xuehai and SP 12 Chongmen, 6 Cun above SP 10 Xuehai.

SP 12 Chongmen
At the lateral end of the inguinal groove, 3.5 Cun lateral to the midpoint of the upper border of the symphysis pubis, lateral to the pulsating external iliac artery.

Zu Jue Yin channel (Liver)

(LR)LV 8 Ququan
On the medial side of the knee, at the medial end of the popliteal crease when the knee is flexed, posterior to the medial epicondyle of the tibia, in the depression of the anterior border of the insertions of the semi-membranous and semi-tendinous muscles.

(LR)LV 9 Yinbao
On the medial side of the thigh, 4 Cun above the medial epicondyle of the femur, between the medial vastus muscle and sartorius muscle.

(LR)LV 10 Zuwuli
On the medial side of the thigh, 3 Cun directly below ST 30 Qichong, at the proximal end of the thigh, below the pubic tubercle and on the lateral border of the long abductor muscle of the thigh.

(LR)LV 11 Yinlian
On the medial side of the thigh, 2 Cun directly below ST 30 Qichong, at the proximal end of the thigh, below the pubic tubercle and on the lateral border of the long abductor muscle of the thigh.

(LR)LV 12 Jimai
Lateral to the pubic tubercle, lateral and inferior to ST 30 Qichong, in the inguinal groove where the pulsation of the femoral artery is palpable, 2.5 Cun lateral to the anterior midline.

Zu Shao Yin channel (Kidney)

KI 11 Henggu
On the lower abdomen, 5 Cun below the umbilicus and 0.5 Cun lateral to the anterior midline.

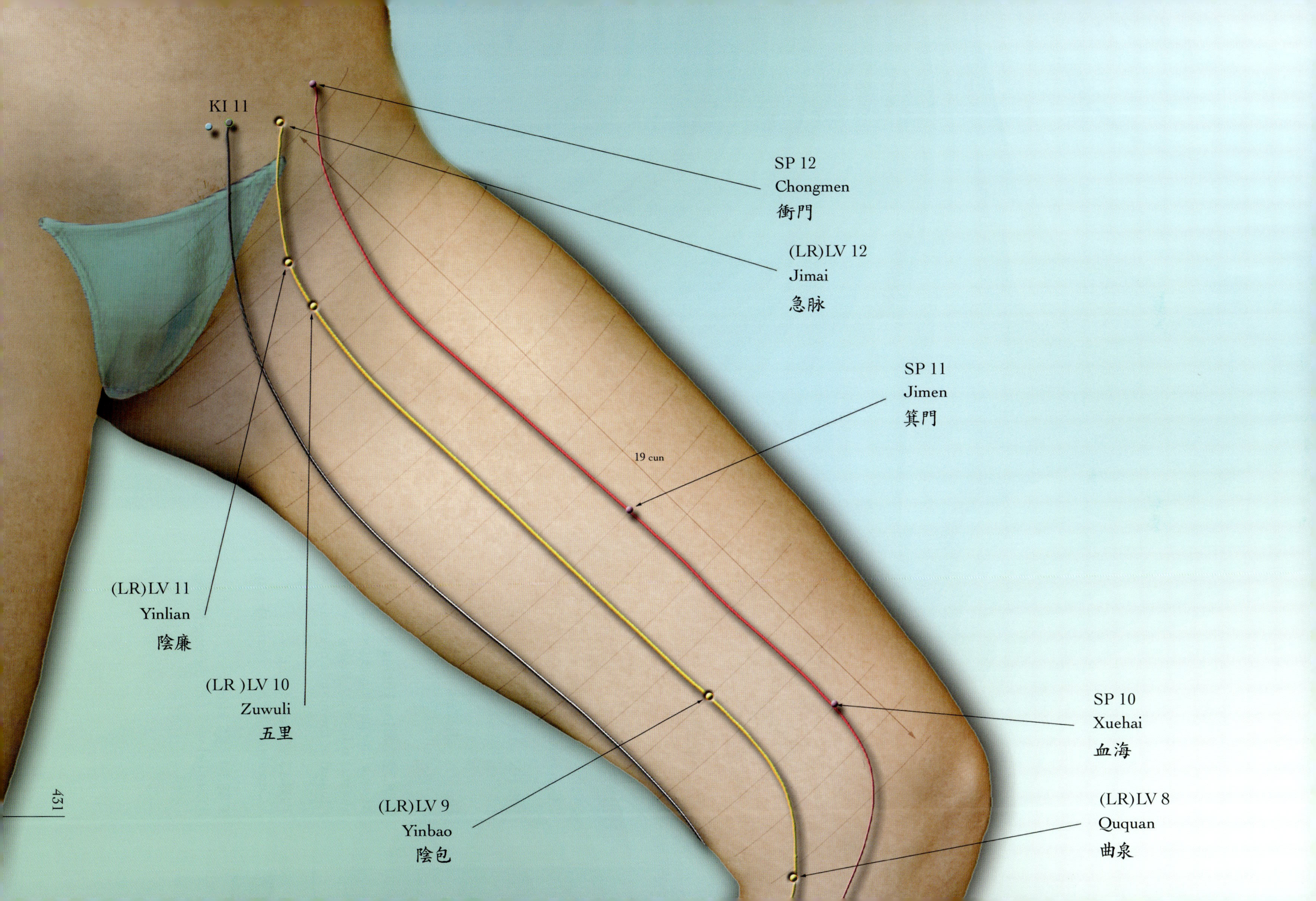

KI 11
SP 12
Chongmen
衝門
(LR)LV 12
Jimai
急脉
SP 11
Jimen
箕門
19 cun
(LR)LV 11
Yinlian
陰廉
(LR)LV 10
Zuwuli
五里
SP 10
Xuehai
血海
(LR)LV 9
Yinbao
陰包
(LR)LV 8
Ququan
曲泉

157 PC Kuangku
1.5 Cun right and left of point ST 34 Xialiao.

160 PC Liaoliao
On the internal side of the knee, on the most prominent part of the condyle of the femur.

161 PC Dalun
On the superiointernal border of the femur, 1 Cun from 160 PC Liaoliao.

162 PC Zuming
On the internal side of the thigh, 2 Cun above point 161 PC Dalun.

163 PC Baichongwo
On the internal side of the thigh, 1 Cun above point SP 10 Xuehai.

164 PC Zuluo
On the internal side of the thigh, 3 Cun from 161 PC Liaoliao.

95 PN Jixia
2 Cun below SP 11 Jimen.

99 PN Jiaoling
2 Cun below LR 10 Zuwuli.

105 PN Hou Xuehai
1.5 Cun below SP 10 Xuehai.

106 PN Jiejian
4 Cun above 105 PN Hou Xuehai.

T 88.12 Ming Huang
Raise a line passing at the midpoint of the internal side of the thigh. The point is situated at the middle of the line.

T 88.13 Tian Huang
3 Cun above T 88.12 Ming Huang.

T 88.14 Qi Huang
3 Cun below T 88.12 Ming Huang.

T 88.15 Huo Zhi
1.5 Cun above T 88.14 Qi Huang, that is to say midpoint between T 88.14 Qi Huang and T 88.12 Ming Huang.

T 88.16 Huo Quan
1.5 Cun below T 88.14 Qi Huang.

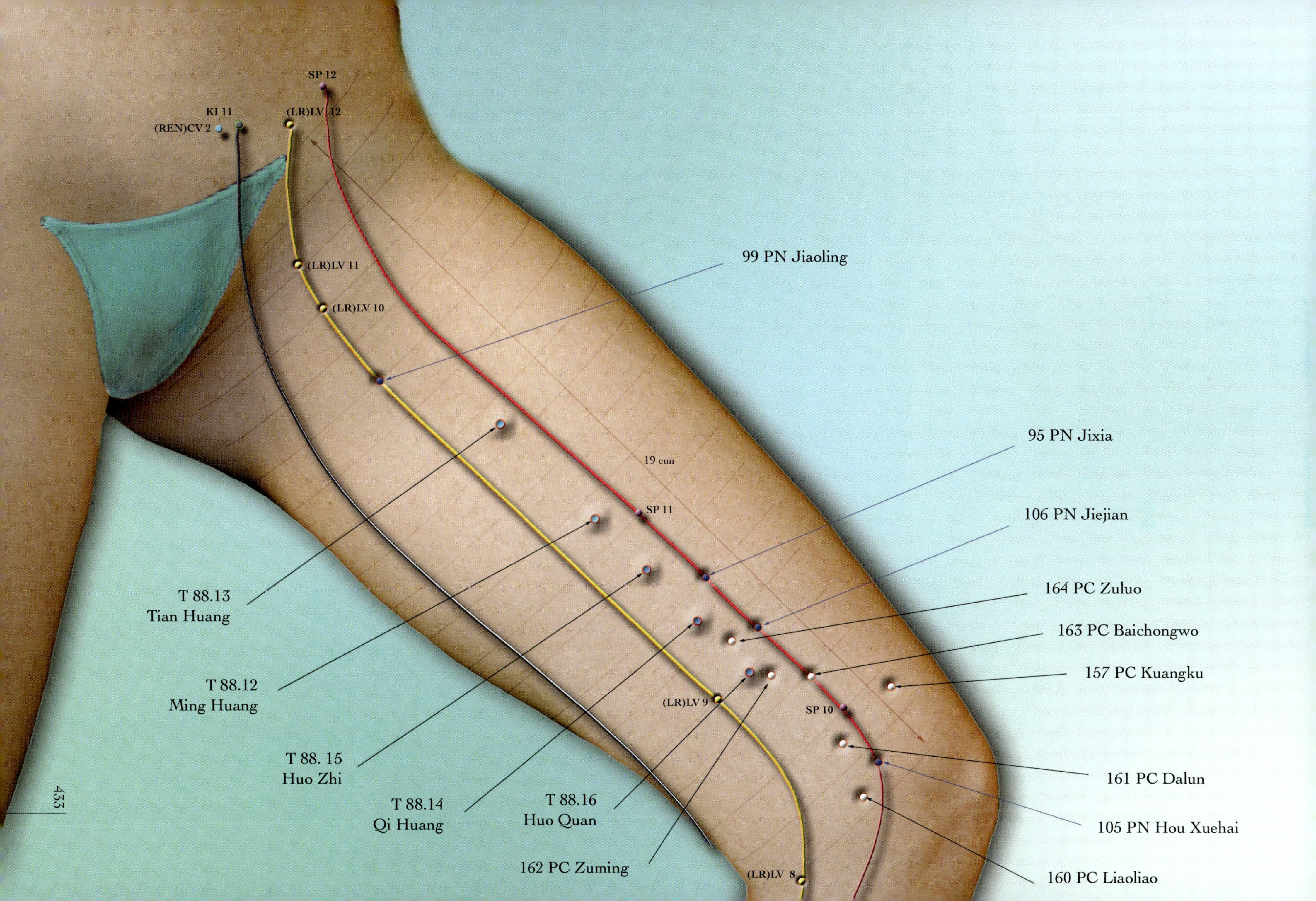
SP 12
KI 11
(REN)CV 2
(LR)LV 12
(LR)LV 11
(LR)LV 10
99 PN Jiaoling
19 cun
SP 11
95 PN Jixia
106 PN Jiejian
164 PC Zuluo
163 PC Baichongwo
157 PC Kuangku
(LR)LV 9
SP 10
161 PC Dalun
105 PN Hou Xuehai
160 PC Liaoliao
(LR)LV 8
T 88.13
Tian Huang
T 88.12
Ming Huang
T 88. 15
Huo Zhi
T 88.14
Qi Huang
T 88.16
Huo Quan
162 PC Zuming

Lateral side

Points of the Main channels

Zu Shao Yang channel (Gall Bladder)

GB 29 Juliao
At the midpoint between the anterosuperior iliac spine and the great trochanter.

GB 30 Huantiao
At the union of the external 3rd and the medial 3rd between the distance of the great trochanter and the sacred hiatus (DU)GV 2 Yaoshu, the patient being in the lateral decubitus position, the leg flexed.

GB 31 Fengshi
At the midpoint the external part of the muscle, 7 Cun above the popliteal crease. When the patient is standing with his hands hanging by his side, the point is where the tip of the middle finger is.

GB 32 Zhongdu
On the external side of the thigh, 2 Cun below GB 31 Fengshi, or at 5 Cun above the popliteal fossa, between the vastus lateralis muscle and the biceps femoris.

GB 33 Xiyangguan
On the outer edge of the knee, 3 Cun above GB 34 Yanglingquan, in the depression located above the tuberosity of the femur..

Extra channel points

Curious points

165 PC Changgu
On the external side of the knee, on the prominence of the condyle of the femur.

166 - 167 - 168 PC Yinwei
There are three points Yinwei. They are situated respectively 1,2,3 Cun above the external extremity of the flex crease of the knee.

169 PC Silian
4 Cun above the external extremity of the flex crease of the knee.

170 PC Wuling
5 Cun above the external extremity of the flex crease of the knee.

171 PC Lingbao
6 Cun above the external extremity of the flex crease of the knee.

New points

108 PN Shang Yangguan
1 Cun above GB 33 Yangguan.

109 PN Shangfengshi
2 Cun above GB 31 Fengshi.

110 PN Qianjian
2.5 Cun above GB 31 Fengshi

X-XZ.55 Jiankua
Below the iliac process, 1.5 Cun above the great trochanter.

X-XZ.56 Kuanjiu
0.5 Cun above the great trochanter.

X-XZ.57 Qiangkua
On the posterior border of the femur, 2 Cun below the great trochanter.

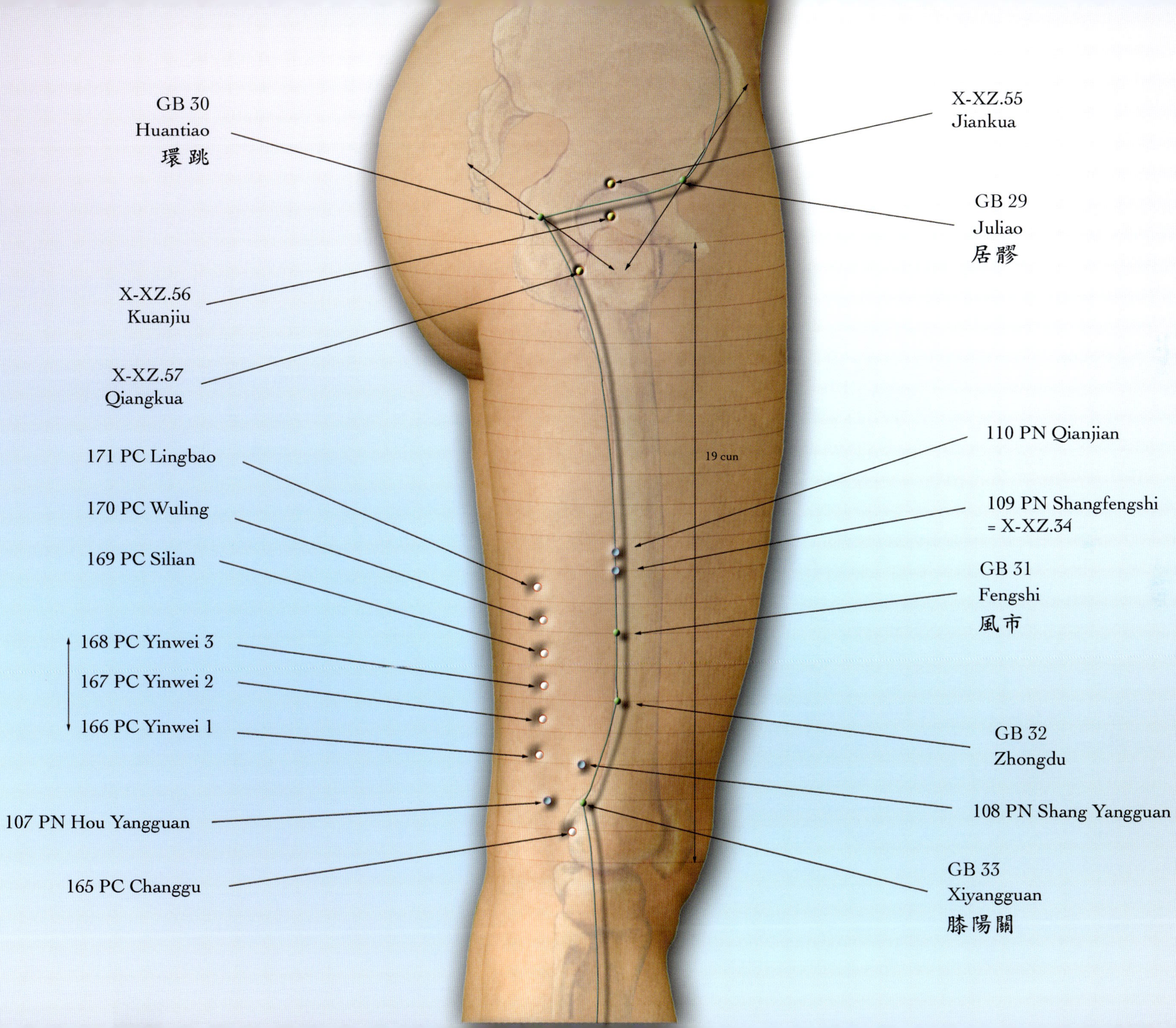
GB 30
Huantiao
環跳
X-XZ.56
Kuanjiu
X-XZ.57
Qiangkua
171 PC Lingbao
170 PC Wuling
169 PC Silian
168 PC Yinwei 3
167 PC Yinwei 2
166 PC Yinwei 1
107 PN Hou Yangguan
165 PC Changgu
X-XZ.55
Jiankua
GB 29
Juliao
居髎
19 cun
110 PN Qianjian
109 PN Shangfengshi
= X-XZ.34
GB 31
Fengshi
風市
GB 32
Zhongdu
108 PN Shang Yangguan
GB 33
Xiyangguan
膝陽關

Posterior side

Points of the Main channels

Zu Tai Yang channel (Bladder)

BL 36 Chengfu
On the posterior side of the thigh, at the midpoint of the inferior gluteal crease.

BL 37 Yinmen
On the posterior side of the thigh and on the line connecting BL 36 Chengfu and BL 40 Weizhong, 6 Cun below BL 36 Chengfu.

BL 38 Fuxi
At the lateral end of the popliteal crease, 1 Cun above BL 39 Weiyang, medial to the tendon of the biceps muscle of the thigh.

BL 39 Weiyang
At the lateral end of the popliteal crease, medial to the tendon of the biceps muscle of the thigh.

BL 40 Weizhong
At the midpoint of the popliteal crease, between the tendons of the biceps muscle of the thigh and the semi-tendinous muscle.

Extra channel points

155 PC Xiwai
At the external extremity of the crease of the knee slightly in front of point BL 39 Waiyang.

100 PN Weishang
2 Cun above BL 40 Weizhong.

101 PN Zhili
4.5 Cun above BL 40 Weizhong.

102 PN Wai Zhili
4 Cun above and 1.5 Cun outside BL 40 Weizhong.

103 PN Yinshang
2 Cun above Yinmen BL 37 Yinmen.

104 PN Yinkang
1.5 Cun inside BL 36 Chengfu.

X-XZ.45 Yinxia
At the midpoint between BL 36 Chengfu and BL 40 Weizhong.

X-XZ.46 Yangkang
1.5 Cun lateral to BL 36 Chengfu.

X-XZ.47 Xin Huantiao
3 Cun lateral to the apex of the coccyx.

X-BY.17 Zuogu
1 Cun below the midpoint between the great trochanter and the apex of the coccyx.

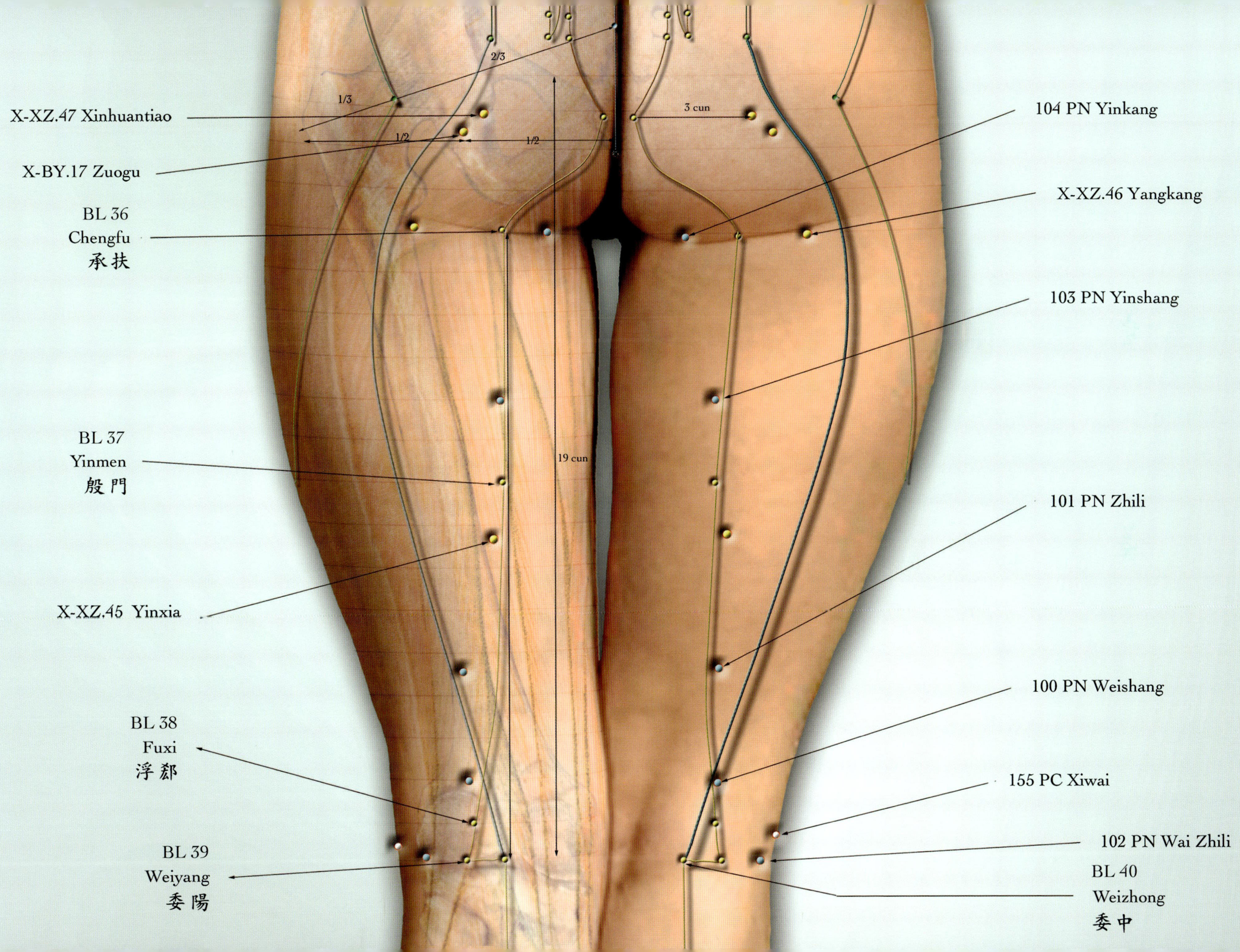

X-XZ.47 Xinhuantiao
X-BY.17 Zuogu
BL 36
Chengfu
承扶
BL 37
Yinmen
殷門
X-XZ.45 Yinxia
BL 38
Fuxi
浮郄
BL 39
Weiyang
委陽
1/3
2/3
1/2
1/2
3 cun
19 cun
104 PN Yinkang
X-XZ.46 Yangkang
103 PN Yinshang
101 PN Zhili
100 PN Weishang
155 PC Xiwai
102 PN Wai Zhili
BL 40
Weizhong
委中

A synthesis of the points of the knee

Anterior side

Points of the Main channels

Zu Yang Ming channel (Stomach)

ST 34 Liangqiu
With the knee flexed, on the anterior side of the thigh and on the line connecting the anterio-superior iliac spine and the superio-lateral corner of the patella, 2 Cun above this corner.

ST 35 Dubi
With the knee flexed, on the knee, in the depression lateral to the patella and its ligament.

ST 36 Zusanli
On the anterio-lateral side of the leg, 3 Cun below ST 35 Dubi, one finger breadth (middle finger) from the anterior crest of the tibia.

Zu Tai Yin channel (Spleen)

SP 9 Yinlingquan
On the medial side of the leg, in the depression posterior and inferior to the medial condyle of the tibia.

SP 10 Xuehai
With the knee flexed, on the inside of the thigh, 2 Cun above the base of the patella, on the prominence that marks the insertion of the quadriceps.

Extra channel points

Curious points

144 PC Xixia
On the lower border of the patella, on the patellar tendon.

145 PC Neixiyang
On the level of the articular interline of the knee, in the two fossas under the patella, on each side of the patellar tendon. The point in the external fossa coincides with ST 35 Dubi.

156 PC Heding
At the midpoint of the upper border of the patella in a fossa. To find the point, have the knee flexed.

157 PC Kuangku
Right and left of point ST 34 Xialiao, 1.5 Cun from that one.

160 PC Liaoliao
On the internal side of the knee, on the most prominent part of the condyle of the femur.

161 PC Dalun
On the superio-internal border of the femur, 1 Cun from 160 PC Liaoliao.

162 PC Zuming
On the internal side of the thigh, 2 Cun above point 161 PC Dalun.

163 PC Baichongwo
On the internal side of the thigh, 1 Cun above point SP 10 Xuehai.

165 PC Changgu
On the external side of the knee, on the prominence of the condyle of the femur.

New points

82 PN Lishang
1 Cun above ST 36 Zusanli.

105 PN Hou Xuehai
1.5 Cun below SP 10 Xuehai.

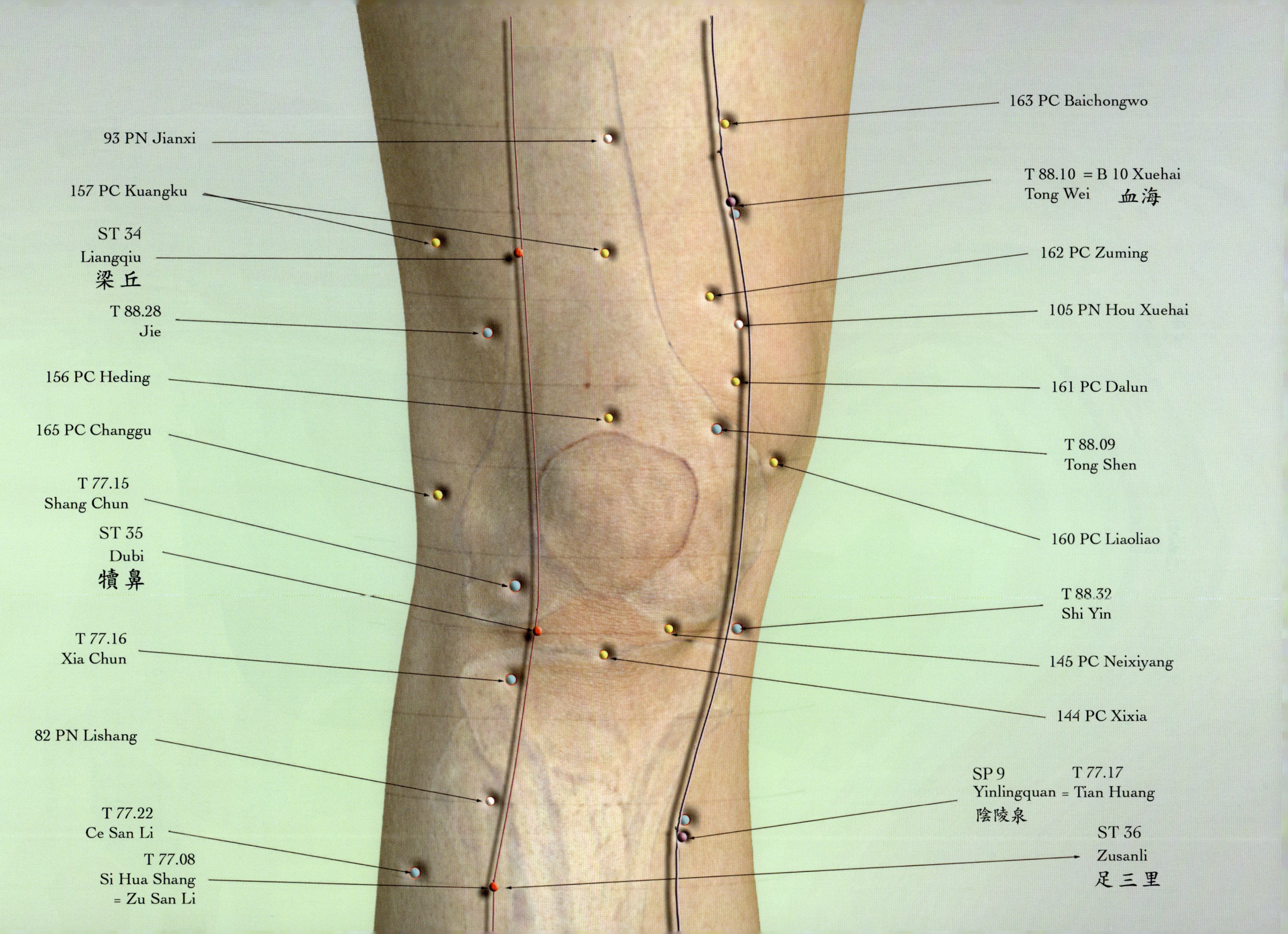

93 PN Jianxi
157 PC Kuangku
ST 34
Liangqiu
梁丘
T 88.28
Jie
156 PC Heding
165 PC Changgu
T 77.15
Shang Chun
ST 35
Dubi
犢鼻
T 77.16
Xia Chun
82 PN Lishang
T 77.22
Ce San Li
T 77.08
Si Hua Shang
= Zu San Li
163 PC Baichongwo
T 88.10 = B 10 Xuehai
Tong Wei 血海
162 PC Zuming
105 PN Hou Xuehai
161 PC Dalun
T 88.09
Tong Shen
160 PC Liaoliao
T 88.32
Shi Yin
145 PC Neixiyang
144 PC Xixia
SP 9
Yinlingquan = Tian Huang
陰陵泉
T 77.17
ST 36
Zusanli
足三里

Points of Master Tong

T 77.08 Si Hua Shang
3 Cun below ST 35 Dubi and one finger breadth outside the anterior border of the tibia. Corresponds to ST 36 Zusanli.

T 77.15 Shang Chun
On the latero-external border of the patella, at the insertion of the patellar ligament.

T 77.16 Xia Chun
1 Cun below T 77.15 Shang Chun.

T 77.17 Tian Huang or Shen Guan
On the inferior border of the internal condyle of the tibia, under the process of the tibia. Corresponds to SP 9 Yinlingquan.

T 77.22 Ce San Li
1.5 Cun latero-externally from T 77.08 Si Hua Shang, that is to say ST 36 Zusanli. This point is situated on the Gall Bladder channel.

T 88.09 Tong Shen
On the superio-internal angle of the patella.

T 88.10 Tong Wei
2 Cun above the superio-internal border of the patella. It is situated near SP 10 Xuehai.

T 88.28 Jie Se
1 Cun above the superior border of the patella and 5 fen on the external side.

T 88.32 Shi Yin
The first point is at the midpoint of the knee joint on its internal side.
The second is 2 Cun below.

Latero-external and posterior side

Points of the Main channels

Zu Shao Yang channel (Gall Bladder)

GB 33 Xiyangguan
On the lateral side of the knee, 3 Cun above GB 34 Yanglingquan, in the tendon of the biceps muscle of the thigh.

GB 34 Yanglingquan
On the lateral side of the leg, in the depression anterior and inferior to the head of the fibula.

Zu Tai Yang channel (Bladder)

BL 38 Fuxi
At the external side of the popliteal crease, 1 Cun above BL 39 Weiyang, medial to the tendon of the biceps muscle of the thigh.

BL 39 Weiyang
On the outer edge of the popliteal fossa and the inner edge of the tendon of the biceps femoris.

BL 40 Weizhong
At the midpoint of the popliteal crease, between the tendons of the biceps muscle of the thigh and the semi-tendinous muscle.

BL 55 Heyang
On the posterior side of the leg and on the line connecting BL 40 Weizhong and BL 57 Chengshan, 2 Cun below BL 40 Weizhong.

Zu Shao Yin channel (Kidney)

KI 10 Yingu
On the medial side of the popliteal fossa, between the tendons of the semi-tendinous and semi-membranous muscles, when the knee is flexed.

Zu Jue Yin channel (Liver)

(LR)LV 7 Xiguan
On the medial side of the leg, posterior and inferior to the medial epicondyle of the tibia, 1 Cun posterior to SP 9 Yinlingquan, at the upper end of the medial head of the gastrocnemius muscle.

(LR)LV 8 Ququan
On the medial side of the knee, at the medial end of the popliteal crease when the knee is flexed, posterior to the medial epicondyle of the tibia, in the depression of the anterior border of the insertions of the semi-membranous and semi-tendinous muscles.

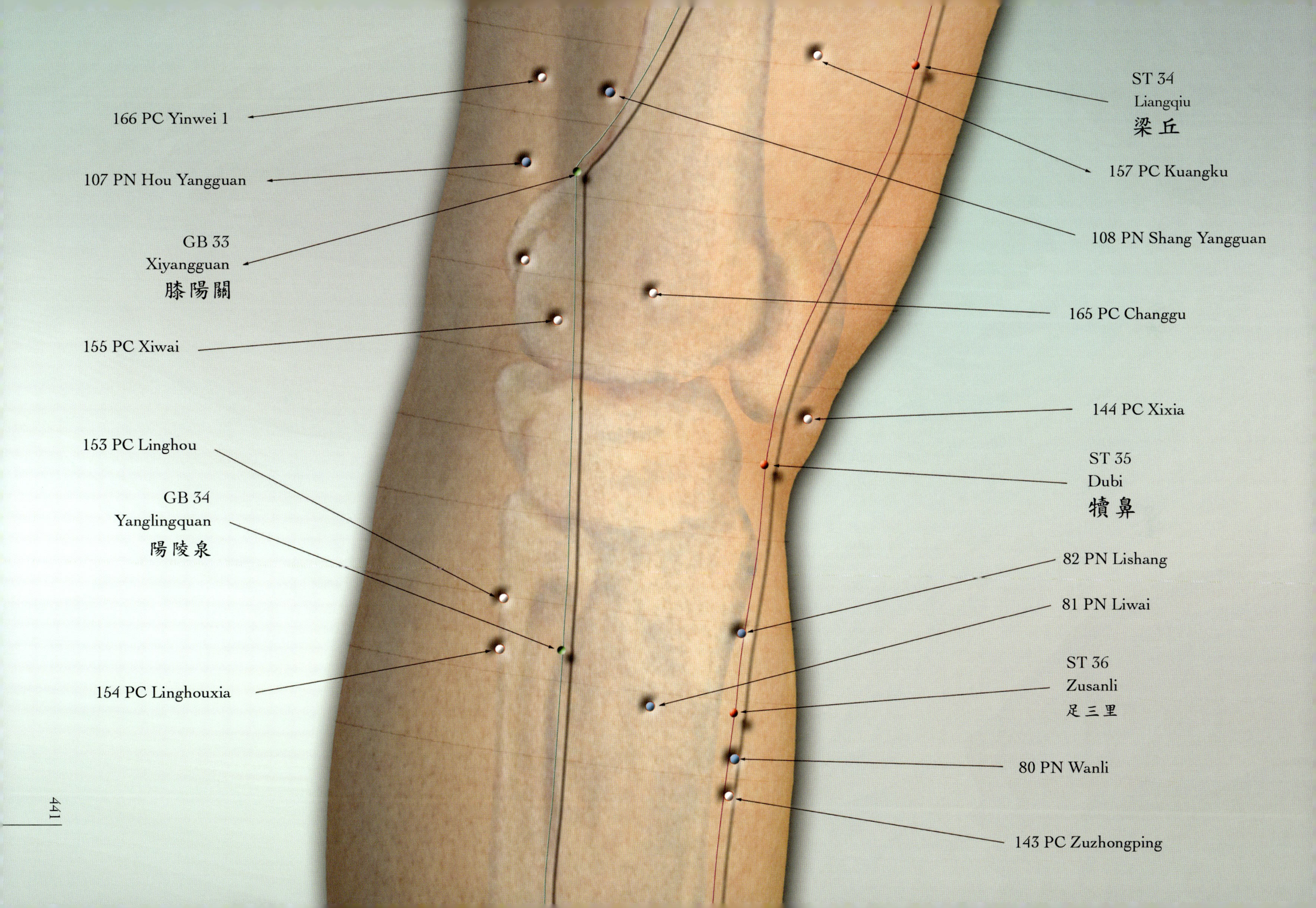

166 PC Yinwei 1
107 PN Hou Yangguan
GB 33
Xiyangguan
膝陽關
155 PC Xiwai
153 PC Linghou
GB 34
Yanglingquan
陽陵泉
154 PC Linghouxia
ST 34
Liangqiu
梁丘
157 PC Kuangku
108 PN Shang Yangguan
165 PC Changgu
144 PC Xixia
ST 35
Dubi
犢鼻
82 PN Lishang
81 PN Liwai
ST 36
Zusanli
足三里
80 PN Wanli
143 PC Zuzhongping

Curious points

143 PC Zu Zhong Ping
1 Cun above ST 36 Zusanli.

153 PC Ling Hou
In a depression, below the head of the fibula, on the posterior border. On the line of GB 34 Yanglingquan, 1 Cun behind it.

154 PC Ling Hou Xia
0.5 Cun below 153 PC Ling Hou.

155 PC Kiwai
On the external extremity of the knee flex crease, slightly in front of point BL 39 Waiyang.

166 - 167 - 168 PC Yinwei
There are 3 Yinwei points. They are respectively 1, 2, 3 Cun above the external extremity of the flex crease of the knee.

New points

80 PN Wanli
3 Cun from 151 PC Wai Huai Jian, that is to say 0.5 Cun below ST 36 Zusanli.

81 PN Li Wai
1 Cun outside ST 36 Zusanli.

100 PN Wei Shang
2 Cun above BL 40 Weizhong.

107 PN Hou Yangguan
1 Cun behind GB 33 Yangguan.

108 PN Shang Yangguan
1 Cun above GB 33 Yangguan

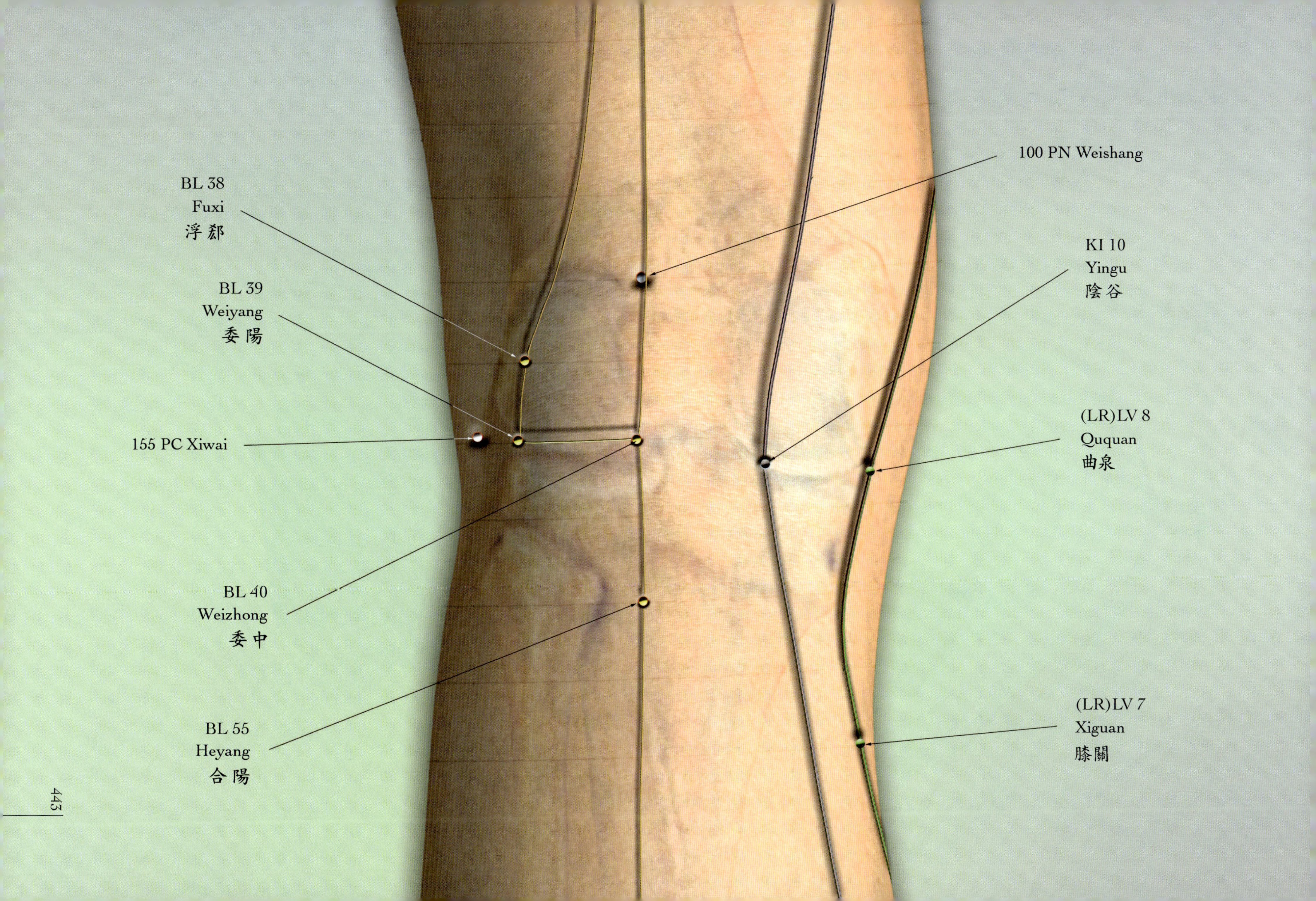
BL 38
Fuxi
浮郄
BL 39
Weiyang
委陽
155 PC Xiwai
BL 40
Weizhong
委中
BL 55
Heyang
合陽
100 PN Weishang
KI 10
Yingu
陰谷
(LR)LV 8
Ququan
曲泉
(LR)LV 7
Xiguan
膝關

The leg

- Anterior, posterior, external, internal side
  - Points of the Main channels
    - Zu Yang Ming (Stomach)
    - Zu Jue Yin (Liver)
    - Zu Tai Yin (Spleen)
  - Extra channel points
    - Curious points
    - New points
    - Points of Master Tong

Anterior side

Points of the Main channels

Zu Yang Ming channel (Stomach)

ST 35 Dubi
With the knee flexed, on the knee, in the depression lateral to the patella and its ligament.

ST 36 Zusanli
On the anterio-external side of the leg, 3 Cun below ST 35 Dubi, one finger breadth (middle finger) from the anterior crest of the tibia.

ST 37 Shangjuxu
On the anterio-external side of the leg, 6 Cun below ST 35 Dubi, one finger breadth (middle finger) from the anterior crest of the tibia.

ST 38 Tiaokou
On the anterio-external side of the leg, 8 Cun below ST 35 Dubi, one finger breadth (middle finger) from the anterior crest of the tibia.

ST 39 Xiajuxu
On the anteriolateral side of the leg, 9 Cun below ST 35 Dubi, one finger breadth (middle finger) from the anterior crest of the tibia.

ST 40 Fenlong
On the anterio-external side of the leg, 8 Cun above the tip of the external malleolus, lateral to ST 38 Tiaokou, and two finger breadths (middle finger) from the anterior crest of the tibia.

ST 41 Jiexi
In the central depression of the crease between the instep of the foot and the leg, between the tendons of the long extensor muscle of the big toe and the long extensor muscle of the toes.

Zu Tai Yin channel (Spleen)

SP 5 Shangqiu
In the depression anterior and inferior to the medial malleolus, at the midpoint of the line connecting the tuberosity of the navicular bone and the tip of the medial malleolus.

SP 6 Sanyinjiao
On the medial side of the leg, 3 Cun above the tip of the medial malleolus, posterior to the medial border of the tibia.

SP 7 Lougu
On the medial side of the leg and on the line connecting the tip of the medial malleolus and SP 9 Yinlingquan, 6 Cun from the tip of the medial malleolus, posterior to the medial border of the tibia.

SP 8 Diji
On the medial side of the leg and on the line connecting the tip of the internal malleonus and SP 9 Yinlingquan, 3 Cun below SP 9 Yinlingquan.

SP 9 Yinlingquan
On the medial side of the leg, in the depression posterior and inferior to the medial condyle of the tibia.

Zu Jue Yin channel (Liver)

(LR)LV 4 Zhongfeng
On the instep of the foot, anterior to the medial malleolus, on the line connecting SP 5 Shangqiu and ST 41 Jiexi, in the depression medial to the tendon of the anterior tibial muscle.

(LR)LV 5 Ligou
On the medial side of the leg, 5 Cun above the tip of the medial malleolus, on the midline of the medial surface of the tibia.

(LR)LV 6 Zhongdu
On the medial side of the leg, 7 Cun above the tip of the medial malleolus, on the midline of the medial surface of the tibia.

(LR)LV 7 Xiguan
On the medial side of the leg, posterior and inferior to the internal epicondyle of the tibia, 1 Cun posterior to SP 9 Yinlingquan, at the upper end of the medial head of the gastrocnemius muscle.

(LR)LV 8 Ququan
On the medial side of the knee, at the medial end of the popliteal crease when the knee is flexed, posterior to the medial epicondyle of the tibia, in the depression of the anterior border of the insertions of the semi-membranous and semi-tendinous muscles.

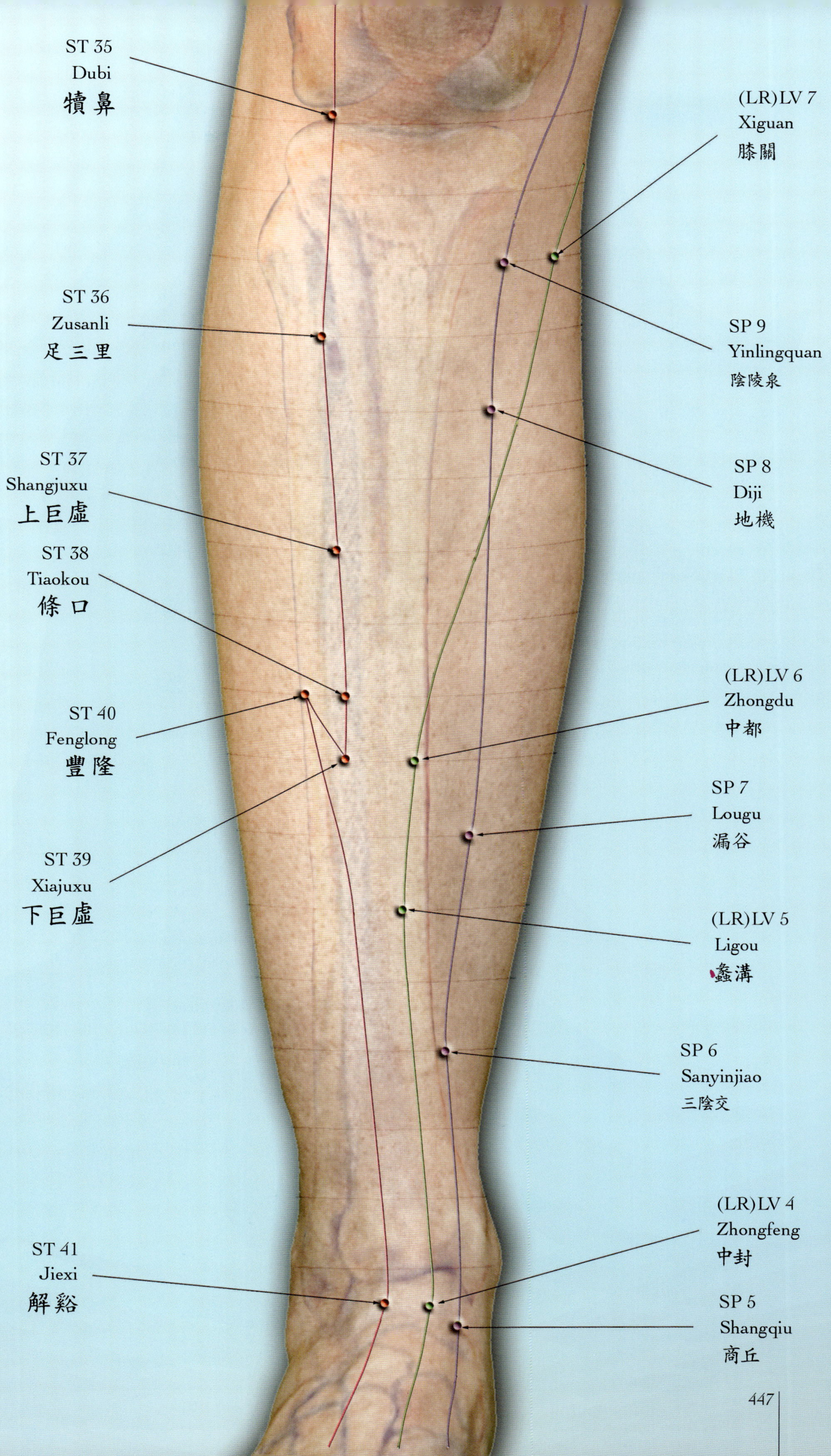

ST 35
Dubi
犢鼻
(LR) LV 7
Xiguan
膝關
ST 36
Zusanli
足三里
SP 9
Yinlingquan
陰陵泉
ST 37
Shangjuxu
上巨虛
SP 8
Diji
地機
ST 38
Tiaokou
條口
ST 40
Fenglong
豐隆
(LR) LV 6
Zhongdu
中都
SP 7
Lougu
漏谷
ST 39
Xiajuxu
下巨虛
(LR) LV 5
Ligou
蠡溝
SP 6
Sanyinjiao
三陰交
ST 41
Jiexi
解谿
(LR) LV 4
Zhongfeng
中封
SP 5
Shangqiu
商丘

Curious points

142 PC Lanwei
On the anterior side of the leg, 2 Cun below ST 36 Zusanli. According to certain writers, it would exist only on the right side.

143 PC Zuzhongping
1 Cun below ST 36 Zusanli.

144 PC Xixia
On the inferior border the patella, on the patellar tendon.

145 PC Neixiyang
On the level of the articular interline of the knee, in the two fossas below the patella, on each side of the patellar tendon. The point of the external fossa coincides with ST 35 Dubi.

New Points

78 PN Naoqing
2 Cun above ST 41 Jiexi, on the external border of the inferior extremity of the tibia.

79 PN Jingxia
3 Cun above ST 41 Jiexi and 1 Cun outside the external border of the extremity of the tibia, on the side of the fibula.

80 PN Wanli
3 Cun from 151 PC Waihuaijian, that is to say 0.5 Cun below ST 36 Zusanli.

81 PN Liwai
1 Cun outside ST 36 Zusanli.

82 PN Lishang
1 Cun above ST 36 Zusanli.

89 PN Juiwai Fan
0.5 Cun below SP 6 Sanyinjiao.

90 PN Dijian
1 Cun below SP 8 Diji.

X-XZ.49 Yiniao
1 Cun above SP 6 Sanyinjiao.

X-XZ.50 Wuanmian
1.5 Cun above SP 6 Sanyinjiao.

X-XZ.38 Liuzhitan
1.5 Cun below PC 142 - J-XZ.13 Lanwei.

X-XZ.59 Erliban
0.5 Cun above ST 36 Zusanli.

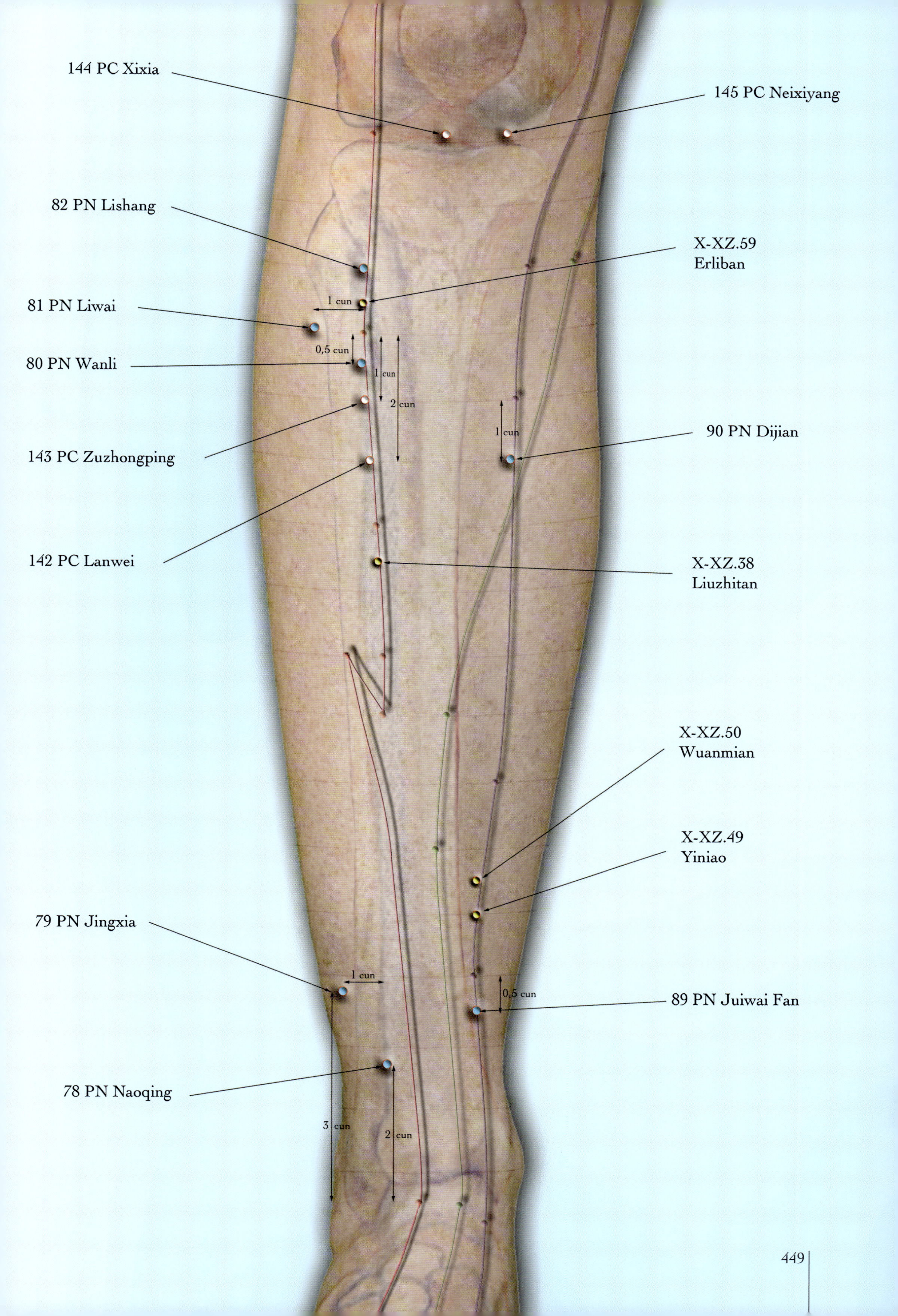
144 PC Xixia
145 PC Neixiyang
82 PN Lishang
X-XZ.59
Erliban
1 cun
81 PN Liwai
0,5 cun
80 PN Wanli
1 cun
2 cun
1 cun
90 PN Dijian
143 PC Zuzhongping
142 PC Lanwei
X-XZ.38
Liuzhitan
X-XZ.50
Wuanmian
X-XZ.49
Yiniao
79 PN Jingxia
1 cun
0,5 cun
89 PN Juiwai Fan
78 PN Naoqing
3 cun
2 cun

T 77.08 Si Hua Shang
3 Cun below ST 35 Dubi and one finger breadth outside the anterior border of the tibia. Corresponds to ST 36 Zusanli.

T 77.09 Si Hua Zhong
4.5 Cun distal from T 77.08 Si Hua Shang
It is situated 5 fen above ST 38 Tiaokou.

T 77.10 Si Hua Fu
2.5 Cun below T 77.09 Si Hua Zhong.
It is situated 1 Cun below ST 39 Xiajuxu.

T 77.11 Si Hua Xia
2.5 Cun below T 77.10 Si Hua Fu.

T 77.12 Fu Chang
1.5 Cun above T 77.11 Si Hua Xia.

T 77.13 Si Hua Li
1.2 Cun internal from T 77.09 Si Hua Zhong.

T 77.14 Si Hua Wai
1.5 Cun external from T 77.09 Si Hua Zhong.

T 77.15 Shang Chun
On the inferior, latero external border of the patella, where the patellar ligament attaches itself.

T 77.16 Xia Chun
1 Cun below T 77.15 Shang Chun.

T 77.17 Tian Huang or Shen Guan
On the inferior border of the internal condyle of the tibia, under the crest of the tibia. Corresponds to SP 9 Yinlingquan.

T 77.18 Tian Huang Fu
1.5 Cun below SP 9 Yinlingquan.

T 77.22 Ce San Li
Situated 1.5 Cun on the latero external side from T 77.08 Si Hua Shang, that is to say ST 36 Zusanli. This point is on the Gall Bladder channel.

T 77.23 Ce Xia San Li
2 Cun below T 77.22 Ce San Li.

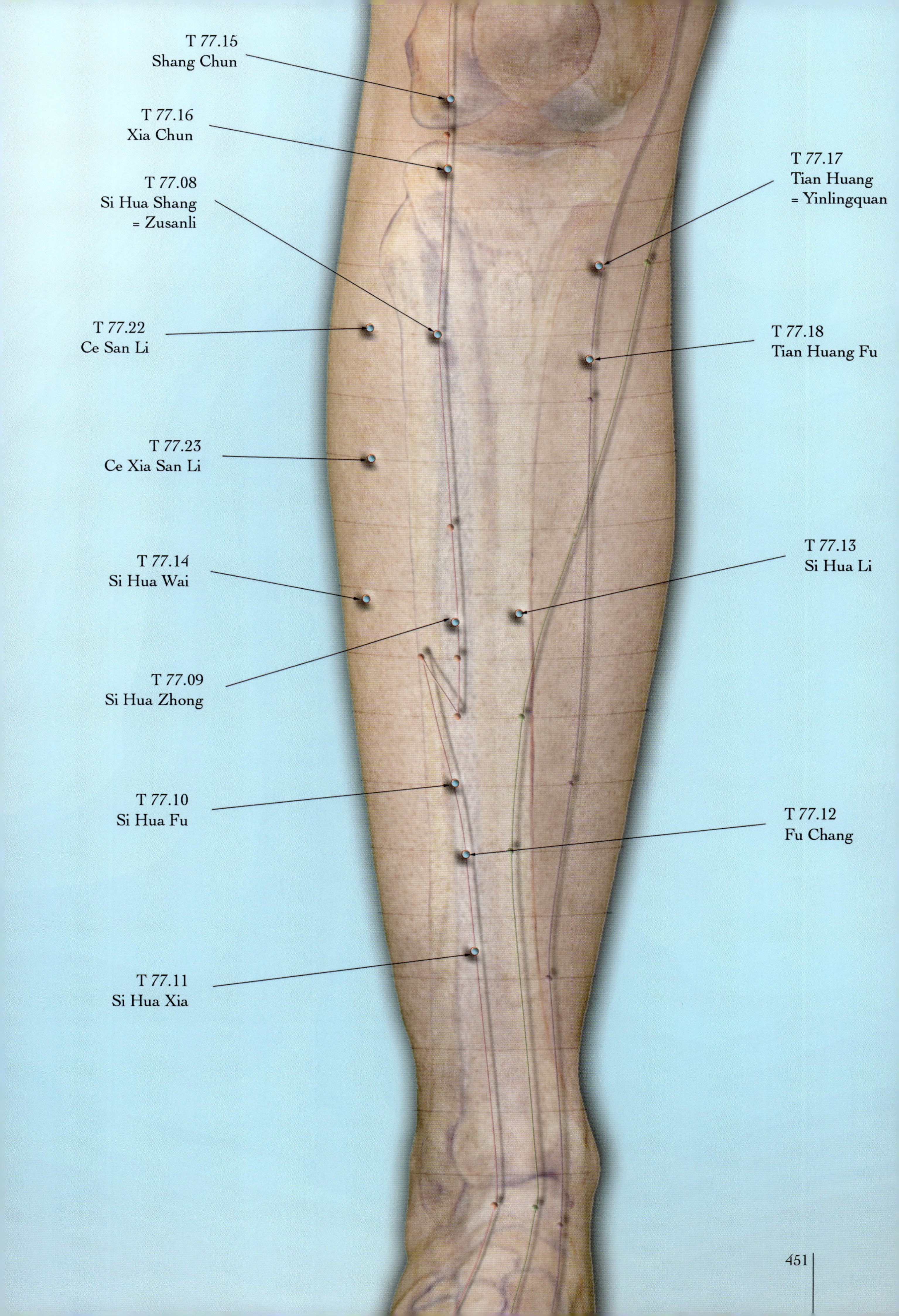
T 77.15
Shang Chun
T 77.16
Xia Chun
T 77.08
Si Hua Shang
= Zusanli
T 77.17
Tian Huang
= Yinlingquan
T 77.22
Ce San Li
T 77.18
Tian Huang Fu
T 77.23
Ce Xia San Li
T 77.14
Si Hua Wai
T 77.13
Si Hua Li
T 77.09
Si Hua Zhong
T 77.10
Si Hua Fu
T 77.12
Fu Chang
T 77.11
Si Hua Xia

Posterior side

Points of the Main channels

Zu Tai Yang channel (Urinary Bladder)

BL 38 Fuxi
At the lateral end of the popliteal crease, 1 Cun above BL 39 Weiyang, medial to the tendons of the biceps muscle of the thigh.

BL 39 Weiyang
At the lateral end of the popliteal crease, medial to the tendon of the biceps muscle of the thigh.

BL 40 Weizhong
At the midpoint of the popliteal crease, between the tendons of the biceps muscle of the thigh and the semi-tendinous muscle.

BL 55 Heyang
On the posterior side of the leg and on the line connecting BL 40 Weizhong and BL 57 Chengshan, 2 Cun below BL 40 Weizhong.

BL 56 Chengjin
On the posterior side of the leg and on the line connecting BL 40 Weizhong and BL 57 Chengshan, at the centre of the gastrocnemius muscle belly, 5 Cun below BL 40 Weizhong.

BL 57 Chengshan
On the posterior midline of the leg, between BL 40 Weizhong and BL 60 Kunlun, in a pointed depression below the gastrocnemius muscle belly when the leg is stretched or the heel is lifted.

BL 58 Fei Yang
On the posterior side of the leg, 7 Cun directly above BL 60 Kunlun and 1 Cun lateral and inferior to BL 57 Chengshan.

BL 59 Fuyang
On the posterior side of the leg, posterior to the external malleolus, 3 Cun directly above BL 60 Kunlun.

BL 60 Kunlun
Posterior to the lateral malleolus, in the depression between the tip of the external malleolus and Achilles tendon

Extra channel points

138 PC Nuxi
On the posterior side of the foot, at the point of intersection of Achilles tendon, on the posterior side of the calcaneum.

139 PC Quanshengzu
On the superior side of the calcaneum, at the midpoint of the tendon.

77 PN Genping
At the midpoint of Achilles tendon, on the line connecting the two malleoluses.

83 PN Luodi
9.5 Cun below the midpoint of the popliteal crease BL 40 Weizhong, that is to say 1 Cun below BL 57 Chengshan.

84 PN Jiuwaifan
1 Cun inside BL 57 Chengshan.

85 PN Jiuneifan
1 Cun outside BL 57 Chengshan.

86 PN Chengjian
At the midpoint of the line connecting BL 57 Chengshan and BL 56 Chengjin.

100 PN Weishang
2 Cun above BL 40 Weizhong.

X-XZ.43 Xiachengshan
0.5 Cun below BL 57 Chengshan.

X-XZ.44 Weixia
4 Cun below and 1.5 Cun laterally and externally from BL 40 Weizhong.

Points of Master Tong

T 77.02 Zheng Zong
2 Cun above T 77.01 Zheng Jin, vertically.

T 77.03 Zheng Shi
2.5 Cun below T 77.04 Bo Qiu.

T 77.01 Zheng Jin
On Achilles tendon between BL 60 Kunlun and KI 3 Taixi.

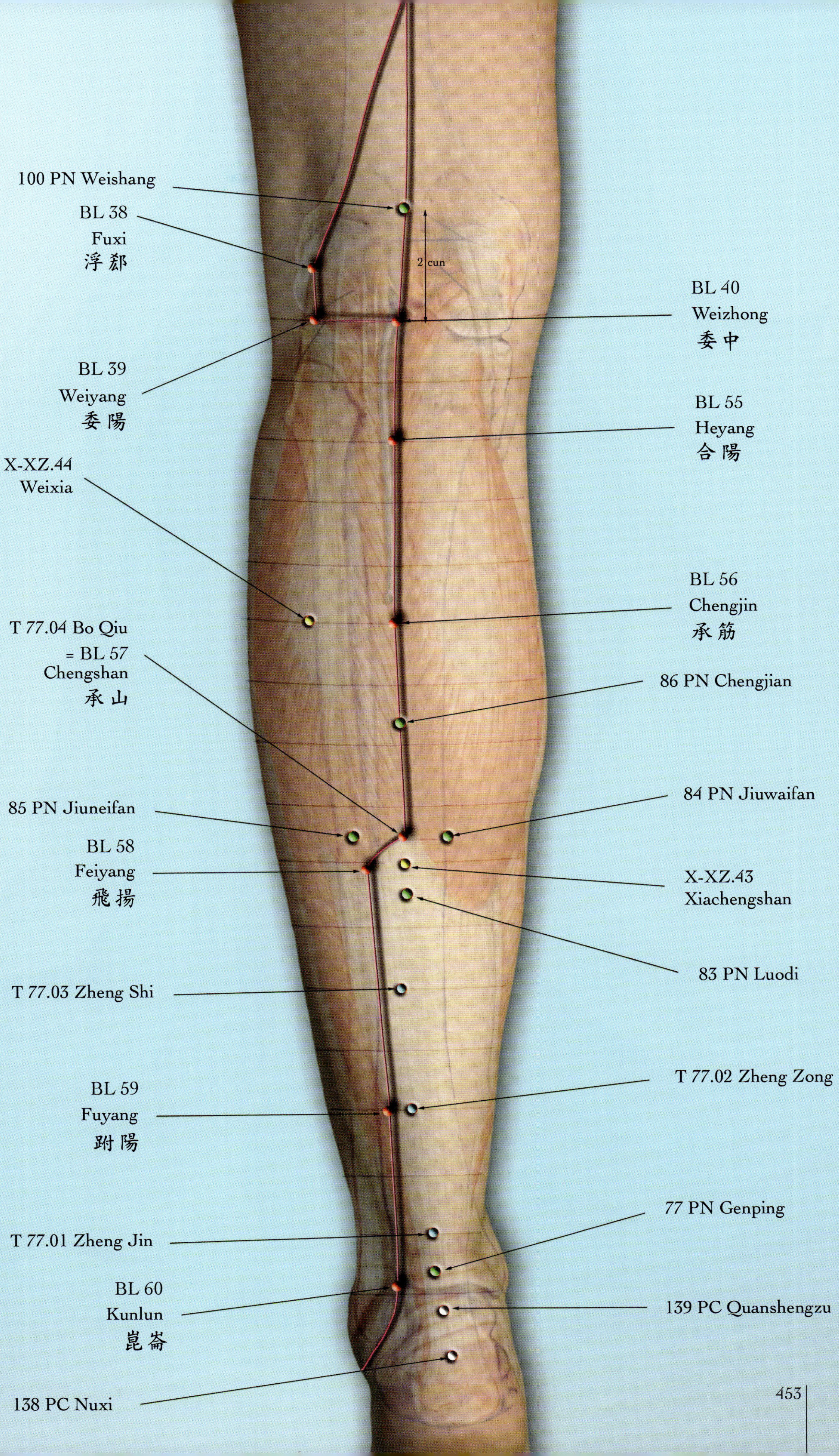
100 PN Weishang
BL 38
Fuxi
浮郄
2 cun
BL 40
Weizhong
委中
BL 39
Weiyang
委陽
BL 55
Heyang
合陽
X-XZ.44
Weixia
BL 56
Chengjin
承筋
T 77.04 Bo Qiu
= BL 57
Chengshan
承山
86 PN Chengjian
85 PN Jiuneifan
84 PN Jiuwaifan
BL 58
Feiyang
飛揚
X-XZ.43
Xiachengshan
83 PN Luodi
T 77.03 Zheng Shi
T 77.02 Zheng Zong
BL 59
Fuyang
跗陽
77 PN Genping
T 77.01 Zheng Jin
BL 60
Kunlun
崑崙
139 PC Quanshengzu
138 PC Nuxi

Latero-external side

Points of the Main channels

Zu Yang Ming channel (Stomach)

ST 35 Dubi
With the knee flexed, on the knee, in the depression lateral to the patella and its ligament.

ST 36 Zusanli
On the anterio-external side of the leg, 3 Cun below ST 35 Dubi, one finger breadth (middle finger) from the anterior crest of the tibia.

ST 37 Shangjuxu
On the anterio-lateral side of the leg, 6 Cun below ST 35 Dubi, one finger breadth (middle finger) from the anterior crest of the tibia.

ST 38 Tiaokou
On the anterio-external side of the leg, 8 Cun below ST 35 Dubi, one finger breadth (middle finger) from the anterior crest of the tibia.

ST 39 Xiajuxu
On the anterioexternal side of the leg, 9 Cun below ST 35 Dubi, one finger breadth (middle finger) from the anterior crest of the tibia.

ST 40 Fenglong
On the anterioexternal side of the leg, 8 Cun above the tip of the external malleolus, outside to ST 38 Tiaokou, and two finger breadths (middle finger) from the anterior crest of the tibia.

ST 41 Jiexi
In the central depression of the crease between the instep of the foot and leg, between the tendons of the long extensor muscle of the big toe and the long extensor muscle of the toes.

Zu Shao Yang channel (Gall Bladder)

GB 34 Yanglingquan
On the external side of the leg, in the depression anterior and inferior to the head of the fibula.

GB 35 Yangjiao
On the external side of the leg, 7 Cun above the tip of the external malleolus, on the posterior border of the fibula.

GB 36 Waiqiu
On the external side of the leg, 7 Cun above the tip of the external malleolus, on the anteriorborder of the fibula and on the level of GB 35 Yangjiao.

GB 37 Guangming
On the external side of the leg, 5 Cun above the tip of the external malleolus, on the anterior border of the fibula.

GB 38 Yangfu
On the external side of the leg, 4 Cun above the tip of the external malleolus, slightly anterior to the anterior border of the fibula.

GB 39 Xuanzhong
On the external side of the leg, 3 Cun above the tip of the external malleolus, on the anterior border of the fibula.

GB 40 Qiuxu
Anterior and inferior to the external malleolus, in the depression external to the tendon of the long extensor muscle of the toes.

Zu Tai Yang channel (Urinary Bladder)

BL 58 Fei Yang
On the posterior side of the leg, posterior to the external malleolus, 7 Cun directly above BL 60 Kunlun and 1 Cun external and inferior to BL 57 Chengshan.

BL 59 Fu Yang
On the posterior side of the leg, posterior to the external malleolus, 3 Cun directly above BL 60 Kunlun.

BL 60 Kunlun
Posterior to the external malleolus, in the depression between the tip of the external malleolus and Achilles tendon.

BL 61 Pucan
On the external side of the foot, posterior and inferior to the external malleolus, directly below BL 60 Kunlun, external to the calcaneum, at the 'junction of the red and white skin'.

BL 62 Shenmai
On the external side of the foot, in the depression directly below the external malleolus.

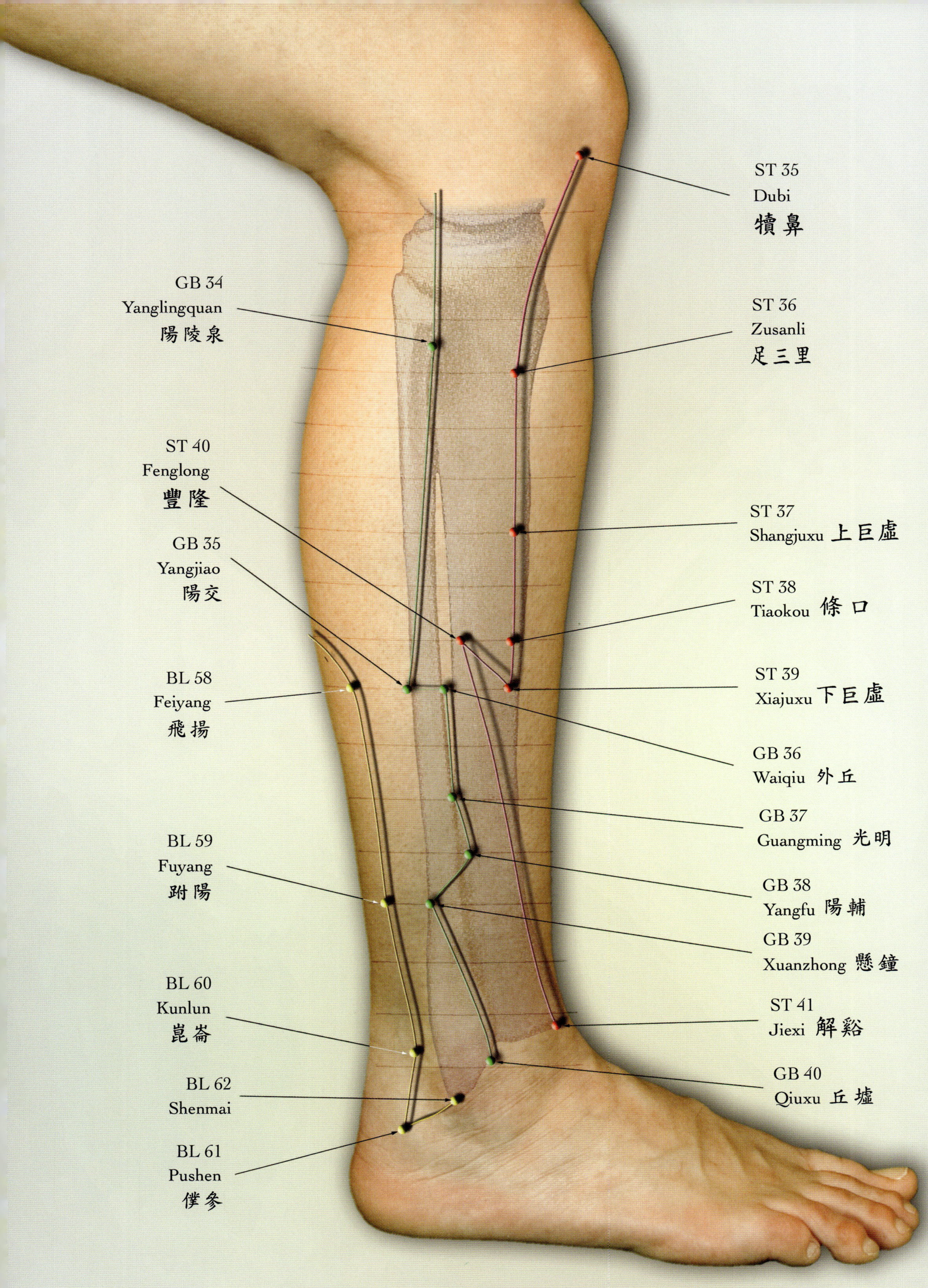
ST 35
Dubi
犢鼻
GB 34
Yanglingquan
陽陵泉
ST 36
Zusanli
足三里
ST 40
Fenglong
豐隆
ST 37
Shangjuxu 上巨虛
GB 35
Yangjiao
陽交
ST 38
Tiaokou 條口
BL 58
Feiyang
飛揚
ST 39
Xiajuxu 下巨虛
GB 36
Waiqiu 外丘
GB 37
Guangming 光明
BL 59
Fuyang
跗陽
GB 38
Yangfu 陽輔
GB 39
Xuanzhong 懸鐘
BL 60
Kunlun
崑崙
ST 41
Jiexi 解谿
GB 40
Qiuxu 丘墟
BL 62
Shenmai
BL 61
Pushen
僕參

Curious points

141 PC
On the external side of the foot, 1 Cun below BL 60 Kunlun.

142 PC Lanwei
On the anterior side of the leg, 2 Cun below ST 36 Zusanli. According to the authors, this point would only exist on the right.

143 PC Zuzhongping
1 Cun below ST 36 Zusanli.

151 PC Waihuaijian
On the prominent part of the external malleolus.

152 PC Danangxue
1.5 Cun below GB 34 Yanglingquan.

153 PC Linghou
In a depression, below the head of the fibula on its posterior border. It is on the line of GB 34 Yanglingquan, 1 Cun behind.

154 PC Linghouxia
0.5 Cun below 153 PC Linghou.

New points

78 PN Naoqing
2 Cun above ST 41 Jiexi, on the external border of the inferior extremity of the tibia.

79 PN Jingxia
3 Cun above ST 41 Jiexi and 1 Cun outside the external border of the extremity of the tibia, on the side of the fibula.

80 PN Wanli
3 Cun from 151 PC Waihuaijian, that is to say 0.5 Cun below ST 36 Zusanli.

81 PN Liwai
1 Cun outside ST 36 Zusanli.

82 PN Lishang
1 Cun above ST 36 Zusanli.

92 PN Lingxia
2 Cun below GB 34 Yanglingquan.

X-XZ.38 Liuzhitan
1.5 Cun below J-XZ.13 Lanwei.

X-XZ.59 Erliban
0.5 Cun above ST 36 Zusanli.

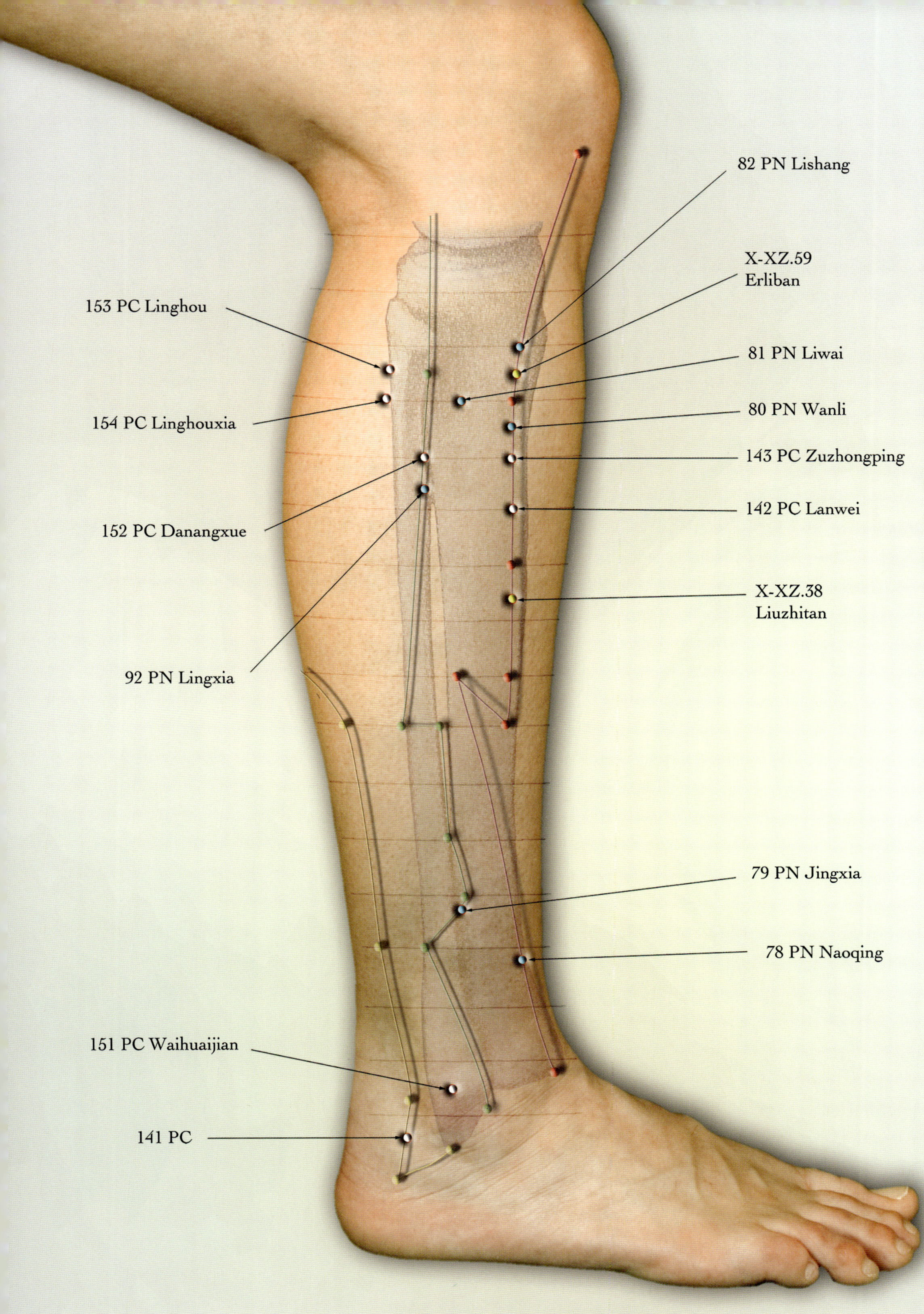
82 PN Lishang
X-XZ.59
Erliban
153 PC Linghou
81 PN Liwai
80 PN Wanli
154 PC Linghouxia
143 PC Zuzhongping
152 PC Danangxue
142 PC Lanwei
X-XZ.38
Liuzhitan
92 PN Lingxia
79 PN Jingxia
78 PN Naoqing
151 PC Waihuaijian
141 PC

T 77.05 Yi Zhong
3 Cun above, and 1 Cun in front of, the apex of the external malleolus, on the posterior border of the fibula.
Also 1 Cun in front of GB 39 Xuan Zhong.

T 77.06 Er Zhong
5 Cun above, and 1 Cun in front of, the apex of the external malleolus, on the posterior border of the fibula.
It is 1 Cun in front of GB 37 Guang Ming.

T 77.07 San Zhong
2 Cun above T 77.06 Er Zhong.

T 77.08 Si Hua Shang
3 Cun below ST 35 Dubi, and one finger breadth outside the anterior border of the tibia.
Corresponds to ST 36 Zusanli.

T 77.09 Si Hua Zhong
4.5 Cun distal from T 77.08 Si Hua Shang.
5 fen above ST 38 Tiaokou.

T 77. 10 Si Hua Fu
2.5 Cun below T 77.09 Si Hua Zhong.
1 Cun below ST 39 Xiajuxu.

T 77.11 Si Hua Xia
2.5 Cun below T 77.10 Si Hua Fu.

T 77.12 Fu Chang
1.5 Cun above T 77.11 Si Hua Xia.

T 77.14 Si Hua Wai
1.5 Cun outside T 77.09 Si Hua Zhong.

T 77.22 Ce San Li
1.5 Cun laterally from T 77.08 Si Hua Shang, that is to say from ST 36 Zusanli. This point is situated on the Gall Bladder channel.

T 77.23 Ce Xia San Li
2 Cun below T 77.22 Ce San Li.

T 77.24 Zu Qian Jin
5 fen outside and 2 Cun below T 77.23 Ce Xia San Li.

T 77.25 Zu Wu Jin
2 cun below T 77.24 Zu Qian Jin.

T 77.26 Qi Hu
Draw a vertical line 1.5 Cun posterior to the prominent part of the external malleolus.

- From the tip of the malleolus, measure 2 Cun above to find the first point.
- The second point is 2 Cun above the first.
- The third point is 4 Cun above the first.

T 77.27 Wai San Guan
Draw a line from the head of the fibula to the inferior border of the external malleolus.
The first point is found at the inferior quarter of the distance between these two extremities, the second at the midpoint of the distance and the third at three quarters.

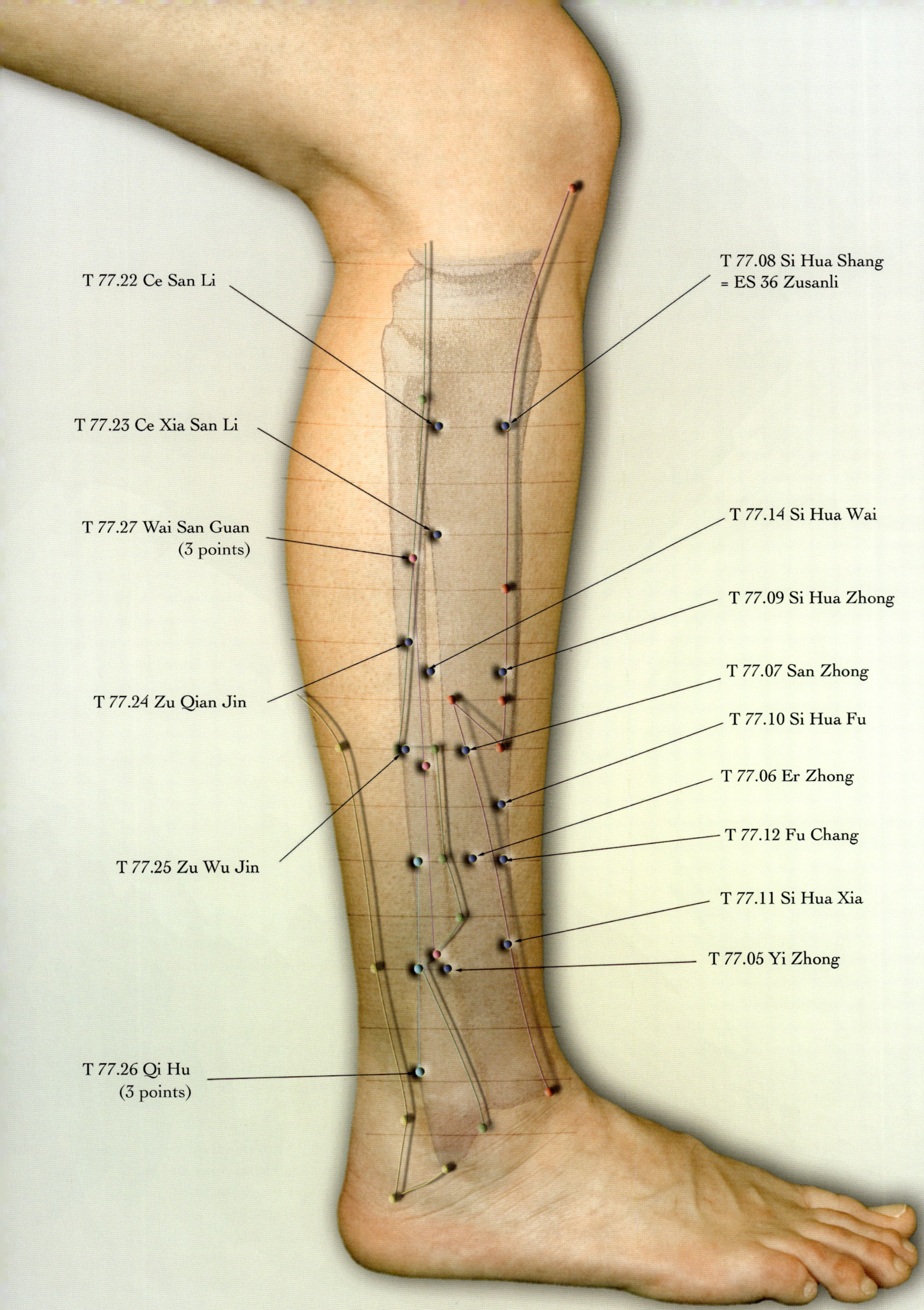
T 77.22 Ce San Li
T 77.08 Si Hua Shang = ES 36 Zusanli
T 77.23 Ce Xia San Li
T 77.27 Wai San Guan (3 points)
T 77.14 Si Hua Wai
T 77.09 Si Hua Zhong
T 77.24 Zu Qian Jin
T 77.07 San Zhong
T 77.10 Si Hua Fu
T 77.06 Er Zhong
T 77.25 Zu Wu Jin
T 77.12 Fu Chang
T 77.11 Si Hua Xia
T 77.05 Yi Zhong
T 77.26 Qi Hu (3 points)

Latero-internal side

Points of the Main channels

Zu Tai Yin channel (Spleen)

SP 5 Shangqiu
In the depression anterior and inferior to the medial malleolus, at the midpoint of the line connecting the tuberosity of the navicular bone and the tip of the medial malleolus.

SP 6 Sanyinjiao
On the medial side of the leg, 3 Cun above the tip of the medial malleolus, posterior to the medial border of the tibia.

SP 7 Lougu
On the medial side of the leg and on the line connecting the tip of the medial malleolus and SP 9 Yinlingquan, 6 Cun from the tip of the medial malleolus, posterior to the medial border of the tibia.

SP 8 Diji
On the medial side of the leg and on the line connecting the tip of the medial malleolus and SP 9 Yinlingquan, 3 Cun below SP 9 Yinlingquan.

SP 9 Yinlingquan
On the medial side of the leg, in the depression posterior and inferior to the medial condyle of the tibia.

Zu Jue Yin channel (Liver)

LR 4 Zhongfeng
On the instep of the foot, anterior to the medial malleolus, on the line connecting SP 5 Shangqiu and ST 41 Jiexi, in the depression medial to the tendon of the anterior tibial muscle.

LR 5 Ligou
On the medial side of the leg, 5 Cun above the tip of the medial malleolus, on the midline of the medial surface of the tibia.

LR 6 Zhongdu
On the medial side of the leg, 7 Cun above the tip of the medial malleolus, on the midline of the medial surface of the tibia.

LR 7 Xiguan
On the medial side of the leg, posterior and inferior to the medial epicondyle of the tibia, 1 Cun posterior to SP 9 Yinlingquan, at the upper end of the medial head of the gastrocnemius muscle.

LR 8 Ququan
On the medial side of the knee, at the medial end of the popliteal crease when the knee is flexed, posterior to the medial epicondyle of the tibia, in the depression of the anterior border of the insertions of the semi-membranous and semi-tendinous muscles.

Zu Shao Yin channel (Kidney)

KI 3 Taixi
On the medial side of the foot, posterior to the medial malleolus, in the depression between the tip of the medial malleolus and Achilles tendon.

KI 4 Dazhong
On the medial side of the foot, posterior and inferior to the medial malleolus, in the depression of the medial side of and anterior to the attachment of Achilles tendon.

KI 5 Shuiquan
On the medial side of the foot, posterior and inferior to the medial malleolus, 1 Cun directly below KI 3 Taixi, in the depression of the medial side of the tuberosity of the calcaneum.

KI 6 Zhaohai
On the medial side of the foot, in the depression below the tip of the medial malleolus.

KI 7 Fuliu
On the medial side of the leg, 2 Cun directly above KI 3 Taixi, anterior to the Achilles tendon.

KI 8 Jiaoxin
On the medial side of the leg, 2 Cun above KI 3 Taixi and 0.5 Cun anterior to KI 7 Fuliu, posterior to the medial border of the tibia.

KI 9 Zhubin
On the medial side of the leg and on the line connecting KI 3 Taixi and KI 10 Yingu, 5 Cun above KI 3 Taixi, medial and inferior to the gastrocnemius muscle belly.

KI 10 Yingu
On the medial side of the popliteal fossa, between the tendons of the semi-tendinous and semi-membranous muscles when the knee is flexed.

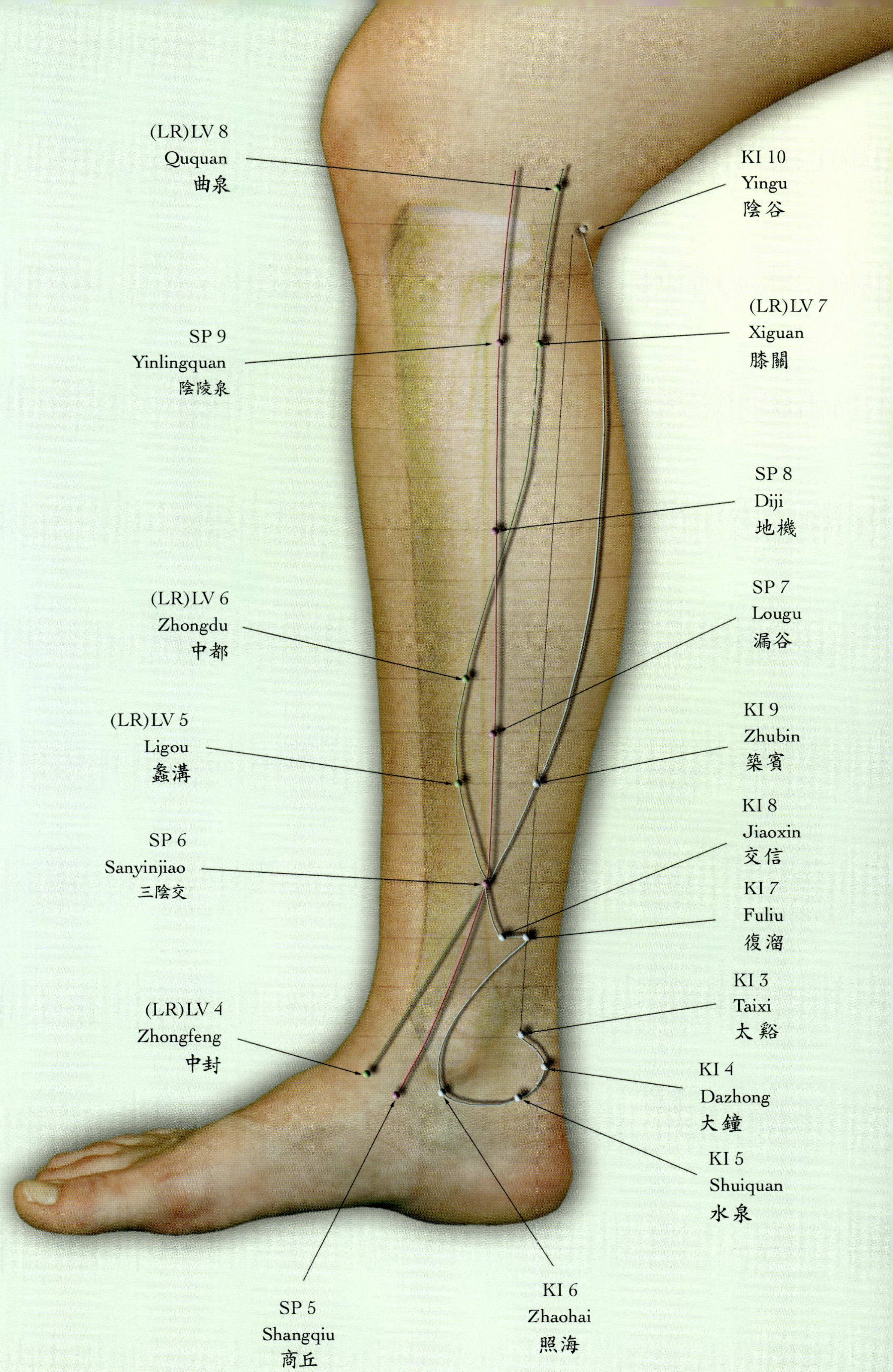

(LR)LV 8
Ququan
曲泉
KI 10
Yingu
陰谷
(LR)LV 7
Xiguan
膝關
SP 9
Yinlingquan
陰陵泉
SP 8
Diji
地機
SP 7
Lougu
漏谷
(LR)LV 6
Zhongdu
中都
KI 9
Zhubin
築賓
(LR)LV 5
Ligou
蠡溝
KI 8
Jiaoxin
交信
SP 6
Sanyinjiao
三陰交
KI 7
Fuliu
復溜
KI 3
Taixi
太谿
(LR)LV 4
Zhongfeng
中封
KI 4
Dazhong
大鐘
KI 5
Shuiquan
水泉
SP 5
Shangqiu
商丘
KI 6
Zhaohai
照海

Curious points

140 PC Taiyinqiao
On the internal surface of the foot, directly below the internal malleolus, in a depression, 0.3 Cun above KI 6 Zhaohai.

146 PC Neihuaijian
On the most prominent part of the internal malleolus.

147 PC Zhizhuanjin
At the midpoint of the superior border of the internal malleolus.

148 PC Shaoyangwei
1.5 Cun above 146 PC Neihuaijian.

149 PC Chengming
3 Cun above KI 3 Taixi.

150 PC Jiaoyi
5 Cun from the crest of the internal malleolus or from 146 PC Neihuaijian.

New Points

87 PN Shangxi
0.5 Cun above KI 3 Taixi.

88 PN Ganyandian
2 Cun above the tip of the internal malleolus.

89 PN Jiuwai Fan
0.5 Cun below SP 6 Sanyinjiao

90 PN Dijian
1 Cun below SP 8 Diji.

Points of Master Tong

T 77.18 Tian Huang Fua
1.5 Cun below SP 9 Yinlingquan.

T 77.19 Di Huang
4 Cun above SP 6 Sanyinjiao, that is to say 7 Cun above the most prominent part of the internal malleolus. It is 1 Cun above SP 7 Lougu.

T 77.20 Si Zhi
3 Cun below T 77.19 Di Huang along the posterior border of the tibia.
1 Cun above SP 6 Sanyinjiao.

T 77.21 Ren Huang
4 Cun below T 77.19 Di Huang on the posterior border of the tibia.
Corresponds to SP 6 Sanyinjiao.

T 88.32 Shi Yin
The first point is situated at the central point of the knee joint in its internal part.
The second is 2 Cun below.

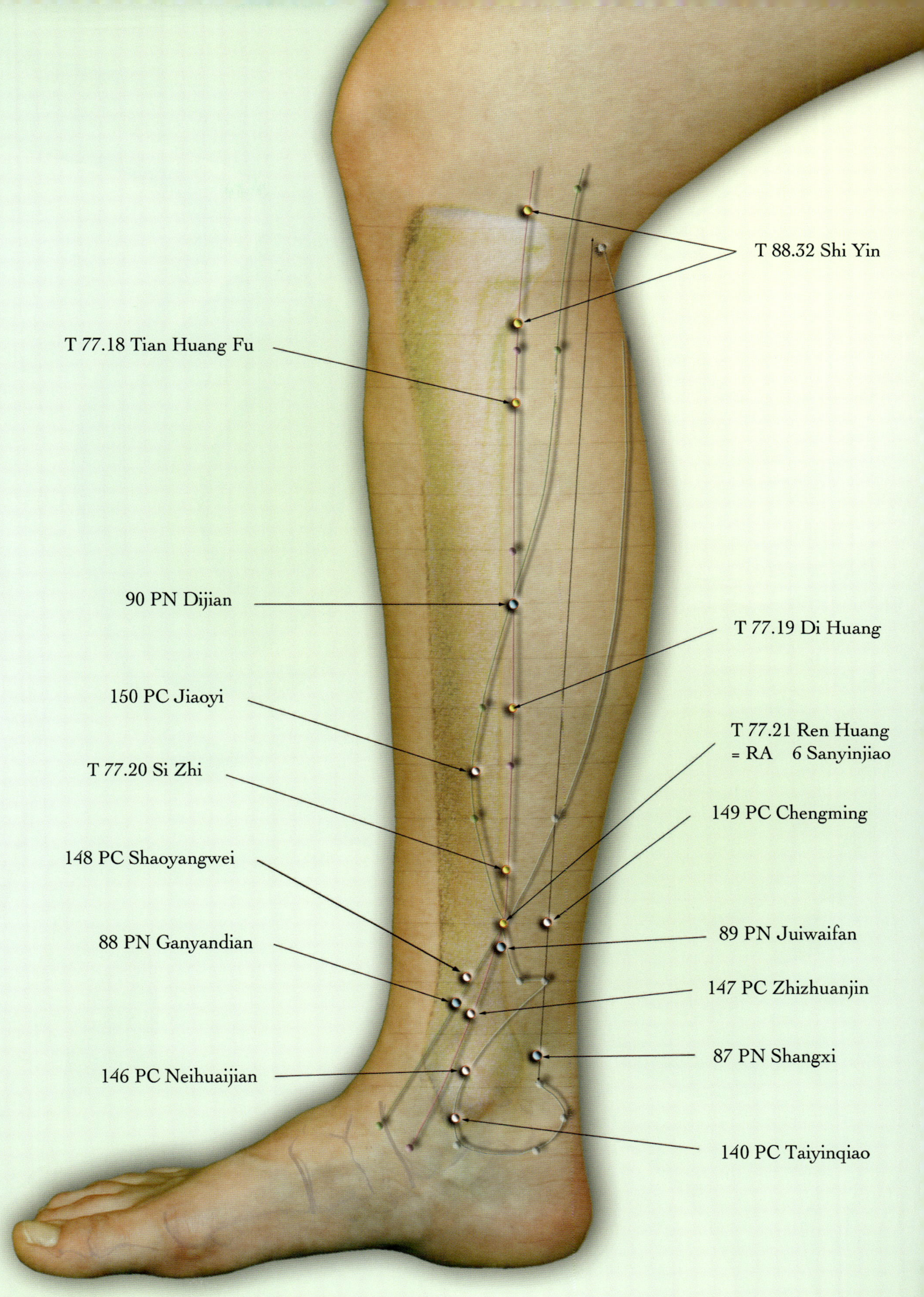
T 88.32 Shi Yin
T 77.18 Tian Huang Fu
90 PN Dijian
T 77.19 Di Huang
150 PC Jiaoyi
T 77.21 Ren Huang
= RA 6 Sanyinjiao
T 77.20 Si Zhi
149 PC Chengming
148 PC Shaoyangwei
88 PN Ganyandian
89 PN Juiwaifan
147 PC Zhizhuanjin
87 PN Shangxi
146 PC Neihuaijian
140 PC Taiyinqiao

The foot

- Anterior, external, internal side, sole
  - Points of the Main channels

    Zu Yang Ming (Stomach)

    Zu Jue Yin (Liver)

    Zu Tai Yin (Spleen)

    Zu Shao Yang (Gall Bladder)

    Zu Tai Yang (Urinary Bladder)
  - Extra channel points

    Curious points

    New points

    Points of Master Tong

Anterior side

Points of the Main channels

Zu Tai Yin channel (Spleen)

SP 1 Yinbai
On the medial side of the distal segment of the big toe, 0.1 Cun from the corner of the toenail.

SP 2 Dadu
On the medial border of the foot, in the depression of the 'junction of the red and white skin', anterior and inferior to the 1st metatarsophalangeal joint.

SP 3 Taibai
On the medial border of the foot, in the depression of the 'junction of the red and white skin', posterior and inferior to the 1st metatarso-phalangeal joint.

SP 4 Gongsun
On the medial border of the foot, anterior and inferior to the proximal end of the 1st metatarsal bone.

SP 5 Shangqiu
In the depression anterior and inferior to the medial malleolus, at the midpoint of the line connecting the tuberosity of the navicular bone and the tip of the medial malleolus.

Zu Jue Yin channel (Liver)

(LR)LV 1 Dadun
On the external side of the distal segment of the big toe, 0.1 Cun from the corner of the toenail.

(LR)LV 2 Xingjian
On the instep of the foot, between the 1st and 2nd toes, at the 'junction of the red and white skin' proximal to the margin of the web.

(LR)LV 3 Taichong
On the instep of the foot, in the depression of the posterior end of the 1st interosseous metatarsal space.

(LR)LV 4 Zhongfeng
On the instep of the foot, anterior to the medial malleolus, on the line connecting SP 5 Shangqiu and ST 41 Jiexi, in the depression medial to the tendon of the anterior tibial muscle.

Zu Yang Ming channel (Stomach)

ST 42 Chongyang
On the dome of the instep of the foot, between the tendons of the long extensor muscle of the big toe and the long extensor muscle of the toes, where the pulsation of the dorsal artery of the foot is palpable.

ST 43 Xiangu
On the instep of the foot, in the depression distal to the commissure of the 2nd and 3rd metatarsal bones.

ST 44 Neiting
On the instep of the foot, at the 'junction of the red and white skin' proximal to the margin of the web between the 2nd and 3rd toes.

ST 45 Lidui
On the external side of the distal segment of the 2nd toe, 0.1 Cun from the corner of the toenail.

Zu Shao Yang channel (Gall Bladder)

GB 40 Qiuxu
Anterior and inferior to the external malleolus, in the depression external to the tendon of the long extensor muscle of the toes.

GB 41 Zulinqi
On the external side of the instep of the foot, posterior to the 4th metatarsophalangeal joint, in the depression external to the tendon of the extensor muscle of the little toe.

GB 42 Diwuhui
On the external side of the instep of the foot, posterior to the 4th metatarso-phalangeal joint, between the 4th and 5th metatarsal bones, medial to the tendon of the extensor muscle of the little toe.

GB 43 Xiaxi
On the external side of the instep of the foot, between the 4th and 5th toes, at the 'junction of the red and white skin', proximal to the margin of the web.

GB 44 Zuqiàoyin
On the external side of the distal segment of the 4th toe, 0.1 Cun from the corner of the toenail.

Zu Tai Yang channel (Urinary Bladder)

BL 63 Jinmen
On the external side of the foot, directly below the anterior border of the external malleolus, on the lower border of the cuboid bone.

BL 64 Jinggu
On the external side of the foot, below the tuberosity of the 5th metatarsal bone, at the 'junction of the red and white skin'.

BL 65 Shugu
On the external side of the foot, posterior to the 5th metatarso-phalangeal joint, at the 'junction of the red and white skin'.

BL 66 Zutonggu
On the external side of the foot, anterior to the 5th metatarso-phalangeal joint, at the 'junction of the red and white skin'.

BL 67 Zhiyin
On the external side of the distal segment of the little toe, 0.1 Cun from the corner of the toenail.

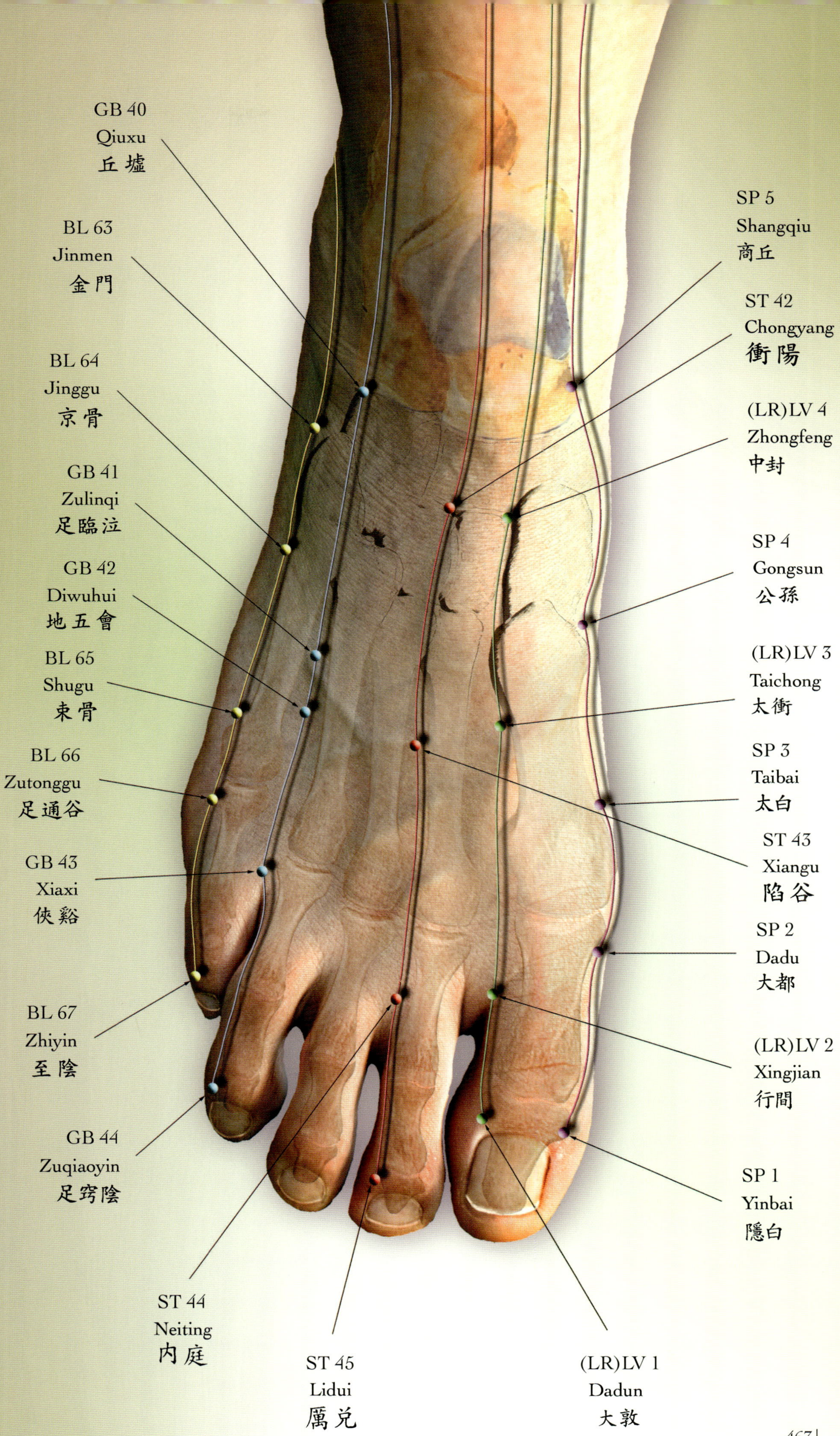
GB 40
Qiuxu
丘墟
BL 63
Jinmen
金門
BL 64
Jinggu
京骨
GB 41
Zulinqi
足臨泣
GB 42
Diwuhui
地五會
BL 65
Shugu
束骨
BL 66
Zutonggu
足通谷
GB 43
Xiaxi
俠谿
BL 67
Zhiyin
至陰
GB 44
Zuqiaoyin
足竅陰
ST 44
Neiting
内庭
ST 45
Lidui
厲兑
(LR)LV 1
Dadun
大敦
SP 5
Shangqiu
商丘
ST 42
Chongyang
衝陽
(LR)LV 4
Zhongfeng
中封
SP 4
Gongsun
公孫
(LR)LV 3
Taichong
太衝
SP 3
Taibai
太白
ST 43
Xiangu
陷谷
SP 2
Dadu
大都
(LR)LV 2
Xingjian
行間
SP 1
Yinbai
隱白

Curious Points

135 PC Xiao Zhijian
At the tip of the 5th toe, 0.1 Cun below the nail beside BL 67 Zhiyin.

136 PC Muzhiheng - Lisanmao
On the dorsal side of the 1st toe, from 0.1 to 0.2 Cun from the midpoint of the interphalangeal line, in the direction of SP 1 Yinbai.

137 PC Bafeng
In the metatarso-phalangeal space, in the interdigital spaces, on the anterior side of the foot. When the foot rests on the sole, the point is found at the end of each interdigital groove. Each foot has 4 points, which makes a total of 8 'winds', among which are ST 44 Neiting, (LR)LV 2 Xingjian, GB 43 Xiaxi.

New Points

76 PN Panggu
1 Cun above the superior surface of the interdigital groove which separates the 3rd from the 4th toe.

Points of Master Tong

T 66.02 Mu Fu
On the dorsal side of the foot, find the centre of the middle phalanx of the 2nd toe.
The point is 3 fen on the lateral external side of that point.

T 66.03 Huo Ying
5 fen from the end of the interdigital groove, between the 1st and the 2nd toe.
Corresponds to (LR)LV 2 Xingjian.

T 66.04 Huo Zhu
Between the 1st and the 2nd metatarsus, along the external border of the 1st matatarsus, where the pulsations of the artery are palpable, 1 Cun behind T 66.03 Huo Ying.
Corresponds to (LR)LV 3 Taichong.

T 66.05 Men Jin
In the distal depression of the joints of the 2nd and 3rd metatarsus.
Corresponds to ST 43 Xiangu.

T 66.06 Mu Liu
In the distal depression of the joints of the 3rd and 4th metatarsus, 1.5 Cun from the space between these toes.

T 66.07 Mu Dou
On the dorsal side of the foot, 5 fen from the space between the 3rd and 4th toe.

T 66.08 Liu Wan Mu Wan
Between the 4th and 5th toe, 5 fen from the commissure.
Corresponds to GB 43 Xiaxi.

T 66.09 Shui Qu
Between the 4th and 5th toe, 1.5 Cun from the groove or 1 Cun from T 66.08 Liu Wan.

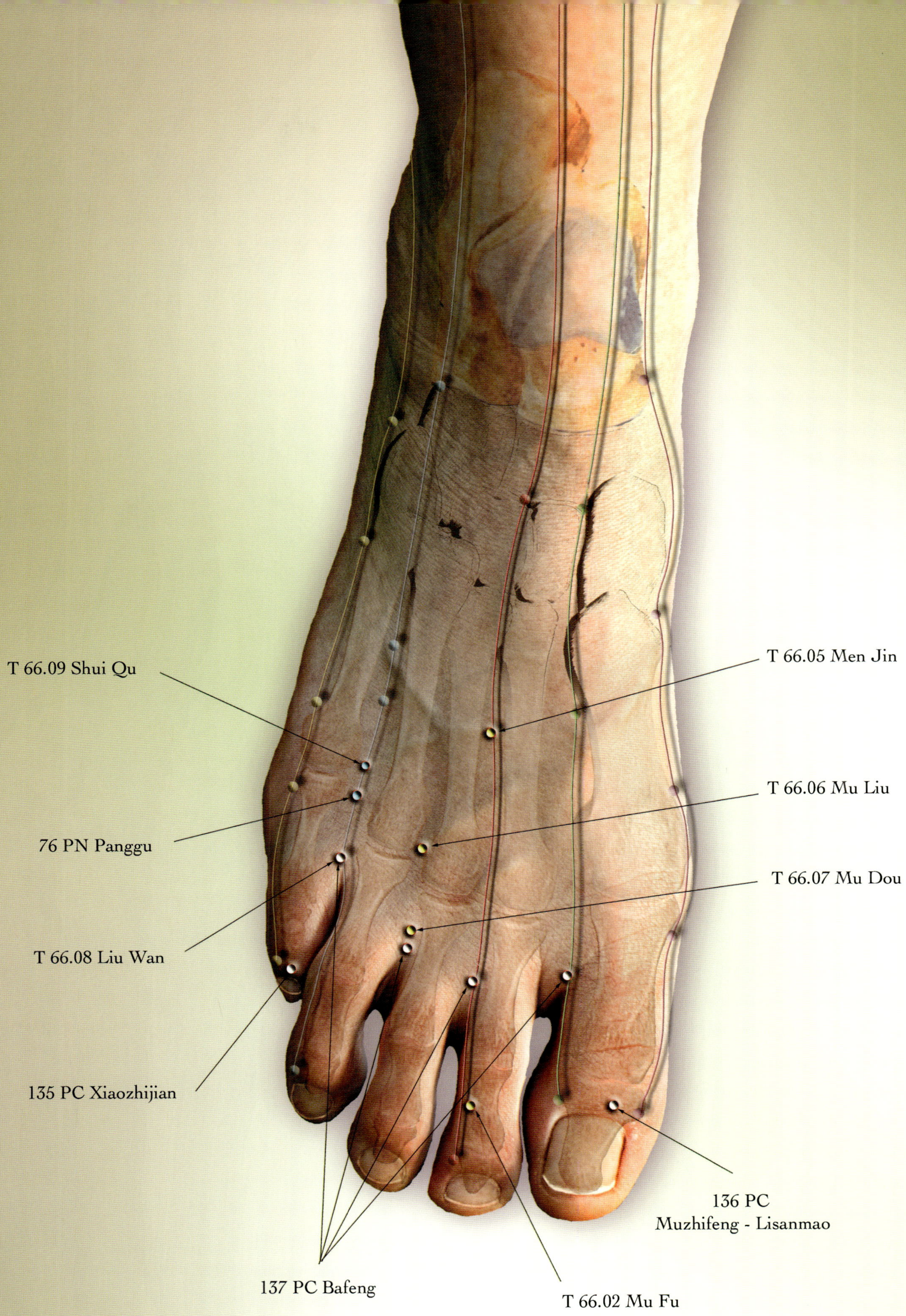
T 66.09 Shui Qu
T 66.05 Men Jin
76 PN Panggu
T 66.06 Mu Liu
T 66.07 Mu Dou
T 66.08 Liu Wan
135 PC Xiaozhijian
136 PC
Muzhifeng - Lisanmao
137 PC Bafeng
T 66.02 Mu Fu

Latero-internal side

Points of the Main channels

Zu Tai Yin channel (Spleen)

SP 1 Yinbai
On the medial side of the distal segment of the big toe, 0.1 Cun from the corner of the toenail.

SP 2 Dadu
On the medial border of the foot, in the depression of the 'junction of the red and white skin', anterior and inferior to the 1st metatarsophalangeal joint.

SP 3 Taibai
On the medial border of the foot, in the depression of the 'junction of the red and white skin', posterior and inferior to the 1st metatarso-phalangeal joint.

SP 4 Gongsun
On the medial border of the foot, anterior and inferior to the proximal end of the 1st metatarsal bone.

SP 5 Shangqiu
In the depression anterior and inferior to the medial malleolus, at the midpoint of the line connecting the tuberosity of the navicular bone and the tip of the medial malleolus.

SP 6 Sanyinjiao
On the medial side of the leg, 3 Cun above the tip of the medial malleolus, posterior to the medial border of the tibia.

Zu Shao Yin channel (Kidney)

KI 2 Rangu
On the medial border of the foot, below the tuberosity of the navicular bone, and at the 'junction of the red and white skin'.

KI 3 Taixi
On the medial side of the foot, posterior to the medial malleolus, in the depression between the tip of the medial malleolus and Achilles tendon.

KI 4 Dazhong
On the medial side of the foot, posterior and inferior to the medial malleolus, in the depression of the medial side of and anterior to the attachment of the Achilles tendon.

KI 5 Shuiquan
On the medial side of the foot, posterior and inferior to the medial malleolus, 1 Cun directly below KI 3 Taixi, in the depression of the medial side of the tuberosity of the calcaneum.

KI 6 Zhaohai
On the medial side of the foot, in the depression below the tip of the medial malleolus.

KI 7 Fuliu
On the medial side of the leg, 2 Cun directly above KI 3 Taixi, anterior to the Achilles tendon.

KI 8 Jiaoxin
On the medial side of the leg, 2 Cun above KI 3 Taixi and 0.5 Cun anterior to KI 7 Fuliu, posterior to the medial border of the tibia.

Zu Jue Yin channel (Liver)

(LR)LV 1 Dadun
On the lateral side of the distal segment of the big toe, 0.1 Cun from the corner of the toenail.

(LR)LV 2 Xingjian
On the instep of the foot, between the 1st and 2nd toe, at the 'junction of the red and white skin', proximal to the margin of the web.

(LR)LV 4 Zhongfeng
On the instep of the foot, anterior to the medial malleolus, on the line connecting SP 5 Shangqiu and ST 41 Jiexi, in the depression medial to the tendon of the anterior tibial muscle.

Extra channel points

X-XZ.48 Chuqixue
0.5 Cun behind KI 2 Rangu.

T 66.10 Huo Lian
On the postero-inferior side of the 1st metatarsus, at the 'junction of the red and white skin'. Corresponds to SP 3 Taibai.

T 66.11 Huo Ju
Before and below the metatarso-cuneiform joint, at the 'junction of the red and white skin'. Corresponds to SP 4 Gongsun.

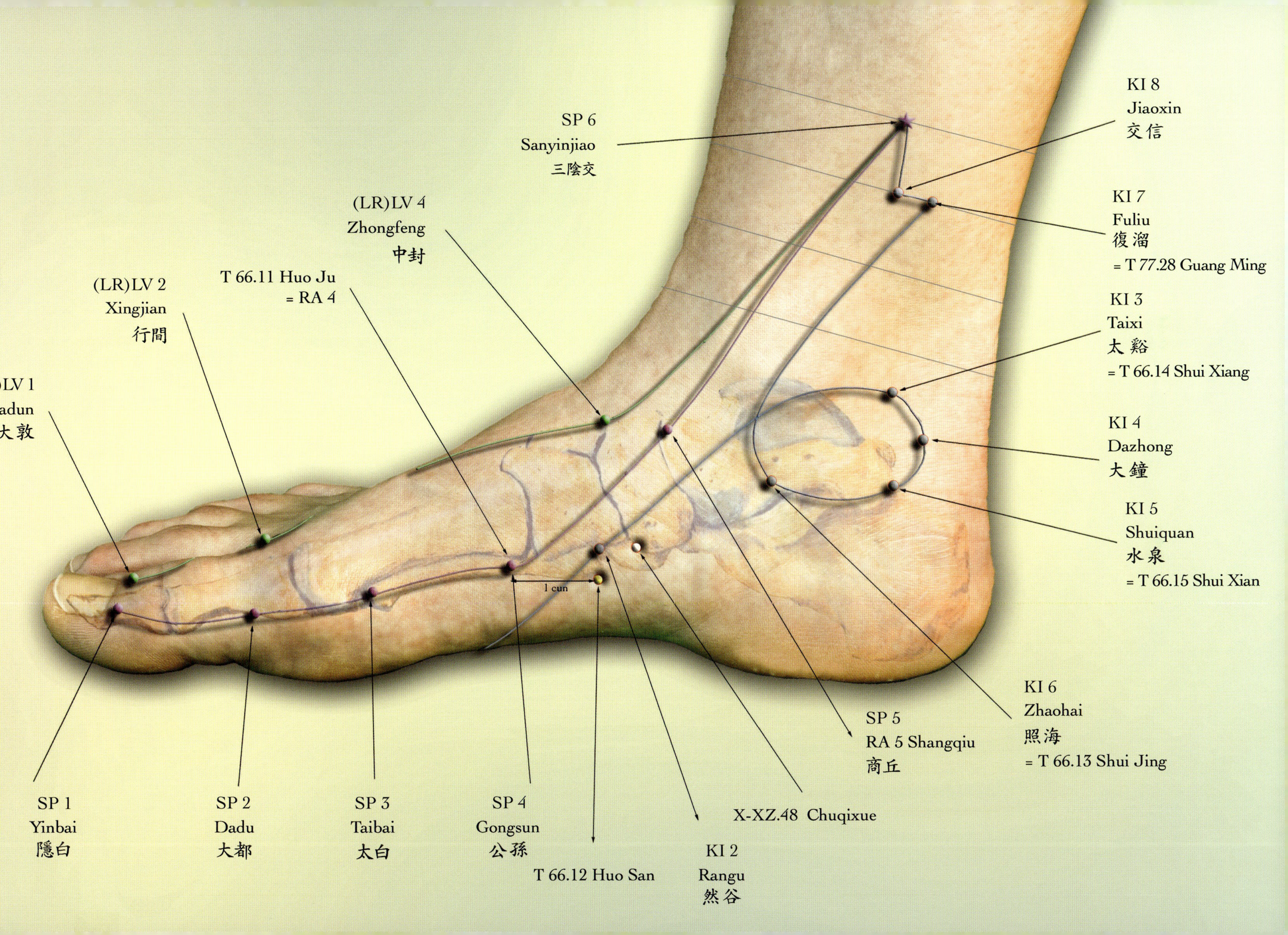

SP 6
Sanyinjiao
三陰交
KI 8
Jiaoxin
交信
KI 7
Fuliu
復溜
= T 77.28 Guang Ming
(LR)LV 4
Zhongfeng
中封
(LR)LV 2
Xingjian
行間
T 66.11 Huo Ju
= RA 4
KI 3
Taixi
太谿
= T 66.14 Shui Xiang
(LR)LV 1
Dadun
大敦
KI 4
Dazhong
大鐘
KI 5
Shuiquan
水泉
= T 66.15 Shui Xian
1 cun
KI 6
Zhaohai
照海
= T 66.13 Shui Jing
SP 5
RA 5 Shangqiu
商丘
SP 1
Yinbai
隱白
SP 2
Dadu
大都
SP 3
Taibai
太白
SP 4
Gongsun
公孫
X-XZ.48 Chuqixue
T 66.12 Huo San
KI 2
Rangu
然谷

T 66.12 Huo San
On the internal side of the foot, in a depression under the scaphoid tubercle, 1 Cun behind SP 4 Gongsun. Corresponds to KI 2 Rangu.

T 66.13 Shui Jing
0.5 Cun below the tip of the internal malleolus. Corresponds to KI 6 Zhaohai.

T 66.14 Shui Xiang
On the posterior tibial artery, right on the posterio-inferior border of the internal malleolus. Corresponds to KI 3 Taixi.

T 66.15 Shui Xian
1 Cun below KI 3 Taixi, in the antero-superior depression on the internal side of the tuberosity of the calcaneum. Corresponds to KI 5 Shuiquan.

T 77.28 Guang Ming
2 Cun above the prominence of the internal malleolus, 0.5 Cun behind the posterior border of the tibia.

Latero-external side

Points of the Main channels

Zu Yang Ming channel (Stomach)

ST 41 Jiexi
In the central depression of the crease between the instep of the foot and the leg, between the tendons of the long extensor muscle of the big toe and the long extensor muscle of the toes.

ST 42 Chongyang
On the dome of the instep of the foot, between the tendons of the long extensor muscle of the big toe and the long extensor muscle of the toes, where the pulsation of the dorsal artery of the foot is palpable.

ST 43 Xiangu
On the instep of the foot, in the depression distal to the commissure of the 2nd and 3rd metatarsal bones.

ST 44 Neiting
On the instep of the foot, at the 'junction of the red and white skin', anterior to the margin of the web between the 2nd and 3rd toe.

ST 45 Lidui
On the external side of the distal segment of the 2nd toe, 0.1 Cun from the corner of the toenail.

Zu Shao Yang channel (Gall Bladder)

GB 39 Xuanzhong
On the external side of the leg, 3 Cun above the tip of the external malleolus, on the anterior border of the fibula.

GB 40 Qiuxu
Anterior and inferior to the external malleolus, in the depression lateral to the tendon of the long extensor muscle of the toes.

GB 41 Zulinqi
On the external side of the instep of the foot, posterior to the 4th metatarso-phalangeal joint, in the depression lateral to the tendon of the extensor muscle of the little toe.

GB 42 Diwuhui
On the external side of the instep of the foot, posterior to the 4th metatarso-phalangeal joint, between the 4th and 5th metatarsal bones, medial to the tendon of the extensor muscle of the little toe.

GB 43 Xiaxi
On the external side of the instep of the foot, between the 4th and 5th toes, at the 'junction of the red and white skin', proximal to the margin of the web.

GB 44 Zuqiaoyin
On the external side of the distal segment of the 4th toe, 0.1 Cun from the corner of the toenail.

Zu Tai Yang channel (Urinary Bladder)

BL 60 Kunlun
Posterior to the external malleolus, in the depression between the tip of the external malleolus and Achilles tendon.

BL 61 Pucan
On the external side of the foot, posterior and inferior to the external malleolus, directly below BL 60 Kunlun, external to the calcaneum, at the 'junction of the red and white skin'.

BL 62 Shenmai
On the external side of the foot, in the depression directly below the external malleolus.

BL 63 Jinmen
On the external side of the foot, directly below the anterior border of the external malleolus, on the lower border of the cuboid bone.

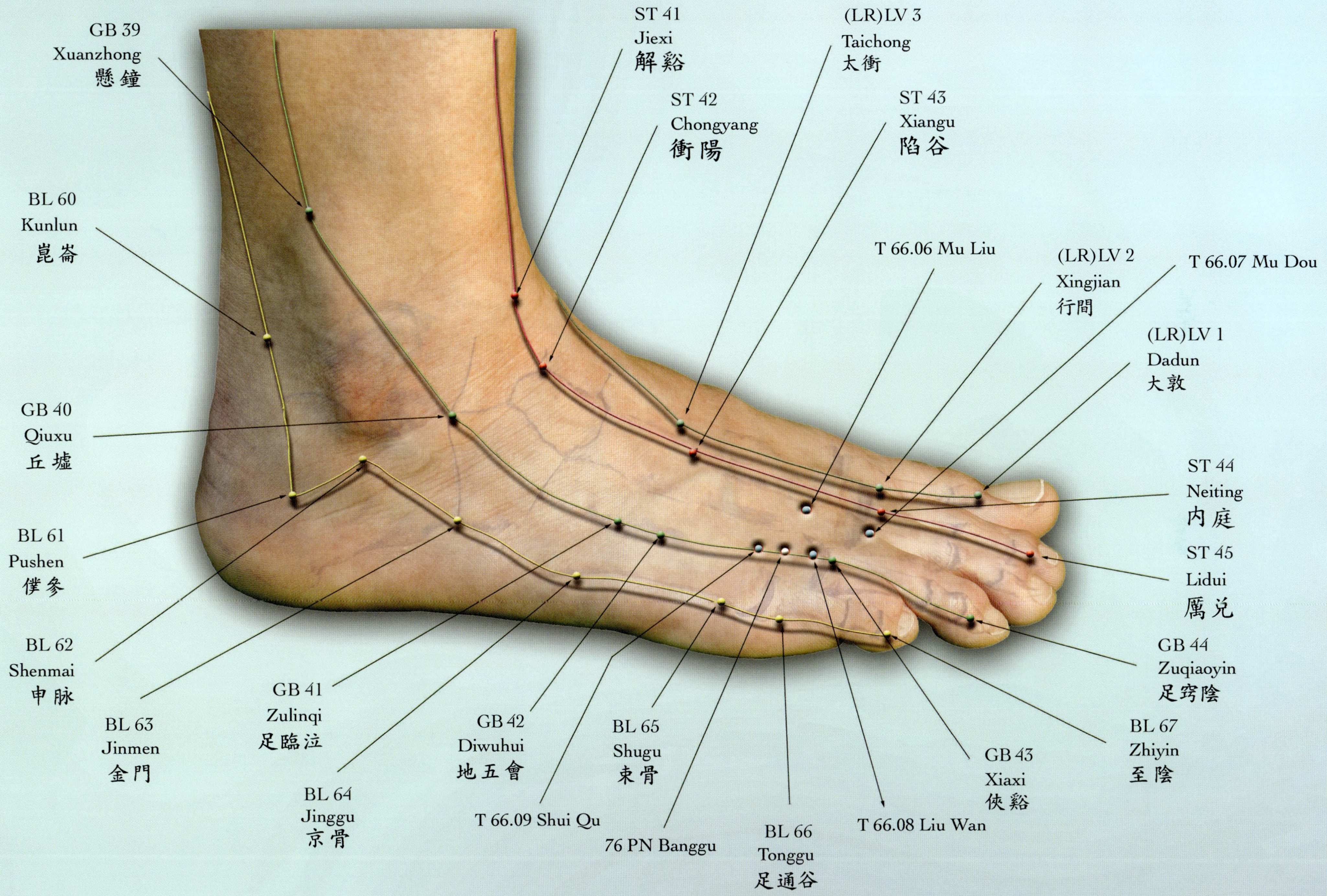

GB 39
Xuanzhong
懸鐘
BL 60
Kunlun
崑崙
GB 40
Qiuxu
丘墟
BL 61
Pushen
僕參
BL 62
Shenmai
申脉
BL 63
Jinmen
金門
GB 41
Zulinqi
足臨泣
BL 64
Jinggu
京骨
GB 42
Diwuhui
地五會
T 66.09 Shui Qu
BL 65
Shugu
束骨
76 PN Banggu
BL 66
Tonggu
足通谷
T 66.08 Liu Wan
GB 43
Xiaxi
侠谿
BL 67
Zhiyin
至陰
GB 44
Zuqiaoyin
足窍陰
ST 45
Lidui
厲兑
ST 44
Neiting
内庭
(LR)LV 1
Dadun
大敦
T 66.07 Mu Dou
(LR)LV 2
Xingjian
行間
T 66.06 Mu Liu
ST 43
Xiangu
陷谷
(LR)LV 3
Taichong
太衝
ST 42
Chongyang
衝陽
ST 41
Jiexi
解谿

BL 64 Jinggu
On the external side of the foot, below the tuberosity of the 5th metatarsal bone, at the 'junction of the red and white skin'.

BL 65 Shugu
On the external side of the foot, posterior to the 5th metatarsophalangeal joint, at the 'junction of the red and white skin'.

BL 66 Zutonggu
On the lateral side of the foot, anterior to the 5th metatarsophalangeal joint, at the 'junction of the red and white skin'.

BL 67 Zhiyin
On the lateral side of the distal segment of the little toe, 0.1 Cun from the corner of the toenail.

Sole of the foot

Points of the Main channels

Zu Shao Yin channel (Kidney)

KI 1 Yongquan
On the sole, in the depression appearing on the anterior part of the sole when the foot is in the plantar flexion, approximately at the junction of the anterior third and posterior two-thirds of the line connecting the base of the 2nd and 3rd toes and the heel.

KI 2 Rangu
On the internal edge of the foot, below the tuberosity of the navicular bone (tarsal scaphoid) and the 'junction of the red skin and white skin'.

Extra channel points

Curious points

130 PC Lineiting
On the sole, in a depression, at the midpoint of the space separating the 2nd and the 3rd metatarsus, 1 Cun below the metatarsophalangeal line.

131 PC Muzhilihengwen Zhiwen (X-XZ.36)
On the sole, at the midpoint of the metatarso-phalangeal line of the 1st toe.

132 PC Qianhouyinzhu
2 points on the sole of the foot.

- The first point is 0.5 Cun in front of KI 1 Yongquan.
- The second point is 0.5 Cun behind KI 1 Yongquan.

133 PC Zuxin
On the sole, 1 Cun behind KI 1 Yongquan.

134 PC Shimian
On the sole, at the midpoint of the heel.

Points of Master Tong

T 55.01 Huo Bao
On the sole, at the midpoint of the joint between the phalanx and the middle phalanx of the 2nd toe.

T 55.02 Hua Gu Yi
On the sole, draw a line between the 1st and the 2nd toe.
- The first point is 5 fen from the space between the toes.
- The second point is 1 Cun from the same space.
- The third point is 1.5 Cun.
- The fourth point is 2.3 Cun from the same space.

T 55.03 Hua Gu Er
On the sole, 1 Cun and 1.5 Cun behind the space between the 2nd and 3rd toe.

T 55.04 Hua Gu San
On the sole, 2 Cun behind the space between the 3rd and 4th toe.

T 55.05 Hua Gu Si
On the internal side of the foot, the point is at the midpoint of the joint between the distal phalanx and the proximal phalanx of the big toe at the junction between the red and white skin.

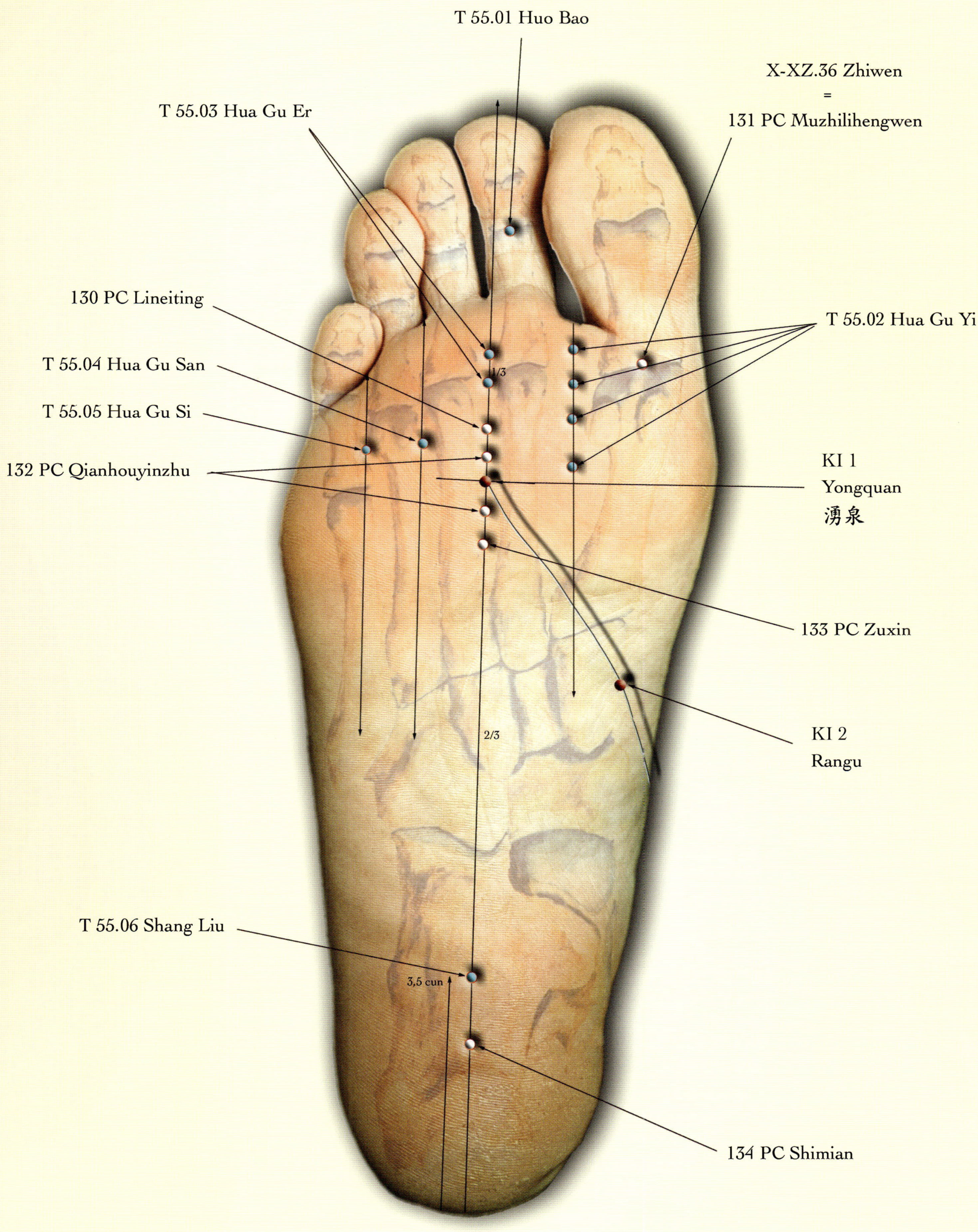
T 55.01 Huo Bao
X-XZ.36 Zhiwen
=
131 PC Muzhilihengwen
T 55.03 Hua Gu Er
130 PC Lineiting
T 55.02 Hua Gu Yi
T 55.04 Hua Gu San
1/3
T 55.05 Hua Gu Si
KI 1
132 PC Qianhouyinzhu
Yongquan
湧泉
133 PC Zuxin
KI 2
Rangu
2/3
T 55.06 Shang Liu
3,5 cun
134 PC Shimian

INDEX

Channel points

By numerical order

By alphabetical order

Extra channel points

Points MN

Points PC

Points PN

Recents points

Points of the Master Tong

Channel points by numerical order

Shu Tai Yin channel: Lung

LU 1	Zhongfu	Meeting the Qi of the Middle Burner.	中府
LU 2	Yunmen	Gateway to the clouds.	雲門
LU 3	Tianfu	Reunion of the celestial Qi.	天府
LU 4	Xiabai	Squeeze the whiteness between them.	俠白
LU 5	Chize	Marsh of the elbow.	尺澤
LU 6	Kongzui	Orifice containing the deepest Qi.	孔最
LU 7	Lieque	A break in the line. A flash.	列缺
LU 8	Jingqu	A constantly flowing rivulet.	經渠
LU 9	Taiyuan	Deep source of life.	太淵
LU 10	Yuji	Limit of the fish belly.	魚際
LU 11	Shaoshang	Small discharge of the Qi of the Lung.	少商

Shu Yang Ming channel: Large Intestine

LI 1	Shangyang	Metal viscus of a Yang channel.	商陽
LI 2	Erjian	Second point in a depression.	二間
LI 3	Sanjian	Third point in a depression.	三間
LI 4	Hegu	Valley formed by the meeting of two bones.	合谷
LI 5	Yangxi	Yang rivulet.	陽谿
LI 6	Pianli	Swerves from its course.	偏歷
LI 7	Wenliu	Keeps the heat.	温溜
LI 8	Xialian	Under the ridge.	下廉
LI 9	Shanglian	Above the ridge.	上廉
LI 10	Shousanli	Three Cun from the hand.	手三里
LI 11	Quchi	Pond of the curve of the elbow.	曲池
LI 12	Zhouliao	Depression near the elbow.	肘髎
LI 13	Shouwouli	Five Cun from the hand.	手五里

LI 14	Binao	Upper part of the arm.	臂臑
LI 15	Jianyu	Hollow of the shoulder.	肩髃
LI 16	Jugu	Big bone.	巨骨
LI 17	Tianding	Celestial three-legged vase.	天鼎
LI 18	Futu	One Fu from the prominence.	扶突
LI 19	Kouheliao	Hollow situated near the mouth.	禾髎
LI 20	Yingxiang	Welcomes perfumes.	迎香

Zu Yang Ming channel: Stomach

ST 1	Chengqi	Receives the tears.	承泣
ST 2	Sibai	Light on all sides.	四白
ST 3	Juliao	Deep hollow.	巨髎
ST 4	Dicang	Clay granary.	地倉
ST 5	Daying	Welcomed by the great artery.	大迎
ST 6	Jiache	Lower jawbone.	頰車
ST 7	Xiaguan	Under the bar.	下關
ST 8	Touwei	In the angle of the head.	頭維
ST 9	Renying	The human being receives Qi.	人迎
ST 10	Shuitu	Water that spurts out.	水突
ST 11	Qishe	Abode of energy.	氣舍
ST 12	Quepen	Chipped bowl.	缺盆
ST 13	Qihu	Gateway to Qi.	氣户
ST 14	Kufang	Shop.	庫房
ST 15	Wuyi	Eaves.	屋翳
ST 16	Yingchuang	Window of the chest.	膺窗
ST 17	Ruzhong	Midpoint of the breast.	乳中
ST 18	Rugen	Lower part of the breast.	乳根
ST 19	Burong	No longer contains.	不容
ST 20	Chengman	Full stomach.	承滿
ST 21	Liangmen	Entrance of the nutriments.	梁門
ST 22	Guanmen	Gateway to the passage.	關門

ST 23	Taiyi	Wide curve.	太乙
ST 24	Huarounmen	Entrance of the essence of the nutriments.	滑肉門
ST 25	Tianshu	Axis of the sky.	天樞
ST 26	Wailing	On the outer side of the hill.	外陵
ST 27	Daju	Large set square.	大巨
ST 28	Shuidao	Way of the waters.	水道
ST 29	Guilai	Way back.	歸來
ST 30	Qichong	Place where the energy goes through.	氣衝
ST 31	Biguan	Joint of the femur.	髀關
ST 32	Futu	Hiding rabbit.	伏兔
ST 33	Yinshi	Yin market.	陰市
ST 34	Liangqiu	Top of the mountains.	梁丘
ST 35	Dubi	Calf's snout.	犢鼻
ST 36	Zusanli	Three Cun from the foot.	足三里
ST 37	Shangjuxu	In the upper part of the vast vacuum.	上巨虛
ST 38	Tiaokou	Large opening.	條口
ST 39	Xiajuxu	In the lower part of the vast vacuum.	下巨虛
ST 40	Fenglong	Abundance and prosperity.	豐隆
ST 41	Jiexi	Rivulet where shoe-laces are tied.	解谿
ST 42	Chongyang	Passage of Qi Yang.	衝陽
ST 43	Xiangu	Bottom of the valley.	陷谷
ST 44	Neiting	Inside the entrance yard.	內庭
ST 45	Lidui	At the other end of the earth.	厲兌

Zu Tai Yin channel: Spleen

SP 1	Yinbai	Hidden whiteness.	隱白
SP 2	Dadu	Large city.	大都
SP 3	Taibai	Beginning of the whiteness.	太白
SP 4	Gongsun	Grandfather and grandson. Yellow emperor.	公孫
SP 5	Shangqiu	Shang hill.	商丘
SP 6	Sanyinjiao	Crossing of the three Yin.	三陰交

SP 7	Lougu	Open valley.	漏谷
SP 8	Diji	Important place on the earth.	地機
SP 9	Yinlingquan	Spring at the foot of the hill.	陰陵泉
SP 10	Xuehai	Sea of blood.	血海
SP 11	Jimen	Gateway to the duster.	箕門
SP 12	Chongmen	Gateway to the crossing.	衝門
SP 13	Fushe	Meet and stay.	府舍
SP 14	Fujie	Knot of the abdomen.	腹結
SP 15	Daheng	Transverse muscle.	大横
SP 16	Fuai	Groan of the abdomen.	腹哀
SP 17	Shidou	Grotto of the nutriments.	食竇
SP 18	Tianxi	Rivulet in the sky.	天谿
SP 19	Xiongxiang	Around the chest.	胸鄉
SP 20	Zhourong	Makes everything splendid.	周榮
SP 21	Dabao	Great steward.	大包

Shu Shao Yin channel: Heart

(HT)HE 1	Jiquan	Supreme source.	極泉
(HT)HE 2	Qingling	Sharp mind.	青靈
(HT)HE 3	Shaohai	Sea of the Shu Shao Yin channel.	少海
(HT)HE 4	Lingdao	Way that leads to the mind.	靈道
(HT)HE 5	Tongli	Communicates with the inside.	通里
(HT)HE 6	Yinxi	Xi point of a Yin channel.	陰郄
(HT)HE 7	Shenmen	Gateway to Shen.	神門
(HT)HE 8	Shaofu	Small workshop.	少府
(HT)HE 9	Shaochong	Important passage of the Shu Shao Yin.	少衝

Shu Tai Yang channel: Small Intestine

SI 1	Shaoze	Small pond.	少澤
SI 2	Qiangu	Anterior valley.	前谷
SI 3	Houxi	Posterior rivulet.	後谿
SI 4	Wangu	Wrist bone.	腕骨

SI 5	Yanggu	Yang valley.	陽谷
SI 6	Yanglao	Helps old age.	養老
SI 7	Zhizheng	Branch that leads to the monarch.	支正
SI 8	Xiaohai	Small sea.	小海
SI 9	Jianzhen	Strength of the shoulder.	肩貞
SI 10	Naoshu	Point of the upper part of the shoulder.	臑俞
SI 11	Tianzong	Celestial ascendant.	天宗
SI 12	Bingfeng	That masters the wind.	秉風
SI 13	Quyuan	Curved wall.	曲垣
SI 14	Jianwaishu	Point of the outer shoulder.	肩外俞
SI 15	Jianzhongshu	Central point of the shoulder.	肩中俞
SI 16	Tianchuang	Window of the sky.	天窗
SI 17	Tianrong	The Qi of the sky reaches the face.	天容
SI 18	Quanliao	Depression of the zygomatic muscle.	顴髎
SI 19	Tinggong	Palace of hearing.	聽宮

Zu Tai Yang channel: Bladder (Urinary)

BL 1	Jingming	Brilliant vessel.	睛明
BL 2	Cuanzhu	Tuft of bamboo leaves.	攢竹
BL 3	Meichong	Assault of the brows.	眉衝
BL 4	Quchai	Deviation and miscarriage.	曲差
BL 5	Wuchu	Located in the middle.	五處
BL 6	Chengguang	Gets the light.	承光
BL 7	Tongtian	That leads to the sky.	通天
BL 8	Luoque	Joins with the brain then withdraws.	絡郤
BL 9	Yuzhen	Jade pillow.	玉枕
BL 10	Tianzhu	Column upholding the sky.	天柱
BL 11	Dazhu	Big shuttle.	大杼
BL 12	Fengmen	Gateway to Feng.	風門
BL 13	Feishu	Acquiescence of the lungs.	肺俞
BL 14	Jueyinshu	Acquiescence of Jueyin.	厥陰俞

BL 15	Xinshu	Acquiescence of the Heart.	心俞
BL 16	Dushu	General acquiescence.	督俞
BL 17	Geshu	Acquiescence of the diaphragm.	膈俞
BL 18	Ganshu	Acquiescence of the Liver.	肝俞
BL 19	Danshu	Acquiescence of the Gall Bladder.	膽俞
BL 20	Pishu	Acquiescence of the Spleen.	脾俞
BL 21	Weishu	Acquiescence of the Stomach.	胃俞
BL 22	Sanjiaoshu	Acquiescence of the Triple Heater.	三焦俞
BL 23	Shenshu	Acquiescence of the Kidney.	腎俞
BL 24	Qihaishu	Acquiescence of (REN)CV 6 Qihai.	氣海俞
BL 25	Dachangshu	Acquiescence of the Large Intestine.	大腸俞
BL 26	Guanyuanshu	Acquiescence of (REN)CV 4 Guanyuan.	關元俞
BL 27	Xiaochangshu	Acquiescence of the Small Intestine.	小腸俞
BL 28	Pangguangshu	Acquiescence of the Bladder.	膀胱俞
BL 29	Zhongliishu	Acquiescence of the central vertebrae.	中膂俞
BL 30	Baihuanshu	Acquiescence of the sphyncter.	白環俞
BL 31	Shangliao	Upper hollow of the bone.	上髎
BL 32	Ciliao	Second hollow of the bone.	次髎
BL 33	Zhongliao	Hollow midpoint of the bone.	中髎
BL 34	Xialiao	Lower hollow of the bone.	下髎
BL 35	Huiyang	Meeting the Yang.	會陽
BL 36	Chengfu	Receives and supports.	承扶
BL 37	Yinmen	Gateway to a prosperous place.	殷門
BL 38	Fuxi	Wavering and crack.	浮郄
BL 39	Weiyang	Outside the delegate.	委陽
BL 40	Weizhong	Midpoint of the flex crease.	委中
BL 41	Fufen	Added branch.	附分
BL 42	Pohu	Abode of Po.	魄户
BL 43	Gaohuang	Life centres.	膏肓俞
BL 44	Shentang	Palace of Shen.	神堂

BL 45	Yixi	Sigh.	譩譆
BL 46	Geguan	Passage of the diaphragm.	膈關
BL 47	Hunmen	Gateway to Hun.	魂門
BL 48	Yanggang	Dominates the Yang.	陽綱
BL 49	Yishe	Abode of Yi.	意舍
BL 50	Weicang	Granary of the stomach.	胃倉
BL 51	Huangmen	Gateway to the membranes.	肓門
BL 52	Zhishi	Chamber of Zhi.	志室
BL 53	Baohuang	Membrane surrounding the uterus.	胞肓
BL 54	Zhibian	On the side of the queue.	秩邊
BL 55	Heyang	Reunion of the Yang.	合陽
BL 56	Chengjin	Receives the muscles.	承筋
BL 57	Chengshan	Upholds the mountain.	承山
BL 58	Feiyang	To rise quickly.	飛揚
BL 59	Fuyang	Yang above the tarsus.	跗陽
BL 60	Kunlun	Kun Lun mountains.	崑崙
BL 61	Pushen	The maid-servant's greeting.	僕參
BL 62	Shenmai	Vessel of Shen hour.	申脉
BL 63	Jinmen	Gateway to the metal.	金門
BL 64	Jinggu	Big bone.	京骨
BL 65	Shugu	Shrunken bone.	束骨
BL 66	Zutonggu	Communicates with the valley.	足通谷
BL 67	Zhiyin	Arrival at Yin.	至陰

Zu Shao Yin channel: Kidney

KI 1	Yongquan	Spring that spurts out.	湧泉
KI 2	Rangu	Valley of fire.	然谷
KI 3	Taixi	Important rivulet.	太谿
KI 4	Dazhong	Big bell.	大鐘
KI 5	Shuiquan	Source of the water.	水泉
KI 6	Zhaohai	Illuminated sea.	照海

KI 7	Fuliu	Returns to its normal course.	復溜
KI 8	Jiaoxin	Meets faithfulness.	交信
KI 9	Zhubin	Built-up shore.	築賓
KI 10	Yingu	Valley of Yin.	陰谷
KI 11	Henggu	Transverse bone.	橫骨
KI 12	Dahe	Great prosperity.	大赫
KI 13	Qixue	Bridge on Qi.	氣穴
KI 14	Siman	Four fullnesses.	四滿
KI 15	Zhongzhu	Pours its Qi into the uterus.	中注
KI 16	Huangshu	Point of the membranes.	肓俞
KI 17	Shanqu	Curve of the Large Intestine.	商曲
KI 18	Shiguan	Stone barrier.	石關
KI 19	Yindu	Yin gathering.	陰都
KI 20	Futonggu	Communicates with the abdomen.	腹通谷
KI 21	Youmen	Pylorus.	幽門
KI 22	Bulang	Veranda of the footsteps.	步廊
KI 23	Shenfeng	Territory of Shen.	廉泉
KI 24	Lingxu	Mountain of the mind.	靈墟
KI 25	Shencang	Divine refuge.	神藏
KI 26	Yuzhong	The splendour is inside.	彧中
KI 27	Shufu	Brings Qi to the organs.	俞府

Shu Jue Yin channel: Pericardium (Master of the Heart)

(MH)PC 1	Tianchi	Lagoon of the sky.	天池
(MH)PC 2	Tianquan	Celestial source.	天泉
(MH)PC 3	Quze	Pond of the curve.	曲澤
(MH)PC 4	Ximen	Crack. Door.	郄門
(MH) PC 5	Jianshi	Minister.	間使
(MH)PC 6	Neiguan	Internal barrier.	內關
(MH)PC 7	Daling	High hillock.	大陵
(MH)PC 8	Laogong	Palace of labour.	勞宮

(MH)PC 9	Zhongchong	Assault of the middle finger.	中衝

Shu Shao Yang channel: Triple Heater or Triple Burner

(TH)TB 1	Guanchong	Assault of the barrier.	關衝
(TH)TB 2	Yemen	Gateway to the liquids.	液門
(TH)TB 3	Zhongzhu	Middle islet.	中渚
(TH)TB 4	Yangchi	Yang marsh.	陽池
(TH)TB 5	Waiguan	Barrier of the Yang.	外關
(TH)TB 6	Zhigou	Crease of the forearm.	支溝
(TH)TB 7	Huizong	Accumulation and reunion.	會宗
(TH)TB 8	Sanyangluo	Reunion of the three Yang.	三陽絡
(TH)TB 9	Sidu	Four important rivers.	四瀆
(TH)TB 10	Tianjing	Well in the sky.	天井
(TH)TB 11	Qinglengyuan	Clear cold abyss.	清冷淵
(TH)TB 12	Xiaoluo	Makes the marshes disappear.	消濼
(TH)TB 13	Naohui	Reunion of the deltoid.	臑會
(TH)TB 14	Jianliao	Hollow of the shoulder.	肩髎
(TH)TB 15	Tianliao	Celestial hollow.	天髎
(TH)TB 16	Tianyou	Side window of the sky.	天牖
(TH)TB 17	Yifeng	Protects from the assaults of the wind.	翳風
(TH)TB 18	Chimai	Vessel of the convulsions.	瘈脉
(TH)TB 19	Luxi	The head finds repose.	顱息
(TH)TB 20	Jiaosun	Tip and small ramifications.	角孫
(TH)TB 21	Ermen	Gateway to the ear.	耳門
(TH)TB 22	Erheliao	Hollow of harmony.	和髎
(TH)TB 23	Sizhukong	Bamboos in a file.	絲竹空

Zu Shao Yang channel: Gall Bladder

GB 1	Tongziliao	Hollow of the pupil.	瞳子髎
GB 2	Tinghui	Reunion of hearing.	聽會
GB 3	Shangguan	Above the bar.	上關
GB 4	Hanyan	Heavy chin.	頷厭

GB 5	Xuanlu	Hanging from the cranium.	懸顱
GB 6	Xuanli	At a short distance from Xuanlu.	懸厘
GB 7	Qubin	Twisted bun.	曲鬢
GB 8	Shuaigu	Commander of all the valleys.	率谷
GB 9	Tianchong	At the conquest of the sky.	天衝
GB 10	Fubai	On the surface and treats the whiteness.	浮白
GB 11	Touqiaoyin	Opening Yin.	頭竅陰
GB 12	Wangu	Perfect bone.	完骨
GB 13	Benshen	Basis of Shen.	本神
GB 14	Yangbai	Yang clarity.	陽白
GB 15	Toulinqui	Close to tears.	頭臨泣
GB 16	Muchuang	Window of the eyes.	目窗
GB 17	Zhengying	Encampment situated in the middle.	正營
GB 18	Chengling	Receives what is immaterial.	承靈
GB 19	Naokong	Hollow of the cranium.	腦空
GB 20	Fengchi	Lagoon of the winds.	風池
GB 21	Jianjing	Well of the shoulder.	肩井
GB 22	Yuanye	Abyss of the armpit.	淵腋
GB 23	Zhejin	Muscles on each side.	輒筋
GB 24	Riyue	The sun and the moon.	日月
GB 25	Jingmen	Gateway to the original energy.	京門
GB 26	Daimai	Belt vessel.	帶脉
GB 27	Wushu	Middle axis.	五樞
GB 28	Weidao	Meeting path.	維道
GB 29	Juliao	Is situated in a hollow of the bone.	居髎
GB 30	Huantiao	To jump into a circle.	環跳
GB 31	Fengshi	Fair of the wind.	風市
GB 32	Zhongdu	Middle canal.	中瀆
GB 33	Xiyangguan	Outside barrier.	膝陽關
GB 34	Yanglingquan	Spring of the central plateau.	陽陵泉

GB 35	Yangjiao	Meeting of the Yang.	陽交
GB 36	Waiqiu	Outer hillock.	外丘
GB 37	Guangming	Bright light.	光明
GB 38	Yangfu	Yang side of the fibula.	陽輔
GB 39	Xuanzhong	Hanging bell.	懸鐘
GB 40	Qiuxu	Between the low and the high hillock.	丘墟
GB 41	Zulinqi	Point of the foot that treats weeping.	足臨泣
GB 42	Diwuhui	Fifth reunion of the earth.	地五會
GB 43	Xiaxi	Deeply embanked rivulet.	俠谿
GB 44	Zuqiaoyin	Point of the foot, passage to Yin.	足竅陰

Zu Jue Yin channel: Liver

(LR)LV 1	Dadun	Emporium.	大敦
(LR)LV 2	Xingjian	Circulation. Cure.	行間
(LR)LV 3	Taichong	Very important passage.	太衝
(LR)LV 4	Zhongfeng	Between two mountains.	中封
(LR)LV 5	Ligou	Woodworm hole.	蠡溝
(LR)LV 6	Zhongdu	Remains in the centre.	中都
(LR)LV 7	Xiguan	Knee joint.	膝關
(LR)LV 8	Ququan	Spring in a curve.	曲泉
(LR)LV 9	Yinbao	Wrapped in Yin.	陰包
(LR)LV 10	Wuli	Fifth phase.	五里
(LR)LV 11	Yinlian	On the side of Yin.	陰廉
(LR)LV 12	Jimai	Fast vessel.	急脉
(LR)LV 13	Zhangmen	Wooden door.	章門
(LR)LV 14	Qimen	End-of-cycle gateway.	期門

Ren Mai vessel: Conception vessel

(REN)CV 1	Huiyin	Meeting of the Yin.	會陰
(REN)CV 2	Qugu	Curved bone.	曲骨
(REN)CV 3	Zhongji	Centre and extremity.	中極
(REN)CV 4	Guanyuan	Contains the original Qi.	關元

(REN)CV 5	Shimen	Stone gateway.	石門
(REN)CV 6	Qihai	Sea of energy.	氣海
(REN)CV 7	Yinjiao	Crossing of the Yin.	陰交
(REN)CV 8	Shenque	Gateway to the palace of Yin.	神闕
(REN)CV 9	Shuifen	Watershed.	水分
(REN)CV 10	Xiawan	Lower gateway to the stomach.	下脘
(REN)CV 11	Jianli	Regularises the inside.	建里
(REN)CV 12	Zhongwan	Middle part of the stomach.	中脘
(REN)CV 13	Shangwan	Upper part of the stomach.	上脘
(REN)CV 14	Juque	Immense gateway to the palace.	巨闕
(REN)CV 15	Jiuwei	Turtle-dove tail.	鳩尾
(REN)CV 16	Zhongting	Central court yard.	中庭
(REN)CV 17	Danzhong	Midpoint of the chest.	膻中
(REN)CV 18	Yutang	Large jade room.	玉堂
(REN)CV 19	Zigong	Crimson palace.	紫宮
(REN)CV 20	Huagai	The emperor's canopy.	華蓋
(REN)CV 21	Xuanji	Ursa Major.	璇璣
(REN)CV 22	Tiantu	Celestial chimney.	天突
(REN)CV 23	Lianquan	Clear spring.	廉泉
(REN)CV 24	Chengjiang	Receives the syrup.	承漿

Du Mai vessel: Governor vessel

(DU)GV 1	Changqiang	Wide and powerful.	長強
(DU)GV 2	Yaoshu	Acquiescence of the loins.	腰俞
(DU)GV 3	Yaoyangguan	Yang barrier.	腰陽關
(DU)GV 4	Mingmen	Gateway to life.	命門
(DU)GV 5	Xuanshu	Hanging axis.	懸樞
(DU)GV 6	Jizhong	Midpoint of the backbone.	脊中
(DU)GV 7	Zhongshu	Middle axis.	中樞
(DU)GV 8	Jinsuo	Contraction of the tendons.	筋縮
(DU)GV 9	Zhiyang	Arrival at Yang.	至陽

(DU)GV 10	Lingtai	Ling rostrum.	靈臺
(DU)GV 11	Shendao	Way of Shen.	神道
(DU)GV 12	Shenzhu	Pillar of the body.	身柱
(DU)GV 13	Taodao	Way of the potters.	陶道
(DU)GV 14	Dazhui	Big vertebra.	大椎
(DU)GV 15	Yamen	Gateway to muteness.	啞門
(DU)GV 16	Fengfu	Where the wind is.	風府
(DU)GV 17	Naohu	Gateway to the brain.	腦户
(DU)GV 18	Qiangjian	Solid place.	強間
(DU)GV 19	Houding	Behind the summit.	後頂
(DU)GV 20	Baihui	Reunion of one hundred vessels.	百會
(DU)GV 21	Qianding	Before the vertex.	前頂
(DU)GV 22	Xinhui	Reunion of the fontanels.	顖會
(DU)GV 23	Shangxing	Star that is up above.	上星
(DU)GV 24	Shenting	Palace of the developped one.	神庭
(DU)GV 25	Suliao	Opening of the whiteness.	素髎
(DU)GV 26	Shuigou	Centre of the being.	水溝
(DU)GV 27	Duiduan	At the extremity of the mouth.	兑端
(DU)GV 28	Yinjiao	Reunion with the gums.	齦交

B

Baihuanshu	BL 30
Baihui	(DU)GV 20
Baohuang	BL 53
Benshen	GB 13
Biguan	ST 31
Binao	LI 14
Bingfen	SI 12
Bulang	KI 22
Burong	ST 19

C

Changqiang	(DU)GV 1
Chengfu	BL 36
Chengguang	BL 6
Chengjiang	(REN)CV 24
Chengjin	BL 56
Chengling	GB 18
Chengman	ST 20
Chengqi	ST 1
Chengshan	BL 57
Chimai	(TH)TB 18
Chize	LU 5
Chongmen	SP 12
Chongyang	ST 42
Ciliao	BL 32
Cuanzhu	BL 2

D

Dabao	SP 21
Dachangshu	BL 25
Dadu	SP 2
Dadun	(LR)LV 1
Dahe	KI 12
Daheng	SP 15
Daimai	GB 26
Daju	ST 27
Daling	(MH)PC 7
Danshu	BL 19
Danzhong	(REN)CV 17
Dazhu	BL 11
Daying	ST 5
Dazhong	KI 4
Dazhui	(DU)GV 14
Dicang	ST 4
Diji	SP 8
Diwuhui	GB 42
Dubi	ST 35
Duiduan	GV 27
Dushu	BL 16

E

Erheliao	(TH)TB 22
Erjian	LI 2
Ermen	(TH)TB 21

F

Feishu	BL 13
Feiyang	BL 58
Fengchi	GB 20
Fengfu	(DU)GV 16
Fenglong	ST 40
Fengmen	BL 12
Fengshi	GB 31
Fuai	SP 16
Fubai	GB 10
Fufen	BL 41
Fujie	SP 14
Fuliu	KI 7
Fushe	SP 13
Futonggu	KI 20
Futu	LI 18
Futu	ST 32
Fuxi	BL 38
Fuyang	BL 59

G

Ganshu	BL 18
Gaohuang	BL 43
Geguan	BL 46
Geshu	BL 17
Gongsun	SP 4
Guanchong	(TH)TB 1

Guangming GB 37
Guanmen ST 22
Guanyuan (REN)CV 4
Guanyuanshu BL 26
Guilai ST 29

H

Hanyan GB 4
Hegu LI 4
Henggu KI 11
Heyang BL 55
Houding GV 19
Houxi SI 3
Huagai CV 20
Huangmen BL 51
Huangshu KI 16
Huantiao GB 30
Huaroumen ST 24
Huiyang BL 35
Huiyin (REN)CV 1
Huizong (TH)TB 7
Hunmen BL 47

J

Jiache ST 6
Jianjing GB 21
Jianli CV 11
Jianliao (TH)TB 14
Jianshi (MH)PC 5
Jianwaishu SI 14
Jianyu LI 15
Jianzhen SI 9
Jianzhongshu SI 15
Jiaosun (TH)TB 20
Jiaoxin KI 8
Jiexi ST 41
Jimai (LR)LV 12
Jimen SP 11
Jinggu BL 64
Jingmen GB 25
Jingming BL 1
Jingqu LU 8
Jinmen BL 63
Jinsuo (DU)GV 8
Jiquan (HT)HE 1
Jiuwei (REN)CV 15
Jizhong (DU)GV 6
Jueyinshu BL 14
Jugu LI 16
Juque (REN)CV 14
Juliao ST 3
Juliao GB 29

K

Kongzui LU 6
Kouheliao LI 19
Kufang ST 14
Kunlun BL 60

L

Laogong (MH)PC 8
Liangmen ST 21
Liangqiu ST 34
Lianquan (REN)CV 23
Lidui ST 45
Lieque LU 7
Ligou (LR)LV 5
Lingdao (HT)HE 4
Lingtai (DU)GV 10
Lingxu KI 24
Lougu SP 7
Luoque BL 8
Luxi (TH)TB 19

M

Meichong BL3
Mingmen (DU)GV 4
Muchuang GB 16

N

Naohu (DU)GV 17
Naohui (TH)TB 13
Naokong GB 19
Naoshu SI 10
Neiguan (MH)PC 6
Neiting ST 44

P

Pangguanshu BL 28
Pianli LI 6
Pishu BL 20
Pohu BL 42
Pushen BL 61

Q

Qianding GV 21
Qiangjian GV 18
Qiangu SI 2
Qichong ST 30
Qihai (REN)CV 6
Qihaishu BL 24
Qihu ST 13
Qimen (LR)LV 14
Qinglengyuan (TH)TB 11
Qingling (HT)HE 2
Qishe ST 11
Qiuxu GB 40
Qixue KI 13
Quanliao SI 18
Qubin GB 7
Quchai BL 4
Quchi LI 11
Quepen ST 12
Qugu (REN)CV 2
Ququan (LR)LV 8
Quyuan SI 13
Quze (MH)PC 3

R

Rangu KI 2
Renying ST 9
Riyue GB 24
Rugen ST 18
Ruzhong ST 17

S

Sanjian LI 3
Sanjiaoshu BL 22
Sanyangluo GB 8
Sanyinjiao SP 6
Shangguan GB 3
Shangjuxu ST 37
Shanglian LI 9
Shangliao BL 31
Shangqiu SP 5
Shangwan (REN)CV 13
Shangxing GV 23
Shangyang LI 1
Shanqu KI 17
Shaochong (HT)HE 9
Shaofu (HT)HE 8
Shaohai (HT)HE 3
Shaoshang LU 11
Shaoze SI 1
Shencang KI 25
Shendao (DU)GV 11
Shenfeng KI 23
Shenmai BL 62
Shenmen (HT)HE 7
Shenque (REN)CV 8
Shenshu BL 23
Shentang BL 44
Shenting (DU)GV 24
Shenzhu (DU)GV 12
Shidou SP 17
Shiguan KI 18
Shimen (REN)CV 5
Shousanli LI 10
Shouwouli LI 13
Shuaigu GB 8
Shufu KI 27
Shugu BL 65
Shuidao ST 28
Shuifen (REN)CV 9
Shuigou (DU)GV 26
Shuiquan KI 5
Shuitu ST 10
Sibai ST 2
Sidu (TH)TB 9
Siman KI 14
Sizhukong (TH)TB 23
Suliao (DU)GV 25

T

Taibai SP 3
Taichong (LR)LV 3
Taixi KI 3
Taiyi ST 23
Taiyuan LU 9

Taodao (DU)GV 13
Tianchi (MH)PC 1
Tianchong GB 9
Tianchuang SI 16
Tianding LI 17
Tianfu LU 3
Tianjing (TH)TB 10
Tianliao (TH)TB 15
Tianquan (MH)PC 2
Tianrong SI 17
Tianshu ST 25
Tiantu (REN)CV 22
Tianxi SP 18
Tianyou (TH)TB 16
Tianzhu BL 10
Tianzong SI 11
Tiaokou ST 38
Tinggong SI 19
Tinghui GB 2
Tongli (HT)HE 5
Tongtian BL 7
Tongziliao GB 1
Toulinqui GB 15
Touqiaoyin GB 11
Touwei ST 8

W

Waiguan (TH)TB 5
Wailing ST 26
Waiqiu GB 36
Wangu SI 4
Wangu GB 12
Weicang BL 50
Weidao GB 28
Weishu BL 21
Weiyang BL 39
Weizhong BL 40
Wenliu LI 7
Wuchu BL 5
Wuli (LR)LV 10
Wushu GB 27
Wuyi ST 15

X

Xiabai LU 4
Xiaguan ST 7
Xiajuxu ST 39
Xialian LI 8
Xialiao BL 34
Xiangu ST 43
Xiaochangshu BL 27
Xiaohai SI 8
Xiaoluo (TH)TB 12
Xiawan (REN)CV 10
Xiaxi GB 43
Xiguan (LR)LV 7
Ximen (MH)PC 4
Xingjian (LR)LV 2
Xinhui (DU)GV 22
Xinshu BL 15
Xiongxiang SP 19
Xiyangguan GB 33
Xuanji (REN)CV 21
Xuanli GB 6
Xuanlu GB 5
Xuanshu (DU)GV 5
Xuanzhong GB 39
Xuehai SP 10

Y

Yamen (DU)GV 15
Yangbai GB 14
Yangchi (TH)TB 4
Yangfu GB 38
Yanggang BL 48
Yanggu SI 5
Yangjiao GB 35
Yanglao SI 6
Yanglingquan GB 34
Yangxi LI 5
Yaoshu (DU)GV 2
Yaoyangguan (DU)GV 3
Yemen (TH)TB 2
Yifeng (TH)TB 17
Yinbai SP 1
Yinbao (LR)LV 9
Yindu KI 19
Yingchuang ST 16
Yingu KI 10
Yingxiang LI 20
Yinjiao (REN)CV 7
Yinjiao (DU)GV 28
Yinlian (LR)LV 11
Yinlingquan SP 9
Yinmen BL 37
Yinshi ST 33

Yinxi (HT)HE 6
Yishe BL 49
Yixi BL 45
Yongquan KI 1
Youmen KI 21
Yuanye GB 22
Yuji LU 10
Yunmen LU 2
Yutang (REN)CV 18
Yuzhen BL 9
Yuzhong KI 26

Z

Zhangmen (LR)LV 13
Zhaohai KI 6
Zhejin GB 23
Zhengying GB 17
Zhibian BL 54
Zhigou (TH)TB 6
Zhishi BL 52
Zhiyang GV 9
Zhiyin BL 67
Zhizheng SI 7
Zhongchong (MH)PC 9
Zhongdu GB 32
Zhongdu (LR)LV 6
Zhongfeng (LR)LV 4
Zhongfu LU 1
Zhongji (REN)CV 3
Zhongliao BL 33
Zhongliishu BL 29
Zhongshu (DU)GV 7
Zhongting (REN)CV 16
Zhongwan (REN)CV 12
Zhongzhu KI 15
Zhongzhu (TH)TB 3
Zhouliao LI 12
Zhourong SP 20
Zhubin KI 9
Zigong (REN)CV 19
Zulinqi GB41
Zuqiaoyin GB 44
Zusanli ST 36
Zutonggu BL 66

Extra channel points PC

By numerical order

1 PC	Sishencong	Reunion of the four gods.
2 PC	Ezhong	Middle of the brow.
3 PC	Yintang	Outline of the brow. Seal of the brow.
4 PC	Shangen	Foot of the mountain.
5 PC	Touquangming	Extremity of the light.
6 PC	Yuyao	Fish back.
7 PC	Yuwei	Fish tail.
8 PC	Qiuhou	Behind the eye.
9 PC	Taiyang	Great Yang.
10 PC	Erjian	Tip of the ear.
11 PC	Lungxue	Point that moves.
12 PC	Hujingmai – Erbei	Venous point of the ear.
13 PC	Yiming	Cures the light. Hidden point that reflects the light. Bright barrier.
14 PC	Shangyingxiang	Above LI 20 Yingxiang. Receives the perfume. Upper point.
15 PC	Jiabi	First and second celestial ones.
16 PC	Biliu	Nasal discharge.
17 PC	Sanjiao	Three burners.
18 PC	Heliao	Malar bone.
19 PC	Dihe	Reunion of the Earth.
20 PC	Jinjin – Yuye	Metal. Jade.
21 PC	Shanglianquan	Above (REN)CV 23 Lianquan.
22 PC	Waijinjin – Waiyuyue	Metal and jade outside.
23 PC	Hongyin	Immense sound.
24 PC	Panglianquan	Beside (REN)CV 23. Lateral source of modesty.
25 PC	Yaxue	Mute point.
26 PC	Biantao	Tonsils. Flattened peach.
27 PC	Luojing	Going down the neck. Stiff neck.
28 PC	Fengyan	Wind and stones. Wind rock.
29 PC	Xinxi	New experience. Known again.
30 PC	Bailiao	One hundred labours, one hundred fatigues.
31 PC	Chonggu	Ancestral bone.
32 PC	Chixue	Red point.
33 PC	Tanchuan	Dyspnoea and cold.
34 PC	Longhan	Dragon chin.
35 PC	Zuoyi – Youyi	Suitable on the right, suitable on the left.
36 PC	Maihua	Plum tree blossom.
37 PC	Shicang	Food shop.
38 PC	Shiguan	Food gate.
39 PC	Hunshe	House of the one who appears.

40 PC	Hangwen	Horizontal line.
41 PC	Qizhong	Central energy. Centre of vital energy ' Chi'.
42 PC	Jingzhong	Central channel.
43 PC	Waisiman	Outside point KI 14 Siman. Four external plenitudes.
44 PC	Jue Yun	
45 PC	Yijing	Spermattorhoea.
46 PC	Qimen	Energy gate.
47 PC	Weibao	Connecting membrane. That attaches the uterus.
48 PC	Changyi	Intestinal residue.
49 PC	Zigong	Uterus. Womb. Palace of the son.
50 PC	Tingtou	Head up.
51 PC	Chuanxi	Dyspnoea.
52 PC	Baizhongfeng	A hundred kinds of winds.
53 PC	Bahua	Eight flowers.
54 PC	Jiujia	Big sensitive point.
55 PC	Liaosen	Nameless.
56 PC	Zhouzhu	Beside the pillar.
57 PC	Juqueshu	Immense void of point Shu. Canal of the grand palace.
58 PC	Liangyan	Measurement of the eye.
59 PC	Qichuan	Asthma.
60 PC	Yinkou	Silver mouth.
61 PC	Bazhuixia	Below the eighth thoracic.
62 PC	Yishu	Point of the spleen.
63 PC	Shubian	Angle of the eyebrow.
64 PC	Zhuogu	Hole in the swimming pool.
65 PC	Jiagu	Reduction of the fracture.
66 PC	Pigen	Land of the curves.
67 PC	Xuechou	Blood – originated sadness.
68 PC	Changfeng	Intestinal winds.
69 PC	Jijupikuai	Collection of curvaceous masses.
70 PC	Zhuzhang	Bamboo cane.
71 PC	Xiajishu	Point Shu below the hip.
72 PC	Huqimen	Behind the curious door.
73 PC	Yaoyi	Suitable to the belt.
74 PC	Yaoyan	Eye of the belt. A look on the lumbar region.
75 PC	Shiqizhuixia	Below the 17th vertebra.
76 PC	Zhonggong	
77 PC	Yaogen	Root of the belt.
78 PC	Couyi – Jiuqi	Branch reserved to the pigeon.
79 PC	Yaoqi	Curious point of the belt.
80 PC	Xiazhui	Low vertebra. Lower vertebra.
81 PC	Yutian	Jade field.
82 PC	Pinxueling	Excellent against anaemia.

83 PC	Tunzhong	Half of the buttock.
84 PC	Huanzhong	Centre of the circle.
85 PC	Huatuojiaji	Dorsal points of Huao Tuo.
86 PC	Shixuan – Sap Sun	Ten announcements. Ten messengers.
87 PC	Muzhijian	Tip of the thumb.
88 PC	Xiaozhijian	Tip of the little finger.
89 PC	Fengquan	Gate of the wind.
90 PC	Jiudianfeng	
91 PC	Muzhijiehengwen	To disperse the knuckle of the thumb.
92 PC	Fangyan	Eye of the typhoon.
93 PC	Mingyan	Clear eye.
94 PC	Sifeng	The four furrows.
95 PC	Shouzhongping	Middle and central part of the hand.
96 PC	Panglaogong	Beside point (MH)PC 8 Laogong.
97 PC	Neiyangchi	Internal Yangchi.
98 PC	Banmen	Trademen's entrance.
99 PC	Sanshang	Three merchants.
100 PC	Dagukong	Void of the big bone. Void between the big bones.
101 PC	Zhongkui – Erjiaoming	Middle of the big ones.
102 PC	Xiaogukong	Void of the little bone.
103 PC	Yiachiamen	First gate of the tower.
104 PC	Erchianmen	Second gate of the tower.
105 PC	Xiaoshixie	To disperse the little finger.
106 PC	Quanjian	Authority and integrity.
107 PC	Baxie	Eight daemons.
108 PC	Luozhen	Mislaid pillow. Fall of the pillow. Stiff neck.
109 PC	Wailaogong	Opposite PC 108.
110 PC	Weiling	Majestic and holy.
111 PC	Jingling	Pure and holy.
112 PC	Shoujinmen	Golden gate of the hand.
113 PC	Dingshu	Anthrax Shu.
114 PC	Erbai	The two white ones.
115 PC	Bizhong	Midpoint of the forearm.
116 PC	Zexia	Lower pond.
117 PC	Zeqian	Front pond.
118 PC	Zhongquan	Central spring.
119 PC	Cunping	One distance from the vessel.
120 PC	Xiawenliu	Lower warm current.
121 PC	Chinao	Interior of the radius.
122 PC	Quyangwei	Treats functional impotence.
123 PC	Sanchi	Three seas.
124 PC	Zhoushu	Point Shu of the elbow.
125 PC	Yeling	Marvellous armpit. Axillary spirit.
126 PC	Tianling	Celestial marvel.
127 PC	Jianyu	Point Shu of the shoulder.

128 PC	Yuqian	In front of the tubercle of the hole.
129 PC	Houye	Behind the armpit.
130 PC	Lineiting	Below ST 44 Neiting.
131 PC	Muzhilihengwen	Transversal line below the first finger.
132 PC	Qianhouyinzhu	Pearls hidden before and behind.
133 PC	Zuxin	Midpoint of the foot.
134 PC	Shimian	Insomnia.
135 PC	Xiaozhijian	Tip of the little toe.
136 PC	Muzhiheng – Lisanmao	Three hairs of the transversal line of the first toe.
137 PC	Bafeng	Eight winds.
138 PC	Nuxi	Woman's knee.
139 PC	Quanshengzu	Spring source that begets the foot.
140 PC	Taiyinqiao	Great Yinqiao.
141 PC		
142 PC	Lanwei	Appendix.
143 PC	Zuzhongping	Median part of the foot.
144 PC	Xixia	Below the knee.
145 PC	Neixiyang	Eye of the knee.
146 PC	Neihuaijian	Tip of the internal section.
147 PC	Zhizhuanjin	To treat dislocation.
148 PC	Shaoyangwei	Little Yangwei.
149 PC	Chengming	To obey orders.
150 PC	Jiaoyi	Good manners.
151 PC	Waihuaijian	Tip of the external section.
152 PC	Danangxue	Point of the gall bladder.
153 PC	Linghou	Behind the tomb. Behind the mound.
154 PC	Linghouxia	Below point 153 PC Linghou.
155 PC	Xiwai	External side of the knee.
156 PC	Heding	Crown of the top. Heron head.
157 PC	Kuangku	Ilium.
158 PC	Shenxi	Kidney system.
159 PC	Quantu	Door of the rabbit hutch.
160 PC	Liaoliao	Extended and empty.
161 PC	Dalun	Great change.
162 PC	Zuming	Bright foot.
163 PC	Baichongwo	Microbial hotbed. Nest of one hundred insects.
164 PC	Zuluo	Basket foot. Cramps of the lower members.
165 PC	Changgu	Formed bone.
166 PC 167PC 168 PC	Yinwei	Yin equilibrium.
169 PC	Silian	Four connections.
170 PC	Wuling	Five marvels.
171 PC	Lingbao	Holy treasure.

Extra channel points PC

By alphabetical order

141 PC

B

Bafeng 137 PC
Bahua 53 PC
Baichongwo 163 PC
Bailiao 30 PC
Baizhongfeng 52 PC
Banmen 98 PC
Baxie 107 PC
Bazhuixia 61 PC
Biantao 26 PC
Biliu 16 PC
Bizhong 115 PC

C

Changfeng 68 PC
Changgu 165 PC
Changyi 48 PC
Chengming 149 PC
Chinao 121 PC
Chixue 32 PC
Chonggu 31 PC
Chuanxi 51 PC
Couyi – Jiuqi 78 PC
Cunping 119 PC

D

Dagukong 100 PC
Dalun 161 PC
Danangxue 152 PC
Dihe 19 PC
Dingshu 113 PC

E

Erbai 114 PC
Erchianmen 104 PC
Erjian 10 PC
Ezhong 2 PC

F

Fangyan 92 PC
Fengquan 89 PC
Fengyan 28 PC

H

Hangwen 40 PC
Heding 156 PC
Heliao 18 PC
Hongyin 23 PC
Houye 129 PC
Huanzhong 84 PC
Huatuojiaji 85 PC
Hujingmai – Erbei 12 PC
Hunshe 39 PC
Huqimen 72 PC

J

Jiabi 15 PC
Jiagu 65 PC
Jianyu 127 PC
Jiaoyi 150 PC
Jijupikuai 69 PC
Jingling 111 PC
Jingzhong 42 PC
Jinjin – Yuye 20 PC
Jiudianfeng 90 PC
Jiujia 54 PC
Jue Yun 44 PC
Juqueshu 57 PC

K

Kuangku 157 PC

L

Lanwei 142 PC
Liangyan 58 PC

Liaoliao 160 PC
Liaosen 55 PC
Lineiting 130 PC
Lingbao 171 PC
Linghou 153 PC
Linghouxia 154 PC
Longhan 34 PC
Lungxue 11 PC
Luojing 27 PC
Luozhen 108 PC

M

Maihua 36 PC
Mingyan 93 PC
Muzhiheng – Lisanmao 136 PC
Muzhijian 87 PC
Muzhijiehengwen 91 PC
Muzhilihengwen 131 PC

N

Neihuaijian 146 PC
Neixiyang 145 PC
Neiyangchi 97 PC
Nuxi 138 PC

P

Panglaogong 96 PC
Panglianquan 24 PC
Pigen 66 PC
Pinxueling 82 PC

Q

Qianhouyinzhu 132 PC
Qichuan 59 PC
Qimen 46 PC
Qiuhou 8 PC
Qizhong 41 PC
Quanjian 106 PC
Quanshengzu 139 PC
Quantu 159 PC
Quyangwei 122 PC

S

Sanchi 123 PC
Sanjiao 17 PC
Sanshang 99 PC
Shangen 4 PC
Shanglianquan 21 PC
Shangyingxiang 14 PC
Shaoyangwei 148 PC
Shenxi 158 PC
Shicang 37 PC
Shiguan 38 PC
Shimian 134 PC
Shiqizhuixia 75 PC
Shixuan – Sap Sun 86 PC
Shoujinmen 112 PC
Shouzhongping 95 PC
Shubian 63 PC
Sifeng 94 PC
Silian 169 PC
Sishencong 1 PC

T

Taiyang 9 PC
Taiyinqiao 140 PC
Tanchuan 33 PC
Tianling 126 PC
Tingtou 50 PC
Touquangming 5 PC
Tunzhong 83 PC

W

Waihuaijian 151 PC
Waijinjin – Waiyuyue 22 PC
Wailaogong – Wailaokong 109 PC
Waisiman 43 PC
Weibao 47 PC
Weiling 110 PC
Wuling 170 PC

X

Xiajishu 71 PC
Xiaogukong 102 PC
Xiaoshixie 105 PC

Xiaozhijian 88 PC
Xiaozhijian 135 PC
Xiawenliu 120 PC
Xiazhui 80 PC
Xinxi 29 PC
Xiwai 155 PC
Xixia 144 PC
Xuechou 67 PC

Y

Yaogen 77 PC
Yaoqi 79 PC
Yaoyan 74 PC
Yaoyi 73 PC
Yaxue 25 PC
Yeling 125 PC
Yiachiamen 103 PC
Yijing 45 PC
Yiming 13 PC
Yinkou 60 PC
Yintang 3 PC
Yinwei 166-167-168 PC
Yishu 62 PC
Yuqian 128 PC
Yutian 81 PC
Yuwei 7 PC
Yuyao 6 PC

Z

Zeqian 117 PC
Zexia 116 PC
Zhizhuanjin 147 PC
Zhonggong 76 PC
Zhongkui – Erjiaoming 101 PC
Zhongquan 118 PC
Zhoushu 124 PC
Zhouzhu 56 PC
Zhuogu 64PC
Zhuzhang 70 PC
Zigong 49 PC
Zuluo 164 PC
Zuming 162 PC
Zuoyi-Youyi 35 PC
Zuxin 133 PC
Zuzhongping 143 PC

Extra channel points PN

In numerical order and translated into English

1 PN	Shangjingming	Superior Jingming.
2 PN	Xiajingming	Inferior Jingming.
3 PN	Jiangming	Blinding light.
4 PN	Jiangming 1	Blinding light n° 1.
5 PN	Jiangming 2	Blinding light n° 2.
6 PN	Jiangming 3	Blinding light n° 3.
7 PN	Jiangming 4	Blinding light n° 4.
8 PN	Shangming	Superior light.
9 PN	Zengming 1	Augmented light n° 1.
10 PN	Zengming 2	Augmented light n° 2.
11 PN	Waiming	External light.
12 PN	Bitong	Stopped nose.
13 PN	Tingxue	
14 PN	Tingling	Excellent hearing.
15 PN	Tingcong	Recuperated hearing.
16 PN	Tingmin	Predisposition to listen.
17 PN	Shangergen	Upper root of the ear.
18 PN	Houtinggong	Posterior SI 19 Tinggong.
19 PN	Houtingxue	Posterior point of hearing.
20 PN	Houtinghui	Posterior GB 2 Tinghui.
21 PN	Hougong	Posterior communication.
22 PN	Qiqian	
23 PN	Yimingxia	Below 13 PC Yiming.
24 PN	Tianting	Celestial listening.
25 PN	Qianzhengxue	Point of correction. To draw and upend.
26 PN	Yanchi	Dangerous lagoon. Pond among the rocks.
27 PN	Anmian 1	Quiet sleep 1.
28 PN	Anmian 2	Quiet sleep 2.
29 PN	Xingfen	Stimulation. Excitement.
30PN	Ronghou	Behind SI 17 Tianrong.
31 PN	Qiangyin	Loud voice.
32 PN	Zhenyin	Raised voice.
33 PN	Xiafutu	Inferior LI 18 Futu.
34 PN	Jingzhong	Midpoint of the neck.
35 PN	Jianei	In the cheek.
36 PN	Shuishang	Above (REN)CV 9 Shuifen.
37 PN	Weile	Stomachal bliss.
38 PN	Zhixie	Antidiarrhoeic.
39 PN	Tituoxue	Supporting door. Raise and support.
40 PN	Chongjian	Between currents.
41 PN	Shuxi	Trail of smiles.

42 PN	Liu Jingzhui Pang	Beside the sixth cervical.
43 PN	Xueya Dian	Point for the arterial tension.
44 PN	Qi Jingzhui Pang	Beside the seventh cervical.
45 PN	Dingchuan	To stop asthma.
46 PN	Wai Dingchuan	Outside 45 PN Dingchuan.
47 PN	Jiehe Xue	Adenitis point. Orifice of Tuberculosis.
48 PN	Weire Xue	Point of stomachal heat.
49 PN	Jiantong Dian	Painful point of the shoulder.
50 PN	Zhongchuan	Specific point of asthma.
51 PN	Pirexue	Point of the heat of the spleen.
52 PN	Shenrexue	Point of the heat of the kidneys.
53 PN	Jianming 5	Blinding light.
54 PN	Kuishang Xue	Point of the ulcer.
55 PN	Weishu	Gastric well-being.
56 PN	Shenji	Kidney spine.
57 PN	Tiao Yue	To jump and kick.
58 PN	Zuogu	Sitting bone.
59 PN	Pangqiang	Beside DM 1 Changqiang.
60 PN	Yatong	Tooth pain.
61 PN	Nuemen	Gate to malaria.
62 PN	Luolingwu	Midway to 108 PC Luozhen.
63 PN	Shang Houxi	Superior SI 3 Houxi.
64 PN	Xishang	Superior (MH)PC 4 Ximen.
65 PN	Luoshang	Superior Luo.
66 PN	Yingxia	Under the eagle.
67 PN	Niushangxue	Point of torsion.
68 PN	Guangzhong	Centre of the barrier.
69 PN		Raise an arm.
70 PN	Taijian	Raise a shoulder.
71 PN	Yingshang	Above the eagle.
72 PN	Jianming	To see clearly. Reinforced brightness.
73 PN	Rushang	Superior flexibility.
74 PN	Sanjian	Three Jian.
75 PN	Zhiping	On the level of the toes.
76 PN	Panggu	Beside the valley.
77 PN	Genping	On the level of the heel.
78 PN	Naoqing	Clear brain.
79 PN	Jingxia	Under the calf of the leg.
80 PN	Wanli	Ten thousand Li.
81 PN	Liwai	Outside ES 36 Zusanli.
82 PN	Lishang	1 Cun above ES 36 Zusanli.
83 PN	Luodi	Path to the land.
84 PN	Jiuwaifan	Corrects the valgus.
85 PN	Jiunei Fan	To come back inside.

86 PN	Chengjian	Supports the middle. Between the supports.
87 PN	Shangxi	Superior KI 3 Taixi.
88 PN	Ganyandian	Liver problems.
89 PN	Juiwai Fan	
90 PN	Dijian	Terrestrial health.
91 PN	Zuyicong	Current useful to the foot.
92 PN	Lingxia	Under the grave.
93 PN	Jianxi	Powerful knee.
94 PN		Four forces.
95 PN	Jixia	Below SP 11 Jimen.
96 PN	Xin Futu	New recumbent rabbit.
97 PN	Qian Fengshi	Anterior GB 31 Fengshi.
98 PN	Meibu	Slow pace.
99 PN	Jiaoling	Miraculous healing. Reinforced inspiration.
100 PN	Weishang	Above BL 40 Weizhong.
101 PN	Zhili	Erect bearing.
102 PN	Wai Zhili	External 101 PN Zhili.
103 PN	Yinshang	Above BL 37 Yinmen.
104 PN	Yinkang	Yin tension. Yin excess.
105 PN	Hou Xuehai	Posterior SP 10 Xuehai.
106 PN	Jiejian	To undo the cold chisel.
107 PN	Hou Yangguan	Posterior GB 33 Yangguan. Yang posterior door.
108 PN	Shang Yangguan	Superior GB 33 Yangguan. Yang superior door.
109 PN	Shangfengshi	Superior market of the wind.
110 PN	Qianjin	To walk and advance.

Extra channel points PN

By alphabetic order

	69 PN
	94 PN
Anmian 1	27 PN
Anmian 2	28 PN
Bitong	12 PN
Chengjian	86 PN
Chongjian	40 PN
Dijian	90 PN
Dingchuan	45 PN
Ganyandian	88 PN
Genping	77 PN
Guangzhong	68 PN
Hou Xuehai	105 PN
Hou Yangguan	107 PN
Hougong	21 PN
Houtinggong	18 PN
Houtinghui	20 PN
Houtingxue	19 PN
Jianei	35 PN
Jiangming	3 PN
Jiangming 1	4 PN
Jiangming 2	5 PN
Jiangming 3	6 PN
Jiangming 4	7 PN
Jianming	72 PN
Jianming 5	53 PN
Jiantong Dian	49 PN
Jianxi	93 PN
Jiaoling	99 PN
Jiehé Xué	47 PN
Jiejian	106 PN
Jingxia	79 PN
Jingzhong	34 PN
Jiunei Fan	85 PN
Jiuwaifan	84 PN
Jixia	95 PN
Juiwai Fan	89 PN
Kuishang Xué	54 PN
Lingxia	92 PN
Lishang	82 PN
Liu Jingzhui Pang	42 PN
Liwai	81 PN
Luodi	83 PN
Luolingwu	62 PN
Luoshang	65 PN
Meibu	98 PN
Naoqing	78 PN
Niushangxue	67 PN
Nuemen	61 PN
Panggu	76 PN
Pangqiang	59 PN
Pirexué	51 PN
Qi Jingzhui Pang	44 PN
Qian Fengshi	97 PN
Qiangyin	31 PN
Qianjin	110 PN
Qianzhengxue	25 PN
Qiqian	22 PN
Ronghou	30 PN
Rushang	73 PN
Sanjian	74 PN
Shang Houxi	63 PN
Shang Yangguan	108 PN
Shangergen	17 PN
Shangfengshi	109 PN
Shangjingming	1 PN
Shangming	8 PN
Shangxi	87 PN
Shenji	56 PN
Shenréxué	52 PN
Shuishang	36 PN
Shuxi	41 PN
Taijian	70 PN
Tianting	24 PN
Tiao Yué	57 PN
Tingcong	15 PN
Tingling	14 PN
Tingmin	16 PN
Tingxue	13 PN
Tituoxue	39 PN
Wai Dingchuan	46 PN
Wai Zhili	102 PN
Waiming	11 PN
Wanli	80 PN
Weilé	37 PN
Weiré Xué	48 PN
Weishang	100 PN
Weishu	55 PN
Xiafutu	33 PN
Xiajingming	2 PN
Xin Futu	96 PN
Xingfen	29 PN

Xishang	64 PN
Xueya Dian	43 PN
Yanchi	26 PN
Yatong	60 PN
Yimingxia	23 PN
Yingshang	71 PN
Yingxia	66 PN
Yinkang	104 PN
Yinshang	103 PN
Zengming 1	9 PN
Zengming 2	10 PN
Zhenyin	32 PN
Zhili	101 PN
Zhiping	75 PN
Zhixié	38 PN
Zhongchuan	50 PN
Zuogu	58 PN
Zuyicong	91 PN

Extra channel points MN

By numerical order

1 MN		Antihypertensor point.
2 MN	Yaotong 2	Lumbar and foot point.
3 MN	Laogong externe	External Laogong.
4 MN	Houtoudian	Vertebral point.
5 MN		Sciatic point.
6 MN		Point of the pharynx.
7 MN		Neck and nape point.
8 MN		Thoracic point.
9 MN		Malleolar point.
10 MN		Shoulder point.
12 MN		Point of the top.
13 MN		Ocular point.
14 MN	Qiantoudian	Forehead point.
15 MN		Vertex point.
16 MN		Point of the reunion of the Yin.
18 MN		Antipyretic point.
20 MN		Tonsil point.
21 MN		Gastro-intestinal point.
22 MN		Anticonvulsive point.
23 MN		Heel point.
24 MN		Odontalgia point.
25 MN		Cough and asthma point.
26 MN		Pollakiuria point.

Extra channel points

Recent points by numerical order

J-BY.3	Liuhua	Six flowers.
J-BY.5	Erzhuixia	Under the second thoracic.
J-BY.6	Huanmen	Gate to the pangs of death.
J-BY.15	Jiegu	Connecting bone.
J-BY.19	Xuefu	Abode of the blood.
J-BY.26	Zhongkong	Intermediate space.
J-BY.37	Jisanxue	Three vertebra holes.
J-SZ.7	Fengyan	Eye of the wind.
J-SZ.32	Zeqian	Resplendent vision.
J-SZ.36	Cirao	Cubitus and radius.
J-SZ.39	Zhoushu	Defile of the elbow.
J-SZ.45	Wuhu	Five tigers.
J-SZ.46	Zhoujian	Bridge of the elbow.
J-SZ.47 = 22 MN	Xiaotianxin	Heart of little heaven.
J-SZ.48	Jianneiling	Internal tomb of the shoulder.
J-SZ.50	Shangbaxie	Above the eight daemons.
J-TJ.12	Erbeijingmaisantiao	Three forks of the vein of the ear.
J-TJ.17	Sanxiao	Contagious smile.
J-TJ.18	Jiachengjiang	Avidity of the receptor of fluids.
J-TJ.32	Muming	Resplendent vision.
J-TJ.33	Neijingming	Inner light of the eyes.
J-TJ.34	Yankou	Mouth of a swallow.
J-TJ.35	Neiyingxiang	Internal scent.
J-TJ.36	Juquan	Source of income.
J-TJ 37	Haiquan	Source of the ocean.
J-TJ.38	Sheshu	Pillar of the tongue.
J-TJ.39	Zhuding	Summit of the pearl.
J-TJ.40	Damen	Big door.
J-XF.21	Eni	Hiccough.
J-XF.22	Xiaoerjixiong	Breast of a small chicken.
J-XF.24	Longmen	Dragon gate.
J-XZ.2	Muzhilihengwen	Transversal crease of the big toe.
J-XZ.13	Lanwei	
X-BY.8	Ganrexue	Vent for the heat of the Liver.
X-BY.11	Shenrexue	Vent for the heat of the Kidney.
X-BY.12	Jianming (Erjianming)	Reinforced brightness.
X-BY.16	Tiaoyue	A jump.
X-BY.17	Zuogu	Ischium.
X-BY.19	Juguxia	Under the big bone.
X-BY.20	Feirexue	Vent for the heat of the Lung.
X-BY.21	Anmian (Sianmian)	Quiet slumber.

X-BY.22	Zhongjiaoshu	Middle burner crease.
X-BY.23	Shenxin	New Kidney.
X-BY.24	Zhantan	Struggle against paralysis.
X-BY.25	Xishang	Above the river.
X-BY.31	Maigen	Root of the vessels.
X-BY.32	Dayan	Blow in the eye.
X-BY.36	Qiaohoushangji	Superior iliac spine.
X-SZ.4	Shanghouxi	Above the posterior river.
X-SZ.9	Gongzhong	Midpoint of the humerus.
X-SZ.10	Jubi	That lifts the arm.
X-SZ.14	Naoshang	Above the shoulder blade.
X-SZ.15	Hubian	Beside the tiger.
X-SZ.16	Zhizhang	Palm of the finger.
X-SZ.17	Neihegu	Inside two valleys.
X-SZ.18	Tongling	Meeting of two valleys.
X-SZ.20	Sanliwai	Outside three miles.
X-SZ.21	Xinquchi	New winding pond.
X-SZ.22	Shangquchi	Above the winding pond.
X-SZ.23	Zhiyang	Eliminates itching.
X-SZ.24	Shenzhou	Extension of the elbow.
X-SZ.25	Xiaxiabai	Inferior flattering appearance.
X-SZ.26	Xiaokuai	Eliminates oedema.
X-SZ.27	Zhitan = 127 PC Jianyu	Controls paralysis.
X-TJ.3	Jianming	Reinforced brightness.
X-TJ.15	Yilong	Barrier of deafness.
X-TJ.16	Houcon	Posterior hearing.
X-TJ.17	Ciqian	In front of the cistern.
X-TJ.24	Ronghou	Behind contention.
X-TJ.26	Zengyin	Augments the sound.
X-TJ.28	Jingzhong	Midpoint of the neck.
X-TJ.31	Tounie	Temple of the head.
X-TJ.32	Dingshen	Fixed mind.
X-TJ.33	Guancai	Abundant door.
X-TJ.34	Xinzanzhu	New crumpled bamboo.
X-TJ.35	Tongming	Shining of the pupil.
X-TJ.36	Tingxiang	Hearing of sounds.
X-TJ.37	Shanglong	Apex of deafness.
X-TJ.38	Tinglongqian	Between hearing and deafness.
X-TJ.39	Yaming	Silent call.
X-TJ.40	Chixia	Under the cistern.
X-TJ.41	Tongerdao	Through the auditory canal.
X-TJ.43	Qiying	Augmentation of vital energy.
X-TJ.44	Shangtianzhu	Above the pillar of heaven.
X-TJ.45	Xiayamen	Under the gate of muteness.

X-TJ.46	Fuyamen	Secondary gate of muteness.
X-TJ.47	Xinyi	New unit.
X-TJ.48	Xiaxinshi	New inferior gratefulness.
X-TJ.49	Zhongjie	Intermediate connection.
X-TJ.50	Dijia	Endemic goitre.
X-TJ.53	Zhiou	Eliminates vomiting.
X-XF.8	Ganshi	Abode of the liver.
X-XF.9	Chuangxinmen	Recent door.
X-XF.10	Tiwei	Turns the Stomach.
X-XF.11	Tongbian	Rumblings
X-XF.12	Xinqixue	Vent of vital energy.
X-XF.13	Xiazhongji	Below the intermediate summit.
X-XF.14	Yeniao	Night urine.
X-XF.17	Yinbian	Beside the genitals.
X-XF.18	Weishangxue	Above the vent of the stomach.
X-XZ.3	Genping	On the level of the ankle.
X-XZ.9	Genjin	Rigid calcaneum.
X-XZ.19	Siqiang	Four forces.
X-XZ.22	Maibu	Long step.
X-XZ.36	Zhiwen	
X-XZ.37	Sili	Four miles.
X-XZ.38	Zhitan (Liuzhitan)	Treatment of paralysis.
X-XZ.40	Tanli	Forces to rise after paralysis.
X-XZ.41	Tankang	Health after paralysis.
X-XZ.42	Qiabinzhong	Between the ileum and the knee.
X-XZ.43	Xiachengshan	Lower prop of the mountain.
X-XZ.44	Weixia	Under command.
X-XZ.45	Yinxia	Under abundance.
X-XZ.46	Yangkang	Excess of Yang.
X-XZ.47	Xinhuantiao	New surrounding circle.
X-XZ.48	Chuqixue	Exit vent of the vital energy.
X-XZ.49	Yiniao	Incontinence.
X-XZ.50	Anmian (Wuanmian)	Peaceful slumber.
X-XZ.55	Jiankua	Fortified hip.
X-XZ.56	Kuanjiu	Acetabulum.
X-XZ.57	Qiangkua	Reinforced hip.
X-XZ.59	Erliban	Two miles and a half.

Extra channel points

Recent points by alphabetic order

A

Anmian (Sianmian) X-BY.21
Anmian (Wuanmian) X-XZ.50

C

Chixia X-TJ.40
Chuangxinmen X-XF.9
Chuqixue X-XZ.48
Ciqian X-TJ.17
Cirao J-SZ.36

D

Damen J-TJ.40
Dayan X-BY.32
Dijia X-TJ.50
Dingshen X-TJ.32

E

Eni J-XF.21
Erbeijingmaisantiao J-TJ.12
Erliban X-XZ.59
Erzhuixia J-BY.5

F

Feirexue X-BY.20
Fengyan J-SZ.7
Fuyamen X-TJ.46

G

Ganrexue X-BY.8
Ganshi X-XF.8
Genjin X-XZ.9
Genping X-XZ.3
Gongzhong X-SZ.9
Guancai X-TJ.33

H

Haiquan J-TJ 37
Houcon X-TJ.16
Huanmen J-BY.6
Hubian X-SZ.15

J

Jiachengjiang J-TJ.18
Jiankua X-XZ.55
Jianming X-TJ.3
Jianming (Erjianming) X-BY.12
Jianneiling J-SZ.48
Jiegu J-BY.15
Jingzhong X-TJ.28
Jisanxue J-BY.37
Jubi X-SZ.10
Juguxia X-BY.19
Juquan J-TJ.36

K

Kuanjiu X-XZ.56

L

Liuhua J-BY.3
Longmen J-XF.24

M

Maibu X-XZ.22
Maigen X-BY.31
Muming J-TJ.32
Muzhilihengwen J-XZ.2

N

Naoshang X-SZ.14
Neihegu X-SZ.17
Neijingming J-TJ.33

Neiyingxiang J-TJ.35

Q

Qiabinzhong X-XZ.42
Qiangkua X-XZ.57
Qiying X-TJ.43
Qiaohoushangji X-BY.36

R

Ronghou X-TJ.24

S

Sanliwai X-SZ.20
Sanxiao J-TJ.17
Shangbaxie J-SZ.50
Shanghouxi X-SZ.4
Shanglong X-TJ.37
Shangquchi X-SZ.22
Shangtianzhu X-TJ.44
Shenrexue X-BY.11
Shenxin X-BY.23
Shenzhou X-SZ.24
Sheshu J-TJ.38
Sili X-XZ.37
Siqiang X-XZ.19

T

Tankang X-XZ.41
Tanli X-XZ.40
Tiaoyue X-BY.16
Tinglongqian X-TJ.38
Tingxiang X-TJ.36
Tiwei X-XF.10
Tongbian X-XF.11
Tongerdao X-TJ.41
Tongling X-SZ.18
Tongming X-TJ.35
Tounie X-TJ.31

W

Weishangxue X-XF.18
Weixia X-XZ.44
Wuhu J-SZ.45

X

Xiachengshan X-XZ.43
Xiaoerjixiong J-XF.22
Xiaokuai X-SZ.26
Xiaotianxin J-SZ.47
= 22 MN
Xiaxiabai X-SZ.25
Xiaxinshi X-TJ.48
Xiayamen X-TJ.45
Xiazhongji X-XF.13
Xinhuantiao X-XZ.47
Xinqixue X-XF.12
Xinquchi X-SZ.21
Xinyi X-TJ.47
Xinzanzhu X-TJ.34
Xishang X-BY.25
Xuefu J-BY.19

Y

Yaming X-TJ.39
Yangkang X-XZ.46
Yankou J-TJ.34
Yeniao X-XF.14
Yilong X-TJ.15
Yinbian X-XF.17
Yiniao X-XZ.49
Yinxia X-XZ.45

Z

Zengyin X-TJ.26
Zeqian J-SZ.32
Zhantan X-BY.24
Zhiou X-TJ.53
Zhitan (Liuzhitan) X-XZ.38
Zhitan = 127 PC Jianyu X-SZ.27
Zhiwen X-XZ.36
Zhiyang X-SZ.23
Zhizhang X-SZ.16
Zhongjiaoshu X-BY.22
Zhongjie X-TJ.49
Zhongkong J-BY.26
Zhoujian J-SZ.46
Zhoushu J-SZ.39

Points of Master Tong by numerical order

DT.01	Fen Zhi Shang	Separate upper branch.
DT.02	Fen Zhi Xia	Separate lower branch.
DT.03	Qi Xing	Seven stars.
DT.04	Wu Ling	Five mountains.
DT.05	Shuang Feng	A couple of phoenixes.
DT.06	Jiu Hou	Nine monkeys.
DT.07	San Jin	Three golden points.
DT.08	Jing Zhi	Jing branch.
DT.09	Jin Lin	Golden forest.
DT.10	Ding Zhu	Upper pillar.
DT.11	Hou Xin	1Behind the heart.
DT.12	Gan Mao San	Influenza n° 3.
DT.13	Shui Zhong	In the water.
DT.14	Shui Fu	Prefecture of the water.
DT.15	San Jiang	Three wide rivers.
DT.16	Shuang He	Double river.
DT.17	Chong Xiao	To push skywards.
T 1010.01	Zheng Hui	Correct reunion.
T 1010.02	Zhou Yuan	About the state.
T 1010.03	Zhou Kun	Big brother of the state.
T 1010.04	Zhou Lun	Mountain of the state.
T 1010.05	Qian Hui	Frontal reunion.
T 1010.06	Hou Hui	Posterior reunion.
T 1010.07	Zhong Shu	General axis.
T 1010.08	Zheng Jing	Repressed tranquillity.
T 1010.09	Shang Li	Above one Li.
T 1010.10	Si Fu Er	4 intestines n° 2.
T 1010.11	Si Fu Yi	4 intestines n° 1.
T 1010.12	Zheng Ben	Correct root.
T 1010.13	Ma Jin Shui	Water of a golden horse.
T 1010.14	Ma Kuai Shui	Water of a fast horse.
T 1010.15	Fu Kuai	Fast intestine.
T 1010.16	Liu Kuai	Fast six.
T 1010.17	Qi Kuai	Fast seven.
T 1010.18	Mu Zhi	Wooden branch.
T 1010.19	Shui Tong	Passage of the water.
T 1010.20	Shui Jin	Gold water.
T 1010.21	Yu Huo	Jade fire.
T 1010.22	Bi Yi	Wing of the nose.
T 1010.23	Zhou Huo	Fire of the state.
T 1010.24	Zhou Jin	Gold of the state.
T 1010.25	Zhou Shui	Water of the state.
T 11.01	Da Jian	Large space.
T 11.02	Xiao Jian	Small space.
T 11.03	Fu Jian	Floating space.
T 11.04	Wai Jian	External space.
T 11.05	Zhong Jian	In the middle of space.
T 11.06	Huan Chao	Back to the nest.

T 11.07	Zhi Si Ma	Four horses of the finger.
T 11.08.1	Zhi Wu Jin	Finger of five ounces of gold.
T 11.08.2	Zhi Qian Jin	Finger of one thousand ounces of gold.
T 11.09	Xin Xi	Heart-type knee.
T 11.10	Mu Huo	Wood fire.
T 11.11	Fei Xin	Lung and Heart.
T 11.12	Er Jiao Ming	Two sharp angles.
T 11.13	Dan	Gall Bladder.
T 11.14	Zhi San Zhong	Three measures in gold.
T 11.15	Zhi Shen	Kidney point of the finger.
T 11.16	Huo Xi	Fire of the knee.
T 11.17	Mu Xue	Grotto of the wood.
T 11.18	Pi Zhong	Oedema of the spleen.
T 11.19	Xin Chang	Regularises the Heart.
T 11.20	Mu Yan	Inflammation of the wood.
T 11.21	San Yan	Three eyes.
T 11.22	Fu Yuan	Retrieved source.
T 11.23	Yan Huang	Yellow eye.
T 11.24	Fu Ke	Special for the woman.
T 11.25	Zhi Xian	Eliminates saliva.
T 11.26	Zhi Wu	Controls what is dirty.
T 11.27	Wu Hu	The five tigers.
T 22.01	Zhong Zi	Double child.
T 22.02	Zhong Xian	Double immortal.
T 22.03	Shang Bai	Superior white.
T 22.04	Da Bai	Big white.
T 22.05	Ling Gu	Spirit of the bone.
T 22.06	Zhong Bai	Middle white.
T 22.07	Xia Bai	Inferior white.
T 22.08	Wan Shun Yi	First discharge of the wrist.
T 22.09	Wan Shun Er	Second discharge of the wrist.
T 22.10	Shou Jie	Relief of the hand.
T 22.11	Tu Shui	Water of the earth.
T 33.01	Qi Men	Gate of the point.
T 33.02	Qi Jiao	Corner of the point.
T 33.03	Qi Zheng	Straight point.
T 33.04	Huo Chuan	Ropes of fire.
T 33.05	Huo Ling	Mound of fire.
T 33.06	Huo Shan	Mountain of fire.
T 33.07	Huo Fu Hai	Sea of the entrails of fire.
T 33.08	Shou Wu Jin	Point of the hand worth five ounces of gold.
T 33.09	Shou Qian Jin	Point of the hand worth one thousand ounces of gold.
T 33.10	Chang Men	Gate to the Intestine.
T 33.11	Gan Men	Gate to the Liver.
T 33.12	Xin Men	Gate to the Heart.
T 33.13	Ren Shi	Erudition of Man.
T 33.14	Di Shi	Erudition of the Earth.
T 33.15	Tian Shi	Erudition of Heaven.
T 33.16	Qu Ling	Mound of the curve.
T 44.01	Fen Jin	Golden division.
T 44.02	Hou Zhui	Posterior vertebra.

T 44.03	Shou Ying	Superior wisdom.
T 44.04	Fu Ding	Superior wealth.
T 44.05	Hou Zhi	Posterior branch.
T 44.06	Jian Zhong	Middle of the shoulder.
T 44.07	Bei Mian	Posterior side.
T 44.08	Ren Zong	Ancestor of Man.
T 44.09	Di Zong	Ancestor of the Earth.
T 44.10	Tian Zong	Ancestor of Heaven.
T 44.11	Yun Bai	White cloud.
T 44.12	Li Bai	White plum.
T 44.13	Zhi Tong	Access branch.
T 44.14	Luo Tong	Descending access.
T 44.15	Xia Qu	Lower curve.
T 44.16	Shang Qu	Upper curve.
T 44.17	Shui Yu	Cure by water.
T 55.01	Huo Bao	Envelope of fire.
T 55.02	Hua Gu Yi	First flower of the bone.
T 55.03	Hua Gu Er	Second flower of the bone.
T 55.04	Hua Gu San	Third flower of the bone.
T 55.05	Hua Gu Si	Fourth flower of the bone.
T 55.06	Shang Liu	Upper tumour.
T 66.01	Hai Bao	
T 66.02	Mu Fu	Wooden woman.
T 66.03	Huo Ying	Hardness of the fire.
T 66.04	Huo Zhu	Master of the fire.
T 66.05	Men Jin	Gate to the large intestine.
T 66.06	Mu Liu	Keeps the wood.
T 66.07	Mu Dou	Wooden spoon.
T 66.08	Liu Wan ou Mu Wan	Finishes off the wood.
T 66.09	Shui Qu	Curve of the water.
T 66.10	Huo Lian	Connection with the fire.
T 66.11	Huo Ju	Fire chrysanthemum.
T 66.12	Huo San	To scatter the fire.
T 66.13	Shui Jing	Crystal water.
T 66.14	Shui Xiang	Minister of the water.
T 66.15	Shui Xian	Immortal water.
T 77.01	Zheng Jin	Correct tendon.
T 77.02	Zheng Zong	Correct ancestor.
T 77.03	Zheng Shi	Correct specialist.
T 77.04	Bo Qiu	To catch the clew.
T 77.05	Yi Zhong	Weight n° 1.
T 77.06	Er Zhong	Weight n° 2.
T 77.07	San Zhong	Weight n° 3.
T 77.08	Si Hua Shang	Four upper flowers.
T 77.09	Si Hua Zhong	Four middle flowers.
T 77.10	Si Hua Fu	Nearly four flowers.
T 77.11	Si Hua Xia	Four lower flowers.
T 77.12	Fu Chang	Viscus intestine.
T 77.13	Si Hua Li	Four internal flowers.
T 77.14	Si Hua Wai	Four external flowers.
T 77.15	Shang Chun	Upper lip.

T 77.16	Xia Chun	Lower lip.
T 77.17	Tian Huang ou Shen Guan	Heaven emperor, or kidney barrier.
T 77.18	Tian Huang Fu	Nearly Heaven emperor.
T 77.19	Di Huang	Emperor of the Earth.
T 77.20	Si Zhi	Four Branches.
T 77.21	Ren Huang	Emperor of Man.
T 77.22	Ce San Li	Beside Zu San Li.
T 77.23	Ce Xia San Li	Outside and below Zu San Li.
T 77.24	Zu Qian Jin	One thousand ounces of lower grade gold.
T 77.25	Zu Wu Jin	Five ounces of lower grade gold.
T 77.26	Qi Hu	Seven tigers.
T 77.27	Wai San Guan	Three barriers outside.
T 77.28	Guang Ming	Bright light.
T 88.01	Tong Guan	Passage across the barrier.
T 88.02	Tong Shan	Passage across the mountain.
T 88.03	Tong Tian	Passage across the sky.
T 88.04	Jie Mei Yi	Sister n° 1.
T 88.05	Jie Mei Er	Sister n° 2.
T 88.06	Jie Mei San	Sister n° 3.
T 88.07	Gan Mao Yi	Influenza n° 1.
T 88.08	Gan Mao Er	Influenza n° 2.
T 88.09	Tong Shen	Passage by the kidney.
T 88.1	Si Ma Zhong	In the middle of a coach and four.
T 88.10	Tong Wei	Passage by the stomach.
T 88.11	Tong Bei	Upper passage by the shoulder.
T 88.12	Ming Huang	Bright yellow.
T 88.13	Tian Huang	Yellow sky.
T 88.14	Qi Huang	Yellow East.
T 88.15	Huo Zhi	Branch of fire.
T 88.16	Huo Quan	Total fire.
T 88.18	Si Ma Shang	Above a coach and four.
T 88.19	Si Ma Xia	Below a coach and four.
T 88.20	Xia Quan	Lower spring.
T 88.21	Zhong Quan	Middle spring
T 88.22	Shang Quan	Upper spring.
T 88.23	Jin Qian Xia	Golden inferior forehead.
T 88.24	Jin Qian Shang	Golden superior forehead.
T 88.25	Zhong Jiu Li	At the middline of the 9 distances.
T 88.26	Shang Jiu Li	Above the 9 distances.
T 88.27	Xia Jiu Li	Below the 9 distances.
T 88.28	Jie	Relief.
T 88.29	Nei Tong Guan	Passage across the internal barrier.
T 88.30	Nei Tong Shan	Internal passage across the mountain.
T 88.31	Nei Tong Tian	Internal passage across the sky.
T 88.32	Shi Yin	Lost voice.
VT.01	Hou Es Jiu	Nine butterflies of the throat.
VT.02	Shi Er Hou	Twelve monkeys.
VT.03	Jin Wu	Five points in gold.
VT.04	Wei Mao Qi	Seven hairs of the stomach.
VT.05	Fu Chao Er Shi San	Twenty three nests of the intestine.

Points of Master Tong

By alphabetic order

B

Bei Mian T 44.07
Bi Yi T 1010.22
Bo Qiu T 77.04

C

Ce San Li T 77.22
Ce Xia San Li T 77.23
Chang Men T 33.10
Chong Xiao DT.17

D

Da Bai T 22.04
Da Jian T 11.01
Dan T 11.13
Di Huang T 77.19
Di Shi T 33.14
Di Zong T 44.09
Ding Zhu DT.10

E

Er Jiao Ming T 11.12
Er Zhong T 77.06

F

Fei Xin T 11.11
Fen Jin T 44.01
Fen Zhi Shang DT.01
Fen Zhi Xia DT.02
Fu Chang T 77.12
Fu Chao Er Shi San VT.05
Fu Ding T 44.04
Fu Jian T 11.03
Fu Ke T 11.24
Fu Kuai T 1010.15
Fu Yuan T 11.22

G

Gan Mao Er T 88.08
Gan Mao San DT.12
Gan Mao Yi T 88.07
Gan Men T 33.11
Guang Ming T 77.28

F

Hai Bao T 66.01
Hou Es Jiu VT.01
Hou Hui T 1010.06
Hou Xin DT.11
Hou Zhi T 44.05
Hou Zhui T 44.02
Hua Gu Er T 55.03
Hua Gu San T 55.04
Hua Gu Si T 55.05
Hua Gu Yi T 55.02
Huan Chao T 11.06
Huo Bao T 55.01
Huo Chuan T 33.04
Huo Fu Hai T 33.07
Huo Ju T 66.11
Huo Lian T 66.10
Huo Ling T 33.05
Huo Quan T 88.16
Huo San T 66.12
Huo Shan T 33.06
Huo Xi T 11.16
Huo Ying T 66.03
Huo Zhi T 88.15
Huo Zhu T 66.04

J

Jian Zhong T 44.06
Jie T 88.28
Jie Mei Er T 88.05
Jie Mei San T 88.06
Jie Mei Yi T 88.04
Jin Lin DT.09

Jin Qian Shang	T 88.24
Jin Qian Xia	T 88.23
Jin Wu	VT.03
Jing Zhi	DT.08
Jiu Hou	DT.06

L

Li Bai	T 44.12
Ling Gu	T 22.05
Liu Kuai	T 1010.16
Liu Wan ou Mu Wan	T 66.08
Luo Tong	T 44.14

M

Ma Jin Shui	T 1010.13
Ma Kuai Shui	T 1010.14
Men Jin	T 66.05
Ming Huang	T 88.12
Mu Dou	T 66.07
Mu Fu	T 66.02
Mu Huo	T 11.10
Mu Liu	T 66.06
Mu Xue	T 11.17
Mu Yan	T 11.20
Mu Zhi	T 1010.18

N

Nei Tong Guan	T 88.29
Nei Tong Shan	T 88.30
Nei Tong Tian	T 88.31

P

Pi Zhong	T 11.18

Q

Qi Hu	T 77.26
Qi Huang	T 88.14
Qi Jiao	T 33.02
Qi Kuai	T 1010.17
Qi Men	T 33.01
Qi Xing	DT.03
Qi Zheng	T 33.03
Qian Hui	T 1010.05
Qu Ling	T 33.16

R

Ren Huang	T 77.21
Ren Shi	T 33.13
Ren Zong	T 44.08

S

San Jiang	DT.15
San Jin	DT.07
San Yan	T 11.21
San Zhong	T 77.07
Shang Bai	T 22.03
Shang Chun	T 77.15
Shang Jiu Li	T 88.26
Shang Li	T 1010.09
Shang Liu	T 55.06
Shang Qu	T 44.16
Shang Quan	T 88.22
Shi Er Hou	VT.02
Shi Yin	T 88.32
Shou Jie	T 22.10
Shou Qian Jin	T 33.09
Shou Wu Jin	T 33.08
Shou Ying	T 44.03
Shuang Feng	DT.05
Shuang He	DT.16
Shui Fu	DT.14
Shui Jin	T 1010.20
Shui Jing	T 66.13
Shui Qu	T 66.09
Shui Tong	T 1010.19
Shui Xian	T 66.15
Shui Xiang	T 66.14
Shui Yu	T 44.17
Shui Zhong	DT.13
Si Fu Er	T 1010.10
Si Fu Yi	T 1010.11
Si Hua Fu	T 77.10
Si Hua Li	T 77.13
Si Hua Shang	T 77.08
Si Hua Wai	T 77.14
Si Hua Xia	T 77.11
Si Hua Zhong	T 77.09
Si Ma Shang	T 88.18

Si Ma Xia	T 88.19
Si Ma Zhong	T 88.1
Si Zhi	T 77.20

T

Tian Huang	T 88.13
Tian Huang Fu	T 77.18
Tian Huang ou Shen Guan	T 77.17
Tian Shi	T 33.15
Tian Zong	T 44.10
Tong Bei	T 88.11
Tong Guan	T 88.01
Tong Shan	T 88.02
Tong Shen	T 88.09
Tong Tian	T 88.03
Tong Wei	T 88.10
Tu Shui	T 22.11
Wai Jian	T 11.04

W

Wai San Guan	T 77.27
Wan Shun Er	T 22.09
Wan Shun Yi	T 22.08
Wei Mao Qi	VT.04
Wu Hu	T 11.27
Wu Ling	DT.04

X

Xia Bai	T 22.07
Xia Chun	T 77.16
Xia Jiu Li	T 88.27
Xia Qu	T 44.15
Xia Quan	T 88.20
Xiao Jian	T 11.02
Xin Chang	T 11.19
Xin Men	T 33.12
Xin Xi	T 11.09

Y

Yan Huang	T 11.23
Yi Zhong	T 77.05
Yu Huo	T 1010.21
Yun Bai	T 44.11

Z

Zheng Ben	T 1010.12
Zheng Hui	T 1010.01
Zheng Jin	T 77.01
Zheng Jing	T 1010.08
Zheng Shi	T 77.03
Zheng Zong	T 77.02
Zhi Qian Jin	T 11.08
Zhi San Zhong	T 11.14
Zhi Shen	T 11.15
Zhi Si Ma	T 11.07
Zhi Tong	T 44.13
Zhi Wu	T 11.26
Zhi Wu Jin	T 11.08
Zhi Xian	T 11.25
Zhong Bai	T 22.06
Zhong Jian	T 11.05
Zhong Jiu Li	T 88.25
Zhong Quan	T 88.21
Zhong Shu	T 1010.07
Zhong Xian	T 22.02
Zhong Zi	T 22.01
Zhou Huo	T 1010.23
Zhou Jin	T 1010.24
Zhou Kun	T 1010.03
Zhou Lun	T 1010.04
Zhou Shui	T 1010.25
Zhou Yuan	T 1010.02
Zu Qian Jin	T 77.24
Zu Wu Jin	T 77.25

Acupuncture points with corresponding pages

Correspondence points – page

Points	Name	Page
LU 1	Zhongfu	4,6,12,112,114,324,374
LU 2	Yunmen	4,12,324,326,374
LU 3	Tianfu	12,316,324
LU 4	Xiabai	12,316,324,326,328
LU 5	Chize	4,12,62,310,316,318,320,324,328,330,336
LU 6	Kongzui	12,316
LU 7	Lieque	4,8,12,260,316
LU 8	Jingqu	12,302,316
LU 9	Taiyuan	4,12,58,302,316
LU 10	Yuji	4,8,12,302,304
LU 11	Shaoshang	4,12,292,294,302
(MH)PC 1	Tianchi	18,24,26,144,190,254,324,370
(MH)PC 2	Tianquan	18,26,324,326
(MH)PC 3	Quze	18,26,254,316,324
(MH)PC 4	Ximen	26,316,318
(MH)PC 5	Jianshi	26,316
(MH)PC 6	Neiguan	22,26,254,316,318
(MH)PC 7	Daling	18,26,96,254,302,304,316
(MH)PC 8	Laogong	18,26,294,296,302,304
(MH)PC 9	Zhongchong	18,26,292,302
(HT)HE 1	Jiquan	32,34,40,42,324
(HT)HE 2	Qingling	42,324
(HT)HE 3	Shaohai	32,42,316,324
(HT)HE 4	Lingdao	42,316,320
(HT)HE 5	Tongli	38,42,316
(HT)HE 6	Yinxi	42,302,316
(HT)HE 7	Shenmen	32,42,78,302,316,318
(HT)HE 8	Shaofu	32,42,302,306
(HT)HE 9	Shaochong	32,42,292,302
LI 1	Shangyang	8,52,62,292
LI 2	Erjian	62,292
LI 3	Sanjian	62,292,300
LI 4	Hegu	8,52,62,292,300
LI 5	Yangxi	62,292,296,310,312
LI 6	Pianli	58,62,310
LI 7	Wenliu	62,310
LI 8	Xialian	62,310

Points	Name	Page
LI 9	Shanglian	62,310
LI 10	Shousanli	62,310,314
LI 11	Quchi	52,62,310,312,314,330,332,336
LI 12	Zhouliao	62,310,330,336
LI 13	Shouwouli	62,330,336
LI 14	Binao	62,336
LI 15	Jianyu	48,52,58,60,62,78,102,204,206,274,276,326,330,336,382,384
LI 16	Jugu	62,274,276,330,332,384
LI 17	Tianding	64,340
LI 18	Futu	10,60,64,84,340,346,360,364,366
LI 19	Kouheliao	64,340
LI 20	Yingxiang	52,64,162,340,346,348,354,356
SI 1	Shaoze	82,292
SI 2	Qiangu	82,292
SI 3	Houxi	82,266,272,292,296,306
SI 4	Wangu	38,82,292,310,296
SI 5	Yanggu	82,292,310
SI 6	Yanglao	82,310
SI 7	Zhizheng	78,82,310
SI 8	Xiaohai	82,310,330
SI 9	Jianzhen	82,330,382
SI 10	Naoshu	84,274,276,278,282,330,334,382,392
SI 11	Tianzong	84,330,382
SI 12	Bingfeng	80,84,90,184,330,382
SI 13	Quyuan	84,102,282,330,382
SI 14	Jianwaishu	84,382
SI 15	Jianzhongshu	84,382
SI 16	Tianchuang	84,340
SI 17	Tianrong	74,84,184,340,346,350,364,366
SI 18	Quanliao	74,84,90,206,340,350,356
SI 19	Tinggong	74,84,90,184,340,346,350,352
(TH)TB 1	Guanchong	100,292
(TH)TB 2	Yemen	100,292
(TH)TB 3	Zhongzhu	100,292,300
(TH)TB 4	Yangchi	22,100,278,292,296,304,310,312
(TH)TB 5	Waiguan	96,100,278,310,312
(TH)TB 6	Zhigou	100,310,314
(TH)TB 7	Huizong	100,310
(TH)TB 8	Sanyangluo	100,310
(TH)TB 9	Sidu	100,310,312
(TH)TB 10	Tianjing	90,100,310,330

Points	Name	Page
(TH)TB 11	Qinglengyuan	100,102,330,334
(TH)TB 12	Xiaoluo	102,330,334
(TH)TB 13	Naohui	102,330,334,336,382
(TH)TB 14	Jianliao	90,102,330,336,382
(TH)TB 15	Tianliao	102,278,282,330,382
(TH)TB 16	Tianyou	24,98,102,340
(TH)TB 17	Yifeng	102,184,340,350,352,360,366
(TH)TB 18	Chimai	102,340
(TH)TB 19	Luxi	102,340
(TH)TB 20	Jiaosun	98,102,184,196,340,344,350
(TH)TB 21	Ermen	102,340,346,350
(TH)TB 22	Erheliao	74,102,184,340,346,348
(TH)TB 23	Sizhukong	90,102,340,356,358
SP 1	Yinbai	110,120,164,466,468,470
SP 2	Dadu	120,470
SP 3	Taibai	120,168,466,470
SP 4	Gongsun	116,120,150,250,252,258,466,470,472
SP 5	Shangqiu	110,120,136,446,460,466,470
SP 6	Sanyinjiao	110,120,126,142,446,448,460,462,470
SP 7	Lougu	120,446,460,462
SP 8	Diji	120,446,448,460,462
SP 9	Yinlingquan	110,120,136,438,440,446,450,460,462
SP 10	Xuehai	120,424,426,430,432,438,440
SP 11	Jimen	120,424,426,430,432
SP 12	Chongmen	120,126,256,378,404,424,430
SP 13	Fushe	110,120,126,254,256,378,404
SP 14	Fujie	110,120, 378,404
SP 15	Daheng	110,112,114,120,254,256,378,404,406
SP 16	Fuai	112,120,254,256,374,378,404
SP 17	Shidou	120,374,378
SP 18	Tianxi	120,374,378
SP 19	Xiongxiang	120,374,378
SP 20	Zhourong	112,114,120,374,378
SP 21	Dabao	116,120,374,378
(LR)LV 1	Dadun	126,136,184,466,470
(LR)LV 2	Xingjian	136,466,468,470
(LR)LV 3	Taichong	136,192,466,468
(LR)LV 4	Zhongfeng	126,136,446,460,466,470
(LR)LV 5	Ligou	132,136,446,460
(LR)LV 6	Zhongdu	136,446,460
(LR)LV 7	Xiguan	136,440,446,460

Points	Name	Page
(LR)LV 8	Ququan	126,136,430,440,446,460
(LR)LV 9	Yinbao	136,424,430
(LR)LV 10	Zuwuli	136,424,426,430,432
(LR)LV 11	Yinlian	136,424,430
(LR)LV 12	Jimai	136,378,404,430
(LR)LV 13	Zhangmen	126,128,130,134,136,184,194,198,284,286,370,378,404
(LR)LV 14	Qimen	112,126,128,130,136,254,256,370,378,404
KI 1	Yongquan	142,150,474
KI 2	Rangu	142,150,258,470,474
KI 3	Taixi	142,150,216,256,258,460,462,470,472
KI 4	Dazhong	142,146,150,250,460,470
KI 5	Shuiquan	150,460,470,472
KI 6	Zhaohai	142,150,258,460,462,470,472
KI 7	Fuliu	142,150,460,470
KI 8	Jiaoxin	150,258,460,470
KI 9	Zhubin	150,254,256,460
KI 10	Yingu	142,148,150,256,440,460
KI 11	Henggu	142,150,250,252,404,424,430
KI 12	Dahe	150,252,404,406
KI 13	Qixue	150,252,404
KI 14	Siman	150,252,404,406
KI 15	Zhongzhu	150,250,252,404
KI 16	Huangshu	152,250,252,404
KI 17	Shanqu	152,252,404
KI 18	Shiguan	152,252,404
KI 19	Yindu	152,252,404
KI 20	Futonggu	152,252,404
KI 21	Youmen	152,250,252,372,404
KI 22	Bulang	152,252,372
KI 23	Shenfeng	152,252,254,372
KI 24	Lingxu	152,252,372,376
KI 25	Shencang	152,252,372
KI 26	Yuzhong	152,252,372
KI 27	Shufu	142,152,250,252,372
ST 1	Chengqi	162,172,174,260,274,276,342,346,348,354,356
ST 2	Sibai	174,276,342,356
ST 3	Juliao	174,274,276,342
ST 4	Dicang	174,274,276,342,346,348,350,356
ST 5	Daying	162,164,174,182,184,274,342
ST 6	Jiache	162,164,174,182,184,342
ST 7	Xiaguan	164,174,184,196,342,344,356

Points	Name	Page
ST 8	Touwei	164,174,196,220,278,282,342,344
ST 9	Renying	118,162,164,172,174,258,274,342,346,360,364,366
ST 10	Shuitu	174,342,364
ST 11	Qishe	164,174,274,342
ST 12	Quepen	10,52,56,58,60,64,74,76,78,90,94,96,98,118,162,164,172,1 74, 184,190,206,258,340,342,370
ST 13	Qihu	174,370
ST 14	Kufang	174,370
ST 15	Wuyi	174,370
ST 16	Yingchuang	174,370,376
ST 17	Ruzhong	174,370
ST 18	Rugen	164,170,174,370,376
ST 19	Burong	174,370,402
ST 20	Chengman	174,402
ST 21	Liangmen	176,402
ST 22	Guanmen	176,402
ST 23	Taiyi	176,402
ST 24	Huaroumen	176,402
ST 25	Tianshu	56,176,402
ST 26	Wailing	176,402
ST 27	Daju	176,402
ST 28	Shuidao	176,402
ST 29	Guilai	176,402
ST 30	Qichong	118,136,162,164,172,176,190,250,378,402,404,424,430
ST 31	Biguan	162,172,176,424,426
ST 32	Futu	158,162,176,424,426
ST 33	Yinshi	176,424
ST 34	Liangqiu	176,424,426,432,438
ST 35	Dubi	176,438,440,446,448,450,454,458
ST 36	Zusanli	56,76,164,176,438,442,446,448,454,456
ST 37	Shangjuxu	56,176,446,454
ST 38	Tiaokou	168,176,446,450,454,458
ST 39	Xiajuxu	76,176,446,454
ST 40	Fenglong	168,176,446,450,454,458
ST 41	Jiexi	136,176,446,448,454,456,460,466,470,472
ST 42	Chongyang	116,164,176,466,472
ST 43	Xiangu	176,466,468,472
ST 44	Neiting	176,466,468,472
ST 45	Lidui	164,176,466,472
GB 1	Tongziliao	74,90,134,184,194,196,344,346,348,350
GB 2	Tinghui	196,344,350
GB 3	Shangguan	90,162,164,196,344

Points	Name	Page
GB 4	Hanyan	90,164,196,344
GB 5	Xuanlu	196,344
GB 6	Xuanli	90,164,196,344
GB 7	Qubin	196,208,344
GB 8	Shuaigu	196,208,344,356
GB 9	Tianchong	196,344
GB 10	Fubai	196,208,344
GB 11	Touqiaoyin	196,208,344
GB 12	Wangu	68,196,206,208,344,352,360,366
GB 13	Benshen	70,196,278,282,344
GB 14	Yangbai	196,278,282,344
GB 15	Toulinqui	196,208,278,282,344,346
GB 16	Muchuang	196,278,282,344
GB 17	Zhengying	198,278,282,344
GB 18	Chengling	198,278,282,344
GB 19	Naokong	198,278,282
GB 20	Fengchi	184,198,206,274,276,278,282,346,360,364,366
GB 21	Jianjing	90,102,184,198,278,282,330,382
GB 22	Yuanye	10,24,40,80,198,378
GB 23	Zhejin	198,378
GB 24	Riyue	112,198,372,378,404
GB 25	Jingmen	198, 378,414
GB 26	Daimai	198,284,286,378,404
GB 27	Wushu	198,284,286,378,404
GB 28	Weidao	198,284,286,378,404
GB 29	Juliao	184,198,274,276,278,282,378,434
GB 30	Huantiao	184,190,194,198,210,378,414,416,434
GB 31	Fengshi	198,434
GB 32	Zhongdu	198,434
GB 33	Xiyangguan	200,434,440,442
GB 34	Yanglingquan	200,206,434,440,442,454,456
GB 35	Yangjiao	200,274,276,278,282,454
GB 36	Waiqiu	200,454
GB 37	Guangming	192,200,454,458
GB 38	Yangfu	200,274,454
GB 39	Xuanzhong	200,454,458,472
GB 40	Qiuxu	132,200,454,466,472
GB 41	Zulinqi	200,284,466,472
GB 42	Diwuhui	200,466,472
GB 43	Xiaxi	200,466,468,472
GB 44	Zuqiaoyin	200,466,472

Points	Name	Page
BL 1	Jingming	74,162,208,220,250,258,266,274,276,342,350,354
BL 2	Cuanzhu	220,342,348,358
BL 3	Meichong	220,342
BL 4	Quchai	220,342
BL 5	Wuchu	220,342,348
BL 6	Chengguang	220,342,348
BL 7	Tongtian	220,342,358
BL 8	Luoque	208,220,358
BL 9	Yuzhen	208,220
BL 10	Tianzhu	148,206,208,210,218,220,226,362
BL 11	Dazhu	74,210,220,226,250,384
BL 12	Fengmen	220,266,384
BL 13	Feishu	220,384
BL 14	Jueyinshu	220,384
BL 15	Xinshu	220,384,386
BL 16	Dushu	220,384,386
BL 17	Geshu	220,384
BL 18	Ganshu	220,384,388,390
BL 19	Danshu	220,384
BL 20	Pishu	220,384
BL 21	Weishu	222,412
BL 22	Sanjiaoshu	222,412
BL 23	Shenshu	148,210,214,222,226,266,284,412,418,420
BL 24	Qihaishu	222,412
BL 25	Dachangshu	222,412,418
BL 26	Guanyuanshu	222,412
BL 27	Xiaochangshu	222,412
BL 28	Pangguanshu	222,412
BL 29	Zhongliishu	222,412
BL 30	Baihuanshu	210,222,412
BL 31	Shangliao	184,222,412
BL 32	Ciliao	222,412
BL 33	Zhongliao	222,412
BL 34	Xialiao	222,412,426,432,438
BL 35	Huiyang	210,222,226,412
BL 36	Chengfu	148,218,222,412,436
BL 37	Yinmen	222,436
BL 38	Fuxi	222,436,440,452
BL 39	Weiyang	94,210,222,436,440,442,452
BL 40	Weizhong	148,210,218,222,224,226,436,440,442,452
BL 41	Fufen	74,210,222,226,384
BL 42	Pohu	222,384
BL 43	Gaohuang	222,384

Points	Name	Page
BL 44	Shentang	224,384
BL 45	Yixi	224,384
BL 46	Geguan	224,384
BL 47	Hunmen	224,384
BL 48	Yanggang	224,384
BL 49	Yishe	224,384
BL 50	Weicang	224,388,414,418
BL 51	Huangmen	224,414
BL 52	Zhishi	224,414
BL 53	Baohuang	224,414
BL 54	Zhibian	224,226,414
BL 55	Heyang	224,226,440,452
BL 56	Chengjin	224,452
BL 57	Chengshan	224,226,240,452,454
BL 58	Feiyang	216,224,452,454
BL 59	Fuyang	224,274,276,452,454
BL 60	Kunlun	216,224,226,276,452,454,456,472
BL 61	Pushen	224,274,276,454,472
BL 62	Shenmai	224,274,454,472
BL 63	Jinmen	224,274,278,282,466,472
BL 64	Jinggu	146,224,466,474
BL 65	Shugu	224,466,474
BL 66	Zutonggu	224,466,474
BL 67	Zhiyin	210,224,466,468,474
(REN)CV 1	Huiyin	250,260,262,266
(REN)CV 2	Qugu	126,134,158,194,260,262,266,402,408
(REN)CV 3	Zhongji	110,126,144,150,158,260,262,402,406
(REN)CV 4	Guanyuan	110,126,144,150,250,260,262,402,406,408
(REN)CV 5	Shimen	150,260,262,402,406
(REN)CV 6	Qihai	260,262,402,406
(REN)CV 7	Yinjiao	150,250,252,260,262,402
(REN)CV 8	Shenque	260,262,402,406
(REN)CV 9	Shuifen	260,262,402,408
(REN)CV 10	Xiawan	110,112,114,164,260,262,402,408
(REN)CV 11	Jianli	112,260,262,402,406
(REN)CV 12	Zhongwan	76,164,260,262,402,406
(REN)CV 13	Shangwan	76,260,262,402
(REN)CV 14	Juque	262,370,372,374,402
(REN)CV 15	Jiuwei	260,262,370,372,374,376,402
(REN)CV 16	Zhongting	260,264,370,372,374
(REN)CV 17	Danzhong	22,94,144,146,170,254,260,264,370,372,374
(REN)CV 18	Yutang	260,264,370,372,374

Points	Name	Page
(REN)CV 19	Zigong	260,264,370,372,374
(REN)CV 20	Huagai	260,264,370,372,374
(REN)CV 21	Xuanji	260,264,364,366,370,372,374,376
(REN)CV 22	Tiantu	254,256,260,264,364,370,372,374
(REN)CV 23	Lianquan	24,148,254,256,260,264,346,364,366
(REN)CV 24	Chengjiang	162,164,260,264,346,348
(DU)GV 1	Changqiang	144,184,266,270,412,418
(DU)GV 2	Yaoshu	270,412,416,418
(DU)GV 3	Yaoyangguan	270,412
(DU)GV 4	Mingmen	270,284,412
(DU)GV 5	Xuanshu	270,412
(DU)GV 6	Jizhong	270,382
(DU)GV 7	Zhongshu	270,382
(DU)GV 8	Jinsuo	270,382
(DU)GV 9	Zhiyang	270,382
(DU)GV 10	Lingtai	270,382
(DU)GV 11	Shendao	146,270,382
(DU)GV 12	Shenzhu	266,270,382
(DU)GV 13	Taodao	210,266,270,382
(DU)GV 14	Dazhui	52,74,90,164,184,198,210,270,282,360,382,386,388
(DU)GV 15	Yamen	270,278,282,360,362,390
(DU)GV 16	Fengfu	198,266,270,274,276,278,282,362
(DU)GV 17	Naohu	208,214,272,330,336,382
(DU)GV 18	Qiangjian	272,362
(DU)GV 19	Houding	272
(DU)GV 20	Baihui	98,128,130,168,208,214,266,272,344,358
(DU)GV 21	Qianding	272,344,358
(DU)GV 22	Xinhui	272,344
(DU)GV 23	Shangxing	272,344
(DU)GV 24	Shenting	164,196,208,220,272,282,342,344
(DU)GV 25	Suliao	266,272,344,356,358
(DU)GV 26	Shuigou	52,64,162,272,340,356
(DU)GV 27	Duiduan	272
(DU)GV 28	Yinjiao	260,266,272

Correspondence
Extra channel points – page

Point	Name	Page
1 PC	Sishencong	358
2 PC	Ezhong	348
3 PC	Yintang	348,358
4 PC	Shangen	348
5 PC	Touquangming	348
6 PC	Yuyao	346,348
7 PC	Yuwei	346,348
8 PC	Qiuhou	346,348,354
9 PC	Taiyang	346,350
10 PC	Erjian	346
11 PC	Lungxue	346
13 PC	Yiming	346,350,352,360,364,366
14 PC	Shangyingxiang	346,348
15 PC	Jiabi	346,348
16 PC	Biliu	346,348
17 PC	Sanjiao	346,348
18 PC	Heliao	346,348
19 PC	Dihe	346,348
20 PC	Jinjin – Yuye	362
21 PC	Shanglianquan	346,364,366
22 PC	Waijinjin – Waiyuyue	346,364,366
23 PC	Hongyin	346,364,366
24 PC	Panglianquan	346,364,366
25 PC	Yaxue	346,360,364,366
26 PC	Biantao	346,364,366
27 PC	Luojing	346,364,366
28 PC	Fengyan	360,366
29 PC	Xinxi	360,362
30 PC	Bailiao	360
31 PC	Chonggu	360
32 PC	Chixue	364,366,376
33 PC	Tanchuan	376
34 PC	Longhan	376
35 PC	Zuoyi – Youyi	376
36 PC	Maihua	406
37 PC	Shicang	406
38 PC	Shiguan	406
39 PC	Hunshe	406
40 PC	Hangwen	406
41 PC	Qizhong	406
42 PC	Jingzhong	406
43 PC	Waisiman	406
44 PC	Jueyun	406
45 PC	Yijing	406

Point	Name	Page
46 PC	Qimen	406
47 PC	Weibao	406
48 PC	Changyi	406
49 PC	Zigong	406
50 PC	Tingtou	406
51 PC	Chuanxi	386
52 PC	Baizhongfeng	386
53 PC	Bahua	386
54 PC	Jiujia	386
55 PC	Liaosen	386
56 PC	Zhouzhu	386
57 PC	Juqueshu	386
58 PC	Liangyan	386
59 PC	Qichuan	386
60 PC	Yinkou	386
61 PC	Bazhuixia	386
62 PC	Yishu	386
63 PC	Shubian	386
64 PC	Zhuogu	386
65 PC	Jiagu	386,416
66 PC	Pigen	386,416
67 PC	Xuechou	386,416
68 PC	Changfeng	386,416
69 PC	Jijupikuai	416
70 PC	Zhuzhang	416
71 PC	Xiajishu	416
72 PC	Huqimen	416
73 PC	Yaoyi	416
74 PC	Yaoyan	416
75 PC	Shiqizhuixia	416
76 PC	Zhonggong	416
77 PC	Yaogen	416
78 PC	Couyi – Jiuqi	416
79 PC	Yaoqi	416
80 PC	Xiazhui	416
81 PC	Yutian	416
82 PC	Pinxueling	416
83 PC	Tunzhong	416
84 PC	Huanzhong	416
85 PC	Huatuojiaji	390
86 PC	Shixuan – Sap Sun	296
87 PC	Muzhijian	296
88 PC	Xiaozhijian	296
89 PC	Fengquan	304
90 PC	Jiudianfeng	304
92 PC	Fangyan	304
93 PC	Mingyan	304
94 PC	Sifeng	304
95 PC	Shouzhongping	304

Point	Name	Page
96 PC	Panglaogong	304
97 PC	Neiyangchi	304
98 PC	Banmen	304
99 PC	Sanshang	294
100 PC	Dagukong	296
101 PC	Zhongkui – Erjiaoming	296
102 PC	Xiaogukong	296
103 PC	Yiachiamen	296
104 PC	Erchianmen	296
105 PC	Xiaoshixie	294
106 PC	Quanjian	294
107 PC	Baxie	296
108 PC	Luozhen	296
109 PC	Wailaogong – Wailaokong	296
110 PC	Weiling	296
111 PC	Jingling	296
112 PC	Shoujinmen	318
113 PC	Dingshu	318
114 PC	Erbai	318
115 PC	Bizhong	318
116 PC	Zexia	318
117 PC	Zeqian	318
118 PC	Zhongquan	296,312
119 PC	Cunping	312
120 PC	Xiawenliu	312
121 PC	Chinao	312
122 PC	Quyangwei	312
123 PC	Sanchi	312
124 PC	Zhoushu	312
125 PC	Yeling	326
126 PC	Tianling	326
127 PC	Jianyu	326
128 PC	Yuqian	326
129 PC	Houye	332
130 PC	Lineiting	474
131 PC	Muzhilihengwen	474
132 PC	Qianhouyinzhu	474
133 PC	Zuxin	474
134 PC	Shimian	474
135 PC	Xiaozhijian	468
136 PC	Muzhiheng – Lisanmao	468
137 PC	Bafeng	468
138 PC	Nuxi	452
139 PC	Quanshengzu	452
140 PC	Taiyinqiao	462
141 PC		456
142 PC	Lanwei	448,456
143 PC	Zuzhongping	442,448,456
144 PC	Xixia	438,448
145 PC	Neixiyang	438,448

Point	Name	Page
146 PC	Neihuaijian	462
147 PC	Zhizhuanjin	462
148 PC	Shaoyangwei	462
149 PC	Chengming	462
150 PC	Jiaoyi	462
151 PC	Waihuaijian	442,456
152 PC	Danangxue	456
153 PC	Linghou	442,456
154 PC	Linghouxia	442,456
155 PC	Xiwai	436,442
156 PC	Heding	426,438
157 PC	Kuangku	426,432,438
158 PC	Shenxi	426
159 PC	Quantu	426
160 PC	Liaoliao	426,432,438
161 PC	Dalun	426,432,438
162 PC	Zuming	426,432,438
163 PC	Baichongwo	426,432,438
164 PC	Zuluo	426,432
165 PC	Changgu	426,434,438
166-167-168 PC	Yinwei	434,442
169 PC	Silian	434
170 PC	Wuling	434
171 PC	Lingbao	434
1 PN	Shangjingming	354
2 PN	Xiajingming	354
3 PN	Jiangming	354
4 PN	Jiangming 1	354
5 PN	Jiangming 2	354
6 PN	Jiangming 3	354
7 PN	Jiangming 4	354
8 PN	Shangming	354
9 PN	Zengming 1	354
10 PN	Zengming 2	354
11 PN	Waiming	354
12 PN	Bitong	354
13 PN	Tingxue	350,352
14 PN	Tingling	350
15 PN	Tingcong	350
16 PN	Tingmin	350,352
17 PN	Shangergen	352
18 PN	Houtinggong	352
19 PN	Houtingxue	352
20 PN	Houtinghui	352
21 PN	Hougong	352
22 PN	Qiqian	360
23 PN	Yimingxia	352
24 PN	Tianting	360
25 PN	Qianzhengxue	350,352

Point	Name	Page
27 PN	Anmian 1	352
28 PN	Anmian 2	360, 366
29 PN	Xingfen	360
31 PN	Qiangyin	364,366
32 PN	Zhenyin	364,366
33 PN	Xiafutu	366
34 PN	Jingzhong	366
35 PN	Jianei	354
36 PN	Shuishang	408
37 PN	Weilé	408
38 PN	Zhixié	408
39 PN	Tituoxue	408
40 PN	Chongjian	408
41 PN	Shuxi	408
42 PN	Liu Jingzhui Pang	360
43 PN	Xueya Dian	360
44 PN	Qi Jingzhui Pang	360,388
45 PN	Dingchuan	388
46 PN	Wai Dingchuan	388
47 PN	Jiehé Xué	360,388
48 PN	Weiré Xué	388
49 PN	Jiantong Dian	388
50 PN	Zhongchuan	388
51 PN	Pirexué	388
52 PN	Shenréxué	388
53 PN	Jianming 5	388
54 PN	Kuishang Xué	388,418
55 PN	Weishu	418
56 PN	Shenji	418
57 PN	Tiao Yué	418
58 PN	Zuogu	418
59 PN	Pangqiang	418
60 PN	Yatong	304
61 PN	Nuemen	296
62 PN	Luolingwu	296
63 PN	Shang Houxi	296
64 PN	Xishang	318
65 PN	Luoshang	312
66 PN	Yingxia	312
67 PN	Niushangxue	312
68 PN	Guangzhong	326
69 PN		332
70 PN	Taijian	332,336
71 PN	Yingshang	332
72 PN	Jianming	336
73 PN	Rushang	336
74 PN	Sanjian	336
76 PN	Panggu	468

Point	Name	Page
77 PN	Genping	452
78 PN	Naoqing	448,456
79 PN	Jingxia	448,456
80 PN	Wanli	442,448,456
81 PN	Liwai	442,448,456
82 PN	Lishang	438,448,456
83 PN	Luodi	452
84 PN	Jiuwaifan	452
85 PN	Jiuneifan	452
86 PN	Chengjian	452
87 PN	Shangxi	462
88 PN	Ganyandian	462
89 PN	Juiwai Fan	448,462
90 PN	Dijian	448,462
92 PN	Lingxia	456
93 PN	Jianxi	426
94 PN		426
95 PN	Jixia	426,432
96 PN	Xin Futu	426
99 PN	Jiaoling	426,432
100 PN	Weishang	426,436,442,452
101 PN	Zhili	436
102 PN	Wai Zhili	436
103 PN	Yinshang	436
104 PN	Yinkang	436
105 PN	Hou Xuehai	426,432,438
106 PN	Jiejian	426,432
107 PN	Hou Yangguan	442
108 PN	Shang Yangguan	434,442
109 PN	Shangfengshi	434
110 PN	Qianjin	434
1 MN	Antihypertensive point	294
2 MN	Yaotong 2	296
3 MN	Laogong external	294
4 MN	Houtoudian	294
5 MN	Sciatica point	294
6 MN	Pharynx point	294
7 MN	Point of the neck and the nape of the neck	294
8 MN	Thoracic point	294
9 MN	Malleolar point	294
10 MN	Shoulder point	294
12 MN	Summit point	294
13 MN	Ocular point	294
14 MN	Qiantoudian	294
15 MN	Vertex point	294
16 MN	Meeting point of the Yin	294
18 MN	Antipyretic point	294
20 MN	Tonsil point	304

Point	Name	Page
22 MN	Xiaotianxin	304
23 MN	Heel point	304
24 MN	Odontalgia point	304
25 MN	Cought and asthma point	304
26 MN	Pollakiuria Point	304
J-BY.3	Liuhua	390
J-BY.5	Erzhuixia	390
J-BY.15	Jiegu	390
J-BY.36	Cirao	318
J-SZ.37	Jisanxue	390
J-SZ.39	Zhoushu	312
J-SZ.47 = 22 MN	Xiaotianxin	304
J-SZ.48	Jianneiling	326
J-TJ.18	Jiachengjiang	350
J-TJ.32	Muming	346
J-TJ.34	Yankou	350
J-TJ.35	Neiyingxiang	350
J-TJ.36	Juquan	362
J-TJ.37	Haiquan	362
J-TJ.38	Sheshu	362
J-TJ.40	Damen	362
J-XF.21	Eni	376
J-XF.22	Xiaoerjixiong	376
J-XF.24	Longmen	406,408
J-XZ.13	Lanwei	448,456
X-BY.8	Ganrexue	390
X-BY.11	Shenrexue	390
X-BY.12	Jianming (Erjianming)	390
X-BY.17	Zuogu	436
X-BY.19	Juguxia	332
X-BY.20	Feirexue	390
X-BY.22	Zhongjiaoshu	390
X-BY.23	Shenxin	418
X-BY.24	Zhantan	418
X-BY.25	Xishang	418
X-BY.31	Maigen	418
X-BY.32	Dayan	418
X-SZ.9	Gongzhong	326
X-SZ.10	Jubi	336
X-SZ.16	Zhizhang	304
X-SZ.17	Neihegu	304
X-SZ.20	Sanliwai	312
X-SZ.21	Xinquchi	332,336
X-SZ.22	Shangquchi	332,336
X-SZ.23	Zhiyang	336
X-SZ.24	Shenzhou	332
X-SZ.25	Xiaxiabai	326
X-SZ.26	Xiaokuai	326

Point	Name	Page
X-SZ.27	Zhitan = 127 PC Jianyu	326
X-SZ.4	Shanghouxi	296
X-SZ.9	Gongzhong	326
X-TJ.3	Jianming	350
X-TJ.8	Tingxue	350
X-TJ.24	Ronghou	350
X-TJ.31	Tounie	350
X-TJ.32	Dingshen	350
X-TJ.33	Guancai	350
X-TJ.35	Tongming	350
X-TJ.36	Tingxiang	350
X-TJ.37	Shanglong	350
X-TJ.38	Tinglongqian	350
X-TJ.40	Chixia	360
X-TJ.41	Tongerdao	362
X-TJ.43	Qiying	364
X-TJ.44	Shangtianzhu	362
X-TJ.45	Xiayamen	362
X-TJ.46	Fuyamen	360
X-TJ.47	Xinyi	362
X-TJ.48	Xiaxinshi	362
X-TJ.49	Zhongjie	362
X-TJ.53	Zhiou	364
X-XF.8	Ganshi	376
X-XF.9	Chuangxinmen	376
X-XF.12	Xinqixue	406
X-XF.13	Xiazhongji	406,408
X-XF.14	Yeniao	408
X-XF.17	Yinbian	408
X-XF.18	Weishangxue	408
X-XZ.19	Siqiang	426
X-XZ.22	Maibu	426
X-XZ.36	Zhiwen	474
X-XZ.38	Zhitan (Liuzhitan)	448,456
X-XZ.40	Tanli	426
X-XZ.41	Tankang	426
X-XZ.42	Qiabinzhong	426
X-XZ.43	Xiachengshan	452
X-XZ.44	Weixia	452
X-XZ.45	Yinxia	436
X-XZ.46	Yangkang	436
X-XZ.47	Xinhuantiao	436
X-XZ.48	Chuqixue	470
X-XZ.49	Yiniao	448
X-XZ.50	Anmian (Wuanmian)	448
X-XZ.55	Jiankua	434
X-XZ.56	Kuanjiu	434
X-XZ.57	Qiangkua	434
X-XZ.59	Erliban	448,456

Correspondence Points of the Master Tong – page

Point	Name	Page
DT.01	Fen Zhi Shang	334,392
DT.02	Fen Zhi Xia	334,392
DT.04	Wu Ling	398
DT.05	Shuang Feng	392
DT.06	Jiu Hou	392
DT.07	San Jin	394
DT.08	Jing Zhi	392
DT.09	Jin Lin	394
DT.10	Ding Zhu	394
DT.11	Hou Xin	396
DT.12	Gan Mao San	396
DT.13	Shui Zhong	420
DT.14	Shui Fu	420
DT.15	San Jiang	420
DT.16	Shuang He	420
DT.17	Chong Xiao	420
T 1010.01	Zheng Hui	358
T 1010.02	Zhou Yuan	358
T 1010.03	Zhou Kun	358
T 1010.04	Zhou Lun	358
T 1010.05	Qian Hui	358
T 1010.08	Zheng Jing	358
T 1010.09	Shang Li	358
T 1010.10	Si Fu Er	358
T 1010.11	Si Fu Yi	358
T 1010.12	Zheng Ben	358
T 1010.13	Ma Jin Shui	356
T 1010.14	Ma Kuai Shui	356
T 1010.15	Fu Kuai	356
T 1010.16	Liu Kuai	356
T 1010.17	Qi Kuai	356
T 1010.18	Mu Zhi	356
T 1010.19	Shui Tong	356
T 1010.20	Shui Jin	356
T 1010.21	Yu Huo	356
T 1010.22	Bi Yi	356
T 1010.23	Zhou Huo	356
T 1010.24	Zhou Jin	356
T 1010.25	Zhou Shui	362
T 11.01	Da Jian	306
T 11.02	Xiao Jian	306
T 11.03	Fu Jian	306
T 11.04	Wai Jian	306

Point	Name	Page
T 11.05	Zhong Jian	306
T 11.06	Huan Chao	306
T 11.07	Zhi Si Ma	298
T 11.08	Zhi Wu Jin	298
T 11.08	Zhi Qian Jin	298
T 11.09	Xin Xi	298
T 11.10	Mu Huo	298
T 11.11	Fei Xin	298
T 11.12	Er Jiao Ming	298
T 11.13	Dan	298
T 11.14	Zhi San Zhong	298
T 11.15	Zhi Shen	298
T 11.16	Huo Xi	300
T 11.17	Mu Xue	306
T 11.18	Pi Zhong	306
T 11.19	Xin Chang	306
T 11.20	Mu Yan	306
T 11.21	San Yan	306
T 11.22	Fu Yuan	306
T 11.23	Yan Huang	306
T 11.24	Fu Ke	300
T 11.25	Zhi Xian	300
T 11.26	Zhi Wu	300
T 11.27	Wu Hu	306
T 22.01	Zhong Zi	306
T 22.04	Da Bai	300
T 22.05	Ling Gu	300
T 22.06	Zhong Bai	300
T 22.07	Xia Bai	300
T 22.08	Wan Shun Yi	306
T 22.09	Wan Shun Er	306
T 22.10	Shou Jie	306
T 22.11	Tu Shui	306
T 33.01	Qi Men	314
T 33.02	Qi Jiao	314
T 33.03	Qi Zheng	314
T 33.04	Huo Chuan	314
T 33.05	Huo Ling	314
T 33.06	Huo Shan	314
T 33.07	Huo Fu Hai	314
T 33.08	Shou Wu Jin	314
T 33.09	Shou Qian Jin	314
T 33.10	Chang Men	320
T 33.11	Gan Men	320
T 33.12	Xin Men	320
T 33.13	Ren Shi	320
T 33.14	Di Shi	320
T 33.15	Tian Shi	320
T 33.16	Qu Ling	320
T 44.01	Fen Jin	328

Point	Name	Page
T 44.02	Hou Zhui	334
T 44.03	Shou Ying	334
T 44.04	Fu Ding	334
T 44.05	Hou Zhi	334
T 44.06	Jian Zhong	334,336
T 44.07	Bei Mian	334
T 44.08	Ren Zong	328
T 44.09	Di Zong	328
T 44.10	Tian Zong	328
T 44.11	Yun Bai	334,336
T 44.12	Li Bai	336
T 44.13	Zhi Tong	334
T 44.14	Luo Tong	334
T 44.15	Xia Qu	334
T 44.16	Shang Qu	334
T 44.17	Shui Yu	334
T 55.01	Huo Bao	474
T 55.02	Hua Gu Yi	474
T 55.03	Hua Gu Er	474
T 55.04	Hua Gu San	474
T 55.05	Hua Gu Si	474
T 55.06	Shang Liu	475
T 66.03	Huo Ying	468
T 66.04	Huo Zhu	468
T 66.05	Men Jin	468
T 66.06	Mu Liu	468
T 66.07	Mu Dou	468
T 66.08	Liu Wan ou Mu Wan	468
T 66.09	Shui Qu	468
T 66.10	Huo Lian	470
T 66.11	Huo Ju	470
T 66.12	Huo San	472
T 66.13	Shui Jing	472
T 66.14	Shui Xiang	472
T 66.15	Shui Xian	472
T 77.01	Zheng Jin	452
T 77.02	Zheng Zong	452
T 77.03	Zheng Shi	452
T 77.05	Yi Zhong	458
T 77.06	Er Zhong	458
T 77.07	San Zhong	458
T 77.08	Si Hua Shang	440,450,458
T 77.09	Si Hua Zhong	450,458
T 77.10	Si Hua Fu	450,458
T 77.11	Si Hua Xia	450,458
T 77.12	Fu Chang	450,458
T 77.13	Si Hua Li	450
T 77.14	Si Hua Wai	450,458
T 77.15	Shang Chun	428,440,450
T 77.16	Xia Chun	450,428

Point	Name	Page
T 77.17	Tian Huang ou Shen Guan	450
T 77.18	Tian Huang Fu	450,462
T 77.19	Di Huang	462
T 77.20	Si Zhi	462
T 77.21	Ren Huang	462
T 77.22	Ce San Li	440,450,458
T 77.23	Ce Xia San Li	450,458
T 77.24	Zu Qian Jin	458
T 77.25	Zu Wu Jin	458
T 77.26	Qi Hu	458
T 77.27	Wai San Guan	458
T 77.28	Guang Ming	472
T 88.01	Tong Guan	428
T 88.02	Tong Shan	428
T 88.03	Tong Tian	428
T 88.04	Jie Mei Yi	428
T 88.05	Jie Mei Er	428
T 88.06	Jie Mei San	428
T 88.07	Gan Mao Yi	428
T 88.08	Gan Mao Er	428
T 88.09	Tong Shen	428,440
T 88.10	Tong Wei	428,440
T 88.11	Tong Bei	428
T 88.12	Ming Huang	432
T 88.13	Tian Huang	432
T 88.14	Qi Huang	432
T 88.15	Huo Zhi	432
T 88.16	Huo Quan	432
T 88.28	Jie	428,440
T 88.29	Nei Tong Guan	428
T 88.30	Nei Tong Shan	428
T 88.31	Nei Tong Tian	428
T 88.32	Shi Yin	440,462
VT.02	Shi Er Hou	376
VT.03	Jin Wu	376
VT.04	Wei Mao Qi	376
VT.05	Fu Chao Er Shi San	408

Main channels

Lung	2,4,6,8,10,16,18,30,32,34,52,56,58,60,126,128,130,142,144,170, 244,292,294,302,316,318,320,324,328,398
Pericardium	4,16,18,22,24,32,94,96,98,144,244,254,278,292,302,304,316, 324,370
Heart	4,10,16,18,20,22,24,30,32,34,38,40,72,76,78,80,82,90,94,96,98, 110,112,114,118,134,142,144,146,170,172,194,218,244,266, 272,292,302,304,316,324,370,398
Large Intestine	4,6,8,10,48,52,56,58,60,84,116,244,274,292,294,310,330,336, 340,384
Small Intestine	34,38,40,68,72,76,80,88,244,274,292,310,330,340,382,386
Triple Heater or Triple Burner	18,48,88,90,96,98,182,184,244,292,310,330,334,336,340,382, 386
Spleen-Pancreas	108,110,114,116,118,126,140,162,164,168,172,244,254,266, 374,378,398,404,406,430,438,446,460,466,470
Liver	110,124,126,128,130,132,134,142,144,182,184,190,192,194, 244,254,370,378,398,404,406,424,430,446,460,466,47
Kidney	110,140,142,144,146,148,208,210,214,216,218,244,250,266, 372,404,406,424,430,460,470,474
Stomach	4,6,52,72,76,110,114,116,118,126,128,130,158,162,164,168, 170,172,244,260,266,274,370,402,406,424,438,446,454,466,47
Gall Bladder	90,126,128,130,132,134,158,180,182,190,192,194,244,274,278, 370,378,404,406,414,440,450,454,458,466,472
Bladder	68,70,140,142,144,146,148,158,180,204,208,210,214,216,218, 244,266,274,278,342,386,392,412,414,416,440,452,454,466,472

Bibliography

Académie de médecine traditionnelle chinoise (Pékin), *Précis d'acupuncture chinoise*, Dangles, 1977.

Aubé Patrick, Delêtre Georges, *L'Acupuncture de A à Z*, J.B. Baillière, 1984.

Auteroche Bernard, Mainville Lucie, Solinas Henri, *Atlas d'acupuncture chinoise*, Maloine, 1990.

Borsarello Jean-François, *Acupuncture*, Masson, 1986.

Borsarello Jean-François, Duron André, Laville-Mery Charles, *Bioénergetique et médecine chinoise*, Maisonneuve, 1973.

Bouchet Alain, Cuilleret Jacques, *Cahiers d'anatomie*, Simep Éditions.

Choain Jean, *La « Voie rationnelle », Tao de la médecine chinoise*, Éditions S.L.E.L., 1957.

Chun Yan Chen, Hammes Michael, C. Kolster Bernard, Yu Lin Lian, *Atlas gráfico de acupuntura*, Könemann, 2005.

Enquin Zhang, *Chinese Acupuncture and Moxibustion*, Shangai College of MTC, 1994.

Latarjet M., Testut L., *Tratado de anatómia humana*, Salvat Editores, 1979.

Localización de los puntos acupunturales, Foreign Languages Press, 1984.

Méridiens [Revue], n° 39–40, A.S.M.A.F., 1977.

Méridiens [Revue], n° 35–36, A.S.M.A.F., 1976.

Niu de Liebenthal Pao-Cheng, *Manual de acupuntura*, El Ateneo, 1988.

Pan Ailian, *Océan d'énergie: étymologie des noms des points d'acupuncture*, Décarie-Maloine, 1993.

Rouvière H., *Anatomie humaine*, Masson, 1967.

Soulié de Morant George, *L'Acupuncture chinoise*, Maloine, 1972.

Standard Meridian Points of Acupuncture, Foreign Languages Press, 1984.

Voisin H., *L'Acupuncture du praticien*, Maloine, 1974.

Wong Ming, *Ling-Shu*, Masson, 1987.

Zhang Jun, Zheng Jin, *Fundamentos de acupuntura*, Foreign Languages Press, 1984.

Zhenguo Yan, *Anatomical Atlas of Acupuncture Points*, Donica Publishing, 2005.

Index

Chapter I • A general view of the channels

The 3 upper Yin

Shu Tai Yin system (Lung) 1

Shu Jue Yin system (Master of the Heart or Pericardium) 15

Shu Shao Yin system (Heart) 29

The 3 upper Yang

Shu Yang Ming system (Large Intestine) 47

Shu Tai Yang system (Small Intestine) 67

Shu Shao Yang system (Triple Heater or Triple Burner) 87

The 3 lower Yin

Zu Tai Yin system (Splenn) 107

Zu Jue Yin system (Liver) 123

Zu Shao Yin system (Kidney) 139

The 3 lower Yang

Zu Yang Ming system (Stomach) 157

Zu Shao Yang system (Gall Bladder) 179

Zu Tai Yang system (Urinary Bladder) 203

A summary of the channels 229

The Main channels 237

Cutaneous territories 243

Marvellous vessels or Extraordinary vessels 249

Chapter II • The points of the different anatomical regions

Points of the hand 291

Points of the forearm 309

Points of the arm 323

Points of the head 339

Chest 369

Back 381

Abdomen 401

Sacro-lumbar region 411

The thigh 423

The leg 445

The foot 465

Index

Channel points - Extra channel points 479

Acupuncture points with corresponding pages 531

Bibliography 557